Handbook of
ENVIRONMENTAL HEALTH AND SAFETY
Principles and Practices

VOLUME I • SECOND EDITION

Handbook of
ENVIRONMENTAL HEALTH AND SAFETY
Principles and Practices

VOLUME I • SECOND EDITION

Herman Koren

LEWIS PUBLISHERS

NATIONAL
ENVIRONMENTAL
HEALTH ASSOCIATION

Library of Congress Cataloging-in-Publication Data

Koren, Herman

    Handbook of environmental health and safety: principles and
practices / Herman Koren. — 2nd ed.
         p.   cm.
    Includes bibliographical references and indexes.
    ISBN 0-87371-272-2 (v. 1). — ISBN 0-87371-414-8 (v. 2).
    1. Environmental health. 2. Environmental engineering.   I. Title.
RA565.K67   1991   90-27423
363.7—dc20   CIP

LEWIS PUBLISHERS, INC.
121 South Main Street, Chelsea, Michigan 48118

PRINTED IN THE UNITED STATES OF AMERICA

Foreword, 1st Edition

Over the years, many books have addressed the subject of environmental health. Here is yet another. The questions naturally arise: Why another book on environmental health? Why at this time? Why this particular book? The first two questions can be answered together in terms of six phenomena characterizing the past few decades: first, of course, is the spectacular growth of the human population. Population growth accentuates environmental problems and hence health problems. When a second caveman moved in with the first, the various waste products of their respective living processes necessarily impinged upon each other. Obviously, the greater the number of people, the greater the environmental problems. The rate of increase in the world's population has more than doubled since the end of World War II, with about 200,000 now being added daily. Concurrent with population growth is greatly increased mobility, with a tendency toward concentration in urban areas further exacerbating environmental problems.

A third phenomenon is the knowledge explosion in science and technology bringing countless benefits to humankind, yet resulting in considerable environmental abuse and an ever-growing catalogue of health hazards. Increasingly, individuals are confronted with risks and hazards. As individuals, they are helpless in the face of the chemical and biological pollution of the air they must breathe, of the water they drink, and of the food they must eat in order to survive. As individuals, they cannot control any of these any more than they can avoid them. A fifth phenomenon is the welcomed expansion of the parameters and capabilities of medical and biological science, facilitating the study of the long-term consequences of exposure to a variety of potential health hazards, including natural and synthetic chemicals in large and minute amounts, audible and inaudible vibrations, and noise. Lastly, there is the very important phenomenon of broad public or consumer education, resulting in a justified insistence on the accountability of researchers, manufacturers, vendors, and regulatory agencies.

Any of these phenomena, and certainly all viewed together, necessarily places the field of environmental health in a new light. This brings us to the third question posed earlier: Why this particular book? The answer is found in its unusual comprehensiveness, as indicated by even a cursory glance at the contents, and by the very solid background of the author, Dr. Koren, in both practice and teaching. In style, this volume is sufficient, readable, and direct, and it is augmented with pertinent reference and bibliographic materials. Because of this fortunate combination of qualities, it should serve not only as an excellent basis for newcomers to the field, but also as a valuable survey and reference book for those already engaged in the important practice of environmental health.

John J. Hanlon
Professor of Public Health
San Diego State University

Formerly Assistant Surgeon General

Foreword, 2nd Edition

Environmental health continues to be a major issue facing our society and the entire world. Citizen interest and demands for a solution to environmental health problems is at an all-time high. Political leaders know that they are being judged by their environmental records. Government and the private sector are initiating additional environmental health measures to more effectively address such environmental concerns as hazardous wastes, toxic chemicals, food protection, radiation protection, air quality, water pollution, safe drinking water, injury prevention, vector control, solid wastes, noise pollution, and occupational health and safety. Increasingly, environmental health professionals are becoming involved in planning decisions relating to transportation, land-use, resource development, and energy productions as the only true methods of preventing environmental health problems.

Global environmental health has finally caught the attention of world leaders. Global warming, acid precipitation, ozone depletion, over-population, desertification, deforestations, and planetary toxification are demanding the attention of properly prepared environmental health professionals in efforts to literally save the world as we know it.

Dr. Herman Koren's revision of his "Handbook of Environmental Health and Safety" is a valuable addition to the literature and will be useful for students and practitioners in the field of environmental health. It is sufficiently comprehensive to effectively address the needs of personnel in government and industry environmental health endeavors.

Larry J. Gordon
Visiting Professor of Public Administration
University of New Mexico
Albuquerque, NM

Formerly President
American Public Health Association
and retired New Mexico Cabinet Secretary for Health & Environment

Preface

This handbook is designed to provide a comprehensive, concise discussion of each of the important environmental health areas, including energy, ecology and people, environmental epidemiology, risk assessment and risk management, environmental law, air quality management, food protection, insect control, rodent control, pesticides, the chemical environment, environmental economics, human disease and injury, occupational health and safety, noise, radiation, recreational environment, indoor environments, medical care institutions, schools and universities, prisons, solid and hazardous waste management, water supply, plumbing, swimming areas, sewage disposal, soils, water pollution control, environmental health emergencies, and nuisance complaints.

Sufficient background material is introduced throughout this book to provide students, practitioners, and other interested readers with an understanding of the areas under discussion. Common problems and potential solutions are described; graphs, inspection sheets, and flow charts are utilized as needed to consolidate or clarify textual material. All facts and data come from the most recent federal government documents, many of which date from 1988 and 1989. Rules and regulations specified will continue to be in effect into the 1990s. For rapidly changing areas in which the existing material used is likely to become dated, the reader is referred to the appropriate sources to update a given environmental health area or portion of an area as needed. This enhances the value of the text by providing basic and current materials that will always be needed and secondary sources that will enable the reader to keep up to date.

This book is neither an engineering text nor a comprehensive text in each area of study. The purpose of this book is to provide a solid working knowledge of each environmental health area with sufficient detail for practitioners. The text can be used in basic courses in environmental health, environmental pollution, ecology, and the environment and people that are offered at all universities and colleges in the United States and abroad. These courses are generally taught in departments of Life Science, Geology, Science Education, Environmental Health, and Health and Safety. For general areas of study, the instructor can omit specific details, such as resources, standards, practices and techniques, and modes of surveillance and evaluation. This same approach may be used by schools of medicine, nursing, and allied health sciences for their students. This text is also suitable for basic introductory courses in schools of public health, environmental health, and sanitary science, as well as junior colleges offering 2-year degree programs in sanitary Science and Environmental Science.

Practitioners in a variety of environmental health and occupational health and safety fields will find this book a handy reference for resolving current problems and for obtaining a better understanding of unfamiliar areas. Practitioners and administrators in other areas, such as food processing, water-quality control, occupational health and safety, and solid and hazardous waste management, will also find this reference book useful.

High school teachers often must introduce environmental health topics in their classes and yet have no specific background in this area. This book could serve as a text in graduate education courses for high school teachers as well as a reference source.

Public interest groups and users of high school and community libraries will obtain an overall view of environmental problems by reading Chapter 1 and the Background and Status, Problems, Potential for Intervention, Resources, and Control sections in each chapter. This volume will also supply a concise reference for administrators in developing nations, for it explains tested controls and provides a better understanding of environmental problems, various standards, practices and techniques, and a variety of available resources.

The material divides easily into two separate courses. Course I would correspond to the content of Volume I and would include Chapter 1. Environment and Humans; Chapter 2. Food Protection; Chapter 3. Food Technology; Chapter 4. Insect Control; Chapter 5. Rodent Control; Chapter 6. Pesticides; Chapter 7. Indoor Environment; Chapter 8. Institutional Environment; Chapter 9. Recreational Environment; and Chapter 10. Occupational Environment.

Course II, corresponding to the content of the Volume II, would include Chapter 1. Air Quality Management; Chapter 2. Solid and Hazardous Waste Management; Chapter 3. Private and Public Water Supplies; Chapter 4. Swimming Areas; Chapter 5. Plumbing; Chapter 6. Private and Public Sewage Disposal and Soils; Chapter 7. Water Pollution and Water Quality Controls; and Chapter 8. Environmental Health Emergencies, Nuisance Complaints, and Special Problems.

Since the problems of the environment are so interrelated, certain materials must be presented at given points in order to give clarity and cohesiveness to the subject matter. As a result, the reader may encounter some duplication of materials throughout the text.

With the exception of Volume I, Chapter 1 and Volume II, Chapter 8, all of the chapters have a consistent style and organization, facilitating retrieval. The introductory nature of Volume I, Chapter 1 and the unusual nature of Volume II, Chapter 8 do not lend themselves to the standard format.

In Volume I, Chapter 1, the reader is introduced to the underlying problems, basic concerns, and basic philosophy of environmental health. The ecologic, economic, energy, toxicologic, and epidemiologic bases provided help the individual to understand his or her relationship to the ecosystem and to the real world of economic and energy concerns; and to understand the relationship between biological, physical, and chemical agents and disease and injury causation. It also provides an understanding of the role of government and the environmental health practitioner in helping to resolve environmental and ecological dilemmas created by humans.

In Volume II, Chapter 8, the many varied facets of environmental emergencies, nuisances, and special problems are discussed. Students may refer to other chapters of the text to obtain a complete idea of each of the problems and the potential solutions.

The general format of Volume I, Chapters 2–10 and Volume II, Chapters 1–7 is as follows:

STANDARD CHAPTER OUTLINE

1. Background and Status (Brief)
2. Scientific, Technological, and General Information
3. Problem
    A. Types
    B. Sources of Exposure
    C. Impact on Other Problems
    D. Disease Potential
    E. Injury Potential
    F. Other Sources of Exposure Contributing to Problems
    G. Economics
4. Potential for Intervention
    A. General
    B. Specific
5. Resources
    A. Scientific and Technical; Industry, Labor, University; Research Groups
    B. Civic
    C. Governmental
6. Standards, Practices, and Techniques
7. Modes of Surveillance and Evaluation
    A. Inspections and Surveys
    B. Sampling and Laboratory Analysis
    C. Plans Review
8. Control
    A. Scientific and Technological
    B. Governmental Programs
    C. Other Programs
    D. Education

9. Summary
10. Research Needs

- The Background and Status section of each chapter presents a brief introduction to, and the current status of, each problem area. An attempt has been made in each case to present the current status of the problem.
- The Problem section is subdivided into several important areas to give the reader a better grasp of the total concerns. To avoid disruption in continuity of the standard outline, the precise subtitles listed may not be found in each chapter. However, the content of the subtitles will be present. The subtitle, Impact on Other Problems, is given as a constant reminder that one impact on the environment may precipitate numerous other problems.
- The Potential for Intervention section is designed to succinctly illustrate whether a given problem can be controlled, the degree of control possible, and some techniques of control. The reader should refer to the Controls section for additional information.
- Resources is a unique section providing a listing of scientific, technical, civic, and governmental resources available at all levels to assist the student and practitioner.
- The section of Standards, Practices, and Techniques is specifically geared to the reader who requires an understanding of some of the specifics related to surveys, environmental studies, operation, and control of a variety of program areas.
- The Modes of Surveillance and Evaluation Section explains many of the techniques available to determine the extent and significance of environmental problems.
- The Control section presents existing scientific, technological, governmental, educational, legal, and civic controls. The reader may refer to the Standards, Practices, and Techniques section in some instances to get a better understanding of controls.
- The Summary presents the highlights of the chapter.
- Research Needs is another unique section that is intended to increase reader awareness to the constantly changing nature of the environment and of the need for continued reading or in-service education on the future concerns of our society.
- The Reference section is extensive and as current as possible. It appears as the last chapter in each volume. It provides the reader with sources for further research and names of individuals and organizations involved in current research.

Author

Dr. Herman Koren, R.E.H.S., M.P.H., H.S.D., is Professor and Director of the Environmental Health Science Program, and Director of the Supervision and Management Program I & II at Indiana State University at Terre Haute. He has been an outstanding researcher, teacher, consultant, and practitioner in the environmental health field, and in the occupational health, hospital, medical care, and safety fields, as well as in management areas of the above and in nursing homes, water and wastewater treatment plants, and other environmental and safety industries for the past 36 years. In addition to numerous publications and presentations at national meetings, he is the author of five books, entitled *Environmental Health and Safety*, Pergamon Press, 1974; *Handbook of Environmental Health and Safety*, Volumes 1 and 2, Pergamon Press, 1980 (now published in updated and vastly expanded format by Lewis Publishers as a 2nd edition); *Basic Supervision and Basic Management*, Parts I & II, Kendall Hunt Publishing, 1987. He has served as a district environmental health practitioner and supervisor at the local and state level. He was an administrator at a 2000 bed hospital. Dr. Koren was on the editorial board of the *Journal of Environmental Health* and the former *Journal of Food Protection*. He is a founder diplomate of the Intersociety Academy for Certification of Sanitarians, a Fellow of the American Public Health Association, a 36-year member of the National Environmental Health Association, founder of the "National" Student National Environmental Health Association, and the founder and advisor of the Indiana State University Student National Environmental Health Association (Alpha chapter). Dr. Koren developed the modern internship concept in environmental health science. He has been a consultant to the U.S. Environmental Protection Agency, the National Institute of Environmental Health Science, and numerous health departments and hospitals, and has served as the keynote speaker and major lecturer for the Canadian Institute of Public Health Inspectors. He is the recipient of the Blue Key Honor Society Award for outstanding teaching and the Alumni and Student plaque and citations for outstanding teaching, research, and service. The National Environmental Health Association has twice honored Dr. Koren with presidential citations for "Distinguished Services, Leadership and Devotion to the Environmental Health Field" and "Excellent Research and Publications."

Acknowledgments

To Boris Osheroff, my friend, teacher, and colleague, for opening the numerous doors needed to obtain the most current information in the environmental health field, and for his many fine suggestions and ideas before and after reading the manuscript; to Ed O'Rourke for his intensive review of the manuscript and his recommendations for revisions and improvements; to Karol Wisniewiski for his review and comments on the manuscript; to Dr. John Hanlon for his encouragement and for helping a young teacher realize his potential; to all the environmental health administrators, supervisors, practitioners, and students who have shared their experiences and problems with me and have given me the opportunity to test many of the practical approaches used in the book; to the National Institute of Occupational Safety and Health, National Institutes of Health, National Institute of Environmental Health Science, U.S. Environmental Protection Agency, U.S. Food and Drug Administration, Cunningham Memorial Library, Indiana State University, Indiana University Library, Purdue University Library, and the many other libraries and resources for providing the material that was used in developing the manuscript; to my wife, Donna Koren, and my student, Evelyn Hutton, for typing substantial portions of the manuscript; to my daughter, Debbie Koren, for helping me organize the materials and for working along with me throughout the night at the time of deadlines to complete the work.

In the second edition, Ryan Hane, one of my recent graduates, has helped as a research assistant. Kim Malone has typed portions of the new manuscript, retyped the entire manuscript, and has been of great value to me. Pat Ensor, Librarian, Indiana State University, has been of considerable value in helping gather large numbers of references in all areas of the book. A very special thanks to my sister-in-law, Betty Gardner, for typing a substantial portion of the new manuscript, despite recurring severe illness. Her cheerfulness during my low periods has helped me complete my work. Finally, thanks to my wife Donna for putting up with my thousands of hours of seclusion in the den, while I was working, and for encouraging me throughout the project and my life with her. She has truly been my best friend.

Contents
Volume I

Contents
Volume II

Chapter 1

ENVIRONMENT AND HUMANS

Health is the avoidance of disease and injury and the promotion of happiness through efficient use of the environment, a properly functioning society, and an inner sense of well-being. *Environmental health* is the art and science of the protection of good health, the promotion of aesthetic values, and the prevention of disease and injury through the control of positive environmental factors and the reduction of potential hazards — physical, biological, chemical, and radiological.

To understand the relationship of the environment to humans and to understand how to protect humans from disease and injury, it will be necessary to discuss the ecosystem, ecosystem dynamics, and energy. Human impact on the environment and the various approaches used to resolve environmental problems, namely, risk assessment, epidemiological, economic, legal, and governmental, will also be discussed. In order to understand abnormal physiology, a brief discussion on normal physiology is included. Finally, it will be necessary to understand the role of professional environmental health practitioners, the skills that they need, and how they work in the ever-growing field of environmental problems.

THE ECOSYSTEM

The Earth is divided into the lithosphere, or land masses, and the hydrosphere, or the oceans, lakes, streams, and underground waters. The hydrosphere includes the entire aquatic environment. Our world, both lithosphere and hydrosphere, is shaped by varying life forms. Permanent forms of life create organic matter and help establish soil; plants cover the land and reduce the potential for soil erosion — the nature and rate of erosion affects the redistribution of materials on the surface of the Earth. Organisms tie up vast quantities of certain chemicals, such

1

as carbon and oxygen; plants and animals, through respiration, release carbon dioxide into the atmosphere — carbon dioxide affects the heat transmission of the atmosphere. Organisms affect the environment and in turn are affected by it.

Two environments, biotic (living environment or community) and abiotic (nonliving environment), combine to form an ecosystem. The ecosystem can also be subdivided by more specific criteria into the following four categories: abiotic, the nutrient minerals that are synthesized into living protoplasm; autotrophic, the producer organisms (largely the green plants), which assimilate the nutrient minerals using energy and combine them into living organic substances; heterotrophic, the consumers, usually the animals that ingest or eat organic matter and release energy; and the heterotrophic reducers, bacteria or fungi, which return the complex organic compounds to their original abiotic state and release the remaining chemical energy. The biotic group in the ecosystem complex is essentially comprised of the autotrophs, or producer organisms that construct organic substances, and the heterotrophs, or consumer or reducer organisms that destroy organic substances. The ecosystem is important when considering the food chain, which is in effect a transfer of energy from plants through a series of organisms that eat and in turn are eaten. Eventually, decay will start the process all over again.

The ecological niche is the combination of function and habitat of each of the approximately 1.5 million species of animals and a half million species of plants on the Earth. There are many interactions between species in the ecosystem, yet a balance is dictated by nature. The law of limiting factors states that a minimum quantity of essentials, such as nutrients, light, heat, moisture, and space, must be available within the ecosystem for survival of the organisms. In some instances where these limiting factors apply or where pesticides or other environmental elements are introduced into the ecosystem, the organism alters itself in order to exist within the new environment. This change is called mutation. Unfortunately, mutation becomes a serious concern in the area of pest control as well as in disease, because the new organism may be highly resistant to effective control and may therefore cause disease, injury, and physical destruction of plants and animals. The ecosystem is always in a dynamic, rather than static, balance. Changes in one part of the ecosystem will cause changes in another.

The Biosphere

The biosphere is that part of the Earth in which life exists. However, this definition is not complete, since spores may commonly be found in areas that are too dry, too cold, or too hot to support organisms that metabolize. The biosphere contains the liquid water necessary for life; it receives an ample supply of energy from an external source, which is ultimately the sun, and within it liquid, solid, and gaseous states of matter interface. All of the actively metabolizing organisms operate within the biosphere. The operation of the biosphere depends on photosynthesis, during which carbon dioxide is reduced to form organic compounds and molecular oxygen. Oxygen, the byproduct of photosynthesis, replenishes the atmosphere and most of the free water, which contains dissolved oxygen.

Ecosystem Dynamics

The ecosystem changes frequently. Several of the cycles that are important and that may be affected by humans include the hydrologic cycle, the carbon cycle, the nitrogen cycle, the phosphorous cycle, and energy flow. The hydrologic cycle is the movement of water from the atmosphere to the Earth and back into the atmosphere. This will be more fully discussed in the chapter on water.

The carbon cycle begins with the fixation of atmospheric carbon dioxide by means of photosynthesis performed by plants and certain microorganisms. During this process carbon dioxide and water react to form carbohydrates, and free oxygen is simultaneously released into the atmosphere. Some of the carbohydrates are stored in the plant, and the rest are consumed by the plant as a source of energy. Some of the carbon that has been fixed by the plants is then consumed by animals, who respire and release carbon dioxide. The plants and animals die, decomposed by action of microorganisms in the soil, and the carbon in their tissues is then oxidized to carbon dioxide and returned to the atmosphere. The carbon dioxide is recycled through the plants and the process repeats itself. Appropriate diagrams on the carbon cycle may be found in the chapter on sewage.

The nitrogen cycle begins when atmospheric nitrogen is fixed or changed into more complex nitrogen compounds by specialized organisms, such as certain bacteria and blue-green algae. Some fixation may occur as a result of lightning, sunlight, or chemical processes, however, the most efficient nitrogen fixation is carried out by biological mechanisms. Other bacteria, fungi, and algae may also play an important role in nitrogen fixation. Basically the atmospheric nitrogen is changed into a nitrate, which is absorbed by plants, eventually becoming a plant protein. The plant protein may decay when the plant dies, change to ammonia, through bacterial action become a nitrite, and through further bacterial action be released as atmospheric nitrogen. The plant protein may also be eaten by animals and become an animal protein. Through decay of the dead animal or breakdown of feces and urine, this protein is changed to ammonia. The ammonia returns to the nitrite stage through bacterial action, and again through bacterial action becomes atmospheric nitrogen. A further description and diagram of the nitrogen cycle will be found in the chapter on sewage.

In the phosphorus cycle, the element moves rapidly through similar stages, becoming locked in sediment or in biological forms such as teeth or bones. The primary sources of phosphorus for agriculture are phosphate rocks and living or dead organisms. Diagrams on the phosphorus cycle may be found in the chapter on sewage disposal.

Food Chain

The cycle of energy flow may also be described as the food web. The food web, or food chain, implies that an organism has consumed a smaller organism and is then consumed by a larger organism. Eventually, the microscopic plants and animals become the food supply for the small fish or animals, which become the food supply for humans. The importance of the food chain is illustrated by biomagnification, in which the impurities found in water are concentrated in the lower forms of life and are reconcentrated substantially during the movement of the impurities through the food chain. For example, whereas people might only get 0.001 µg of mercury in drinking water, they might get 30–50 µg of mercury by consuming fish that have been bioconcentrating mercury.

Energy Cycles

Solar energy is absorbed by the Earth and is eventually reradiated into space as heat. The heat is distributed over the surface of the Earth through circulation caused by the atmosphere and the oceans. Diurnal changes or changes occurring in a 24-hr period due to wind, temperature, and humidity are important near ground level and at high levels in the atmosphere. In certain localities land masses and sea breezes may affect the overall heat and weather patterns. However, total heat movement is not affected significantly by local conditions.

About 30% of the solar energy entering the atmosphere is deflected or scattered back toward outer space because of the atmosphere, the clouds, or the Earth's surface. This portion of the solar energy is lost. About 50% of the incoming radiation reaches the ground or ocean, where it is absorbed as heat. The properties of the surface that receive the energy will determine the thickness of the layer over which the available heat will be distributed. In the oceans the surface wave motions effectively distribute the heat over a 300-ft layer of air. On land the energy transferred downward into the ground occurs very slowly through the process of molecular heat conduction. The penetration distance is very small. About 20% of the incoming solar radiation is absorbed as it goes through the atmosphere. In the upper atmosphere, oxygen and ozone molecules absorb an estimated 1–3% of the incoming radiation. This absorption occurs in the ultraviolet range and, therefore, limits the penetrating radiation to wavelengths longer than 300 nm. This absorption is very important, since it is the main source of energy for the movement of air in the upper atmosphere. This absorption also shields people on Earth from damaging effects of ultraviolet radiation. Most of the rest of the 20% is absorbed by water vapor, dust, and water droplets in the clouds. The energy that sustains all living systems is fixed in photosynthesis, as previously mentioned. Part of the energy fixed in plants and animals has been compressed over millions of years and is the source for stored energy, namely, coal, oil, and natural gas.

Current Ecosystem Problems

People and advanced technology affect the ecological niche by interfacing with the intricate function and habitat of various species of animals and plants. Air pollution, toxic waste, pesticides, raw sewage, solid wastes, and medical wastes create an environmental pressure that is detrimental to all life forms, including people. From the destruction of the Rain Forest of Brazil, because of the excessive clearing of trees, to the potential destruction of forests in the Northern Hemisphere due to acid rain, a huge number of ecosystems may be eliminated. Nature's diversity offers many opportunities for finding new pharmaceuticals, genetic mapping, genetic engineering, and the potential power to improve crops. We are wasting our greatest natural resource on which we depend for food, oxygen, clean water, energy, building materials, clothing, medicine, sociological well-being, and countless other benefits.

Biological diversity includes two related concepts, genetic diversity and ecological diversity. Genetic diversity is the amount of genetic variability among individuals in a single species, whereas ecological diversity is the number of species in a community of organisms. An organism's ability to withstand the challenge of varying physical problems, chemical problems, parasites, and competition for resources is largely determined by its genetic makeup. The more successful organisms that survive pass on the genes to the next generation. An example of this would be in the case of climactic changes where plants have smaller, thicker leaves in order to lose less water in areas that are becoming arid. Genetic diversity is important in developing new crops to meet new conditions. This is how disease-resistant crops can be utilized to sustain life for people. Genetic variation among species is greater than within the species. It is difficult to determine which of these species are of extreme importance and will be essential to people in the future as new supplies of food, energy, industrial chemicals, and medicine. It is anticipated that the human population will increase by nearly 50% in the next 20 years, with much of this increase occurring in very poor nations. The resources needed to feed these individuals will be substantial, and many new forms of food will have to be developed in order to avoid starvation.

There appears to be adequate sources of fossil fuel for the immediate future. However, these fossil fuels, when burnt, cause severe pollution. Eventually, fossil fuel sources will be depleted. It will then be necessary to utilize plants that can produce energy-rich materials, such as soybean oil. At present this process is very expensive.

Coastal waters contain several major natural systems. Since they directly abut the land, they are affected by activities on the land. Pollutants coming from coastal sewage treatment plants, industrial facilities, the settling of air contaminants, and erosion of land, many hundreds of miles upstream, can cause considerable problems to the coastal waters. Oil slicks are extremely hazardous to ecosystems.

Trace metals can be concentrated in the food chain and, thereby, become a hazard to people eating sea life. Heavy metals, such as mercury, lead, and cadmium, have been found in coastal waters.

Molluscs become indicators of environmental contamination and environmental quality. These shellfish, which live in the mud or sand bottoms of the aquatic ecosystems, accurately reflect some of the important characteristics of the water adjacent to the land. These organisms may show a direct adverse effect when they accumulate chemical substances or microorganisms. They are good for systematic monitoring, because they are stationary in nature, except for the early egg and larva stages, and filter large quantities of water in order to feed themselves. Since they are of significant commercial importance, there is reliable quantitative data available on harvesting and other aspects of the molluscs' life cycle.

In the past, the EPA sponsored studies of the accumulations of organic chlorine pesticides in shellfish. Residues of DDT were found for years after the use of the chemical was banned in 1972. The Mussel Watch Program has shown elevated levels of polychlorinated biphenyls (PCBs). High-level concentrations of silver, cadmium, zinc, copper, and nickel have been detected in various bodies of water. Mussels have also shown the presence of radionuclei as a result of weapon testing programs and also from effluent that has come from nuclear reactors. Records of commercial catches of clams and oysters indicate that the number of molluscs in coastal waters has been steadily declining over the years. This may be due to overfishing, loss of habitat, and deterioration due to pollution and natural disasters. It has been shown that the building of a sewage treatment plant has the potential to adversely affect shellfish habitats by altering water salinity.

The continued loss of tropical forests is the single greatest threat to the preservation of biological diversity. Much of Central America's ecosystems may be affected by these problems.

Wetlands have exceptional value because of their location at the land/water interface. They provide ecological, biological, economic, scenic, and recreational resources. Many wetlands are among the most productive ecosystems on the planet, especially estuaries and mangrove swamps. At least one half of the biological production of the oceans occurs in the coastal wetlands. From 60 to 80% of the world's commercially important marine fish either spend time in the estuaries or feed upon the nutrients produced there. Coastal wetlands help protect inland areas from erosion, flood, waves, and hurricanes. The wetlands are an important economic resource, because they provide cash crops, such as timber, marsh hay, wild rice, cranberries, blueberries, fish, and shellfish.

Desertification is the process that occurs in semiarid lands where sterile sandy desert-like ecosystems are created from grasslands, brushlands, and sparse open forests. This desertification can be stimulated beyond natural forces by overgrazing of animals and by misuse of the area. This can lead to mass starvation and death.

ENERGY

Energy problems came to the attention of the American public and the world in 1973 as a result of the Arab oil embargo and the actions of the Organization of Petroleum Exporting Countries (OPEC). Unfortunately, in the years preceding these dramatic events, the United States was buying an increasing amount of petroleum abroad and utilizing more natural gas than was readily available. As a result of these shortages, the President authorized the allocation of scarce fuels. By May 1973 the Office of Emergency Preparedness was reporting widespread problems of gasoline stations closing due to lack of fuel. Voluntary guidelines were announced and a variety of emergency actions taken including the lowering of the speed limit to 55 mi/hr. Project Independence was conceived with the goal of self-sufficiency in energy production by 1980. Congress established the Federal Energy Administration in December 1973, and the President created the Federal Energy Office to deal with the immediate crisis. Funding was provided for additional research and development in a variety of energy areas.

Unfortunately, by 1978 the amount of fuel being purchased abroad was escalating. The variety of programs at the federal level were so confusing, complex, and spread out over a variety of agencies that the administration in Washington, DC requested the establishment of a new federal Department of Energy with an administrator of cabinet rank. It was hoped that the new department would bring organization to the chaotic mass of confusing energy problems and move the country toward self-sufficiency.

Conservation and development of new energy sources or the expansion of existing sources were two major phases of energy programs. Energy conservation was of critical importance to the United States; yet despite predictions of severe shortages, conservation attempts had not been entirely successful and great amounts of energy were still wasted. Greater energy efficiency was needed in cars and in electrical and gas equipment. Better insulated homes and a general reduction in energy usage was essential. The United States had to become less dependent on oil and gas as sources of energy.

Oil and gas in 1972 accounted for nearly 78% of U.S. energy consumption. By 1980, oil and gas remained the greatest sources of energy used, despite the fact that the United States had about a 300-year supply of coal available. Although in 1979, Alaska was providing new oil, the United States was more dependent than ever on foreign oil. It had been recommended that natural gas be deregulated at the wellhead, enabling the gas companies to provide adequate funds to further research. However, it was understood that no matter how many sources of oil and gas were developed, eventually the supplies would be depleted; therefore, it was necessary to examine other kinds of energy sources. These sources included oil shale, nuclear energy, geothermal energy, solar energy, fusion energy, and coal. Oil shale was available, but at rather high costs. Additional research was necessary to determine how best to extract the oil from the shale both economically and without creating surface environmental problems.

Nuclear power could certainly provide a continuing source of available energy to the country. However, many permits had been delayed because of consumer groups concerned with the possibility of nuclear accidents and with the disposal of the radioactive wastes. The threat of terrorist groups obtaining wastes from the fusion process and utilizing these materials to terrorize the world was also of great concern. In 1973, the U.S. nuclear electrical generating capacity was 20,000 MW, or over 5% of the nation's total electrical capacity.

Geothermal energy is energy produced by tapping the Earth's heat. It was being used at the geyser sites in Oregon and in California, where the Pacific Gas and Electric Company employed geothermal steam as a power source for a 400-MW electrical generating facility. Geothermal resources include dry steam, hot steam, and hot water. At that time, because of cost factors, only dry geothermal heat was used to drive electric turbines.

Solar energy was both economically and technologically feasible. The benefits of capturing the sun's energy had been long recognized. It was predicted that if research and development programs were successful, solar energy could provide, by the year 2020, 25% of the total U.S. energy requirements. Funding for solar energy research had increased and considerable research was needed. The 1979 federal budget included $137,500,000 for solar research and development. Techniques under investigation included direct thermal applications, solar electric applications, and fuels from biomass. Direct thermal application consisted of solar heating for cooling of buildings and hot water supply, and the use of solar heat for agriculture and industrial processes. Agricultural applications were the drying of food, crop drying, lumber drying, heating and cooling of greenhouses, and heating of animal shelters. The solar electrical applications included research on the generation of electricity from windmills, solar cells (photovoltaic), solar thermal electric systems, and ocean thermal energy conversion. Theoretically, these systems were well understood; however, their applications were not yet economically feasible. The fuels produced from biomass suggested large-scale use of organic materials, such as animal manure, field crops, crop waste, forest crops and waste, and marine plants and animals. The conversion process was technically feasible. During World War II, much of the liquid fuel of France consisted of methanol produced from wood. It is hoped that by the year 2000 biomass will supply the equivalent of 1.5 million barrels of oil per day. Processes requiring study included fermentation to produce methane and alcohol, chemical processes to produce methanol, and pyrolysis to convert organic waste material to low British thermal unit (Btu) gaseous fuels and oils.

Coal was the one major source of energy that could certainly be utilized to reduce our reliance on other nations. The most serious deterrent was the many contaminants, particularly sulfur, found in coal. Sulfur could be removed by utilizing new technologies in coal mining and conversion. Since coal made up 85% of the U.S. fossil energy resource base, it was wise to utilize it as well and as quickly as possible. New coal technology under development promised two very important changes in the potential pollution areas: the reduction of sulfur oxide in particulate matter from direct burning of coal, and the provision of liquid and gaseous fuel substitutes for domestic oil and gas production. The emphasis in research had shifted from conventional combustion to direct fuel to fuel conversion, that is, coal gasification and coal liquification. Sulfur could be extracted even when coal was directly burned using fluidized beds in which crushed coal was injected into the boiler near its base. Coal was fluidized by blowing air uniformly through a grid plate. Sulfur dioxide was then removed by injecting the crushed coal with limestone particles less than $1/_8$ in. in diameter. The limestone particles were noncombustible and would accept the sulfur dioxide. This system also reduced the nitrogen oxides below emission standards for utility boilers of 0.7 lb of nitrogen dioxide per million Btu. The nitrogen dioxide could actually be reduced to 0.2–0.3 lb per million Btu. The gasification process involved the reaction of coal with air, oxygen, steam, or a mixture of these that yielded a combustion product containing carbon monoxide, hydrogen, methane, nitrogen, and little or no sulfur. The energy of the product ranged from 125–175 Btu per standard cubic foot for a low Btu gasification process to about 900–1000 Btu per standard cubic foot for a high Btu gasification process. The high Btu gasification process was comparable with the energy content of natural gas. The liquification transformed the coal into a liquid hydrocarbon fuel and simultaneously removed most of the ash and sulfur. Hydrogen was frequently added in many of the direct catalytic liquification techniques.

THE ENERGY DILEMMA

The 1980s

The trauma of the 1970s led to a more active role by government in the 1980s to try to resolve energy problems in the United States. The Energy Security Act and the Crude Oil Windfall Act of 1980 gave the President additional power to encourage the development of oil and gas reserves and a synthetic fuels industry. Since coal was our greatest fuel resource, there was an attempt to substitute coal for oil in various industries. Energy was used more efficiently because of a variety of conservation programs. The federal conservation and solar programs were passed by Congress, and additional funds were allotted for weatherization assistance for low-income people and tax breaks for others who made their homes more efficient. Funds were provided by the Solar Energy and Conservation Bank to finance the use of solar energy systems. Solar wind and geothermal tax credits were made available for residential users. Funds were provided for research and development in the use of energy sources other than fossil fuels. The biomass program and the alcohol tax credits provided 1.6 billion dollars thru 1990 to promote the conversion of grain, farm residues, and other biomasses to alcohol fuels. Sixteen and one half billion dollars were allotted through 1990 to improve transportation efficiency by enhancing public transportation programs. A fuel economy program was put in place and funds were provided for enforcement of the 55 mi/hr speed limit.

Auto efficiency standards were raised from 20 mi/gal for the 1980 model year vehicle to 27.5 mi/gal by 1985. A gas guzzler tax was added to new vehicles with mileage below 15 mi/gal. A building energy conservation code and an appliance efficiency standard code was established to improve fuel conservation and efficiency. As a result of all this effort, the growth of energy usage slowed substantially from the previous years.

By 1984 the national energy picture had continued to brighten substantially. Prices of gasoline had dropped and the price of crude oil had dropped substantially. This progress was due in part to the decontrol of the domestic petroleum markets and the removal of restrictions on energy use. Potential new sources of energy were being developed in the oceans abutting the United States. This area is larger than the country's land mass and includes many of the deep water areas, as well as the Outer Continental Shelf. Further, the OPEC nations were in disarray because of the long war between Iran and Iraq, and the propensity for cheating on the amount of oil that each country was permitted to produce in order to maintain the high fuel prices. Continued conservation of energy because of the many permanent changes made in energy use over the years helped keep the price of oil at lower levels.

In 1984, 1608 geothermal leases were in effect, in which 30 were capable of production and 12 actually were in production.

There was an estimated 2 trillion barrels (bbl) of shale oil in the Green River formation in northwestern Colorado, northwestern Utah, and southwestern Wyoming. Of this amount, it was estimated that 731 billion bbl of oil could be extracted commercially. It was also estimated that in Utah, there were approximately 2 billion bbl of oil recoverable with current technology. Approximately 100–200 million bbl could be recovered by surface mining. The question was not the availability of these energy resources, but the cost of developing them.

Since the Three Mile Island debacle of March 1979, there have been numerous public protests about the building and use of nuclear power as a source of energy. There still is a need for facilities for the permanent disposal of the high radioactive waste that is generated during the production of electricity in the nuclear power plants. Until very recently, in 1988, it was assumed that the nuclear waste was being handled appropriately on site. However, it has been disclosed that there have been numerous leaks of nuclear waste from the nuclear power plants over the decades. A prime example of this is the serious problems occurring in Ohio, which must be corrected to prevent disease. In 1990 the cleanup of these sites were estimated to cost 31.4 billion dollars.

The 1990s

In 1988, the United States consumed more energy than in any previous year in its history. The demand for petroleum was about 17 million bbl/day which was the highest point since 1980. However, the strong underlying energy efficiency trend continued, except for the use of energy in the transportation sector. The projection for energy use by the year 2000 is 90 quads, with the current usage of 79.4 quads in 1988. A quad is equivalent to one quadrillion British thermal units. The U.S. domestic oil production, which was on the rise in the 1980s will start to decline again, and it is anticipated that by the year 1994 over half of the oil used by the United States will be imported. This may once again leave the country vulnerable to the demands and needs of OPEC, wars, and other catastrophes, such as the running aground of the Exxon Valdez in Prince William Sound, Alaska. This ecological debacle temporarily cut the amount of oil coming from Alaska down to 20% of its normal flow. Although oil flow recovered and prices then declined, one wonders about future catastrophes and their effect.

Because of lower oil prices during 1989, U.S. exploration decreased sharply. In 1989, there were only 740 drilling rigs, whereas in 1981 there were 4500 functioning drilling rigs. There will be a need for a long lead time to simply get ready to do further exploration for oil in the United States. The environmental and conservation groups will attempt to limit further oil exploration in Alaska because of the incident that occurred in March of 1989.

On August 2nd, 1990, Iraq invaded Kuwait. By the end of February 1991, Iraq had set on fire over 600 oil wells in Kuwait. Although there is currently an over supply of oil now that Operation Desert Storm is successfully completed, there will be a drop in oil production. The restoration of oil production from Kuwait will be a long, complicated, and expensive process.

The demand for natural gas could expand between 10 and 15% before the year 2000. This would mean gas production in excess of 20 trillion ft^3/year again. Electricity, despite increased efficiency in usage, is projected to rise at a yearly average 3.2%. Since there are not any newly ordered nuclear power plants to be on line by the year 2000, fossil-fuel plants must continue to provide most of the generating capacity.

Despite the hope of the early 1970s, nuclear power has never achieved anywhere near the amount of electrical output that had been anticipated. This is in part due to the concern of citizens, a variety of lawsuits, and in a large part due to the Three Mile Island incident in 1979, which still has not been fully resolved. The shoddy construction and need to obtain multiple permits, such as in Indiana, have further aggravated the situation. The U.S. Nuclear Regulatory Commission (NRC) has listed a series of unresolved safety issues. These include systems interactions, seismic design criteria, emergency containment, containment performance, station black out, shutdown decay, heat-removal requirements, seismic qualification of equipment in operating plants, safety implications of control systems, hydrogen control measures and effects of hydrogen burns on safety equipment, high-level waste management, low-level waste management, and uranium recovery and mill tailings.

With the increasing problems related to oil and gas, as well as nuclear power, electric utilities and others may turn increasingly to coal as a source of fuel and power. This occurs at a time when the issue of acid rain is gaining momentum. The current administration in the White House will find it hard to maintain the line against mandated actions related to acid rain. The damage caused by acid rain is increasing, as well as the pressure from Canada to do something about it. Although new energy technologies related to coal, as well as other sources, have been put pretty much into a holding pattern, it seems that it will now be necessary to explore these further and attempt to utilize other means of securing energy.

While it is true that the Clean Air Act of 1990 puts numerous additional constraints on the use of coal as a source of energy, at some time in the future it may be economically feasible to clean up the coal and use this abundant energy resource.

HEALTH IMPLICATIONS OF NEW ENERGY TECHNOLOGIES

Coal, which is our most abundant fossil fuel, presents some of the gravest threats to human health and the environment. Coal miners are exposed to carcinogens and also suffer from pneumoconiosis, black lung disease, silicosis, emphysema, tuberculosis, and bronchitis. Polycyclic aromatic hydrocarbons (PAHs) are found in mixtures within fossil fuels and the byproducts of fossil fuels. The fossil fuel byproducts break out into two major groupings: coal-tar, which comes from the combustion or distillation of coal, and polycyclic aromatic hydrocarbons. Coal tar products can cause skin and lung cancer through the cutaneous and respiratory route.

Human exposure to complex mixtures of PAHs has been extensive, and carcinogenic effects following long-term exposure to them have been documented. Coke-oven emissions and related substances, such as coal-tar, have long been associated with excess disease. The first observation of occupational cancer, that is, scrotal cancer among London chimney sweeps, was made by Percval Pott in 1775. More recently, lung and genitourinary cancer mortality have been associated with coke-oven emissions. Human tumorigenicity has also been associated with the exposure to creosote. *Creosote* is a generic term that refers to wood preservatives derived from coal-tar creosote or coal-tar oil, and includes extremely complex mixtures of liquid and solid aromatic hydrocarbons.

Exposure to many other complex chemical mixtures that include PAHs has been associated with human disease. These mixtures have been found in cigarette smoke, roofing tar emissions, and shale oils.

Bituminous coal is the starting point for coal liquification and coal gasification. The mere process of breaking up the coal's structure appears to release a large number of carcinogenic compounds. Some of these compounds include benz(a)anthracene, chrysene, and benzopyrene. In the coal gasification process, where coal is exposed to molecular oxygen and steam at temperatures of 900°C or higher, much of the hydrocarbon structure is destroyed, and therefore, the

carcinogenic compound level is drastically reduced. However, at these temperatures the hazards due to carbon monoxide increase, as well as the potential hazards
due to other toxic agents such as hydrogen sulfide and other carbon-nitrogen
products that may be released to the environment.

Coal liquification occurs in the temperature range of 450–500°C. This is the
most economical temperature range to produce pumpable liquid. During the initial
conversion reaction, hydrogen is added to the coal, which produces a wide range
of compounds containing condensed aromatic configurations. Many of these are
carcinogenic. A potential also exists for the production of other toxic compounds,
including carbon monoxide.

Oil shale is a generic name for the sedimentary rock that contains substantial
organic materials known as kerogen. The kerogen content of oil shale may vary
between 5 and 80 gal of oil per equivalent ton. The extraction of the oil from the
oil shale occurs when the shale is heated to 350–550°C under an inert atmosphere.
The products include oil vapor, hydrocarbon gases, and a carbonaceous residue.
There are major concerns related to the retorting shale process. (1) The production
of one ton of spent shale per barrel of oil. (2) The volume of the shale increases
by more than 50% during the crushing process and a considerable amount of alkali
minerals contained in the oil shale could reach into the groundwater supply. (3)
Since a large amount of water is needed for controlling dust and reducing the
alkalinity of spent shale, this would create a problem in areas where water is in
short supply.

During the processing of the oil shale, organic materials containing nitrogen,
oxygen, and sulfur are produced in large quantities. Further, waste gases containing hydrocarbons, hydrogen sulfide, sulfur dioxide, and nitrogen oxides are
formed.

Geothermal energy is a general term that refers to the release of the stored heat
of the Earth that can be recovered through current or new technologies. The only
available technologies are those that extract heat from hydrothermal convection
systems, where water or steam transfer heat from the deep parts of the system to
areas where this energy can be tapped. The amount of pollutants found in the
energy source varies dramatically from area to area. In any case, geothermal fluids
may contain arsenic, boron, selenium, lead, cadmium, and fluorides. They also
may contain hydrogen sulfide, mercury, ammonia, radon, carbon dioxide, and
methane. Wastewater management is an environmental problem if the spent
liquids are put into surface water. They may contaminate the surface water and
also cause an elliptical dish-shaped depression, such as occurred in New Zealand.
If the liquids are injected under high pressure back into the geological formation,
they may enhance seismic activities. Another problem that is related to the
production of geothermal energy is noise. Some noise levels reach as high as 120
dBA. Blowouts can occur and produce accidents with potential injuries. Airborne
emissions and noise can affect human health. The most significant potential
problem is hydrogen sulfide gas. Hydrogen sulfide is a very toxic gas that

produces immediate collapse with respiratory paralysis at concentrations above 1000 ppm by volume. Serious eye injuries can occur at 50–100 ppm by volume. Further, hydrogen sulfide has a very offensive odor that can be detected by the human nose at very low concentrations.

The photovoltaic solar cell is an efficient, direct energy-conversion device. The sun falls on this device and liberates electrons, which flow as an electrical current. Solar cells are used for powering most satellites. The major health hazards involved relate to the mining and the various steps needed in the production of the silicon cells. Other cells utilized include cadmium sulfide and gallium arsenide. The health hazards involved relate to the production and use of these chemicals in the occupational setting. If satellite solar power stations come on line in the future, there will need to be an extensive research effort to determine if there will be potential biohazards due to the microwave beams being brought back to Earth from the solar space stations.

Biomass is plant materials and animal waste used as a source of fuel. These plants and animal materials include forest products and residues, animal manure, sewage, municipal wastes, agriculture crops and residues, as well as aquatic plants. The values of biomass is that there is a continuous renewable source of energy. The problems of using biomass for energy purposes are many. They include the following: (1) firewood, used for residential heating, which contributes to air pollution, indoor pollution, and fire hazards; and (2) ethyl alcohol, which can be added to gasoline, in a 1:9 ratio to make gasohol, and comes from major farm crops. Overfarming to produce these crops could contribute to excessive soil loss and nutrient loss. Further, during times of drought, the grains may be needed for food for people or feed for animals.

Health hazards may be associated with the emissions from biomass combustion residuals, biomass gasification residuals, biomass liquification, and air and water pollutants. The burning of biomass creates the same primary air pollutants as in the burning of coal. Unburned hydrocarbons, sulfur oxides, and high-fugitive dust levels are created. Organic acids and minerals that affect water quality may be found in boiler-water treatment chemicals, in leachates from ash residues, and in biomass storages piles. Residential wood burners, because of incomplete combustion, release carbon monoxide and unburned hydrocarbons, which may include photochemically reactive chemicals as well as carcinogens. Emissions from biomass gasification may originate from the process stack, waste ponds, storage tanks, equipment leaks, and storage piles. These pollutants include oxides of nitrogen, hydrogen cyanide, hydrocarbons, ammonia, carbon monoxide, and particulates. The process water and condensates may also contain phenols and trace metals. The tars produced by the thermochemical decomposition of organic substances may contain PAHs, some of which are known carcinogens. Anaerobic digestion may lead to the production of odors, hydrogen sulfide, and ammonia gases. The effluent contains large quantities of biochemically oxygen-demanding materials, organic acids, and mineral salts.

Nuclear power plants have the potential of providing abundant supplies of electricity without contributing substantial amounts of pollutants to the environment. The industry, however, has failed to deliver on the promise because the costs of making nuclear energy safe have spiraled out of control. The nuclear cycle starts with the mining of uranium and the resulting risk of lung cancer in miners attributable to the alpha radiation from the decay of radon-22 daughter products. Further, there are injuries associated with the mining and quarrying of uranium. During the milling of uranium ore, there are small airborne releases of radon and the residues or "tailings" still contain most of the radioactive species of the original ore and require careful disposal. Next, the uranium oxide is converted to uranium hexafluoride. This enrichment increases the percentage amount of fissionable uranium-235. Finally, fuel rods are fabricated during the processing of the uranium after it has been mined. (The potential amount of worker radiation exposure that comes from gaseous solid wastes is small.) The reactor then produces energy. Finally, after the fuel rods are spent, the irradiated rods are removed from the reactor cooler and stored for long periods of time to permit decay of the short-lived fission products before reprocessing. Radioactive water and waste must be contained.

Engineers can build reactors that are safer than those now in operation. The basic technology has been available for more than 20 years. This technology has been ignored in favor of water-cooled reactors, which have already been proven in nuclear submarines. However, these reactors are particularly susceptible to the rapid loss of coolant, which led to the accidents at Three Mile Island and Chernobyl. All nuclear reactors split large atoms into smaller pieces and thereby release heat. It is necessary to keep the core of nuclear fuel from overheating and melting into an uncontrollable mass that may breach the containment walls and release radioactive material. One way to prevent a meltdown is to make sure that the circulating coolant, which is water, will always be present in adequate quantities. To prevent mechanical failures from interrupting the transfer of heat, most reactors employ multiple backup systems. This technique is known as "defense in depth." The problem with this technique is that it can never be 100% safe against a meltdown.

The U.S. Department of Energy wanted to use a new strategy in Idaho Falls, Idaho. They wanted to build a series of four small-scale modular reactors that use fuel in such small quantities that their cores could not achieve meltdown temperatures under any circumstances. The fuel would be packed inside tiny heat-resistant ceramic spheres and cooled by inert helium gas. The whole apparatus would be buried below ground. The main problem is that the smaller units produce less electrical output and, therefore, are less economical initially. However, over time the units may become more efficient.

During the early morning hours on March 28, 1979, the most serious accident in the history of U.S. nuclear power took place. A series of highly improbable events involving both mechanical failure and human error led to the release of a considerable amount of radioactivity and the evacuation of preschool children and pregnant women within 5 mi of the Three Mile Island nuclear power plant, located near Middletown, Pennsylvania. The accident occurred at unit #2 of the complex. The accident was initially triggered by the failure of a valve and the subsequent shutting down of a pump supplying feed water to a steam generator. This in turn led to a "turbine trip" and a shutdown of the reactor. This in itself was not the problem, because of the backup systems that were used anticipate such failures. However, a pressure relief valve failed to close, which led to the loss of substantial amounts of reactor coolant, usually in quench tanks. Auxiliary feed water pumps were nonoperational because of closed valves, in violation of the Nuclear Regulatory Commission (NRC) regulations. The operator failed to respond promptly to the stuck relief valve. There were faulty readings on the control room instruments, which led the operators to turn off the pumps for the emergency core cooling system. The operator shut down the cooling pumps more than an hour after the emergency began. Two days after the initial problems, a large bubble of radioactive gas, including potentially explosive hydrogen, was believed to have formed in the top of the reactor. The President of the United States established a 12-member committee to review the accident and make recommendations. The major findings of the commission after a 6-month study were (1) the accident was initiated by mechanical malfunction and made worse by a series of human errors; (2) the accident revealed very serious shortcomings in the entire government and private sectors systems used to regulate and manage nuclear power; (3) the NRC was so preoccupied with the licensing of new plants that it had not given adequate consideration to overall safety issues; (4) although the training of power plant operators at Three Mile Island conformed to NRC standards, it was extremely inadequate; (5) the utility owning Three Mile Island and the power plant failed to acquire enough information on safety to make good judgements; (6) there was an extremely poor level of coordination and a lack of urgency on the part of all levels of government after the accident occurred. The NRC had not made mandatory, state emergency or evacuation plans.

Other potential emergency situations have occurred in the past. December 12, 1952, the accidental removal of four control rods at an experimental nuclear reactor at Chalk River, Canada near Ottawa led to a near meltdown of the reactor's uranium core. A million gallons of radioactive water accumulated inside. Fortunately, there were no accident-related injuries. On October 7, 1957, a fire occurred in the reactor north of Liverpool, England. Like the Chernobyl facility, the Windscale Tile #1 plutonium production plant used graphite to slow down neutrons emitted during nuclear fission. Two hundred square miles of countryside were contaminated. Officials banned the sale of milk from the cows grazing the area for more than a month. It was estimated that at least 33 cancer deaths could be traced to the effects of the accident. On January 3, 1961, a worker's error in removing control rods from the core of an SL-1 military experimental reactor near Idaho Falls, Idaho caused a fatal steam explosion, and three servicemen were killed. On March 22, 1975, a worker using a lighted candle to check for air leaks in the Brown Ferry reactor near Decatur, Alabama touched off a fire that damaged electrical cables connected to the safety system. Although the reactors coolant water dropped to dangerous levels, no radioactive material escaped into the atmosphere. On March 8, 1981, radioactive wastewater leaked for several hours from a problem-ridden nuclear power station in Tsuruga, Japan. Workers who mopped up the wastewater were exposed to radiation. The public did not become aware of the accident for 6 weeks until radioactive materials were detected. On January 4, 1986, one worker at the Kerr-McGee Corporation, a uranium processing plant in Gore, Oklahoma, died from exposure to a caustic chemical that formed when an improperly heated, overfilled container of nuclear materials burst. Some of the radiation flowed out of the plant. More than 100 people went to local hospitals.

April 26, 1986, the worst disaster in history relating to a nuclear plant occurred in Chernobyl, Russia. A loss of water coolant seemed to trigger the accident. When the water circulation failed, the temperature in the reactor core soared over 5000°F, causing the uranium fuel to begin melting. This produced steam that reacted with the zirconium alloy of the fuel rod to produce explosive-type hydrogen gas. Apparently, a second reaction produced free hydrogen and carbon monoxide. When the hydrogen combined with the oxygen, it caused an explosion, which blew off the top of the building and ignited the graphite. Next, a dense cloud of radioactive fission products went off into the air as a result of the burning graphite. In subsequent days, radioactive materials were found in Finland, Sweden, Norway, Denmark, Poland, Romania, and Austria. In Italy, freight cars loaded with cattle, sheep, and horses from Poland and Austria were turned back because of the concern due to abnormally high levels of radiation found in many of the animals. Britain cancelled a spring tour of the London Festival Ballet, which was to go to the Soviet Union. In West Germany, officials insisted that children be kept out of sand boxes to avoid contamination. Slight amounts of radiation were found in Tokyo, in Canada, and in the United States. In the immediate vicinity of the reactor and up to 60 mi^2 from the reactor, the topsoil may be contaminated for decades. The residents of Kiev, 80 mi from the disaster and the surrounding areas, were told to wash frequently and to keep their windows closed. They were warned against eating lettuce and swimming outdoors. Water trucks were used to wash down streets to wash away radioactive dust. It will be many years before there is an accurate determination as to the number of people who will die as a result of this catastrophe. Approximately 100,000 people will have to be monitored for the rest of their lives for signs of cancer. In 1990, large populated areas surrounding the reactor site in the Ukraine and in nearby Belorussia remain contaminated with high levels of radioactivity. The cost of the cleanup could run as high as $358 billion dollars.

It should be understood that the nation will have to make a myriad of difficult decisions concerning the conservation and use of fuel. Obviously the burning of carbonaceous fuels has significantly contributed to air pollution and accompanying health problems. The energy problem has affected our entire economy by the purchase of energy from abroad. An adequate program of development and use of energy sources within our own country will provide us with the needed energy and yet prevent the many potential hazards that may occur as a result of the burning of fossil fuels. This nation must respond to the new energy crisis of the 1990s in the same way that it responded to every other crisis in its history.

Environmental Problems Related to Energy

The major environmental problems related to energy are caused by pollutants created by fossil fuels, destruction of the natural environment by removal or spillage of fossil fuels, and the effect of the continued rise in cost of fossil fuels on the economy, and therefore, on our way of life. The pollutants will be detailed in the chapter on air pollution. With an increased need for fossil fuels comes an increase in the destruction of the natural environment. Improper fuel removal, fuel storage, or fuel transportation degrades the environment, destroys the aesthetic value of many of our most beautiful areas, and causes destruction of fish, wildlife, and plant life. The economics of energy is extremely important, since the world continues to shift huge sums of money from industrialized societies to oil-producing nations. If the industrialized nations are unable to pay for the energy being provided, mass unemployment and recessions may result. In the Third World, the nations burdened by huge costs for energy will be faced with the choice of literally shutting down their countries or allowing their people to starve because of the tremendous financial drain. The problems related to health, aesthetics, and economics are critical, since they are totally interrelated, and therefore must be considered as a unit.

TRANSPORT AND ALTERATION OF CHEMICALS IN THE ENVIRONMENT

Humans are part of the flow of energy; in the biosphere we interact with thousands of plants and animals. Because of our power and productivity, we have the ability to alter the ecosystems of the Earth, of which we are a part, in a beneficial or harmful manner. Technological advancement since the early 1800s has been revolutionary rather than evolutionary. In the great movement forward to improve our own lifestyle, we have destroyed many of the natural ecosystems and severely polluted much of the air, water, and soil. To better understand the nature of the environmental alterations, it is important to understand the role of chemicals in the environment and how they are transported and altered.

Much of the environment is made up of chemicals, the largest amount of which are occurring naturally and have few detrimental effects on humans. However, we change the form, distribution, and concentration of these chemicals and synthesize

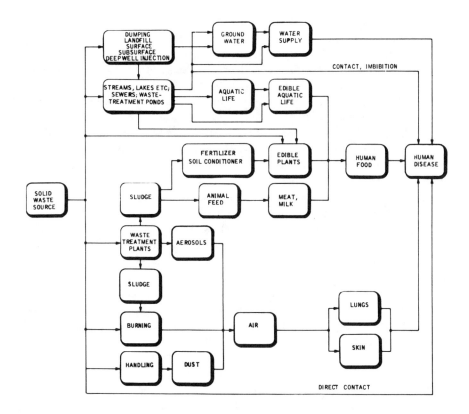

Figure 1.1.   Chemical waste/human disease pathways (postulated).

or produce, either as products or as byproducts, chemicals that are naturally found in the environment, as well as chemical compounds that are not. It is necessary to understand the movement of those chemical compounds that are not natural and to understand the movement of all of these chemicals through the environment, their concentrations, the degree of human exposure, and the potential hazards to human health (Figure 1.1).

Chemicals or pollutants reach us through the air, water, food, skin, and at times through a combination of these routes. We may continuously, intermittently but repeatedly, or sporadically be exposed. We may be exposed to chemicals present in substantial or in minute quantities. In some instances, chemicals present in minute quantities are more hazardous than those in large quantities. There is also direct intentional exposure through chemicals added to foods and cosmetics, and those involved in the preparation of drugs. Good sense dictates limiting the level of exposure of any chemicals that may cause adverse responses in the individual. Chemicals producing eutrophication in lakes and aiding the growth of toxin-producing fungi should be contained. *Aspergillus flavus* produces an aflatoxin in cotton seeds that may be harmful to people. Aflatoxin production can be prevented or reduced by changing the cotton-planting processes. Other toxic substances derive from nuclear fallout, fly ash from power plants, emissions from smelters or refineries, or airborne agricultural chemicals. Many of these substances enter the food chain through crops and eventually affect humans.

Once a pollutant is released into the environment, the chemicals are transported and transformed in a variety of complex ways. They are carried by air, water, or with solid particles; transformed by chemical or biochemical reactions; and diluted, diffused, or concentrated by physical or biological processes. Human health is endangered not only at the point of use or discharge of a specific chemical, but also at varying points in the ecosystem. Unfortunately, many of the fundamental changes occurring in the environment are poorly understood.

Reconcentration, specifically bioconcentration, is a process involving potentially harmful human exposure to chemicals. Plants and animals accumulate certain chemicals at levels higher than those of the ambient environment. Chemicals may also be absorbed by air- or waterborne particles that are inhaled and concentrated in the lungs.

Chemicals are transported widely throughout the environment by air, water, biota (biological life), vaporization, sublimation, diffusion, and leaching. For some substances, only passive transport and diffusion occurs as air or water moves past the substance. Some substances diffuse upward, where they are degraded by ultraviolet light, or diffuse downward and are adsorbed onto the surfaces of suspended particulate matter. Other substances are dissolved in water droplets and returned to Earth by rainfall. Weather conditions may set up cycles of elimination of toxic substances from air onto land and then into water, and eventually they are reintroduced into the air. Where sources of pollution are in close proximity to sources of water usage, an extensive and potentially hazardous series of chemicals may be transported directly into the water source and cause significant health problems (Figure 1.2).

Dispersion of Contaminants

Contaminants tend to spread continuously within the environment. The characteristics of the medium affect dispersion of the contaminants. When contaminants are dispersed within the atmosphere, it is called diffusion, which may be of a convective or turbulent manner. Convection is the circulatory motion that occurs in a fluid at a nonuniform temperature owing to the variation of its density and the action of gravity. Turbulence in the atmosphere is a complicated phenomenon that depends upon the wind velocity, the direction of the wind, the altitude, and the friction of the air over the surface of the Earth, the temperature of the air, the pressure in the air and the presence or absence of bodies of water, mountains, and

Chemicals in the Environment

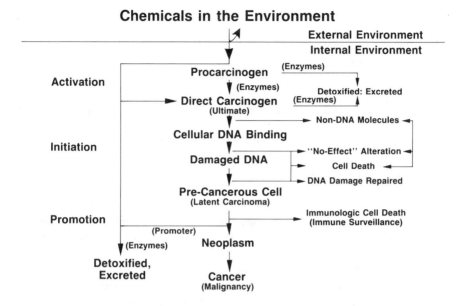

Figure 1.2. Chemicals in the environment. *Source: Environmental Toxicology and Risk Assessment: An Introduction, Student Manual* (Chicago, IL: United States Environmental Protection Agency, Region V), Visual 3.8.

flat areas. Dispersion from ground-level sources are seen almost immediately downwind from the source. Dispersion from tall smoke stacks or any elevated discharge may be found for considerable distances from the source.

Dispersion in the hydrosphere is much more varied and complex than dispersion in the atmosphere. The dilution volumes and mixing characteristics, as well as the rate of transport of the contaminants, vary with rivers, lakes, estuaries, coastal waters, and oceans. The contaminants in the water may further contaminate the land, the groundwater supply, and the air.

The environment of the soil is very complex with regards to the dispersion of contaminants. The dispersion depends upon the specific nature of the chemicals, the type of soil, and other factors, such as moisture, pH, and temperature. Most chemical contaminants do not move readily through the soil once they enter it, unless they channel into a groundwater supply. Gases may also easily diffuse through air pockets and channels in the soil. Soil water is the major means of moving chemicals downward to the lower horizons of the soil or upward to the point of evaporation. Water may also move the contaminants in a lateral manner or help in surface runoff depending upon the slope of the soil. The more rapidly water percolates through the soil, the more readily the chemicals may diffuse into the pores between the soil particles. If a chemical is of a higher molecular weight, such as DDT and PCBs (halogenated organic compounds), little movement will occur through the soil. Further, ammonium and phosphate ions are strongly bound to soil particles, while nitrates are not, and therefore, readily leach through the soil.

Alteration of Contaminants

Contaminants are removed from the atmosphere by a variety of means. The contaminants may either be in the original form or may have altered during the dispersion process. Particles are removed from the atmosphere by gravity, impacting on objects, and through rain, snow, or sleet. The larger the particle, the more rapidly it will settle out. The particles may attract gases, vapors, and/or viruses and bacteria. These gases and vapors may also be dissolved in moisture, or through chemical reactions in the atmosphere they may be converted into other gases or particles. Chemicals found in water may be bioconcentrated in fish or plant life, deposited on surfaces, or turned into other products by the self-purification process that is related to chemical oxidation. Physical self-purification may occur when the chemicals are bound to suspended particles in water.

The persistence of chemicals in the soil is a complex function, related to physical, chemical, and biological factors. The major means of removal of contaminants in soil are by degradation due to microbial action, chemical degradation, evaporation, volatilization from the surface, and uptake by vegetation.

In order to get some better understanding of transport and alteration of chemicals in the environment, seven chemicals have been chosen for discussion. These seven all have been identified on the CERCLA (Superfund) National Priorities List. They are part of a large group that pose the most significant potential threat to human health, as determined by the Agency for Toxic Substances and Disease. These chemicals are chrysene, chloroform, cyanide, lead, 2,3,7,8-tetrachloro-dibenzo-p-dioxin (2,3,7,8-TCDD), tetrachloroethylene, and trichloroethylene.

Chrysene is found in the environment from natural sources, such as forest fires and volcanos, and from man-made sources. Chrysene, like all PAHs, is formed during high-temperature pyrolytic processes. Combustion is the major source of environmental chrysene. Virtually all direct releases into the environment are to the air, although small amounts may be released to the water and land. Chrysene is removed from the atmosphere by photochemical oxidation and dry or wet deposition onto land or water. It is very persistent in either soil or sediment. Biodegradation is a slow process. The half-life of chrysene in the soil is estimated to be 1000 days.

Incomplete combustion of carbonaceous materials is the major source of chrysene in the environment. Residential heating and open burning are the largest combustion sources. The inefficient combustion process creates uncontrolled emissions, with heating in the home as the greatest contributor to chrysene emissions. Hazardous waste sites can be concentrated sources of PAHs on a local scale and, therefore, may contain larger quantities of chrysene. Chrysene in the atmosphere is expected to be associated primarily with particulate matter, especially soot. Chrysene in aquatic systems is expected to be strongly bound to suspended particles or sediments. Chrysene is expected to be strongly sorbed to surface soils.

Chloroform is a colorless or water-white liquid used to make fluorocarbon-22. It is also used as a pesticide and as a solvent in the manufacture of pesticides and dyes. Chloroform is discharged into the environment from pulp and paper mills, pharmaceutical manufacturing plants, chemical manufacturing treatment plants, and from the chlorination of wastewater. Most of the chloroform released ends up in the air. It may be transported for long distances before being degraded by reacting with photochemically generated hydroxyl radicals. Significant amounts of chloroform from the air may be removed by precipitation. However, the chloroform may reenter the air by volatilization. When released to soil, chloroform will either volatilize rapidly from the surface or leach readily through the soil to the groundwater, where it may persist for relatively long periods of time. It is estimated that its half-life in the atmosphere is 70–79 days. However, when chloroform is found in photochemical smog, its half-life is 260 days.

Volatilization is the primary mechanism for removal of chloroform from water. Its half-life in water may vary from 36 hr to 10 days. Although chloroform has been found to adsorb strongly to peat moss and less strongly to clay, typically it will volatilize rapidly in either dry or wet soil.

Although cyanides are naturally occurring substances found in a number of foods and plants and produced by certain bacteria, fungi, and algae, the greatest amount found in the environment comes from industrial processing. Hydrogen cyanide is used primarily in the production of organic chemicals. Cyanide salts are used primarily in electroplating and metal treatment. The major sources of cyanide released to water are reported to be discharges from metal finishing industries. The major source of the cyanide released to air is vehicle exhausts, and the major sources of cyanide released to soil appear to be disposal of cyanide waste in landfills and the use of cyanide-containing road salt. Cyanide released to air appears to be transported over long distances before reacting with photochemically generated hydroxyl radicals. The half-life of cyanide is 334 days. In water cyanide may volatilize, or if in an alkali metal salt form, it may readily disassociate into anion and cations. The resulting cyanide ion may then form hydrogen cyanide or react with metals to form metal cyanides. Insoluble metal cyanides are expected to adsorb to sediment and possibly bioaccumulate in aquatic organisms.

In soil, cyanides may occur in the form of hydrogen cyanide, alkali metal salts, or immobile metallocyanide complexes. At soil surface with a pH less than 9.2, it is expected that hydrogen cyanide will volatilize. In subsurface soil, where the cyanide present is at a low concentration, it is probably biodegradable. In soil with a pH less than 9.2, hydrogen cyanide is expected to be highly mobile, and where cyanide levels are toxic to microorganisms such as in landfills or in spills, the cyanide may reach the groundwater.

Lead is a naturally occurring element that may be found in the Earth's crust and in all parts of the biosphere. It is released into the atmosphere by a variety of industrial processes and from leaded gasoline. The deposition of lead found in both soils and surface waters generally comes from the atmosphere. Lead is transferred continuously among air, water, and soil by natural chemical and physical processes, such as weathering, runoff, precipitation, dry deposition of dust, and stream/river flow. Long distance transport of up to about 600 mi may occur. Lead is extremely persistent in both water and soil. In the atmosphere, lead exists primarily in the particulate form. The chemistry of lead in the water is highly complex, because it may be found in a multitude of forms. The amount of lead that remains in solution depends upon the pH of the water and the dissolved salt content. In most soils, the accumulation of lead is primarily a function of the rate of deposition from the atmosphere. Most lead is maintained strongly in the soil and very little is transported into surface water or groundwater.

The chemical 2,3,7,8-TCDD is commonly called dioxin, which is an inaccurate colloquial name. Except when it is used as a reference standard, it is not intentionally manufactured by any industry. It is inadvertently produced in very small amounts as an impurity during the manufacture of certain herbicides and germicides. It is also produced during the incineration of municipal and industrial wastes. The environmental fate of 2,3,7,8-TCDD is not understood with certainty for air, water, and soil.

Tetrachloroethylene is a nonflammable liquid solvent widely used for dry cleaning fabrics and textiles, and for metal-degreasing operations. It is used as a starter chemical for the production of other chemicals. It is also called perchloroethylene, perc, PCE, perclene, and perchlor. Most of the tetrachloroethylene used in the United States is released into the atmosphere by evaporative losses. The dry cleaning industry is the major source. When the chemical is released in water, it rapidly volatilizes and returns to the air. However, rainwater will dissolve it, and once again it falls to the land. The usual transformation process for tetrachloroethylene in the atmosphere is the result of the reaction with photochemically produced hydroxyl radicals. The degradation products of this reaction include phosgene and chloroacetyl chlorides (phosgene is a severe respiratory irritant).

Trichloroethylene is used as a solvent for removing grease from metal parts. It is released into the atmosphere by evaporative losses. It may also be released into the environment through gaseous emissions from waste-disposal landfills, and it may leach into groundwater from waste-disposal landfills. It is dissolved in rainwater. The dominant transformation process for trichloroethylene in the atmosphere is its reaction with sunlight-produced hydroxyl radicals. The degradation products of this reaction include phosgene, dichloracetyl chloride, and formyl chloride.

Toxic Chemicals and the Environment

The EPA is authorized to control the risks of over 65,000 existing chemical substances that are used today, as well as to do a review on new chemicals before they are put into the marketplace. This huge number of chemicals are essential to our lifestyle in the 1990s and into the future. Most of these chemicals are not harmful if used properly. Others can be extremely harmful if individuals are exposed for even a short period of time. The health effects may range from cancer to birth defects to the destruction of various parts of the environment. Some examples of these problems include the following:

1. exposure to asbestos and the potential for causing cancer and lung disease
2. contamination of food or water with the pesticide ethylene dibromide
3. contamination of water and shellfish with PCBs
4. buildup in wildlife of DDT, aldrin, and chlordane
5. other chemicals, such as benzene and a group of other chlorinated hydrocarbons may cause cancer, birth defects, and other severe health problems

As a result of a major need to control our chemical environment, the Congress has passed a series of laws. These include the following:

1. Toxic Substances Control Act, which authorizes the EPA to regulate the production, use, or disposal of chemicals
2. The Federal Insecticide, Fungicide and Rodenticide Act, which authorizes the EPA to register all pesticides and to remove unreasonable hazardous pesticides from the marketplace
3. The Federal Food Drug and Cosmetic Act, which authorizes the EPA to cooperate with the FDA to establish tolerance levels for pesticide residues on food and in food products
4. The Resource Conservation and Recovery Act, which authorizes the EPA to identify hazardous wastes, and to regulate their generation, transportation, treatment, storage, and disposal
5. The Comprehensive Environmental Response Compensation and Liability Act, which requires the EPA to designate hazardous substances that can present substantial danger to the public and also authorizes the cleanup of sites contaminated with these substances
6. The Clean Air Act, which authorizes the EPA to set emission standards to limit the release of hazardous air pollutants
7. The Clean Water Act, which requires the EPA to establish a list of toxic water pollutants and set appropriate standards to avoid health problems

8. The Safe Drinking Water Act, which requires the EPA to set drinking water standards to protect the public health of our society from hazardous substances
9. The Marine Protection Research and Sanctuaries Act, which regulates ocean dumping of toxic contaminants
10. The Asbestos School Hazard Act, which authorizes the EPA to provide loans and grants to schools to help abate severe asbestos hazards
11. The Asbestos Hazard Emergency Response Act, which requires the EPA to establish a comprehensive regulatory system for controlling asbestos hazards in schools
12. The Emergency Planning and Community Right-to-Know Act, which requires states to develop programs to respond to a hazardous chemical release and also requires industries to report the presence of hazardous chemicals and the release of hazardous chemicals

The potential problem of hazardous chemicals in the environment is in many ways tied to the fact that we are using organic compounds that come from oil. The rest of the chemicals that are inorganic, such as ammonia, chlorine, and metals, may also be hazardous to us. These chemicals are used in preserving food, in transportation, communication, and in a variety of many other kinds of systems that are needed for our lifestyles.

ENVIRONMENTAL PROBLEMS AND HUMAN HEALTH

This volume concentrates on problems related to human health, rather than to the various other ecosystems coexisting on Earth. Numerous publications are available to the reader interested in exploring the environment and the human effect on the ecosystem as it relates to various animals and plants.

The decay of the environment and the resultant effect on human health grows daily. The sun is shrouded by the smoke of industry and dwellings. Our lungs are blackened, irritated, and corroded by air pollutants. Emphysema and lung cancer are increasing sharply. Our eyes tear. Our noses are offended by acrid, noxious doses of chemicals, overflowing sewage, fumes, and other obnoxious odors. We are deafened by the noise of traffic, industry, construction, and jet aircraft.

Pesticides essential to food production and protection against the severe epidemics of the past, such as malaria, yellow fever, and plague, are misused and frankly abused. Too many pesticides are being applied carelessly by uninformed or indifferent individuals. Research on the ultimate effects of pesticides on humans and their environment has not kept pace with the sharp increase in usage in the last 20 years. Over three billion pounds of pesticides are applied annually in the United States alone.

We continue to use dangerous, misbranded, or adulterated food produced under unsanitary conditions despite several governmental enforcement acts, including the Pure Food and Drug Act of 1906, the Food, Drug, and Cosmetic Act of 1938, and its numerous amendments in the 1950s, 1960s, 1970s, and 1980s. *S. aureus* is present in most supplies of pooled raw milk. *Salmonella* organisms are found in 15–30% of raw dressed poultry and in many commercial egg products. *C. perfringens* is present on more than half of the red meat sold. *C. botulinum*, type E, is frequently found in raw fish. *S. aureus* is present on the mucous membranes or skin of 30–50% of the population. Chemical residues, pesticides, and additives are components of food that may cause subtle problems. Although the overall incidence of food-borne disease is unknown, it is estimated from statistics of actual outbreaks that 5–10 million cases of gastroenteritis occur each year and that 200,000–1,000,000 of these cases are salmonellosis. In smaller communities, food-service programs are either inadequate or totally lacking.

The residential environment for 6 million poor American families is the hazardous blighted slums of 19th century England. The healthful, pleasant, attractive, comfortable housing of the make-believe world of television is replaced in reality by tin and tarpaulin shacks; crowded, ramshackle houses; and tenements overflowing with garbage, flies, roaches, rodents, and sewage. Walls and floors deteriorate, lights fail, windows disintegrate, heating is poor, and ventilation is almost nonexistent. Lead poisoning, tuberculosis, and infectious diseases are endemic in this environment.

In the suburban areas, inadequate planning and poor land utilization have led to the subdivision nightmare, where children play in overflowing sewage and houses crack, settle, and are subject to flooding. Home accidents related to housing have sharply increased during the last 30 years.

Insects and rodents spread disease, cause annoyance, destroy crops, and ruin property. During the outbreaks of bubonic plague in Europe in the 14th and 15th centuries, an estimated 25 million people died of this rat-borne disease. In the 1980s, bubonic plague was still present in southwestern United States. Yellow fever affected 23,000 out of a population of 37,000 people in Philadelphia in 1793. Four thousand people died. In addition, malaria, encephalitis, spotted fever, tick fever, tularemia, and rickettsial pox have occurred regularly during the 19th and 20th centuries. In 1959, an epidemic of eastern encephalitis occurred in New Jersey, and in 1962 an epidemic of St. Louis encephalitis occurred in the St. Petersburg-Tampa Bay area of Florida. In 1975 and 1976, St. Louis encephalitis spread to many states. In 1986, Houston, Texas experienced its largest outbreak since 1980. It has also reoccurred in Long Beach, California. Further, from 1984 to present, there has been a substantial amount of Lyme's disease from ticks. Tens of thousands of rat bites occur each year, notably to very young children.

Many individuals spend part of each day or most of their lives in a variety of institutions. Each institution, particularly if it provides sleeping accommodations, is in effect a small community, with all of the environmental health problems discussed in this book, and the added problem of a rapidly shifting mobile population that is highly susceptible to disease and accidents.

Noise, or unwanted sound, causes temporary or permanent hearing loss, physical and mental disturbances, breakdowns in the reception of oral communications, reduced efficiency in performing work-related tasks, irritability, disruption of sleep and rest, and increased potential for accidents. Steady exposure to 90 dB of sound can cause eventual hearing loss. Heavy traffic can reach 90 dB, and the noise of jet planes, rock-and-roll bands, motorcycles, power mowers, auto horns, heavy construction, and farm equipment exceed this level.

Occupational hazards occur in all industries. In 1986, 10,700 workers died of occupational hazards and 1,800,000 were the victims of disabling injuries. Billions of dollars are spent each year on worker's compensation cases for injuries and work-related illnesses. These problems are caused by accidents and various health hazards, such as toxic chemicals, including dusts, gases, fumes, mists, vapors; physical agents, including noise, pressure, ionizing radiation, and severe temperature variations; biological hazards, including insects, bacteria, yeasts, fungi, and viruses; and other hazards, including unusual work-related posture, boredom, fatigue, repetitive motion, and monotony.

Low and high levels of radiation may adversely affect human health by causing genetic damage, burns, destruction of tissue, reduced life span, and cancer. Old X-ray and medical fluoroscopic machines may contain inadequate devices to filter out unneeded or stray radiation. Improper use, storage, and/or disposal of radioactive material and wastes could lead to serious potential hazards.

Outdoor recreation is a complex and rapidly expanding area of the economy. Americans spend over 30 billion dollars a year on recreation and over 100 million Americans vacation each year. The environmental health problems resulting from these activities are enormous. They include all areas discussed in these books, plus the problems of mass migrations; inadequate housing, water, and sewage facilities; and a high potential for accidents.

Soil is the single most important natural factor in choosing a site for construction of dwellings and for developing an effective, operating, on-site sewage disposal system. Unfortunately, lack of knowledge in this area, poorly designed tests, and inadequate supervision have led to severe settlement cracks in houses, flooded basements, flooded homes, slippage of hillsides and houses, soil erosion, contaminated or inadequate water supplies, and overflowing sewage. The disease and accident potential is substantial.

Solid wastes generate enormous economic, aesthetic, social, and health problems. Our affluent society, improved technology, new packaging methods, and disposable items have increased the amount of solid waste per person per day from 2.65 lbs in 1960 to 3.58 lbs in 1986 and growing. At this rate, our society will become inundated by waste unless changes in waste production and disposal are made immediately. Unsatisfactory storage, collection, and disposal of solid waste lead to insect and rodent problems, air pollution, offensive odors, accidents, fires, explosions, contamination of the water supply, degradation of the landscape, destruction of fish, and conversion of bodies of water into open sewers. Hazardous waste has complicated the problem enormously.

Water, our most precious resource, is used to sustain life, support the growth of food, develop business and industry, and provide recreation. Despite these essential human needs, individuals, municipalities, industries, commercial establishments, and agriculture continue to pollute our water supply. Historically many of our worst epidemics have been caused by contaminated water. Although in 1990 the epidemics are gone, water-borne outbreaks of disease still occur and the potential for disease outbreaks continues. In addition, a new hazard has been added: chemical pollution. Mercury has caused food poisoning in Japan and Sweden. Oil spills have caused the deaths of birds and fish, and have befouled once lovely beach areas. Insecticides have caused millions of fish to die and could well cause serious harm to people. Fertilizers are helping to destroy our streams and lakes by transforming them into open sewers.

Potable water used for drinking and swimming may be unsafe because of physical, chemical, and bacteriological hazards. Physical deficiencies exist because of inadequate groundwater sources, inadequate design of treatment plants, inadequate disinfection capacity, inadequate system capacity, improper design of swimming pools, and inadequate training of water plant and swimming pool operators. Poor, outmoded, unsafe plumbing can easily lead to contamination of potable water by nonpotable water through cross-connections and submerged inlets.

PHYSIOLOGY

Physiology is the basic biomedical science dealing with function and living organisms. To understand function it will also be necessary to understand something about anatomy. A major concern of physiology is how environmental factors influence the function of individuals. The intracellular processes in people proceed properly if the fluid environment surrounding each cell is maintained in a nearly constant state. Temperature, oxygen supply, acidity, and nutrients must be at a nearly constant level. The respiratory, circulatory, alimentary, and excretory systems help maintain this internal environment of the body. The external environment that the individual is subjected to varies considerably. There are wide ranges of temperature, humidity, ionizing and nonionizing radiation, pressure, toxic gases, particles, and a variety of chemicals, as well as microorganisms, in this environment that can affect people. All of these factors alone or in combination can elicit a biological response that can be harmful to the individual. In order to understand the adverse physiological reactions caused by chemicals, physical agents, and microorganisms, it is necessary to have some knowledge of the cell and a select group of body systems.

The Cell

The cell is the basic unit of all living organisms. The cell consists of the following: cytoplasmic membrane, endoplasmic reticulum (ER), Golgi apparatus, mitochondria, lysosomes, ribosomes, nucleus, and nucleoli. The cytoplasmic membrane is the boundary of the cell and maintains its integrity. The endoplasmic reticulum serves as the cell's circulatory system. The Golgi apparatus synthesizes carbohydrates and then combines with protein to produce glycoprotein. The mitochondria function in catabolism and ATP (adenosine triphospate molecules) synthesis. The lysosomes are the cell's digestive system. The ribosomes synthesize proteins. The nucleus dictates how protein synthesis occurs. Therefore, it directs the other cell activities, such as active transport, metabolism, growth, and heredity. The nucleoli are essential in the formation of ribosomes. The nucleus stores, transcribes, and transmits genetic information. This information is stored in the DNA molecules present in the nucleus. The sequence of the DNA base pairs is transcribed into the sequence of base pairs in RNA molecules. This RNA is known as the messenger RNA. The messenger RNA then acts as a bearer of a coded message that is translated in the ribosome into the specific proteins they synthesize. The DNA and RNA process are very complex mechanisms.

Metabolism consists of catabolism, which is a decomposition process, in which relatively large food molecules are broken down to yield smaller molecules and energy, and anabolism, in which synthesis occurs, using energy to build relatively small molecules into larger molecules. Enzymes, hormones, and antibodies are all produced during the process of anabolism. Enzymes catalyze both catabolic and anabolic chemical reactions. There are intracellular enzymes that are synthesized within the cell and also function within the cell. There are extracellular enzymes that are synthesized in cells and function outside of the cells. Energy changes accompany metabolic reactions. Catabolism releases energy as heat or chemical energy. Chemical energy released by catabolism is first converted into ATP molecules. The ATP molecules, which are easily broken, then release the energy, which can be utilized in anabolism.

The Blood

The primary function of blood is the transportation of various substances to and from the body cells and the exchange of materials, that is, oxygen, nutrients, and waste materials, between the respiratory, digestive, and excretory organs. Its secondary function contributes to the homeostasis of fluid volume, pH, and temperature. It is also necessary for cellular metabolism and defense against microorganisms. (Homeostasis is the maintaining of the internal environment of the body in a relatively uniform manner.) The blood consists of the erythrocytes or red blood cells, leukocytes or white blood cells, and thrombocytes or platelets. The red blood cells transport oxygen and carbon dioxide. The white blood cells are important in the immune response. The platelets initiate blood clotting and homeostasis, which is the arrest of bleeding. The plasma is the liquid part of the blood minus its cells.

The immunological response starts when bacteria or viruses invade the body. Scavengers, including neutrophils, are among the first line of defense at the site of the infection. Neutrophils are produced in the bone marrow. They survive for a few days. Next, the complement system, which is a group of at least 20 proteins circulating in the blood, stick to the microorganism, setting off a chain reaction that eventually will destroy it. Macrophages, which are long-lived scavengers, migrate through the body and engulf foreign matter as well as cellular debris. They signal other cells in the immune system to confront and destroy the microorganisms. A macrophage, after ingesting the microbe, shows specific markers on its surface. These markers or antigens signal other immune system cells, called helper T-cells, to come forth. T-cells have their own set of receptors that can recognize specific antigens of a given microbe. The T-cell grows and divides. The helper T-cells start to reproduce when a protein, which is called interleukin-1, is released by a macrophage. The helper T-cells then produce a variety of interleukins that activate other T-cells and B-cells. They also produce gamma interferon, which activates the macrophages. The B-cells that are stimulated by the helper T-cells divide and mature into plasma cells. The plasma cells produce antibodies that are directed against the specific antigens. (Antibodies are proteins that recognize and bind to a specific microbe.) This either stops the microbe from going further or it makes the microbe more vulnerable to macrophages and neutrophils. Antibodies also activate the complement system. It takes about 1 week for the immune system to be at its highest level of functioning. Killer T-cells recognize and destroy virus-infected cells or cancer cells. They use lethal proteins that punch holes in the cell's membrane and cause it to rupture. The suppressor T-cells probably work by sending out chemical signals that either slow or stop the immune reaction after the organisms have been destroyed. This keeps the body from attacking itself. Memory cells are types of B- and T-cells that circulate through the body after an infection has occurred. The next time the antigens are released, the memory cells start the process of destroying the antigens. Cancer cells, whether they are created by chemicals, radiological agents, or possibly viruses, are surrounded by killer T-cells because they were attracted by the surface antigens. These cells release chemicals that break down the membrane of the cancer cells and cause the cells to die.

The Lymphatic System

The lymphatic system is a specialized part of the circulatory system. It consists of a moving fluid, lymph. The lymph comes from the blood and tissue fluid, and returns to the blood. Lymph is a clear, watery fluid that is found in the lymphatic vessels. The interstitial (intercellular) fluid fills the spaces between the cells. The clear watery fluid, in fact, is complex and organized material. It makes up the internal environment of the body. Both lymph and interstitial fluids closely resemble blood plasma in composition. The main difference is that they contain a lower percentage of proteins than does the plasma. The lymphatic system, like the blood system, can carry contaminants throughout the body. The lymphatics return water and proteins from interstitial fluid to the blood from where they came. Lymph nodes have two major functions. They filter out injurious substances and phagocytose them. Unfortunately, sometimes the number of microorganisms entering the lymph nodes are greater than the phagocytes can destroy, and therefore the node becomes infected. Also, cancer cells that break away from a malignant tumor may enter the lymphatics, travel to the lymph nodes, and set up new sites of cancer growth. The lymphatic tissue of the lymph nodes forms lymphocytes and monocytes, which are the nongranular white blood cells and plasma cells. This is called hemopoiesis.

Membranes

Membranes are thin sheets of tissues covering or lining the various parts of the body. Four important membranes are mucous, serous, synoval, and cutaneous (skin). The mucous membranes line the cavities of passageways of the body that open to the exterior. This includes the lining of the mouth, digestive tract, respiratory tract, and genitourinary tract. Mucous membranes protect the underlying tissue, secrete mucus, and absorb water, salts, and other solutes. The serous and synoval membranes line the closed cavities of the body. The pleura is the serous membrane that lines the thoracic cavity. The mucous film of this lining consists of a superficial gel layer that traps the inhaled particles. The ciliated mucous membrane contains goblet cells. The cilia propel the mucus upward in the respiratory tract to the pharynx, which causes a coughing reflex. Another serous membrane is the pericardium, the sac in which the heart lies. The synovial membrane has smooth, moist surfaces that protect against friction. It lines the joints, tendon sheaths, and bursae.

The cutaneous membrane (skin) protects the body against various microorganisms, sunlight, and chemicals. The skin helps to maintain the normal body temperature by regulating the amount of blood flowing through it and by sweat secretion. Skin consists of an epidermis, dermis, and accessory organs of the skin, such as the hair, nails, and skin glands. The very top layer of the epidermis is composed of dead cells and is virtually waterproof. Because of the innumerable microscopic nerve endings throughout the skin, the body is kept informed of changes in the environment. Finally, beneath the dermis is the subcutaneous fatty tissue.

Glands

The two main kinds of multicellular glands are the endocrine or ductless and the exocrine or duct glands. Endocrine glands secrete into the blood capillaries, whereas the exocrine glands secrete into the ducts that open on the surface of the epithelium. The endocrine glands include the pituitary gland, pineal gland, thyroid gland, parathyroid gland, thymus gland, adrenal glands, pancreas (cells of islands of Langerhans), and ovaries or testes.

Nervous System

The nervous system is a major interface between humans and the environment. It is involved in maintaining homeostasis. It controls posture and body movements. It is the center of subjective experience, memory, language, and the thought process that is peculiar to human activity. The fundamental unit of the nervous system is the neuron, which consists of a cell body, dendrites, and the axon, which may be several feet in length. The dendrites conduct impulses to the cell body of the neuron. Receptors that receive the stimuli that initiate conduction are the distal ends of dendrites of sensory neurons. (Distal is the furthest point away from the point of origin.) The neuron axon conducts impulses away from the cell body. It can vary in diameter from about 2 μm down to 1 μm. In general, the larger diameter of the axon, the more rapid the impulses will be conducted. About 10% of the cells in the nervous system are neurons; the rest are supporting elements.

The connections between the nerve cells are called synapses. They play a key role in the transmission of impulses in the nervous system. A single cell may be connected to as many as 15,000 other cells by means of synapses. The transmission of impulses occur from the depolarization of the cell membrane brought about by chemical or mechanical events. The nerve impulse or action potential is a self-propagating impulse of electrical negativity that travels along the surface of the neuron's cytoplasmic membrane. A stimulus to the neuron greatly increases the membrane's permeability to sodium ions. Sodium ions rush into the cell at the point of stimulation. The membrane becomes depolarized at the point of stimulus. This occurs for an instant. As more sodium ions come into the cell, they produce an excess of positive ions inside the cell and leave an excess of negative ions outside the cell. The negatively charged point of the membrane sets up a local current, with the positive point adjacent to it. The local current acts as the stimulus. Within a fraction of a second, the adjacent point on the membrane becomes depolarized and its potential reverses from positive to negative. The cycle keeps repeating itself until the electrical current, in a wave motion, moves the full length of the neuron.

Many nerve impulses travel a route called the reflex arc. The reflex arc consists of two or more neurons arranged in a series that conduct impulses from the periphery or from the outside portion to the central nervous system. The receptors are stimulated, and the nerve impulse flows into the spinal cord and brain and back out to the effectors, which aid in the carrying out of the intended action. The synapse is the place where the nerve impulses are transmitted from one neuron to another neuron. For instance, if there is an environmental change, such as a rise in temperature, that is a stimulus to the receptor neuron, it then goes along the afferent nervous pathway to carry the action potentials to the central nervous system. Within the spinal cord or brain, there is an integrating center where the action potentials are now sent out along different pathways from the central nervous system and the effector is activated. The body will now make some adaptation to the temperature change, if it is possible.

The synapse is the place where the nerve impulses are transmitted from one neuron to another neuron. A synapse consists of a synaptic knob, a synaptic cleft, and the cytoplasmic membrane of the neuron dendrite or cell body. Since an action potential cannot cross the synaptic clefts, a chemical mechanism operates, where the chemical, which is called a neurotransmitter, is released from the synaptic knobs into the synaptic cleft. The neurotransmitter molecules diffuse rapidly across the microscopic width of the synaptic cleft and bind to specific protein molecules called neurotransmitter receptors. This leads to an opening of channels in the membrane through which sodium ions diffuse into and potassium ions diffuse out of the interior of the postsynaptic neuron. The message now moves forward.

There are three functional classes of neurons. First are the afferent neurons, which frequently are connected to sensory receptors, such as touch, taste, smell, and sight. They transmit information in the form of a coded action potential from the peripheral nervous system to the central nervous system (CNS). Second, the efferent neurons transmit action protentials from centers of the spinal cord and brain to the skeletal muscle, smooth muscle, or secretory cells in the periphery. Third, the interneurons make up about 97% of the total neurons and provide the vast number of connections in the central nervous system. The sensory receptors are either responsive or show a lack of responsiveness to different kinds of energy in the environment.

Several different compounds serve as neurotransmitters. Acetylcholine is released at neuromuscular junctions with skeletal muscle cells. Acetylcholine is also released at neuromuscular junctions with smooth muscle cells, cardiac muscle cells, and in neuroglandular junctions. Norepinephrine is released at other neuroglandular junctions and at some neuromuscular junctions with smooth muscle and cardiac muscle cells. The enzyme acetylcholinesterase rapidly halts the stimulation of muscle cells by acetylcholine. The enzyme hydrolyzes acetylcholine to acetate and choline.

The nervous system is composed of the central nervous system and the peripheral nervous system. The CNS consists of the brain and spinal cord. The cranial nerves, spinal nerves, and ganglia make up the peripheral nervous system. The nervous system is also classified by the effectors innervated, that is, the somatic nervous system and the autonomic nervous system. The somatic nervous system consists of the brain, spinal cord, cranial nerves, and spinal nerves. It innervates the skeletal muscles. The autonomic nervous system is composed of the autonomic or visceral motorneurons, which innervate the cardiac muscle, smooth muscle, and glandular epithelial tissues.

The autonomic system is divided into two parts by function. The sympathetic part is an emergency system that greatly increases sympathetic impulses to most visceral effectors during stress and prepares the body for the use of maximum energy. It increases the secretion of epinephrine by the adrenal medulla, which in turn increases and prolongs the effects of norepinephrine. The parasympathetic nervous system regulates many visceral effects under normal conditions. Its neurotransmitter, acetylcholine, stimulates the digestive juices, insulin secretion, and contraction of the smooth muscle of the digestive tract. The autonomic system does not function autonomously or independently of the CNS.

The CNS is intimately related to the endocrine system. The two systems perform the same general function for the body of communication, integration, and control. Whereas the nervous system sends nerve impulses conducted by neurons from one specific structure to another, the endocrine system sends tiny quantities of chemical messengers, known as hormones. Nerve impulses produce rapid, short-lasting responses, whereas hormones produce slower and longer lasting responses. The cerebral cortex sends impulses into the hypothalamus, which controls the production and secretion of six separate hormones of the interior pituitary. The hypothalamus secretes the antidiurectic hormone (ADH), which regulates the reabsorption of water into the kidney. The hypothalamus also controls the autonomic nervous system, which in turn controls the secretion of epinephrine.

The Respiratory System

The respiratory system's function is to distribute air and to act as a means of exchange of gases, so that oxygen may be supplied to body cells and carbon dioxide may be removed from them. The circulatory system is jointly responsible for meeting the respiratory needs of the body, since it provides a mechanism for oxygen to reach all of the body cells. The nose, pharynx, larynx, trachea, bronchi, and bronchioles act as a conduit for air into the alveoli, where the actual gas exchange takes place with the blood capillaries. The nose also filters out impurities and warms, moistens, and chemically examines the air for substances that might become irritating to the mucous lining of the respiratory tract. It is also an organ of smell, and acts as an aid in phonation (the production of vocal sounds, especially speech). The pharynx is a corridor for the respiratory and digestive tracts. Both air and food pass through the structure before reaching the appropriate tubes. It is also important in phonation. The larynx or voice box is another part or the corridor to the lungs. It protects the airway from solids or liquids during swallowing. It is also the organ of voice production. The trachea is a continuation of the corridor through which air passes to the lungs. The bronchi is a continuation of the trachea and aids in the distribution of air to the interior of the lungs. The alveoli are surrounded by networks of capillaries. This is where the main function of the lungs occurs, that is, gas exchange between the air and the blood. The lungs contain the bronchus, bronchioles, alveoli, pleura (sac around the lungs), and alveolar sacs (Figure 1.3).

The conducting portion of the respiratory system provides little resistance to

Conductive Airways of the Respiratory System, with Details of the Ciliated Epithelial Lining

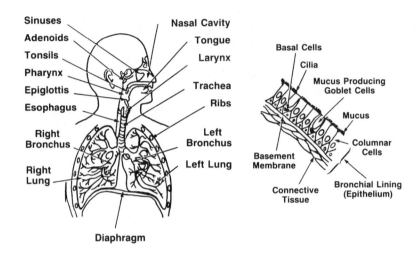

Figure 1.3.   Conductive airways of the respiratory systems, with details of the ciliated epithelial lining. *Source*: *Environmental Toxicology and Risk Assessment: An Introduction, Student Manual* (Chicago, IL: United States Environmental Protection Agency, Region V), Visual 2.6.

the movement of gases to the alveolar surface. Further, this area conditions the air and protects the lungs from the largest of infectious or potentially toxic particles in the air. The convoluted, moist, and richly vascular mucosa of the nose protects nose-breathers from inhaling particles larger than 5–10 μ in diameter. Soluble gases may be removed by absorption. The air is warmed and moistened, or cooled when the conditions are hot and dry. The trachae-bronchial system is lined with cilia that constantly force mucus toward the larynx. The mucus carries with it the microorganisms and particles that have been trapped in it, as well as the macrophages that move out of the alveoli with material that have been scavenged from the alveolar surface. When the mucus reaches the pharynx it is usually swallowed or may be spit out. These pulmonary mechanisms help prevent problems in the respiratory system due to the inhalation of dusts, fumes, and other materials that can cause disease or injury. The coughing mechanism helps in the removal of these materials.

Pulmonary ventilation is brought about by changes in the size of the thorax (chest). The contraction of the diaphragm and chest-elevating muscles leads to an expansion of lungs that decreases the alveolar pressure and creates an inspiration. Expiration is a passive process that begins when pressures change. This occurs when the muscles are relaxed, which leads to a decrease in size of the thorax, a decrease in the size of the lungs, and a pressure gradient in the alveoli that is greater than the atmosphere.

The exchange of gases in the lungs take place between the alveolar air and venous blood that flows to the lung's capillaries. The alveolar-capillary membranes allow gases to move back and forth. Oxygen enters the blood because the PO_2 of the alveolar air is greater than the PO_2 of the venous blood. At the same time, the PCO_2 of the venous blood is much higher than the PCO_2 of the alveolar blood. This causes carbon dioxide to flow from the venous blood to the alveolar blood. Once the oxygen enters the blood, about 2.5% is transported as a solute in the blood, whereas about 97.5% is transported as oxyhemoglobin in the red blood cells. The carbon dioxide is transported in the blood as a true solute in small quantities. A large amount of the carbon dioxide is transported as bicarbonate ions in the plasma. A moderate amount of the carbon dioxide is transported in the red blood cells as carbaminohemoglobin.

Carbon dioxide is the major regulator of respiration. An increase in blood carbon dioxide to a certain level may stimulate respiration, and a decrease may cause a decreased level of respiration. If the oxygen level of blood is low, this also may stimulate respiration to a certain point.

The Digestive System

The digestive tract is largely a tube that is open at both ends. In a real sense, the contents of the gastrointestinal tract is exterior to the body (Figure 1.4). In order to get into the tissues or cells, the ingested materials have to go through an extremely acidic environment in the stomach, as well as being broken apart by enzymes. Some of the large molecules, such as cellulose, are unaltered and therefore are excreted in the feces. Almost all of the digestion and absorption of food and water take place in the small intestine. The three basic food groups—the carbohydrates, proteins, and fats—are broken down in different ways. The carbohydrates, most of which are ingested in the form of starch, are split into disaccharides by the enzyme amylase, found in the saliva and in the pancreas. They are then split into monosaccharides by enzymes in the small intestinal mucosa. The sugar molecules are then transported into the blood. The proteins are broken down further into peptides and into free amino acids, which are transported across the intestinal cells. Most of the fat digestion occurs in the small intestine from the combined actions of pancreatic lipase and bile salts secreted by the liver. The bile salts act chiefly as emulsifying agents. Fatty acids are resynthesized to triglycerides in the intestinal cells. They are then secreted into the lymphatics as small lipid droplets.

The function of the digestive system is to accept raw materials in the form of different foods, minerals, vitamins, and liquids, and to prepare them for absorption into the capillaries or lymphatics for distribution through the body.

The main organs of the digestive system are the mouth, pharynx, esophagus,

The Gastrointestinal Tract

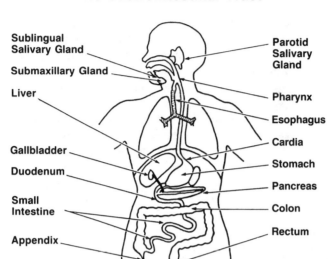

Figure 1.4.   The gastrointestinal tract. *Source: Environmental Toxicology and Risk Assessment: An Introduction, Student Manual,* (Chicago, IL: United State Environmental Protection Agency, Region V), Visual 2.12.

stomach, and intestine. The accessory organs that are open to the main organs are the salivary glands, liver, gallbladder, and pancreas.

The stomach serves as a food reservoir that expands and contracts depending on the amount of food present. The glands secrete most of the gastric juice, which is fluid make up of mucus, enzymes, and hydrochloric acid. The epithelial cells that form the surface of the gastric mucosa secrete mucus. The chief cells (zymogenic cells) secrete the enzymes of gastric juice. The parietal cells secrete hydrochloric acid and are also thought to produce a protein known as intrinsic factor. When the muscular coat of the stomach contracts, it churns up food and breaks it into small particles, which allows for mixing with the gastric juice. In time it moves the gastric content into the duodenum. The stomach, in a limited quantity, absorbs water, alcohol, and certain drugs. It also produces the hormone gastrin in cells that are in the pyloric region.

The small intestine, which is about 20 ft in length, has three major divisions: the duodenum, the jejunum, and the ileum. The function of the small intestine is to complete the digestion of food, to absorb the end products of digestion into the blood and lymph, and to secrete hormones that help control the secretion of pancreatic juice, bile, and intestinal juice. The intestinal juice contains mucus from the intestine, digestive enzymes from the pancreas, and bile from the liver.

The lower part of the alimentary canal is called the large intestine. It is divided into the cecum, colon, and rectum. The main functions of the large intestine are the absorption of water, secretion of mucus, and elimination of the wastes of digestion.

The peritoneum is a large, continuous sheet of serous membrane that lines the walls of the entire abdominal cavity and forms the serous outer coat of the organs.

The liver, which is the largest gland in the body, is one of the most vital organs. Liver cells detoxify a variety of substances. They secrete about a pint of bile a day. The liver carries on a number of important steps in the metabolism of proteins, fats, and carbohydrates. Liver cells store iron, vitamins A, B_{12}, and D. Poisonous substances that enter the blood from the intestine are circulated to the liver, where a series of chemical reactions occur that turn the substances into nontoxic compounds. The bile secreted by the liver is made up of bile salts, bile pigments, and cholesterol. The bile salts that are formed in the liver from the cholesterol are the most essential part of the bile that aids in the absorption of fats.

The gallbladder is a storage organ for the bile, which is produced by the liver. When digestion is going on in the stomach and intestines, the gallbladder contracts, ejecting the concentrated bile into the duodenum.

The pancreas is composed of two different types of glandular tissue, one exocrine and one endocrine. Embedded between the exocrine units of the pancreas are clusters of exocrine cells called islands of Langerhans or pancreatic islets. The pancreas secretes the digestive enzymes found in pancreatic juice. Its beta cells secrete insulin, and its alpha cells secrete glucagon.

The last step in the digestive process is defecation. The stomach is emptied in about 2–6 hr after a meal, depending upon what is eaten. Chyme, which is gastric juices mixed with food to form a milky white material, is ejected about every 20 sec into the duodenum. Defecation is brought about by a reflex that is caused by stimulation of the receptors in the rectal mucosa. Constipation occurs when the material in the lower colon in the rectum moves at a rate that is slower than normal. This causes extra water to be absorbed from the fecal mass and thereby hardens the stool. Diarrhea occurs when chyme moves too quickly through the small intestine, and therefore, not enough water and electrolytes are absorbed. It is the loss of water, leading to dehydration and loss of electrolytes, that makes diarrhea in infants such a serious problem.

The Urinary System

The urinary system consists of the kidneys, ureters, bladder, and urethra. The functions of the kidneys are to excrete urine, which eliminates various toxins and metabolic wastes. They also help maintain fluid, electrolytes, acid-base balance, and the proper level of blood pressure in the body. The ureters collect urine and drain it into the bladder. The bladder is a reservoir for storing urine and also acts as the organ that expels the urine through the urethra which is the passageway for expulsion from the body.

THE EPIDEMIOLOGICAL APPROACH TO ENVIRONMENTAL HEALTH PROBLEMS

Epidemiology is the study of the factors that determine the occurrence and frequency of disease in a population. Epidemiology is more a method than a body of knowledge drawing upon the knowledge and skills of clinical medicine, microbiology, pathology, zoology, sociology, anthropology, and toxicology. Although originally the epidemiological technique was applied only to disease and disease causation, more recently it has been applied to accident causation and injuries, or diseases caused by chemicals and environmental factors.

Two major categories encompass epidemiology. Descriptive epidemiology is the study of the amount and distribution of disease within a given population by person, place, and time; analytical epidemiology is the study of causes determining the relatively high or low frequency of disease in specific groups. Age, sex, ethnicity, race, social class, occupation, and marital status are the principal personal variables in epidemiological research.

Age is the most important determinant. Mortality and morbidity rates often show a relationship to age. Acute respiratory infections occur with highest frequency among the very young, and upper respiratory infections, in general, are of greatest concern to the very young and the very old. Sex is another important epidemiological determinant. The mortality rate for emphysema and respiratory cancer is higher in males than in females, whereas the frequency of occurrence of diabetes is higher in females than in males. Ethnic groups and racial groups exhibit definite differences in morbidity and mortality; blacks have a substantially higher age-adjusted death rate from hypertensive heart disease, accidents, and tuberculosis. Cancer of the cervix is higher in blacks than in whites. Whites have a higher death rate from suicide and leukemia. Social class is an important determinant in the epidemiological study of a specific disease. The death rate for unskilled workers is considerably higher than for the professional classes; infant mortality rates follow the same pattern. Schizophrenia occurs more frequently in the lowest social classes of the urban community. Occupation influences susceptibility to given diseases. Pulmonary fibrosis is found among workers exposed to free silica; cancer is found among workers exposed to aniline dyes.

The physical and biological environment influences the occurrence of disease in a given area. Multiple sclerosis is known to occur most frequently in northern climates; malaria and yellow fever are most frequently found in southern climates; lung cancer is more prevalent in urban areas.

Time is another extremely important variable in the study of disease occurrence. If it is possible to determine the cycle of diseases, there may be clues concerning the mutation of various organisms and also the potential effect of various chemicals on arthropods and on human beings. When the time element is determined in an outbreak of disease, it may be possible to quickly recognize the cause of the outbreak. Time is particularly important in outbreaks of food-borne disease. It is also important in determining degrees of exposure. Time, temperature, and concentration are the keys to many environmental health problems and often to the resolution of these problems or to the protection of the individual.

Descriptive epidemiology systematically summarizes the basic data available on health and the major causes of disease and death. These data permit the evaluation of trends in health and the comparison of these trends from subgroup to subgroup and from country to country. Descriptive epidemiology provides a basis for the planning and evaluation of health services and for the identification of health problems.

Analytical epidemiology seeks to discover the causes of problems by formulating hypotheses based on actual observations of existing diseases, conducting studies of outbreaks of disease, and evaluating the data to determine the accuracy of original hypotheses. The classic example of this method was the determination of the epidemiology of cholera by John Snow, the English physician, who realized through observation that cholera was spread through water from contaminated sources.

Germ Theory of Disease

John Snow hypothesized that a specific microorganism caused cholera, yet this concept was not generally accepted during his time. The germ theory of disease was postulated on the basis of Louis Pasteur's research. Pasteur demonstrated that microorganisms caused fermentation and that these microorganisms were not spontaneously generated but could be found in the air. His work was followed by Robert Koch, who first isolated the organisms causing tuberculosis and Asiatic cholera. Koch introduced scientific rigor to the proof of the cause of disease. Koch's postulates were that the parasite had to be shown to be present in every case of the disease by isolation and pure culture; it could not be found in other diseases; once the parasite was isolated, it had to be capable of reproducing the disease in experimental animals; the parasite had to be recovered from the diseased animal. Unfortunately, Koch's postulates did not apply to viruses, since they could not be cultivated in pure culture and were host specific. His postulates also did not apply well to certain pathogenic organisms that infected a host but did not necessarily produce a recognizable disease.

Infectious Disease Causation

In order to understand the spread of infectious or communicable disease, it is necessary to understand the host-parasite relationship and the reservoir of disease. The reservoir of disease is defined as the living organisms or inanimate objects in which infectious agents live and multiply. A parasite is the microorganism causing a specific disease. A host may be either people or animals. The chain of infection of a given disease is usually illustrated as follows:

In the environmental health field, the practitioner attempts to halt the transmission of disease by eliminating the vector, which may be an arthropod, or controlling or purifying the vehicle, which may be an inanimate object — water, air, or solid waste. Generally, the reservoir of disease is difficult to control, with the exception of those diseases where hosts can be "vaccinated." Eliminating the culex mosquito, which in turn eliminates St. Louis encephalitis, is an example of breaking the chain of infection at the vector state. Purification and chlorination of water supplies, which destroys the microorganisms causing cholera, is an example of halting the spread of infection at the vehicle stage.

Humans are not only the recipients, but often the carriers, of disease. Carriers discharge microorganisms into the environment through feces and urine, and from the skin, nose, or pharynx. Since carriers may not exhibit symptoms of disease, they are dangerous in the transference of disease. Carriers may have a subclinical case of a disease, or may be in the incubation or in the convalescent period of disease. Some examples of diseases transmitted through carriers are polio, chicken pox, measles, hepatitis, diphtheria, salmonella, and staphylococcus infections. Disease may be transmitted directly from person to person through kissing, sexual intercourse, coughing, sneezing, direct contact of the skin, or through indirect transmission by means of vehicles or vectors. Some examples of vehicles, also called fomites, include bedding, toys, surgical instruments, surgical dressings, contaminated food or water, and needles. Indirect transmission of disease may occur when dust and droplet nuclei enter the airstream through coughing, sneezing, talking, dusting, walking, or through a variety of housekeeping or medical procedures. Q fever, histoplasmosis, tuberculosis, psittacosis, staphylococcus infections, and other diseases are transferred in this manner. Disease may be spread from a common source or from secondary sources. Generally the initial outbreak of an epidemic spreads from a common source; the secondary outbreak, which may occur days or weeks later, spreads from infected secondary sources, or the secondary environment, which became infected during the primary outbreak of disease. The scope of an outbreak of disease depends upon the virulence of the organism, its pathogenicity, the number of unprotected individuals at risk, various environmental factors, and the intervening factors used to control the disease. Virulence refers to the degree to which an infectious agent can produce serious illness. Pathogenicity refers to the ability of an agent to cause disease in an infected host.

The cause of the spread of infectious disease is generally complex, even when the primary cause, or agent, is known. The many contributing factors, or secondary causes, including the environmental factors, sharply influence the rate of disease occurrence. When the disease has run a specific course and new cases are not occurring, and if this is a spontaneous type of leveling off of the disease process, it is recognized that an equilibrium has been established between the parasite and the host, and this equilibrium will last until new hosts and/or a more virulent form of the organism appears, or the environmental factors are changed.

Chemical Dose-Response Curve

In order to determine the effects of a chemical on the health and well-being of the individual and society, it is necessary to know the nature of the chemical, the time period over which the chemical may react with the individual, and the quantity of the chemical present. Large quantities of chemicals may act over a short time period and cause acute disease or injury. Small quantities of chemicals can act over long time periods to cause chronic effects or such conditions as carcinogenesis, mutagenesis, teratogenesis, and other toxicities.

Chemicals are increasingly becoming a major concern to the environmental health practitioner, since the level of infectious and communicable diseases is dropping and the level of environmental contaminants that may cause disease or injury is increasing sharply. To understand the chemical problem, it is necessary to understand the dose-response curve. The dose-response curve is a relationship between the potential toxicity inherent in a given chemical and the kind of symptoms exhibited when the chemical interacts with the living biological system. A chemical may be extremely toxic, but safe in given situations in which no contact with the living organism occurs. However, a chemical may be mildly toxic yet problematic if it comes into constant contact with the living organism. The dose-response curve is determined by subjecting laboratory animals to a given chemical under specific conditions. These conditions, and the error attributed to the conditions, must also be incorporated in the dose-response curve. The dose-response curve refers to that dose (LD_{50}) producing death in 50% of the animals under experimental conditions.

Toxicity is the ability of a substance to produce injury. Toxicology is the science of poisons. In chemical toxicology, the potential hazards involved in the use of chemicals, the manner in which chemicals affect the living organism, the mode by which chemicals enter the body, the type of toxicity that may occur, and the methods useful in measuring potential hazards from toxic substances are investigated. Epidemiology, the branch of medicine dealing with epidemic diseases, is also concerned with the serious problem of chemicals and uses the information and techniques of the toxicologist, along with a vast number of other techniques, to determine the potential for disease and injury. The epidemiologist attempts to extrapolate acceptable exposure limits from the dose-response data to reduce the chemical hazard to humans.

Environmental Factors in Disease Causation

The environment is enormously complex and constantly changing. Humans contribute to the alteration of the environment and to the transportation of toxic agents. Substances in the environment enter the internal and external tissues of an individual through inhalation, ingestion, skin contact, or actions on various sensory organs. Environmental contaminants found in the air include sulfur oxides, carbon monoxide, and photochemical oxidants. Water contaminants include inorganic and organic chemicals, microorganisms, weeds, silt, and other materials. Food contaminants include inorganic and organic materials purposely or accidentally added to food, such as chemical preservatives, pesticides, and fertilizers. Physical factors in the environment that may affect health include ionizing radiation, ultraviolet and visible light, infrared radiation, microwaves, noise, vibration, ultrasonics, heat, cold, humidity, extremes of barometric pressure, geography, climate, geology, and varying seasons of the year. Environmental contaminants causing disease or injury come from waste disposal, recycling of wastes, and disposal of wastes into the air, water, and land. In addition, natural toxicants of biological or chemical origin may be found in the air, water, and food. These environmental factors cause disease or injury when the susceptible host is exposed and when the condition is severe enough to produce a dose response causing harm to the living organism.

Environmental Contaminants and Disease

Cancer caused by environmental contaminants is becoming a major problem in our society today. Cancer attributed to occupational or other environmental factors is higher among adult males because of a higher proportion of males employed in industry. More than half of the cancers observed in the third national cancer survey were caused or suspected to have been caused by an external agent. However, in evaluating cancer problems, it is important to recognize that cancer is also associated with socioeconomic status, cigarette smoking, diet, alcohol, and radiation.

There has been a sharp increase in morbidity and mortality for bronchitis, pulmonary emphysema, and to a lesser extent for asthma. These three conditions, which are collectively termed *chronic obstructive pulmonary disease*, are found more frequently among smokers than nonsmokers. A higher mortality rate exists among the low-income urban population compared with the upper-income or rural population. Studies indicate that environmental conditions contribute to the increase in these diseases.

A sharp increase in cardiovascular conditions, particularly coronary heart disease, has been attributed to diet, cigarette smoking, and problems of the environment. Clinical experimental evidence suggests that exposure to carbon monoxide exacerbates symptoms of angina. There appears to be a strong statistical correlation between the season of the year, the yearly temperature change, and mortality attributed to coronary heart disease. Although infectious agents contribute to this problem, the environment is also involved. Numerous, although controversial, publications concerning the relationship between water hardness and cardiovascular disease have appeared. Some association appears to exist between the level of trace metals found in water and the susceptibility to sudden death from arrhythmia. It is also suspected that carbon disulfide and freon gas contribute to arrhythmia in the occupational environment.

Many investigators are now studying the effect of environmental conditions on the outcome of pregnancy. During pregnancy the internal organism is more vulnerable to problems, since the detoxification mechanisms are altered and changes occur in protein, carbohydrate, and lipid metabolism. Concern exists today regarding teratogens, or substances producing abnormal variants, such as congenital malformations and spontaneous abortions. While further research is needed in this area, certain chemicals are conclusively embryopathic, including methyl mercury, aminopterin, thalidomide, iodine deficiency, steroid hormones with androgenic activity, and carbon monoxide. Certain chemicals are suspected of affecting human prenatal development, including cortisone, vitamin A deficiency, and diethylstilbestrol.

Infants are susceptible to methemoglobinemia caused by the ingestion of nitrates or nitrites with drinking water. Chemicals may also affect the CNS during postnatal development, affecting particularly the cerebrum. Lead poisoning in children produces irreparable brain damage and permanent retardation and eventual death. Long-term exposure to air pollutants by infants and children impairs respiratory health and aggravates chronic bronchitis, asthma, and other lung conditions. From 1971 to 1982, 399 waterborne outbreaks of disease occurred in the United States. The frequency of occurrence of waterborne disease is increasing. In addition to waterborne infections, an increasing number of deaths occurred due to chemical poisoning through the ingestion of water. In the reuse of water, problems are created by the presence of microorganisms and toxic minerals, which ordinarily are not found in the initial pure water supply. Some laboratory data suggest that mutations occur as a result of environmental pollutants. Although this is not conclusively demonstrated, concern exists, and mutation must be considered one of the potential disease problems related to environmental contaminants.

Epidemiological Investigations and Environmental Health

An epidemiological investigation is a systematic study of available current data, past records and data, and on-site investigations, when feasible. Epidemiological investigations vary with the type of study being conducted. Specific techniques will be discussed in the chapters on water and food. A typical epidemiologic study consists of a preliminary review of material, further investigations, analysis of the data, a hypothesis concerning cause, and in some cases, the testing of the hypothesis through the review of new data. In the initial gathering of materials, standardized forms must be used to obtain consistent and complete information. The data are verified through diagnosis of the specific diseases, conditions, and injuries. The investigator determines the time, place, and individuals involved, and plots the cases by time and condition. The investigator asks the questions: how?, why?, what?, when?, where?, and who?. A test hypothesis concerning the cause of the outbreak or the cause of the disease is theorized; additional cases are sought that either fit or do not fit the pattern. If the cases do fit the pattern, the hypothesis is strengthened. Comprehensive, accurate analysis of data is essential in determining the causes of diseases, conditions, or injuries. One of the principal problems facing the investigator is failure on the part of victims to report the disease to the proper authorities, which results in an incomplete picture of disease conditions. Determining the actual cause, or causes, of a given condition can be problematic, however, the trained investigator, in collaboration with the environmentalist and medical nursing team, uses available data in an attempt to control the problem. The work of an epidemiologist is similar to that of a detective: it is long and tedious, and frequently gives poor results. However, where the results are satisfactory, the population is saved from debilitating or death-causing diseases.

Epidemiological analysis of health data is made by age, race, sex, place, and time. Reports include incidence, prevalence, and case fatality, or morbidity of a given disease or condition. Variation by time and place reflects exposure and response to different environmental pollutants. Epidemiological analysis also includes variations in pulmonary function, cholinesterase levels, and potential toxic effects based on blood analysis of various chemicals.

Epidemiological methods are used to compare exposed and controlled groups. For example, the amount of lung disease in a specific group of workers is compared to the general population. The greatest problem in this type of study is controlling the significant variables, such as age, sex, economic status, ethnic group, and smoking habits for the two groups.

Another technique is a longitudinal study of a specific group over a period of time to determine the rate and degree of occurrence of a specific disease within varying time periods. This technique is particularly effective in determining the acute effects of exposure to given environmental pressures. The determination of community exposures, including not only the occupational but other environmental settings, is difficult, since groups are usually poorly defined, vary within the geographic area, and receive different levels of exposure to substances.

In attempting to study the relationship between health and environmental exposure in a community, the subjects must be selected either by geographic comparisons, comparisons over time, or dose-response relationships. In geographic comparison, the exposure is defined by a geographic area, for example, health problems experienced by individuals living in heavily air-polluted areas. A problem with this type of study is that a single measure or combinations of measures of exposure is used to represent the entire group exposure. In interpreting data from this type of study, it is important to account for possible difference in factors other than those under study. These differences include migration, age, and ethnic group.

In time-comparison studies, a specific time period is compared to health effects over that time period. For instance, a community and its health problems are evaluated for a given period and then are reevaluated after a month, a year, or longer. If the time study period is less than a year, seasonal variation must be introduced. An attempt is made to understand differences in health effects in the population between the initial time period and the subsequent one. Occasionally geographic and time comparison methods are combined, producing a more reliable comparison than if the two were made separately.

Dose-response relationships for individuals are frequently studied. For example, exposure to carbon monoxide is determined for an individual by using personal monitoring equipment and analyzing the degree of carboxyhemoglobin in the individual compared to the quantity of carbon monoxide present in the environment.

In a surveillance study, available data are collected and analyzed, and departures from the norm are noted regarding the severity of environmental problems and health effects. This type of study is used to take corrective action as quickly as possible. In all data collection and epidemiological studies of environmental problems, recognition and control or elimination of variables contributing to illness are essential. Without this, data are of little or no value. In examining environmental influences on health, rare occurrences of illnesses, such as the sudden occurrence of liver cancer in the United States, are important indicators. The outbreak of liver cancer aided in the discovery of the problems related to occupational exposure of vinyl chloride.

Death certificates and morbidity data records from state compensation agencies provide information regarding occupational or other exposures to given environmental factors. This information is of value in developing the initial epidemiological study for a given environmental factor and its relationship to poor health. Other record sources include cancer registers, union records, social security records, and interviews with surviving employees or next of kin. The National Center for Health Statistics is another source of data, and is involved in epidemiological analysis by utilizing data available through various other components of the Department of Health and Human Services, the Social Security Administration, and other branches of government and industry. Although death certificates are required in all states, a National Death Index is needed that would provide the epidemiologist with information on a yearly basis. Great Britain and the states of California and Washington analyze death by occupation. This would be invaluable epidemiological data if provided on a national scale by the National Center for Health Statistics. The National Health Interview survey is a source of data on mortality, but until recently inadequate information on morbidity was provided. Better morbidity data are needed. A study that should provide better data is the National Health Examination Survey, developed to determine the possible role of water hardness on cardiovascular disease. The national census could be used to provide mortality and morbidity data for men and for women by occupation and social class. Each yearly census contributes an additional source of information that should be analyzed for environmental problems. The state health departments also gather and provide data on environmental health problems.

At present, data are available from the Environmental Protection Agency, the National Institute of Occupational Safety and Health, and the Food and Drug Administration. As can be seen, adequate data are available in some areas but are lacking in others. However, there is a tremendous need not only to provide data, but also to coordinate data and to make them available to epidemiological, medical, and environmental support services.

Environmental Epidemiology

The establishment of contamination criteria and/or exposure limits can be done through epidemiological studies or toxicological studies. The epidemiological studies are concerned with the exposure of populations at their occupational settings or through the contamination of food, drinking water, or air. These studies provide a statistical association between the levels of contaminants and the reported effects. The toxicological studies are of groups of animals exposed intentionally in a variety of controlled laboratory experiments. In this situation, it is possible to define the doses, frequency of application, and metabolic pathways, areas of storage, and types and amounts of biological damage created by the chemical agent. The advantage of using epidemiological data for establishing safe limits of human exposure to chemicals is that you do not have to have the compounding factor of interpreting animal data and extrapolating from this data. The epidemiological studies try to determine whether or not correlations exist between the frequency or prevalence of a disease or health condition in humans and some specific factor, such as the concentration of a toxic chemical in the environment. Another advantage to this type of determination is that typically large numbers of humans are involved in the study and the exposure levels are usually subclinical. Therefore, the data are directly relevant to the study and determination of the dosage. The disadvantages are (1) the human exposure to chemicals is fortunately very limited, (2) controlled studies of exposure in humans are neither feasible nor legal, and (3) the application of the available epidemiological data may be limited because of its quality or because there are many risk factors involved in the study that cannot be truly limited or measured. These risk factors might relate to the ages of the individuals, their educational backgrounds, occupational histories, prior smoking habits, drugs or alcohol consumed, dietary practices, general health, and sexual and racial factors. Another major problem is determining what is the health endpoint in a clear and concise manner. Death, of course, is a reliable and a definable statistic, but how do you measure the level of good health or poor health related to a combination of chemicals to which the individual has been exposed? Further, there are a group of chronic nonspecific diseases with nonspecific causations that may occur. Even if measurements are made of the appropriate chemicals, are they accurate and do they show the overall population exposure to the chemicals? The most effective types of epidemiological studies are of working populations where known groups of individuals are exposed to chemicals.

The two classes of experimental design in epidemiology are descriptive and analytical. The two types of descriptive studies include the case study and ecological study. The case study provides information for a single individual about the relationship of an exposure and disease. Many times case studies are the starting point for in-depth investigations. The ecological study is used to determine how a single factor is distributed in the population. The ecological study describes the disease in terms of prevalence, incidence, and mortality rates for the population. The ecological study compares the trends of disease in two or more populations or over time. It is used to generate hypotheses about the impact of an agent on the disease found in human populations. These hypotheses are then used to help design the analytical studies. The two types of analytical studies are the retrospective and prospective. In the retrospective study, the population being studied is determined from death certificates, hospital records, and other sources of data indicating morbidity. A control group that is free of the disease is then chosen for comparison. The exposure history of the study group has been assessed and the causal relationship between exposure to the chemical or other substance and the level of disease is established.

In the prospective design, a cohort (a group) of disease-free patients are followed over time, and the development of disease is monitored. Of the pairs of individuals one half are exposed to the test agent while the other half is not exposed. The development of disease is measured in both groups, and a determination is made of the causal relationship between the agent and disease. Other problems with the cohort study relate to diseases that rarely occur, the sample size of the study, design management problems when the disease has a long latency period, and death of the subjects.

To get around some of the difficulties that have been mentioned, the Environmental Epidemiology Branch of the National Cancer Institute did comprehensive studies of 3056 counties in the United States for a 20-year period between 1950 and 1969. They were trying to do a human risk assessment for cancer. From the studies and maps that were drawn, they were able to identify cancer hot spots. They found Salem County, New Jersey had the highest rate of bladder cancer mortality in the United States. They then did ecological studies and generated maps of industries in that state. The maps included the total population of the county employed in a specific area. Eighteen major industrial categories were examined. They were able to use existing vital records and census data that eliminated extensive data collection. In the comparison rates of mortality and morbidity, it was critical to adjust for confounding factors. (A confounding factor or confounder is a risk factor for the disease under study.) If you do not adjust for the confounders, then it is possible to get a mistaken assessment of the risk for disease as a result of specific agents. In cancer, age always is a confounder. Further, many cancers are sex or race specific. Smoking is a confounder in many cancers, including lung and bladder, when the individual is also exposed to environmental or industrial sources. Another problem is called bias. (Bias is a systematic error in the design, conduct, or analysis of a study that causes a mistaken estimation of the relationship of the study factor to the disease.) Some of the biases include a nonrespondent bias, a diagnostic suspicion bias, an exposure suspicion bias, and a family information bias.

Epidemiological investigations, when adjusted for bias and the confounding factors, can be of enormous importance because they provide information about humans under actual conditions to exposure to a specific agent. The surveillance systems used in epidemiology need to be set up in a very careful manner, since these systems vary widely in methodology, scope, objectives, and characteristics. The strength of the evaluation depends on the evaluator, how each of the items in the framework flow, and how they are established and assessed.

The public health significance of the health event must be established. In the concern about a few persons being exposed to highly toxic chemicals, a cluster of cases occurring, or a general outbreak of a specific disease, the health event may be measured by (1) the total number of cases, incidence, and prevalence; (2) the case-fatality ratio, which is an index of severity; (3) the mortality rate; (4) the index of loss productivity; (5) the index of the years of potential life loss; (6) the medical costs involved; and (7) the degree of preventability. Next, what are the objectives of the system? Describe the health event under surveillance and state a case definition for each health event. Draw a flow chart of the system. Describe the components and operation of the system, including (1) what is the population under surveillance? (2) what is the period of time of data collection? (3) what information is collected? (4) who provides the information? (5) how is it transferred? (6) how is it stored? (7) who analyzes the information? (8) how is this data analyzed and how often? (9) how often are reports disseminated and distributed and to whom are they sent? and (10) what is the level of usefulness of the data from the surveillance system?

Each system has to be simple, flexible, acceptable, sensitive (collects considerable amounts of data), predictive of problems, representative, and timely.

Epidemiological studies that are adjusted for bias and confounding factors and meet the guidelines previously stated may be extremely beneficial in providing information about humans under actual conditions of exposure to specific agents. A well-designed, properly controlled study can provide competent results in health areas and the determination of risk assessment. The weight of evidence needed to determine whether or not something is potentially carcinogenic in humans or whether other kinds of health problems exist will be based on a combination of epidemiological studies, long-term animal studies, pharmacokinetic studies, and relevant toxicological studies.

RISK ASSESSMENT AND RISK MANAGEMENT

Risk is the possibility of loss or injury. It is the potential realization of unwanted consequences of an event. Both the probability of the event occurring and the magnitude of the consequences are involved in the term risk. Risk is a function of the hazard involved, the dose-response from the person, and the situation in which people are exposed to the substance.

Risk Assessment

Risk assessment is the determination of what the problems are. Risk management is the process of deciding what to do about the problems. Risk assessment is made up of four steps: hazard identification, dose-response assessment, exposure assessment, and risk characterization. Hazard identification involves the gathering of data on the substance, including information about the link between the substance and adverse health effects. This step determines that such a link exists. Unfortunately when the hazard identification is based on the extrapolation from animal data to humans, it is difficult at times to reasonably assert that the substance is a significant hazard. Once it is determined that a chemical is likely to cause a specific health effect in humans, it is necessary to establish the relationship between the amount of exposure (dose) and the effect that is produced. It is not entirely clear that safe levels or thresholds exist for toxic chemicals, since short exposures at low doses may cause damage to health in the future. This is a very complex area. It is easy enough to establish a health endpoint when the individual is subject to high doses of a chemical and an immediate result is seen in the body.

Exposure assessment is a determination of how much of the chemical is available to the individuals in a specific area. Its two components include an analysis of the path of transportation and an assessment of the impact of transformation, as well as the available dose and an analysis of the population characteristics. The best way of measuring human exposure is to do direct measurement or monitoring of the ambient conditions. The degree of exposure may vary from chemical to chemical. Human data typically are quite limited because of the types of monitoring, the expense, and the time required to gather information. Modeling often is used as a substitute. In this technique, data are fed into the computer on pollutant releases, release characteristics, meteorology, hydrology, geography, and other information. The exposed population is estimated through the use of census data, etc. There is considerable uncertainty about these estimates. Risk characterization is the exposure time, times the potency, times the unit or the individual at risk. This step requires an evaluation of the information from the first three steps. For noncarcinogens, the margin of safety (MOS) is estimated by dividing the experimental NOAEL (No Observed Adverse Effect Level) by the estimated daily human dose. For carcinogens, the risk is estimated at the human dose by multiplying the actual human dose by the risk per unit of dose projected in dose-response modeling. It may be a range of risk produced by using different models and assumptions about dose-response curves and the relative susceptibility of humans and animals.

Hazard assessment has been based on the underlying principles that animal bioassays are indicative of probable human response and that there is no threshold of response to carcinogens. Technical guidelines have been established by various agencies in risk assessment as they relate to carcinogens.

Toxic substances can lead to adverse effects in the liver, kidneys, lung, and other body systems. These systems have certain thresholds at which they will respond to the chemicals. This further complicates the problem of risk assessment.

Most risk assessments are for individual chemicals. However, in reality, people are subjected many times to chemical mixtures that would alter the risk assessment picture considerably. Are there carcinogenic or additive effects to the mixtures of the chemicals?

The exposure assessment is a combination of field monitoring, mathematical modeling, measurement of actual concentrations of chemicals in tissues, and laboratory modeling data. A complication in this exposure assessment is determining how many people are exposed to the chemicals through the air, soil, water, drinking water, or food. It is then important to calculate the rate of uptake of the chemicals through breathing, eating, drinking, or absorption through the skin. Then it is essential to track the chemicals through the body and through the process of metabolism.

To summarize risk analysis it is necessary to (1) determine the source of the chemical and how long it is released to the environment; (2) the pathways through which the chemical travels through the environment, such as the air, water, food, etc.; (3) the behavior of the substances in the body and the pathways that the chemical takes during metabolism, as well as the toxicity of the metabolite products; (4) estimate the concentration or dose at the specific site in the organs and how long the organ is exposed to the chemical; (5) determine the persistence of the substance in the environment and how long it can be taken into a human being and then passed on to the specific organ; (6) determine the relationship between the dose the individual gets and the effect on the individual; and (7) estimate the risk to the exposed population (see discussion of industrial toxicology in Chapter 10 for further information).

Risk Management

Risk management helps set priorities. Many of the priorities set for the federal agencies, especially the EPA, have been decided by Congress. Because of potential emergencies, it is best that the agencies can take a hard look at what the potential problems are and decide what is of the most pressing nature. Risk management produces a basis for balanced analysis and decision making. Because we are exposed to a complex, highly dilute mixture of chemicals that come through the air, water, land, and food, it is frequently difficult to determine which area must receive immediate attention. Rational decisions can be made on the basis of good scientific data. Risk management produces more efficient and consistent risk reduction policies. By doing complete evaluations, it is possible to determine how best to approach risk management. Risk management is the complex of judgement and analysis that uses the results of risk assessment to produce the decisions that are necessary to bring about an environmental action.

Inherent in risk management is the idea of comparability. The EPA has a variety of goals, some of which may be in conflict with others. For instance, deep ocean dumping of sewage sludge may reduce the human health risk in comparison with incineration or land spreading, but may have an adverse effect on marine ecosystems. Pollution control, and therefore the reduction of risk, is an incremental process. Consistency must be used here because of the cost involved. At some point when increasing the effectiveness of the process, the cost needed to improve the process by a small increment could be beyond reasonable levels of expenditure. Risk management programs have been set up by the EPA for a variety of problem areas, such as hazardous waste, air pollution, water pollution, etc.

CANCER, MUTAGENS, AND TERATOGENS

Cancer is a malignant tumor of potentially unlimited growth that expands locally by invasion and systemically by metastasis. A mutagen is an agent that brings about a change in the body's genetic material. The mutation is a relatively permanent change in hereditary material involving either a physical change in chromosome relations or a biochemical change. A teratogen is a teratogenic agent. Teratogenic is related to or the cause of developmental malformations and monstrosities.

Carcinogenesis

Cancer is a general term for a group of related diseases that cause uncontrolled growth of certain types of tissues. Cancer and organ toxicity differ in that cancer results from abnormalities in the reproduction and growth of cells, rather than in the alteration of cell structure or function. Chemical carcinogenesis generally means the induction by chemicals of neoplasms not usually observed, the induction by chemicals of more neoplasms than are generally found, or the earlier induction by chemicals of neoplasms than commonly observed. Chemical carcinogens undergo biotransformation, as with any other similarly structured pharmacological agent. Biotransformation is the transformation of chemical compounds within a living system. The response to the chemical carcinogen varies with the species, strain, and sex of the experimental animal. Chemical carcinogens interact with other environmental agents that sometimes enhance and other times decrease their effect. Chemical carcinogens have a persistent and delayed biological effect, may be more effective in divided doses than in an individual large dose, and may have distinct interactions with host genetic elements. Carcinogens may either be direct initiators of the genetic change or act only as a promoter in combination with another initiator. At times, one carcinogen may be required to cause a second carcinogen to promote a response. They would then be called co-carcinogens. One of the major problems in determining the possibilities of carcinogenic activity are the long periods of latency, which may extend for many years. Typically, time periods for the ultimate production of the cancer may take from 15 to 40 years after the onset of the exposure and prior to disease manifestation. The site of the development of a malignant tumor may vary considerably. It is based either on the portal of entry or on the site where biotransformation takes place. Asbestos causes cancer in the lung. Azo dyes produce tumors or cancer in the liver, which is the site of biotransformation. Radium produces cancer in the bone, where the radium is stored. The aromatic amines produce cancer in the bladder, the site at which excretion takes place.

The testing, assessing, and regulation of carcinogens is carried out by a variety of agencies, based on a large number of health, safety, and environmental laws that have been passed in the last 12 years. These agencies have issued guidelines and policies on how they intend to identify, evaluate, and regulate carcinogens. The Office of Technology Assessment has been given the task of gathering together all the information and techniques used by these agencies in making their decisions. There are many important issues in assessing potentially carcinogenic chemicals. They are based on interpretation of test data and the use of assumptions. These assumptions are derived from theories about how cancer is caused and decisions about what is considered appropriate public policy. The OTA has identified four important kinds of assumptions: (1) assumptions used when the data are not available for a particular case; (2) assumptions that are potentially testable but have not yet been tested; (3) assumptions that probably cannot be tested because there are certain experimental limitations; (4) assumptions that cannot be tested because there are ethical considerations.

There are four kinds of evidence that may be used for qualitatively identifying carcinogens. They are epidemiological studies, long-term animal bioassays, short-term tests, and structure-activity relationships.

Epidemiological studies are used to collect information about human exposures and diseases. Reports of individuals or clusters of cases help to generate hypotheses for later study. Many of the chemicals now known to be human carcinogens were first identified by physicians. Larger epidemiological studies were then devised and conducted. Descriptive epidemiological studies correlate risk factors, exposures, and diseases that are causes of death in specific groups. They are useful in establishing hypotheses for further study in providing clues about the hazards. Analytical epidemiological studies are used in comparing populations between a group exposed to the agent and a group that is not exposed. In case-control studies, the comparison is made between people with a given disease and those who do not have the disease.

Long-term animal bioassays are laboratory studies in which animals are exposed to suspected carcinogens for long periods of time (about 2 years). The animals are then examined for tumors. The tissues of animals of those who survive and those who die are compared.

Short-term tests are used to examine genetic change in laboratory cultures of cells, in humans or other animals. These tests may take days or weeks to conduct.

Structure-activity relationships (SARs) are used to determine the chemical structures of substances and carcinogenicity. This is a comparison that helps to predict whether or not chemicals of a class closely related to carcinogens may also be carcinogenic.

The political considerations related to carcinogenesis are considerable. In regulatory proceedings, industry, labor, environmental groups, public interest organizations, and government may voice opinions that are frequently and substantially different. These groups place different values on the harm caused by the unnecessary regulation of a chemical that may later turn out to be safe. Even when accepting the value of animal data, there are many discussions on whether a particular animal study is reliable and if the data apply appropriately to people.

The Food and Drug Administration (FDA) was the first agency to establish guidelines for toxicity. The FDA, under the 1958 Food Additives Amendment to the Food, Drug and Cosmetic Act, which included the "DELANEY Clause", prohibited the intentional use of food and color additives that were determined to be carcinogenic in animals or humans. The "DELANEY Clause" does not apply to all food ingredients, because some were federally sanctioned prior to the 1958 amendment and some were considered to be generally safe. In the 1970s the FDA began using quantitative assessments for certain environmental contaminants found in food. By the 1980s, the FDA began applying these techniques to food and color additives. Much discussion has centered around what is an approved analytical technique to determine if the food and color additives are contaminated with small amounts of carcinogenic impurities, or when the additive itself has been determined to be carcinogenic. In 1985, it was decided that an approved analytical technique could be defined as one that could detect residue contents as low as the level associated with the upper portion of human risk estimate of one cancer for every one million persons exposed. This technique requires a risk assessment to estimate what residue levels correspond to this risk level.

The Environmental Protection Agency (EPA) began to develop carcinogenic assessment guidelines during the regulatory proceedings on the suspension and cancellation of several pesticides. The attorneys for the EPA summarized the expert testimony and developed summaries that were called *cancer principles*. There was considerable criticism of these cancer principles. In 1977, the Environmental Defense Fund petitioned the EPA to establish a policy on classifying and regulating air pollutants that were carcinogenic. In response to a court order to assess the hazards and risks of a large group of substances related to the Clean Water Act, the EPA set a methodology for assessing human risk. In 1984, the EPA published a proposed revision of its Carcinogen Assessment Guidelines, and also published proposed guidelines for assessing exposure to agents for mutagenicity and the developmental toxicants. It also published proposed guidelines for risk assessments of chemical mixtures. The final version of these guidelines were published in 1986.

The Carcinogen Assessment Guidelines published in the Federal Register described the general framework to be used in assessing carcinogenic risks and some of the principles to be used in evaluating the quality of the data and in making judgements concerning the nature and magnitude of the risk of cancer from suspected carcinogens. The various steps of risk assessment include hazard identification, dose-response assessment, exposure assessment, and risk characterization. The policy also presents a *weight-of-the-evidence* classification system. The Environmental Protection Agency lists five groups. (1) Group A includes human carcinogens. This group is only used when there is sufficient evidence from epidemiological studies to support an association between exposure from agents and cancer. (2) Group B is the probable human carcinogens. In this group the weight of the evidence of human carcinogenicity as based on epidemiological studies is limited. However, the weight of evidence based on animal studies is sufficient. (3) Group C is the possible human carcinogens. In this group, the evidence of carcinogenicity in animals is limited and there is also an absence of human data. (4) Group D is not classified as to human carcinogenicity. Here, there is inadequate human and animal evidence of carcinogenicity. (5) In Group E, there is evidence of noncarcinogenicity for humans.

The Consumer Product Safety Commission (CPSC) published carcinogen guidelines in 1978. The guidelines were challenged in court and were thrown out. Subsequently, CPSC decided to use the guidelines adopted by the Inter Agency Regulator Liaison Group (IRLG).

The Occupational Safety and Health Administration (OSHA) and numerous other agencies have written regulations on carcinogens. There have been a lot of arguments by various scientific groups, and there have been numerous legal challenges of a variety of carcinogen guidelines.

Most of the policies established by the various federal agencies declare that well-conducted positive epidemiological studies provide conclusive evidence for carcinogenicity. Some of the factors used in evaluating epidemiological studies include the strength of association, the level of statistical significance, information on dose-response relationship, biological plausibility, temporal relationships, accuracy of exposure and cause-of-death classification, adequacy of follow-up, and a determination of the amount of time needed for latent effects to show up. A "nonpositive" study cannot be used to indicate an absence of a carcinogenic hazard. It can be used to establish upper limits of risk. There were nonpositive studies for arsenic, benzene, coke-oven emissions, petroleum refinery emissions, and vinyl chloride. Subsequently, these chemicals were proven to be carcinogenic. Long-term animal bioassay data have been accepted as a predictor for human beings. Substances shown to be carcinogenic in animals are presumed to present carcinogenic risk to humans. The reason for this conclusion is twofold. One, a number of chemicals that were first identified as animal carcinogens were subsequently confirmed as human carcinogens. Two, all chemicals accepted as human carcinogens, if properly studied in animals, have been shown to be carcinogenic in at least one species of animals. Although this cannot establish that all animal carcinogens also cause cancer in humans, in the absence of data on humans, it is biologically plausible. However, one must consider the amount of substance that is necessary to cause the potential carcinogenic effect in people and then determine if the chemical should be banned from use. Further, the route of administration in animal studies must be taken into account when determining the potential health effects in people. A big question is what do you do when you have conflicting animal data or conflicting animal and human data?

Quantitative risk assessment relates to dose-response determination, exposure estimation, and risk characterization. Quantitative estimation is a serious problem because often a series of untestable assumptions must be made from an extrapolation of animal to human cases. The exposure levels are deliberately set at high quantities to maximize the probability of detecting a carcinogenic effect. Extrapolating from a high concentration to a low concentration, even when using human data, can be difficult. Therefore, it can be seen how difficult it must be to reach appropriate conclusions when using animal data. The relationship between the dose or exposure of an agent and the biological response of the human is one of the most fundamental in the fields of toxicology and epidemiology. The dose-response curve may have several different shapes, ranging from a straight line to different curves. The data from animals usually represent dose levels substantially higher than the range of human exposure. An important question is whether or not threshold limits exist for exposure to carcinogens. These limits would be set at a level where there are no possible effects from the carcinogenic agent.

The development of cancer takes place in stages. These are typically called initiation, promotion, and progression. Initiation involves an alteration in the cell's genetic material. This alteration of genetic materials can remain latent (that is, without apparent disease showing) for years. Promotion involves the expression of genetic information and the transformation of latent initiated cells into tumors. Progression consists of the growth of the tumors and the development of metastasis, which is the transfer of cancerous cells to other tissues. Some chemicals are primarily initiators. Other chemicals act only as promoters. Some chemicals are both initiators and promoters and are called complete carcinogens. It is possible that the mechanism of promotion involves an alteration in body chemistry, cellular growth, and repair, and other processes. *Pharmacokinetics* is a term that includes the variety of biological and chemical reactions that may occur in the body when exposed to a drug or chemical. The drug or chemical can be absorbed into the body, metabolized into other substances, distributed to other organs and tissues, or removed from the body. During any of these actions or reactions, various organs of the body may be involved and may also be damaged.

Mutagenesis

Mutagenesis is brought about by a change in genetic material that was transmitted during cell division. This change can cause deleterious effects. If the change occurs in either the sperm and/or egg cells, the mutagenesis may occur in future generations. The effect of the mutagen may either be long term or delayed, however, high doses may cause toxicity to occur rapidly. Mutagenesis has a number of characteristics in common with carcinogenicity. The carcinogen may bring about its effects through the proliferation of normal cells, rather than loss of function. The same thing may occur with the mutagen. Apparently, there is an absence of a threshold of a given chemical that may cause the mutagenesis, as well as carcinogenicity. The Environmental Mutagen Society in 1975 revealed that a variety of active compounds in all chemical classes could cause mutagenesis. There is a clear and present hazard for the human population now, as well as for succeeding generations. It is necessary to understand how gene mutations occur, how chromosome breaks and rearrangements occur, as well as nondisjunction (nondisjunction is the failure of two chromosomes to separate subsequent to metaphase in meiosis or mitosis, so that one daughter cell has both chromosomes and the other has neither of the chromosomes). Also, it is necessary to understand mitotic recombination and sister-chromatid change. Unscheduled DNA synthesis and differential growth-inhibitory effects and repair-deficient strains need also to be considered. A variety of test systems need to be used in order to determine if any given chemical is mutagenistic. Further monitoring studies of populations at great risk contribute significant information to determine what might be mutagenic and what might not. Genetic hazards to humans are typically predicted by bringing together the chemical and biological data generated from various tests and animal models. The degree of hazard or risk is usually a function of the perceived relevance of a test system that provides the positive responses and the total body of evidence that is available. Some known or suspected chemical mutagens include DDT, sodium arsenate, 2,4-D, cadmium sulfate, 2,4,5-T, some lead salts, dioxin, nitrite, ozone, and benzene.

Teratogenesis

Teratology is the study of malformations induced during the development of an animal fetus from conception to birth. The teratogen causes a change to occur during the early embryonic development of the fetus, which results in anatomical defects, other functional defects, or biochemical developmental errors. Teratogens specifically cause malformations and are part of a class of developmental toxicants. Teratogens may also cause the death of the embryo or fetus. Teratogenesis is usually considered to be a chronic effect, although the toxicity that caused it appeared over a short period of time during the developmental stage. The severity of the condition relates to the kind of exposure, the amount of the exposure, and when the exposure occurs during the developmental process. Drugs and chemicals considered to be teratogenic include androgenic hormones (a male hormone), thalidomide, organic mercury, dioxin, 2,4,5-T, esters, cadmium sulfate, sodium arsenate, and phenylmercuric acetate. It is also possible that various heavy metals, such as cadmium, copper, lead, and selenium, may be teratogenetic agents.

ENVIRONMENTAL HEALTH PROBLEMS AND THE ECONOMY

The environment and the economy are not mutually exclusive; together they form an integral system. As the public demands a higher quality environment, the economy must adjust to meet this need. In the past, Americans have placed a high priority on convenience and consumer goods. There seemed to be no need for concern over the air, water, or the Earth. The economic system was established such that incentives were given to individuals to better their position in our society without taking into account the problem of environmental degradation. The rich endowment of natural resources; the large amounts of investment, technology, and education; the influx of skilled immigrants; and the political and economic systems in the United States have combined to produce an economy with enormous material wealth. American business and industry, through its technological developments, produce huge amounts of goods and services. Consumers continually demand new products, and business and industry have accommodated consumers. Old products are discarded as quickly as new ones become available.

However, a new demand for better environmental quality and cleaner surroundings has emerged in the United States. The American public is demanding cleaner water, cleaner air, relief from noise and congestion, and a more aesthetically pleasing environment. Concurrently, the supposedly unlimited supply of water and air is befouled with huge quantities of pollutants, creating severe health, economic, and aesthetic problems. Exotic chemicals have entered the environment and moved through the food chain to a point where they are becoming a potentially dangerous health hazard.

During the last 10 years the production of synthetic organic chemicals, containers, packaging, and electronic equipment has increased sharply. As production has increased, consumer demand has increased. Unfortunately, although the products are available to the consumer at reasonable prices, these prices do not fully represent the cost to our society, since they do not include the cost of waste of our natural resources or the cost of environmental degradation. Due to the unfair economic incentive for use of raw materials vs reuse of recycled materials, there is further destruction of our environment. The dumping of wastes into the air, water, and land must end. The burden of the cost of the environmental cleanup and environmental improvement must be borne by all. Pollution costs include higher prices due to damage to crops, materials, and properties, and due to disease and injury. The actual costs of control measures are difficult to assess, since one must take into account abatement programs, expenditures by business and industry, expenditures by government, research and development, monitoring devices, and a variety of equipment and service purchases.

Dollars spent improving the environment are an investment in life and human society, although the benefits of these investments may not be readily seen. Imposing environmental controls results in the reallocation of resources, which in the short run may cause adverse economic effects, such as higher prices, temporary unemployment, and plant dislocations. The decrease in medical bills, the increase in recreational activities, the decrease in damage to materials, and the provision of a better society must be weighed against the temporary cost increases. Marginal firms that fail to meet these costs will be forced to close. The ultimate good to the greatest number of people should be the deciding factor in determining what should or should not be done to improve the environment. However, in any given area where changes are undertaken, adequate additional funds must be made available if the area is likely to suffer economic dislocation. In some cases where health is not endangered by pollution, it may be best to carefully and cautiously make temporary exceptions to certain standards in an effort to ease the strain on local economies. These exceptions should only be made with the greatest of care.

RISK-BENEFIT ANALYSES

In all areas, but particularly with regard to chemicals, it is necessary to compare the social benefits of a given environmental substance to the risk it causes to our society. Many of the pesticides are potentially extremely dangerous, and yet without the availability of pesticides the country and the world would be rampant with disease and would probably suffer from famines and starvation. It is therefore necessary to evaluate each of the environmental substances to determine the relative safety of the substance, the relative efficiency in production of the substance compared to the use of natural resources, and the relative impact on the environment at various stages of manufacture, use, or disposal. In all cases, risk-benefit analyses must be completely considered. The benefits include the value of the use of the substance to the consumer, its aesthetic value, the conservation of natural resources and energy resulting from the production of the substance, and the economic impact of employment, regional development, and balance of trade in the production of the substance. The risks include the adverse effect on health; the potential for death occurring as a result of use of the substance; environmental damage to air, water, land, wildlife, vegetation, aesthetics, and property; and the misuse of natural resources and energy. Establishing risk-benefit analyses for each substance utilized within the environment is difficult. However, with analyses of this type, society can determine the necessary trade-offs between potential hazard and employment, health, gross national product, balance of trade, conservation of natural resources, and environmental quality.

Benefit Cost Analysis

The four case studies that will be discussed include the following: lead in gasoline, air toxics issues, incineration at sea, and municipal sewage sludge reuse and disposal.

Lead in Gasoline

In the case of the lead-in-gasoline rule, a careful quantitative analysis was used. Lead in gasoline has been regulated by the EPA for more than 10 years. There was a concern that leaded gasoline was being used in cars that required unleaded gasoline. This would affect the pollution control catalysts. There was a further concern that the lead was a potential risk to children's health. Although previous rules had reduced lead in gasoline, in March 1984, the EPA set a very tight phasedown schedule. The final rule issued in March 1985 was more stringent than the original proposal. It required that lead in gasoline be reduced from 1.1 g per leaded gallon to 0.5 g per leaded gallon by July 1985, and to 0.1 g per leaded gallon by January 1986. Not all of the lead was removed from leaded gasoline, because of the potential effect it would have on older engines. Since the 1920s, refineries have added lead to gasoline as an inexpensive way of boosting octane. With the removal of lead, refineries had to do additional processing or use other additives, which tended to raise the cost of gasoline. The EPA used a computer model of the refining industry that had been developed by the Department of Energy to estimate the cost of the rule. The computer model included various equations that showed how different inputs could be turned into different end products at varying costs and the constraints on the industry's capacity. To estimate the cost of the rule, the EPA first ran the model specifying the then current lead limit of 1.0 g per leaded gallon and computed the cost of meeting the demand for the refined products. It then ran the model specifying the lower lead limits and recomputed the overall cost. The difference between the original and anticipated lead limits was the estimated cost of the tighter standard. Based on that analysis, the EPA estimated that the rule would cost less than 100 million dollars in the second half of 1985 to meet the 0.5 g per leaded gallon standard and approximately 600 million dollars in 1986 to meet the 0.1 g per leaded gallon standard. They then ran extensive tests using more pessimistic assumptions, such as an unexpected high demand for high-octane unleaded gasoline, increased downtime for equipment, and a reduced availability of alcohol additives. The 0.1 g per leaded gallon rule met virtually all conditions. It was extremely unlikely that a combination of these conditions would occur at the same time.

The benefits of the rule were estimated for the maintenance of good health in children and the improvement in educational efforts, compared to the efforts of children who were exposed to lead. Further, benefits were estimated based on the reduction in damage caused by excessive emissions of pollutants from vehicles using the wrong gasoline, the impacts on vehicle maintenance, and the impacts on fuel economy. The EPA also used existing data to study the relationship between levels of lead in the blood and blood pressure to make some estimate of the health effects that adults would have from this rule. In each category, the EPA first estimated the impact of reduced lead in physical terms. In the case of children's health effects, it used statistical studies relating to lead in gasoline and blood lead levels projected in children. It also estimated the cost to people and society and the environment by allowing the lead levels to stay the same vs. the reduced lead levels. It was determined that the benefits of the lead rule exceeded the cost by a 3:1 ratio and that if the potential benefits in reductions in blood pressure, illness, and death were to be counted, the ratio would rise to 10:1.

Air Toxics

Risk assessment technique were used to determine the environmental hazards of toxic air emissions. A study was made to determine how much of the air toxics problems could be controlled by using existing EPA programs. A new toxic strategy was then developed and put into place. It has three main parts: (1) direct federal regulation of significant nationwide problems, (2) state and local control of significant pollutant problems that are national in nature, and (3) an increased study of geographical areas that are subjected to particularly high levels of air pollution. In the initial study, 42 air toxic compounds were evaluated, basically for their potential to cause cancer. A determination was made of how many people were exposed across the country. An estimate of the risk associated with each compound in this national exposure was determined. The estimate was of long-term cancer incidents (a 70-year time frame) associated with these compounds. The analysis suggested that the air toxics problem was complex, caused by many pollutants and sources, varied significantly from city to city, and even varied within a city.

Apparently, preexisting EPA air toxic strategies that focused on regulating each pollutant as part of national emissions standards were too narrowly set up to be effective. A comparison of air quality data for 1970 and 1980, however, showed that significant reductions in national cancer incidents related to air pollution had occurred as a result of the reduction of air pollutants. Air toxic problems were examined in detail in Baltimore, Baton Rouge, Los Angeles, Philadelphia, and Phoenix. Fourteen compounds that were identified in the original study as being of the greatest concern for cancer risk for the general population and how these compounds could be controlled were studied. Again, cancer was the indicator of health effects, since other health effect indicators were more difficult to determine. Emission sources were identified in each of the five cities for these 14 compounds. Human exposure modeling was used to determine the degree of risk experienced by the people of these selected areas and the cancer incidents expected from this risk. The assumption was made that there would be a full implementation of the criteria of the pollutant programs.

The greatest incidents of cancer are associated with vehicles or area sources, such as numerous small pollution sources, like wood stoves. Point-source pollution sources, which are large, relatively identifiable sources of pollution, such as utilities and steel plants, appear to account for little of the total incidents.

All this data suggests that the health problems of air toxics may require a targeted strategy to make the best use of available resources. The data also suggest that additional focus must be made on small and nontraditional sources in order to get the greatest benefit for the health of the individuals. Additional attention needs to be given to wood smoke, vehicles, waste oil burning, and gasoline stations. The risk assessment here has helped to highlight the important areas where the greatest amount of good results will occur from the smallest amount of dollar expenditures.

Incineration at Sea

Industry generates more than 70 billion gal of hazardous waste each year in the United States. This waste must be safely managed through treatment, storage, and disposal. Because the EPA is restricting the use of management practices through permitting, regulation, and enforcement programs, there is a need to determine how best to get rid of this hazardous waste. Incineration is a technique that is being used to help destroy this hazardous waste. Public opinion, however, is strongly against the issuing of permits for new incinerators.

Between 1974 and 1982, the EPA issued permits for four series of burns conducted by the incinerator ship Vulcanus. Three of the burns were in the Gulf of Mexico and one in the Pacific Ocean. Public opposition to incineration at sea has intensified greatly. The EPA undertook a number of studies to determine whether incineration on land or the oceans is more desirable.

The EPA developed a risk assessment study that compared the human and environmental exposure most likely to occur from releases of land-based incineration vs. ocean incineration. The study used existing information and added new analysis of the emissions, transport, fate, and alternate effects. Both systems used land transportation, transfer and storage operations, and final incineration. The ocean system also included an ocean transportation step. For the purposes of the study, the waste was assumed to be a combination of 35% PCBs by weight and 50% ethylene dichloride (EDC) by weight. These two chemicals were used to simplify the waste stream. The analysis was to determine the statistically expected amount of pollutant released from accidental spills and air emissions. The transportation and handling accounted for less than 15% of the expected releases, while the incineration accounted for about 85%.

During land transportation, there were two types of potential losses. They were vehicular accidents and spills from the containers in route. Vehicular accidents were expected to occur on the average of once every 4–5 years and container failure once every 3–4 years. The analysis of transfer and storage of the waste was considered and three types of releases were anticipated: (1) spills when unloading the waste from tank trucks, (2) spills from equipment at the waste transfer and storage facilities, (3) and fugitive emissions in transfer and storage. Spills from transfer and storage components are infrequent events that are estimated to occur at about 0.50% per year and from the tank trucks, loading of waste to ship in the ocean system is approximately 0.002% per year. It was recognized that spills of this low quantity would likely be contained at the facility.

About 320 voyages of incineration ships have been made in the North Sea since 1972, and no casualties or spills have occurred. The spill rates from ships were based on worldwide historical data. It was estimated that the frequency of all spills for the Vulcanus would be about 1 per 12,000 operating years.

The study estimated and compared the possible human health and environmental effects due to incinerator releases and fugitive releases from the transfer and storage equipment.

The analysis of human health risk was based on the most exposed individual, who lives at the location, and the persons at the highest overall risk due to air concentrations from the incinerator stack and transfer storage facility. This risk is based on the cancer potential for an individual who has 70 years of continuous exposure. The risk on land-based incineration was 3 chances in 100,000, whereas the risk on ocean burning was in a range of 1 in 1,000,000 to 6 in 10,000,000. The relative risk on land for incineration of PCB waste is about 40 times more than on the ocean. For EDC waste, the ratio of land/ocean risk is about 30:1.

The conclusions of the incineration study were as follows: (1) incineration, whether on land or at sea, is a valuable and environmentally sound option; (2) there is no clear preference for ocean or land incineration as it relates to human health risk and the environment; (3) future demands will significantly exceed the capacity for disposal of hazardous waste; (4) continuing research is needed to improve current knowledge of combustion processes and effects; (5) the EPA needs to improve its public communications effort in the area of hazardous waste management.

Municipal Sludge Reuse and Disposal

Sludge management is an intrical part of municipal sewage disposal. A determination was made of the cost effectiveness of the disposal of sludge through various media. Municipalities generate about $6\frac{1}{2}$ million dry tons of wastewater sludge a year. Sludge production is expected to double to approximately 13 million dry tons a year by the year 2000.

The five major sludge use/disposal options currently available are land application, distribution and sale of sludge products, landfill, incineration, and ocean disposal.

Municipal sewage sludge contains over 200 different substances, such as toxic metals, organic chemicals, and pathogenic organisms.

There are many regulations controlling the disposal of sludge. However, the

EPA, in order to achieve better results, is consolidating existing sludge management authorities under the Clean Water Act. The Sludge Policy requires a consideration of risks, benefits, and costs of all sludge use and disposal practices. Since there is no methodology available to evaluate and compare the various impacts of sludge disposal, the EPA established the Integrated Environmental Management Division (IEMD) to conduct a major study of human health risks and costs related to the different disposal options used for sludge. The purpose of this approach is to identify high-risk areas and to develop a profile of the disposal options that provide a cost-effective way to reduce these risks. The high-risks disposal options include the exposure route, contaminants, and wastewater treatment plant types.

The IEMD model has four key components. They are (1) model plants, (2) risk profiles, (3) cost, and (4) cost-effectiveness analysis. There are three model plant sizes, each with one or two different types of sludge. These sludge types vary by the amount of industrial liquid waste added to the plants. The risk profiles are based on the four exposure routes of inhalation, groundwater, surface water, and ingestion through food, as they relate to potential carcinogenicity and renal effects. Costs have been estimated for each of the disposal options associated with each of the model plants. Cost-effectiveness analysis are made on a dollar basis for switching from one disposal option to another. The effectiveness of the reduction in the potential for disease is determined. Three major results have come from the IEMD analysis of sludge disposal. (1) There are aspects of sludge management that present very little risk and, therefore, provide very large gains if these aspects could be regulated properly. (2) Some options may be better, with lower cost and lower risk, while others may be worse, with some cost reduction but an increase in risk. (3) With some refinement and the use of site-specific data, the model may be used as an effective decision-making tool at the regional or local level.

The EPA is attempting to use its very best judgement based on scientific findings and projection of data to develop good risk management programs. Considerable additional research will be needed in many areas in order to ultimately reach a proper management program that will reduce the risk to the health of people in a cost-effective manner.

ENVIRONMENTAL HEALTH PROBLEMS AND THE LAW

The environmental health practitioner works for a unit of the local, state, or federal government, or, if working for industry, is constantly dealing with various levels of government. (For a detailed discussion on the functions of government, the reader is encouraged to refer to one of the many publications on local and state government.) For the purposes of this book, a brief discussion on federal-state relationships, legislative procedure, pressure groups, local government, and state and local finances is included.

The governmental system in a democratic society performs two distinct functions. It provides a foundation for debate on issues and a vehicle for the solution of problems. It also provides a service and regulatory function, since the individual in a complex society cannot provide for all of his or her needs. These needs include adequate police, fire, and health protection. For the society to operate properly, the individual or group must adhere to rules and regulations formulated and enforced by the elected government.

According to the U.S. Constitution and Amendments thereto, federal powers include the control of interstate and foreign commerce, conduct of foreign policy, and national defense. The federal government cannot levy direct taxes, other than income taxes, except in proportion to the population of the states. The federal government cannot abridge civil rights. The state is sovereign. The powers not delegated to the federal government or prohibited by the constitution are reserved for the states or the people. The state powers include the administration of elections, the establishment and operation of local government, education, intrastate commerce, the creation of corporations, police force, and the promotion of health, safety, and welfare. Although the federal government cannot dictate organization and administration of programs, or the establishment of policy, it has gained some control via its use of funds. In order for a state to receive federal funds, it must adhere to certain requirements established by the federal government for the use of these funds. These requirements include preparation and submission of plans, approval of plans by a central federal agency, the establishment of necessary state agencies, provision for matching state funds, and the supervision and auditing of state programs by the appropriate federal agencies.

Public policy is a result of the interaction of groups and individuals. The formulated policy is not necessarily ideal for the public. Major pressure groups — including business, industry, agriculture, labor, medicine, and religion, try to influence the legislature to pass laws beneficial to their own self-interest. These groups also pressure administrative agencies to make or change decisions on policies that are not in the best interest of the public. These groups compete with each other for public support. They use sizable sums of money to influence opinions on legislation. Citizens groups properly educated and stimulated by trained professional public health workers could use the techniques of the pressure groups to establish public policy that would be beneficial to the citizens and would improve the environment.

The state is divided into local government, including counties, townships, villages, cities, and boroughs. Local government functions may include all the functions of the state government, such as the powers of taxation, establishment of budgets, licensing, and the administration of health, safety, and environmental programs. The local government is expected to deliver quality service and to protect the public. It should prevent the duplication of efforts, budgets, and facilities of the state government. The environmentalist must recognize that although concern for the environment is foremost in his or her mind, the county commissioners or township supervisors may be more concerned with other programs, such as road building, recreational facilities, police and fire service, etc. The environmentalist has to compete for budget monies with each of the other operating departments.

State and local governments obtain their funds for operation from taxes levied on sales, automobile licensing, gasoline, corporations, personal income, alcohol and tobacco, gross receipts, property, and establishment licensing. In many cases, tax rates become oppressive and yet income derived from taxes is inadequate to properly finance the necessary programs for the community. Then it becomes necessary for the federal government to supply funds; in this way, the federal government exercises control in some states even at the township level.

Law and Public Health

Public health policies will be discussed briefly, since the environmental health practitioner should have some understanding of the law to function in his or her capacity as a government official. (For more detailed discussion, many books are available describing current laws relating to the environment.)

Law is the rule of civil conduct prescribed by the supreme power in a state commanding what is right and forbidding what is wrong. Law should represent the community's desires or commands, apply to all members of the community, be backed by the full power of the government, and provide the administration of justice under the law for all people. The purpose of law is to protect by the regulation of human conduct of the individual from other individuals, groups, or the state, and vice versa. Statutory laws are basically legislative acts passed by a legislative body; common laws are established customs of the community; equity is a decision by a judge in a given situation where rules and regulations have been violated; and administrative laws are rules and regulations in a given area established by an agency authorized by the legislature. The value of administrative laws is that the rules and regulations established can be changed readily when new scientific data are available. If regulation were part of a statute, the legislature would have to go through the complex process of rewriting the act.

Public health law is that body of statutes, regulations, and precedents that protects and promotes individual and community health. Public health law is founded on the Preamble to the Constitution, which ordains that the government shall "... promote the general welfare," and Section 8, Article 1 of the Constitution, which "... provide(s) for the common defense and general welfare." The interpretation of these clauses and other clauses of the Constitution by the U.S. Supreme Court has established the legal basis for public health. The law of eminent domain empowers the state with the authority to seize, appropriate, or limit the use of property in the best interests of the community. This power is reflected in the establishment of zoning and land use regulations that are so essential to the preservation of a good environment.

Nuisance laws, originating in the Middle Ages, are used repeatedly in environmental health work. Basically they state that the use of private property is unrestricted only as long as it does not injure another person or his property. If an injury occurs, a nuisance exists. Large numbers of public health officials use nuisance laws to eliminate problems caused by sewage, solid waste, air pollution, and insects and rodents.

The public health law owes its effectiveness to the police power of the state. In times of great stress, such as severe fires, floods, and outbreaks of diseases, the private property of an individual might be summarily appropriated, used, or even destroyed if the ultimate relief, protection, or safety of the community demands that such action be taken. The legislature can and does delegate police power to an administrative agency for use in the event of an emergency. If a public official fails to use the delegated police power, he or she is guilty of nonfeasance of office. The use of this police power is determined by the chief administrative office, usually the State Secretary of Health, in consultation with the governor of the state.

Administrative law is that series of rules, regulations, and standards needed to implement statutes. The development and use of sound rules, regulations, and standards based on strong scientific data puts the burden of proof on the defendant, who has to show that the rules have not been fairly applied, rather than that they have limited value. Through judicial presumption, the judge presumes that the laws are correct, because they are based on sound scientific criteria or on the judgment of a group of experts. The judge only decides if the law is applied fairly.

Licensing is an important means of control and enforcement of environmental standards. If an individual or business does not meet the standards, a license is denied or revoked. An individual operating without a license is subjected to severe penalties.

Environmental health practitioners must be conscious of those actions that may occur in the enforcement of environmental health laws. These actions are (1) misfeasance, which is the performance of a lawful action in an illegal or improper manner; (2) malfeasance, which is wrongful conduct by a public official; and (3) nonfeasance, which is an omission in doing what should be done in an official action.

Environmental Law

For historical purposes, the Rivers and Harbors Act of 1899 was passed by Congress to prevent the illegal discharge or deposit of refuse or sewage into navigable waters of the United States. This law was used before 1972 to penalize some corporations for polluting the water.

The National Environmental Policy Act of 1969 signed into law on January 1, 1970 (Public Law 91-190), amended by PL 94-52 and PL 94-83, in 1975 set forth the continuing responsibility of the federal government to

1. Fulfill the responsibilities of each generation as trustee of the environment for succeeding generations.
2. Assure safe, healthful, productive, aesthetically, and culturally pleasing surroundings for all Americans.
3. Attain the widest range of beneficial uses of the environment without degradation, risk to health or safety, or other undesirable and unintended consequences.
4. Preserve important historic, cultural, and natural aspects of our national heritage, and maintain, wherever possible, an environment that supports diversity and variety of individual choice.
5. Achieve a balance between population and resource use that will permit high standards of living and a wide sharing of life's amenities.
6. Enhance the quality of renewable resources and approach the maximum attainable recycling of depletable resources.

The National Environmental Policy Act (NEPA) also states that all federal agencies shall

1. Utilize a systematic, interdisciplinary approach ensuring the integrated use of the natural and social sciences and the environmental design arts in planning and in decision making when planning may be an impact on the human environment.
2. Identify and develop methods and procedures, in consultation with the Council on Environmental Quality, ensuring that other environmental amenities and values be given appropriate consideration in decision making, along with economic and technical considerations.
3. Include in every recommendation or report on proposals for legislation and other major federal actions significantly affecting the quality of the human environment, a detailed statement by the responsible official on
   a. the environmental impact of the proposed action
   b. any adverse unavoidable environmental effects should the proposal be implemented
   c. alternatives to the proposed action
   d. relationship between local short-term uses of the environment and the maintenance and enhancement of long-term productivity
   e. any irreversible and irretrievable commitments of resources involved in the proposed action should it be implemented
4. Study, develop, and describe appropriate alternatives to recommended courses of action in any proposal involving unresolved conflicts, as well as alternative uses of available resources.
5. Recognize the worldwide and long-range character of environmental problems and, where consistent with the foreign policy of the United States, lend appropriate support to initiatives, resolutions, and programs designed to maximize international cooperation in anticipating and preventing a decline in the quality of the world environment.
6. Make available to states, counties, municipalities, institutions, and individuals advice and information useful in restoring, maintaining, and enhancing the quality of the environment.

7. Initiate and utilize ecological information in the planning and development of resource-oriented projects.
8. Establish the Council on Environmental Quality, whose functions are as follows:
   a.   to assist and advise the President in the preparation of the environmental quality report, which will be submitted to the Congress starting in 1970 (this report is an excellent digest of various environmental health issues)
   b.   to gather timely and authoritative information concerning the conditions and trends in the quality of the environment, currently and for the future
   c.   to review and appraise the various programs and activities of the federal government
   d.   to develop and recommend to the President, national policies to foster and promote the improvement of environmental quality
   e.   to conduct investigations, studies, surveys, research, and analyses relating to ecological systems and environmental quality
   f.   to document and define changes in the natural environment
   g.   to report at least once every year to the President on the state and condition of the environment
   h.   to make and furnish studies, reports, and recommendations with respect to matters related to policy and legislation

Since 1970 when the NEPA Act was passed and a Council on Environmental Quality was established, Congress has passed a large number of laws and amendments to the laws, related to a variety of environmental issues that potentially affect the health of people and affect the environment and/or ecosystems. These laws include the Clean Air Act (CAA), the Comprehensive Environmental Response, Compensation, and Liability Act (CERCLA), the Consumer Product Safety Act (CPSA), the Emergency Planning and Community Right-to-Know-Act, the Energy Supply and Environmental Coordination Act, the Federal Insecticide, Fungicide, and Rodenticide Act (FIFRA), the Food, Drug, and Cosmetic Act Amendments (FDCA), the Occupational Safety and Health Act (OSHA), the Noise Control Act, the Radon Gas and Indoor Air Quality Research Act, the Resource Conservation Recovery Act (RCRA), the Super Fund Amendments and Reauthorization Act (SARA), and the Toxic Substances Control Act (TSCA).

There have been a number of laws related to water and water pollution that have been enacted since the original Rivers and Harbor Acts of 1899. These laws include the Clean Water Amendments of the Federal Water Pollution Control Act (CWA), the Coastal Zone Management Act, the Deep Water Port Act, the Marine Protection, Research, and Sanctuaries Act, the National Ocean Pollution Planning Act, the Outer Continental Shelf Lands Act, the Port and Tanker Safety Act, the Shore Protection Act, the Safe Drinking Water Act (SDWA), the Water Resources Planning Act, and the Water Resources Research Act.

The major amendments to the Clean Air Act were in 1977, updated in 1983, and again a major amendment was passed in 1990. There had been a substantial debate for the last 8 years concerning how to change the law. The congressional findings included: (1) that the predominant part of the nation's population was located in rapidly expanding metropolitan and other urban areas that generally cross the boundary lines of local jurisdictions and expand into multiple states; (2) that the growth and the amount of complexity of air pollution brought about by urbanization, industrial development, and increased use of motor vehicles has resulted in growing dangers to the public health and welfare, crops, property, and air and ground transportation; and (3) that the prevention and control of air pollution at its source was the primary responsibility of the state and local governments and that federal financial assistance and leadership was necessary for the improvement of air quality.

The law provides an elaborate federal-state scheme for controlling conventional pollutants, such as ozone and carbon monoxide. The 1990 amendments create tighter controls on tailpipe exhaust, reduction of acid rain, reduction of nitrogen oxides, reduction of air toxics, which may be carcinogenic, and protection of the ozone layer by phasing out chlorofluorocarbons, carbon tetrachloride, methylchloroform, and hydrochlorofluorocarbons.

The Comprehensive Environmental Response Compensation and Liability Act of 1980 (CERCLA), also known as Super Fund was designed to handle the problems of cleaning up the existing hazardous waste sites in the United States. The act, which was originally passed in 1980, was updated numerous times until 1988. The hazardous waste problems range from spills that need immediate attention to hazardous waste dumps that are leaking into the environment and posing long-term health and environmental hazards.

The Emergency Planning and Community Right-to-Know Act of 1986 established state emergency response commissions, emergency planning districts, emergency planning committees, and comprehensive emergency plans. A list of extremely hazardous substances has been prepared and published.

The Energy Supply and Environmental Act of 1974 was updated in 1978. The purposes of this act were to provide for a way to assist in meeting the needs of the United States for fuels in a consistent, practical manner, and to protect and improve the environment. It allowed for coal conversion or coal derivatives to be used in place of oil in power plants. This, of course, can contribute to greater levels of air pollution.

The Consumer Product Safety Commission Act (CPSA) of 1970, updated in 1984, established the Consumer Product Safety Commission as an independent regulatory agency. The CPSA gives the commission the power to regulate consumer products and to oppose unreasonable risks of injury or illness. They also regulate consumer products, except for foods, drugs, pesticides, tobacco and tobacco products, motor vehicles, aircraft and aircraft equipment, and boats and boat accessories. The law authorized the commission to publish consumer product safety standards in order to reduce the level of unreasonable risks. The commission has recalled hair dryers containing asbestos because of potential hazards. They also may regulate carcinogens.

The Federal Insecticide, Fungicide, and Rodenticide Act (FIFRA) was originally passed by Congress in 1947 and was last updated in 1988. It provides for the registration of new pesticides; the review, cancellation, and suspension of registered pesticides; and the reregistration of pesticides. It is concerned with the production, storage, transportation, use, and disposal of pesticides. It also includes areas of research and monitoring.

The Food, Drug and Cosmetic Act, which is an amendment to the Food and Drug Act of 1906, and the Federal Food, Drug, and Cosmetic Act of 1938, has been further amended numerous times. It was the first federal statute to regulate food safety. The amendment of 1938 established the general outlines for the authority of the FDA. Food additives, food contaminants, naturally occurring parts of food or color additives to food, drugs, or cosmetics, as well as potential carcinogens, have been regulated by various parts of the law. In 1958, the Food Additives Amendment, better known as the Delaney clause, stated "... that known additives shall be deemed to be unsafe if found to induce cancer when ingested by man and/or animal, or if it is found, after tests which are appropriate for the evaluation of food additives, to induce cancer in man or animal ... ."

The Noise Control Act of 1972 was amended last in 1978. The findings of Congress were as follows: (1) that inadequately controlled noise presents a growing danger to the health and welfare of the population of the United States; (2) that the major sources of noise include transportation vehicles and equipment, machinery, appliances, and other products used in commerce; (3) that, although the primary responsibility for control of noise belongs to the state and local governments, federal action is needed to deal with major noise sources in commerce.

Since aircraft contribute considerable amounts of noise to the environment, two laws were passed and further amended to reduce these noise sources. The first law was entitled "An Act to Require Aircraft Noise Abatement Regulation." This law was signed in 1968 and amended several times until it became the Quiet Communities Act of 1978. The Congress decided to control and abate aircraft noise and sonic booms. The second law was the Aviation Safety and Noise Abatement Act of 1979, enacted in 1980 and amended in 1982. The purpose of the act was to provide assistance to airport operators to prepare and carry out noise compatibility programs, to provide assistance to assure continued safety in aviation, and for other purposes. It helped to establish a single system of measuring noise, which was reliable, and also a single system for determining the exposure of individuals to noise at airports.

The Occupational Safety and Health Act of 1970 established the Occupational Safety and Health Administration or OSHA, and the National Institute for Occupational Safety and Health (NIOSH). OSHA sets and enforces regulations to control occupational health and safety hazards, including exposure to carcinogens. NIOSH is responsible for research; for various evaluations, publications, training; and for the regulation of carcinogens by supporting epidemiological research and recommending changes in health standards to OSHA.

The Occupational Health and Safety Act provides three mechanisms by law for setting standards to protect employees from hazardous substances. The act initially authorized OSHA to adopt the health and safety standards already established by federal agencies or adopted as national consensus standards. This authority was for the first 2 years in 1972 and 1973. It also authorizes OSHA to issue emergency temporary standards (ETS) that require employers to take immediate steps to reduce workplace hazards. The ETS may be issued by OSHA when they determine employees are exposed to grave danger and that the emergency standard is necessary. Although the public has not had a chance to comment on a standard, because of the nature of the potential hazard, it must be enforced. However, the final standard must be issued within 6 months of the emergency standard. The third way that OSHA sets standards is to issue new permanent exposure standards and to modify or revoke existing ones. However, the informal rulemaking that goes along with modifying or revoking existing standards is subject to court review. OSHA's approach to rulemaking can result in requirements in monitoring and medical surveillance, workplace procedures and practices, personal protective equipment, engineering controls, training, recordkeeping, and new or modified permissible exposure limits (PELs). The permissible exposure limits are the maximum concentration of toxic substances allowed in the workplace air.

Through 1988, NIOSH has published 35 occupational safety and health guidelines with technical information about chemical hazards for workers. In addition, they publish the criteria documents (CD), alerts, current intelligence bulletins (CIB), health and safety guides (HSG), symposium or conference proceedings, NIOSH administrative and management reports, scientific investigations, data compilations, and other worker-related booklets. The CDs are recommended occupational safety and health standards for the Department of Labor. Usually, as part of the recommended standard is a recommended exposure limit (REL). The CIBs relate important public health information and recommend protective measures to industry, labor, public interest groups, and the academic world. The HSGs provide basic information for employers and employees to help them have a safe and healthful work environment.

The Radon Gas and Indoor Air Quality Research Act of 1986 included the following findings by Congress: (1) high levels of radon gas pose a serious health threat in structures in certain areas of the country; (2) certain scientific studies suggest that exposure to radon, including naturally occurring radon and indoor air pollutants, may cause a public health risk; (3) existing federal radon and indoor air pollution research programs are fragmented and underfunded; (4) there is a need for adequate information concerning exposure to radon and indoor air pollutants; and (5) this need should be met by appropriate federal agencies.

The Resource Conservation and Recovery Act of 1976 (RCRA) was updated through 1988. This law replaced the previous Solid Waste Disposal Act. The Used Oil Recycling Act of 1980 amended the RCRA Act and then was incorporated into the main text of the act.

The RCRA provides for regulating the treatment, transportation, and disposal of hazardous waste. Hazardous waste is defined as solid waste that may cause death or serious disease, or may present a substantial hazard to human health or the environment if it is improperly treated, stored, transported, or disposed. Solid waste includes solid, liquid, semisolid, or contained gaseous materials from a variety of industrial and commercial processes. This definition excludes solid or dissolved materials found in domestic sewage or related to irrigation, industrial discharges subject to the Clean Water Act, or mining wastes. RCRA requires that the EPA develop and issue criteria for identifying the characteristics of hazardous wastes. Defining characteristics of hazardous wastes are (1) it poses a present or potential hazard to human health and environment when it is improperly managed; and (2) it can be measured by a quick, available, standardized test method, or it can be reasonably detected by generators of solid wastes through their knowledge of their wastes: ignitability, corrosivity, reactivity, and extraction procedure toxicity. RCRA also includes the regulation of underground storage tanks that may be used for a variety of storage processes. Further, it includes a program in medical waste tracking.

The Superfund Amendments and Reauthorization Act of 1986 (SARA) of CERCLA established an Alternative or Innovative Treatment Technology Research and Demonstration Program. Its function is to promote the development, demonstration, and use of new or innovative treatment technologies and to demonstrate and evaluate new, innovative measurement and monitoring technologies. The SARA amendments require that an information dissemination program be established along with the research efforts. The three types of technologies include (1) available alternative technology, such as incineration; (2) innovative alternative technology, which is any fully developed technology for which cost or performance information is incomplete and the technology needs full-scale field testing; and (3) emerging alternative technologies that are in their early stage of development and the research has not yet fully passed the laboratory or pilot testing phase.

The Toxic Substances Control Act (TSCA) was enacted in 1976 and amended through 1986. TSCA was enacted by Congress to allow for the regulation of chemicals in commerce, as well as before they even entered commerce. This policy includes (1) that chemical manufacturers and processors are responsible for developing data about the health and environmental effect of their chemicals; (2) that the government regulate chemical substances that pose an unreasonable risk of injury to health or the environment and act promptly on substances that pose imminent hazards; and (3) that regulatory efforts should not unduly hinder industrial innovation. TSCA is directed at hazardous substances, wherever they occur. They do not have to be within a special environment. As long as they may cause a danger to the public — including a significant risk of cancer, genetic mutations, birth defects, or other potential serious hazards — they may be restricted or even banned by the EPA. TSCA permits the EPA to regulate new or existing toxic substances by requiring the testing of new or existing chemicals and by requiring their restriction in production and use, or even outright banning of substances that pose an unreasonable risk to health or the environment. The Asbestos Emergency Response Program is part of TSCA.

There are 11 separate acts of Congress related to water quality and water pollution. These range from the Clean Water Act to proper management of the oceans near the continental United States and the improvement of water quality.

The Federal Water Pollution Control Act was amended by the Clean Water Act of 1977 and was amended numerous times through 1988. The initial act was passed in 1948 and was then referred to as the Federal Water Pollution Control Act. In 1972, Congress set the goals of achieving fishable, swimmable waters by 1983 and prohibiting the discharge of toxic pollutants in toxic amounts by 1985. In the 1977 amendments, Congress endorsed a new means for regulating toxic pollutants, which had been developed to settle a lawsuit between environmental organizations and the EPA. In 1987, Congress continued its emphasis on the control of toxic pollutants. Although the Clean Water Act Provisions are less directly related to human health than the provisions of the Safe Drinking Water Act, it still aims at regulating human exposure to carcinogens and other toxic materials. An important part of the CWA is the National Pollution Discharge Elimination System (NPDES), which creates permits for direct discharges into the waters. It is lawful to discharge a pollutant only if the discharge is in compliance with the NPDES permit, which has to be issued by the EPA or by states whose permit programs are approved by the EPA. To obtain a permit, a facility must have the following information submitted: (1) a list of pollutants that must be regulated according to federal or state law, along with a permissible amount of each pollutant that may be discharged per unit of time; (2) monitoring requirements and schedules for implementing the pollution concentration requirements; and (3) special conditions regarding the pollutants that the agencies feel should be imposed on the polluters. These may include additional testing and procedures for spills of pollutants into the water.

The Coastal Zone Management Act of 1972 was updated through 1986. Congress determined that it was in the national interest to have effective management, beneficial use, protection, and development of the coastal zone. The coastal zone included the fish, shellfish, and other living marine resources and wildlife found in this ecologically fragile area. It considered the increasing and competing demands of economic development, population growth, and harvesting of the fish and shellfish.

The Deep Water Port Act of 1974 was amended through 1984. Congress decided to authorize and regulate the location, ownership, construction, and operation of deep-water ports in waters beyond the territorial limits of the United States. The act provides for the protection of the marine and coastal environment beyond these territorial limits.

The Marine Protection Research and Sanctuaries Act of 1972 was amended through 1988. Congress determined that unregulated dumping of material into the ocean waters endangered human health, welfare, and amenities, and the marine environment, ecological systems, and economic potential. It decided that it was the policy of the United States to regulate the dumping of all types of materials into ocean waters and to prevent or strictly limit the dumping of any material into the ocean waters that would adversely effect health, welfare, and the ecosystem. It regulated the transportation of the material and dumping at sea.

The National Ocean Pollution Planning Act of 1978 was updated through 1988. Congress determined that the activities of people in the marine environment can have a profound short-term and long-term impact on such an environment and can greatly affect ocean and coastal resources. It stated that a comprehensive federal plan should be developed for ocean pollution research and for the development and monitoring of the material that had been dumped, the fate of the material, and the effects of the pollutants on the marine environment.

The outer Continental Shelf Lands Act of 1953 was updated through 1988. The *Outer Continental Shelf* means all the submerged lands lying seaward and outside of the areas of lands beneath navigable waters. This includes the subsoil and seabed that is attached to the United States. The Outer Continental Shelf is of vital importance to the United States, since it is a national resource reserve of a variety of potential minerals, oil, etc. Congress decided that since the exploration, development, and production of the minerals of the Outer Continental Shelf will have significant impact on the coastal and noncoastal areas of the United States and on other affected states, it was in the national interest that this exploration be controlled by the federal government. The Secretary of the Interior, under this act, has the authority to authorize the leasing of the Outer Continental Shelf and to enforce safety, environmental, and conservation laws and regulations, and to formulate and promulgate such regulations, as necessary, to prevent problems from occurring.

The Port and Tanker Safety Act of 1978 was updated through 1986. Congress decided that the navigation and vessel safety and protection of the marine environment were matters of major national importance. It said that increased vessel traffic in the nation's ports and waterways created a substantial hazard to life, property, and the marine environment, and that there was an increased need for supervision of the vessel and port operations. It was particularly interested in the handling of dangerous articles and substances on, or immediately adjacent to, the navigable waters of the United States. It stated that advanced planning is critical in determining proper, adequate, and protective matters for the Nation's ports and waterways, and the marine environment.

The Shore Protection Act was passed in 1988. Congress stated that a vessel may not transport municipal or commercial wastes in coastal waters without a permit from the Secretary of Transportation and without displaying a number or other marking on the vessel, as prescribed by the Secretary. The permit included the name, address, and telephone number of the vessel owner and operator, its transport capacity, and its history of cargo transportation during the previous year, including wastes. The Secretary of Transportation had the right to enforce regulations concerning loading, securing, offloading, and cleanup. The secretary can also refuse the permit.

The Safe Drinking Water Act (SDWA) of 1974 was updated through 1988. It was originally passed to ensure a safe drinking water supply. The Clean Water Act was designed to control water pollution, but it did not provide authority to regulate polluted water discharged into nonnavigable waters, such as groundwater, which often is a source of drinking water. The SDWA is used primarily to regulate water provided by public water systems, and it contains several provisions that may be used to regulate hazardous substances, including carcinogens in drinking water. The SDWA is more directly concerned in protecting human health than the CWA. Under the SDWA, the EPA regulates contaminants that may have an adverse effect on the health of people. It then establishes the steps that the agency must go through, over time, to protect drinking water. The EPA published national interim drinking regulations in 1975. The regulations were to protect the health of the people to the extent feasible using current technology treatment techniques and other means that the administrator determined were generally available. Congress also required that the EPA request a National Academy of Sciences study to determine the potential adverse health effects of contaminants in the water and to help establish recommended maximum contaminant levels (RMCLs). These RMCLs were the recommended maximum level to be set for contaminants to prevent the occurrence of any known or anticipated adverse effect. It had to include an adequate margin of safety, unless there was no safe threshold. In that case, the recommended maximum contaminant level could be set at 0. The RMCLs are not enforceable health goals but are used as guidelines for establishing enforceable drinking water standards. The MCLs or maximum contaminant levels were to be as close to the RMCLs as feasible. The enforcement of the MCLs rested with the states, whereas EPA sets the MCLs (see 1986 SDWA Amendments).

The Water Resources Planning Act of 1965 was updated through 1988. The act stated that in order to meet the rapidly expanding demand for water throughout the nation, it is declared to be the policy of the Congress to encourage the conservation, development, and utilization of water and related land resources of the United States on a comprehensive and coordinated basis by the federal government, states, counties, and private enterprises with the cooperation of all of the various agencies and governments, individuals, and businesses involved.

The Water Resources Act was passed in 1984. The Congress declared that the existence of an adequate supply of water of good quality for the production of materials and energy for the nation's needs and for the efficient use of the nation's energy and water resources is essential to national economic stability and growth and to the well-being of the people. It also stated that the management of water resources is closely related to maintaining environmental quality and social well-being, and that there should be a continuing national investment in water and related research in technology that was commensurate with growing national needs. This research and development of technology should include the development of a technology for the conversion of saline and other impaired waters to a quality usable for municipal, industrial, agricultural, recreational, and other beneficial uses.

How a Bill Becomes a Law

Although a passage of bills in the state legislatures and the federal government varies somewhat, they all follow an approximate route from the introduction of the bill to the final passage by both houses of the state government or the federal government. They are then signed or vetoed by the governor of the particular state or by the President of the United States. In order to get a better understanding of this process of how a bill becomes a law, the procedure used in the State of Indiana will now be discussed. Recognize that this path varies somewhat with each of the kinds of government throughout the country. For a bill to become a law in Indiana, it must follow a prescribed path with carefully planned steps. The path has many detours, lost turns, stumbling points, barriers, and frustrations. The reason for all these problems is because of the unique system of checks and balances that we find in government in the United States. The process in Indiana is as follows: (1) a representative or senator decides to introduce legislation; (2) a legal specialist draws up a bill; (3) the bill is introduced and assigned to a committee; (4) the committee approves the bill or amends the bill and approves it or rejects it; (5) the house or senate votes to accept the committee recommendation; (6) the bill is printed; (7) the bill remains on each representative's desk or senator's desk for 24 hours and then it is placed on the House or Senate calendar for a second reading; (8) the bill is ordered "engrossed" or reprinted to show amendments; (9) the bill passes a second reading. Amendments may be added at this time; (10) the third reading of the bill occurs; (11) speeches are made for or against the bill; (12) the vote is taken (in Indiana, the constitutional majority needed for either approval or rejection is 51 votes in the House or 26 votes in the Senate); (13) the bill is now delivered to the opposite house of the state or federal government, where either a senator or representative has promised to sponsor it; (14) the bill goes through a similar process in the opposite house, and it may also be amended here; (15) if the bill is amended, it must be approved by the House-Senate Conference Committee; (16) it is then submitted to the house and senate for passage; (17) when passed, it is signed by the House Speaker and the President of the Senate; (18) the Attorney General checks its constitutionality; (19) it is signed by the governor or vetoed and sent to the Secretary of State; (20) if it is signed, it becomes an enrolled act and is printed and bound in the "Act of Indiana"; (21) if it is vetoed, it goes back to the house and senate, where a constitutional majority would be a two-thirds vote in order to overrule the veto of the governor; and (22) it is either finally overruled or, if it has been enrolled, it is an enrolled act; it now takes effect when distributed among the circuit court clerks of all the counties of the state.

As stated previously, although the procedures in the federal government and the other states vary, the approach to becoming a law is about the same.

ENVIRONMENTAL IMPACT STATEMENTS

The National Environmental Policy Act (NEPA), which established the council on environmental quality (CEQ), requires each federal agency to prepare a statement of environmental impact in advance of each major action, recommendation, or report on legislation that may significantly affect the quality of the human environment. An environmental impact statement is the heart of a federal administrative process designed to ensure achievement of national environmental goals. Each statement must access in detail the potential environmental impact of a proposed action. All federal agencies are required to prepare statements for matters under their jurisdiction. The impact statement is called an EIS. The purpose of the statement is to disclose the environmental consequences of the proposed action and to be continually conscious of environmental considerations.

The actions that the statement cover must be major and environmentally significant. The actions may be a precedent for much larger actions, which may have considerable environmental impact, or they may alter the future course of other environmental actions.

Each environmental impact statement must include (1) a detailed description of the proposed action, including information and technical data adequate to permit a careful assessment of environmental impact; (2) discussion of the probable impact on the environment including any impact on ecological systems and any direct or indirect consequences that may result from the action; (3) any adverse environmental effects that cannot be avoided; (4) alternatives to the proposed action that might avoid some or all of the adverse environmental effects, including an analysis of costs and environmental impacts of these alternatives; (5) an assessment of the cumulative, long-term effects of the proposed action, including its relationship to short-term use of the environment vs. the environment's long-term productivity; (6) any irreversible or irretrievable commitment of the resources that might occur from the action or that would curtail beneficial use of the environment; and (7) a final impact statement must include a discussion of problems and objections raised by other federal, state, and local agencies, private organizations, and individuals during the draft statement's review process.

A draft statement must be prepared and circulated for comment at least 90 days before the proposed action. A final statement must be made public at least 30 days before the proposed action. Any agency unable to meet these requirements must consult with the CEQ.

NEPA requires each federal agency to consult with and obtain the comments of any other agency — state, local, as well as federal — that has jurisdiction by law or special expertise related to the environmental impact.

In its guidelines for preparing statements, CEQ lists the agencies that must be consulted in the following areas: air quality and air pollution control; weather modification; environmental aspects of energy generation and transmission; toxic materials; pesticides; transportation and handling of hazardous materials; coastal areas, including estuaries, waterfowl refuges, and beaches; historic and archaeological sites; floodplains and watersheds; mineral land reclamation; parks, forests, and outdoor recreational areas; noise control and abatement; chemical contamination of food products; food additives and food sanitation; microbiological contamination; radiation and radiological health; sanitation and waste systems; shellfish sanitation; transportation and air quality; transportation and water quality; congestion in urban areas; housing and building displacement; environmental effects with special impacts on low-income neighborhoods; rodent control; urban planning; water quality and water pollution control; marine pollution; river and canal regulation and stream channelization; and wildlife.

The guidelines also require that these statements be made available for public comment. Many individual agencies ask for such comments from interested parties and private organizations.

The impact statement procedure allows the public an opportunity to participate in federal decisions as they affect the human environment. The statements are announced in the *Federal Register*, and many agencies have other procedures to reach interested citizens.

A study of environmental impact statements for a 6-year period made by the CEQ indicated the following findings and recommendations: (1) the impact statement process can and should be more useful in agency planning and decision making; (2) agencies with major EIS responsibilities should support high-level, well-staffed offices charged with implementing NEPA and the EIS process effectively; (3) guidelines from CEQ are necessary to help agencies draw up EISs for broad federal programs or groups of projects; (4) procedures that agencies use to notify other federal, state, and local agencies and the public of important EIS actions need improvement; (5) to improve EIS review, agencies in consultation with CEQ should clearly define their expertise and jurisdiction for purposes of commenting on EISs; (6) the EIS process has contributed much to interagency coordination in environmental matters, but this coordination can be strengthened further; (7) the quality and content of EISs need continuing improvement; (8) several special issues arising from the EIS process require attention and remedy (among these are procedures for impact statements and federal permit actions); and (9) the adequacy of public participation in the EIS process needs thorough study and evaluation.

The environmental impact statements and their resulting conclusions can and have been challenged in the courts. Each year, there are numerous cases in the courts concerning the statements. It is, therefore, necessary that the statements be completed in a thorough and comprehensive manner.

Cases have been filed in widely varying situations. Calvert Cliff's Coordinating Committee vs the NEC was an early landmark case that set the direction of federal agencies' NEPA responsibilities. The NEC was required to revise its activities to systematically analyze environmental impacts. Since then courts have handed down numerous decisions expanding federal agency responsibilities under NEPA. The court ruled in a case involving the Army Corp of Engineers that the findings by the Council on Environmental Quality indicated that inadequate considerations was given to the safety and water quality effects of a project being carried out by the Army Corp of Engineers. The court required the corp to revise its impact statement and ordered the lower court to review this revised statement in light of CEQ's findings. Thus, for the first time, CEQ's findings as the principal overseer of the NEPA process were considered by a federal court.

Half of all of the environmental impact statements have been written about road-building actions undertaken by the Federal Highway Administration. The statements have resulted in significant planning changes. The second largest number of statements have been prepared for watershed protection and flood-control projects. Statements have also been prepared in coal development and in strip mining and its effect on existing use of land and water. As a result of the evaluation and the review process of environmental impact statements, several federally sponsored projects have been suspended or modified in recent years. Many states have followed the federal approach and have passed environmental impact statement laws.

THE ENVIRONMENTAL HEALTH FIELD

The environmental health field is composed of individuals whose efforts are directed toward controlling, preserving, or improving the environment in order that people may have optimum health, safety, comfort, and well-being now and in future generations. Given the complexity of the environment, a variety of disciplines and career categories have developed, including engineering, comprised of sanitary, safety, mechanical, and chemical; education; science, comprised of generalists or specialists in limnology, medicine, etc.; environmental health, comprised of generalists or specialists in three basic types of categories: (1) testing, sampling, routine inspections, and distribution of public information, (2) investigation consultation, planning, and education; and (3) supervision, administration planning, enforcement, and public relations.

The Environmental Health Practitioner

The environmental health practitioner is an applied scientist and educator who uses the knowledge and skills of the natural, behavioral, and environmental sciences to prevent disease and injury and to promote human well-being.

The environmental health practitioner makes inspections; conducts special studies; samples air, water, and food; reviews plans; acts as an educator, public relations officer, and community organizer; plans programs; acts as a consultant to civic groups, business, industry, and individuals; and enforces environmental and public health laws. The environmental health practitioner is involved in a multitude of program areas, including accident prevention, air pollution control, communicable diseases, environmental emergencies, food processing and food handling, outbreaks of food-borne infections and poisonings, hazardous substances control, housing, indoor environment, insect and rodent control, institutional environment, meat, milk, migrant labor camps, noise control, nuisance abatement, occupational health, planning, product safety, radiation control, recreational sanitation, public and private sewage, solid and hazardous waste management, swimming pool sanitation, water pollution, and public and private water supply.

The environmentalist of today and the future must be a highly skilled, well-trained generalist possessing a multitude of competencies as an effective member of the health team. These competencies are divided into 6 general and 12 specific areas as follows:

General Competencies

A. General Science

1. knowledge of general inorganic and organic chemistry
2. knowledge of general biology
3. knowledge of general microbiology
4. knowledge of general college math, including algebra, trigonometry, and basic statistics
5. knowledge of physics (mechanics and fluids)
6. knowledge of epidemiological principles
7. knowledge of risk assessment techniques

B. Communications and Education

1. knowledge of various communications, verbal and written
2. knowledge of how to work with people
3. knowledge of the use of audiovisual aids
4. knowledge of group dynamic techniques and group processes
5. knowledge of interviewing techniques
6. knowledge of teaching and learning principles
7. understanding of the need for public information and proper relationships with the news media
8. understanding of how to work with and motivate community organizations including the use of small group dynamic techniques
9. knowledge of use of databases

C. Planning and Management

1. knowledge of techniques needed to establish a program in any of the environmental health areas
2. knowledge of electronic data processing techniques and their application
3. knowledge of techniques used to establish priorities
4. ability to design surveys and survey forms
5. ability to use survey techniques to determine the extent of given environmental health problems

6. ability to interpret survey findings
7. ability to determine the advisability of legal action and when to initiate it

D. General Technical Skills

1. sound knowledge of the principles of learning and possession of skills in training, testing, evaluation, and use of aids in various areas of environmental health
2. knowledge of how to use inspectional techniques to determine environmental health problems
3. knowledge of a variety of sampling techniques related to air, water, food, hazardous chemicals, etc.
4. ability to accurately collect samples, complete record forms, and interpret result of lab samples in light of surveys made
5. ability to use survey and field test instruments such as light meters, chlorine test kits and fly grids, lead and radon samples

E. Administrative and Supervisory Skills

1. knowledge of environmental and public health laws, regulations, ordinances, codes, and their application
2. knowledge of supervisory techniques used in environmental health programs
3. knowledge administrative techniques used in the management of environmental health programs
4. understanding of the significance of the application of existing environmental and public health laws
5. understanding of the systems approach to the analysis of environmental health problems
6. understanding of the vital role of continued maintenance in the permanent resolution of environmental control problems
7. understanding of the relationship between health departments, other public agencies, voluntary agencies, business, and industry
8. understanding of basic principles of economics and how they relate to the existence of environmental health problems and the potential for successful environmental health programs
9. understanding of overall health problems and health priorities
10. knowledge of risk management techniques

F. Professional Attitudes

1. desire to work with people and to use basic environmental health sciences to resolve environmental health problems
2. sense of obligation to fulfill the requirements of the job and to carry out assigned duties in a professional manner
3. approach recipients of services with a cooperative attitude
4. display courtesy in personal relationships with fellow employees
5. accept constructive criticism from employees, peers, and the public
6. sense of dedication to the environmental health profession reflected in par- ticipation in continuing education

7. control emotions and perform in a mature manner during periods of stress
8. desire to communicate public health principles

Specific Competencies

A. Air

1. knowledge of the different air pollutants and their sources
2. knowledge of the relationship of weather conditions to air pollution
3. knowledge of effects of air pollutants on the biosphere
4. understanding of the relationship of air pollution to topography
5. knowledge of microflow of air
6. knowledge of functional operation of air pollution control devices
7. knowledge of preventive measure in air pollution control
8. knowledge of corrective measure in air pollution control
9. knowledge of the practical applications of air pollution control procedures and techniques
10. knowledge of the principles of combustion engineering
11. knowledge of air, air sampling techniques, and the ability to conduct air sampling
12. ability to implement surveys to clarify and identify the extent of problems
13. ability to evaluate results of surveys in light of long-range and short-range programs within the community
14. ability to design and implement cost-benefit analysis of control programs
15. knowledge of air toxics

B. Environmental Chemicals

1. knowledge of potential chemical contaminants of food
2. knowledge of potential chemical contaminants of potable water supplies
3. knowledge of transport requirements for hazardous chemicals
4. knowledge of the techniques and procedures for identifying environmental chemicals
5. knowledge of the means of disposal for environmental chemicals
6. understanding of decontamination of objects or substances that have been contaminated with environmental chemicals
7. knowledge of field tests used to determine the presence and concentration of environmental chemicals
8. knowledge of detergent and disinfectant chemistry
9. ability to evaluate detergents in an "in-use" situation
10. knowledge of economic poisons and how they affect the human ecology of the region
11. knowledge of principles and practices in the application of economic poisons
12. ability to formulate poisons in proper dilutions
13. knowledge of bait formulations used in pest control
14. understanding of the safety features needed to prevent accidents with environmental chemicals
15. knowledge of disinfectant detergents and their use

C. Environmental Injuries

1. knowledge of public health and ecological aspects of environmental injuries problems
2. knowledge of the instrumentation, material, and procedures involved in determining the causes of accidents
3. knowledge of epidemiological techniques used for studying accident problems
4. ability to motivate voluntary corrective action on the part of the public
5. ability to evaluate accidents and their causes

D. Food

1. knowledge of food technology and its relationship to health
2. knowledge of principles of food manufacturing, processing, and preservation
3. knowledge of food-borne diseases and their control
4. knowledge of epidemiological techniques and procedures
5. knowledge of design, location, and construction of food establishments and their equipment
6. knowledge of principles of food establishment operations, housekeeping, and maintenance
7. knowledge of equipment design, operation, maintenance, and cleaning techniques
8. knowledge of methods of motivating industrial management to understand, accept, and carry out its responsibilities in the food environment, personnel training, and personal supervision
9. knowledge of legal requirements of food technology
10. knowledge of inspection, survey techniques, and significance of data
11. knowledge of the examination and licensure of food establishment managers
12. knowledge of techniques used by different cultural and ethnic groups in food growing and preparation
13. knowledge of institutional food-handling practices
14. ability to obtain public support for food programs
15. knowledge of characteristic and properties of milk
16. knowledge of dairy bacteriology
17. knowledge of milk production and processing
18. knowledge of legal standards of food and milk composition
19. knowledge of techniques used to investigate dairy farms
20. knowledge of milk processing operations and control
21. ability to inspect pasteurization plants

E. Insects and Rodents

1. understanding of the epidemiology of vector-borne diseases
2. understanding the natural habitat and control of common insects of public health and economic significance
3. knowledge of basic life cycles of insects and rodents of public health significance
4. ability to identify a variety of insects of public health or economic significance in the field

5. understanding of environmental factors related to vector control
6. ability to identify scope of field problems and to determine control activities required
7. understanding of advantages and limitations of insecticides and their effect on the ecology of the region
8. understanding of the operation of sprayers and other pest control equipment
9. knowledge of the epidemiology of rodent-borne diseases
10. understanding of environmental procedures used in rodent control
11. understanding of biological control of rodents
12. understanding of chemical control of rodent ectoparasites
13. understanding of relationship of environmental health personnel to pest control operators
14. understanding of the production, transportation, storage, use, and disposal of pesticides

F. Noise

1. knowledge of public health and ecological effects of noise on the individual and community
2. knowledge of instrumentation and procedures involved in noise measurements
3. knowledge of existing laws pertaining to nuisances and noise abatement
4. knowledge of practical applications of control measures
5. ability to implement surveys designed to define the extent of noise problem
6. ability to evaluate results of surveys and to establish long-range and short-range goals for control
7. knowledge of work-related noise stress

G. Population and Space Utilization

1. understanding of the population explosion and its effect on the present and future needs of our society
2. understanding of the health hazards related to congestion
3. understanding of individual space needs
4. understanding of the effects of different cultures on population control
5. understanding of the use of community planning and zoning on space utilization
6. understanding of establishment of priorities for the proper use of existing space

H. Radiation

1. knowledge of radiation theory and principles
2. knowledge of dangers of radiation
3. knowledge of use of radiation and radioisotopes
4. knowledge of effects of radiation
5. knowledge of safety precautions
6. knowledge of monitoring techniques and instrumentation used in radiation detection
7. knowledge of techniques of storage and disposal of radioactive materials

8. knowledge of techniques of transportation of radioactive materials
9. knowledge of techniques of decontamination
10. knowledge of legal requirements of transportation, use, storage, and disposal of radioactive materials

I. Indoor Environment

1. knowledge of cultural, economic, and sociological aspects of individual and multiple dwelling units
2. knowledge of housing conditions needed for health, comfort, and well-being
3. knowledge of impact of transportation on housing
4. knowledge of real-estate laws and prevailing practices
5. knowledge of the various agencies involved in supervision and licensing of community shelters
6. knowledge of techniques used to evaluate individual and multiple dwelling units
7. knowledge of local, state, and federal housing programs
8. understanding of zoning laws and their effect on the use of individual and multiple dwelling units
9. understanding of the relationship of minority groups and poverty to housing use
10. knowledge of indoor air-pollution problems

J. Solid Wastes

1. knowledge of the types of solid waste generated in the community
2. knowledge of the types of waste generated by common industrial processes
3. knowledge of various methods of storage, collection, and disposal of solid waste
4. knowledge of public health and ecological aspects of solid wastes
5. knowledge of the use of systems analysis in waste disposal management
6. knowledge of economics of solid-waste disposal
7. ability to evaluate the results of solid-waste surveys and to establish long-range and short-range goals
8. ability to implement surveys to determine the extent of the solid waste problems
9. ability to design, implement, and evaluate programs related to waste disposal vs. public health problems

K. Hazardous Waste

1. knowledge of the health and safety concerns related to hazardous waste sites
2. knowledge of the effects of exposure to toxic chemicals in a hazardous waste site
3. knowledge of the route of entry to the body of hazardous chemicals, including inhalation, skin absorption, ingestion, and puncture wounds (injection)
4. understanding of the potential health effects due to acute and chronic exposure to various chemicals at the hazardous waste site
5. knowledge of the symptoms of exposure to hazardous chemicals, such as burning, coughing, nausea, tearing eyes, rashes, unconsciousness, and death
6. knowledge of the potential chemical reactions that may produce explosion, fire, or heat

7. understanding of the psychological effects of oxygen deficiency in humans related to an increase in specific chemicals in the immediate environment
8. understanding of the health effects of ionizing radiation related to alpha radiation, beta radiation, gamma radiation, and X-rays
9. knowledge of techniques used to dispose of radioactive material
10. understanding of the potential types of hospital and research facility waste that may cause biological hazards for the individual and may be spread through the environment
11. knowledge of the various safety hazards that may be found at hazardous waste sites
12. an understanding of the potential electrical hazards that may occur from overhead power lines, downed electrical wires, and buried cables that have been sub-jected to potential damage from hazardous waste situations
13. an understanding of the psychological effects that may occur to the individual at the hazardous waste site due to heat stress or cold exposure

L. Water and Liquid Wastes

1. knowledge of water sources
2. knowledge of potable drinking water quality and standards (physical, chemical, biological, and radiological)
3. knowledge of water-borne diseases and how they are transmitted
4. knowledge of sampling and testing of potable water
5. interpretation of laboratory analysis of water samples
6. knowledge of legal aspects of water quality control
7. knowledge of different types of water usage
8. understanding of the protection and selection of individual water supplies
9. understanding principles of water treatment
10. knowledge of physical and biological composition of sewage, including common and exotic industrial wastes
11. knowledge of types of industrial wastes and their significance
12. knowledge of the effects of sewage discharge on water quality
13. understanding of the epidemiology of sewage-associated diseases
14. knowledge of the technology and basic engineering principles related to water flow
15. understanding the principles of individual sewage disposal
16. knowledge of principles of municipal sewage treatment
17. knowledge of small sewage treatment units
18. knowledge of the measurement of absorptive quality of soils
19. knderstanding principles of nonwater sewage disposal
20. understanding of techniques used in problems of emergency situations related to water and sewage
21. understanding the techniques and potential hazards of sludge disposal

SUMMARY

Humans, who alone have the intellectual capacity to improve their life through science, share the Earth with numerous other biosystems, some of which affect health. Humans contribute to the destruction of biosystems and to their own health problems through degradation of the environment by means of solid waste, air, water, and land pollutants. The pollutants are transported biologically or physically through the environment and cause a variety of environmental problems that may eventually destroy much of the Earth as we know it today. People through use of skill, good sense, and planning can avoid this destruction and provide for a wholesome, safe environment for future generations.

The problems of the environment are numerous and complex, and can only be resolved by determining the cause, means of prevention, and necessary controls of specific hazards and combined hazards through the use of epidemiological research, special study techniques, risk assessment, and risk management techniques. Not only the ecosystem, the food chain, the growth of population, and the energy cycle must be understood, but also the use and abuse of energy and the impact of humans on their own environment. An economic determination as to the benefit-risk of each type of environmental impact must be made backed by reasonable decisions concerning these impacts. Further, the legal system and the trained environmental health practitioners, along with all levels of government, industry, and concerned citizenry, must be employed to voluntarily improve the environment of all living organisms.

It is certainly clear that individual or collective human decisions influence our environment to a considerable degree and that the quality of our life and the time and manner of our death are related to these decisions. It is not only important to eliminate specific environmental components that cause or contribute to disease or injury, but it is essential to develop a preventive strategy that will eliminate many of the environmental problems before they occur.

In this chapter, the general problems affecting the environment and some of the kinds of primary concerns have been discussed. In the succeeding chapters, each environmental health area will be treated separately, and the specific environmental problem will be approached from the background and status of the problem; necessary scientific, technological, and general information available; the source, scope, and potential for disease related to the problem; the potential for intervention; the resources that may be utilized; the standards, practices and technique available; the modes of surveillance and evaluation available; specific controls; and research needs. (See Preface for a description of how best to use this book and Volume II.)

THE FUTURE

The environmental health practitioner, specialist, and scientist of the future must have appropriate knowledge of health and the environment and recognize the people-made pressures that have been placed upon natural resources, both living and nonliving. These pressures are far more severe than have ever been previously suspected. There is substantial evidence that potentially health-threatening groundwater contamination is a problem of increasing concern in the United States and that toxic chemicals in hazardous waste dumps and other ground storage may pose serious health and environmental threats and create public health problems. Toxic chemicals are present in the air, water, and in the workplace and are growing in quantity and complexity. Many examples of potential damage to the ecosystem, as well as to people, have been cited and will be cited in the succeeding chapters.

Short-term research and short-term horizons have been the techniques used for trying to resolve long-term problems. There is concern, not only in government, but also in the private sector. It is, therefore, necessary to have a better understanding of the interdisciplinary activities that will be needed to carry out the long-term research that will determine the kinds of risks that are occurring and the kinds of risk management that will be necessary in the future.

Research

New research techniques include molecular epidemiology, which is based upon the measurement in the exposed individual of the interaction of a toxic chemical or its derivative with a tissue constituent or a tissue alteration resulting from exposure to the chemical. It provides an indirect measure of individual exposure. Recent research has determined means for detecting and measuring the interaction of a foreign chemical with easily accessible normal human constituents, such as chemical carcinogen interactions with deoxyribonucleic acid (DNA). It is now possible, at times, to detect a few altered DNA molecules out of millions of cells. Altered chromosomes can be determined, and important advances will be occurring in detecting and quantitating human exposure to foreign chemicals.

Susceptibility to chemical toxicants varies widely in the human host population. The host factors are genetic diversity, current and prior disease, sex, and age. There are approximately 2000 genetically identifiable human diseases. Genetic conditions are likely to enhance the risk to individuals of developing environmentally or occupationally associated adverse health effects. The extent of the risk is unknown. The extent of the risk of enhancement is unknown. Considerable study is needed in genetic diversity-susceptibility and biological mechanisms.

Exposure to humans, other animals, and the environment of pollutants or mixtures of substances is not clearly understood. There are no valid general rules for determining the presence of synergistic activity in the mixtures of chemicals to which these individuals or the environment is subjected to.

The emerging technology of biotechnology and microelectronics are of considerable interest and potential concern. Biotechnology deals with genetic engineering, which hopefully will produce new medical products, agricultural products, chemicals, and other products. Microelectronics involves the production of microelectronic shifts with the use of a variety of virtually unstudied chemicals, such as gallium arsenide, silicon, and halogenated hydrocarbon solvents.

There is inadequate data available for the physical, chemical, and ecological variations in fresh waters, oceans, and the atmosphere over extended periods of time. It is, therefore, difficult to determine what is a natural change over a long period of time and what is caused by human activity. There is a need to understand a normal range of variations in ecosystems and how they can deal with the pollutants that are currently flowing into them and that will flow into them in the future.

As pollutants move from one environmental medium to another, the rate of transfer is not understood. Quantity assessments are necessary, and the results of these movements of pollutants need to be understood. At what rates are aerosols formed from chemically reactive organic pollutants and what effect do these aerosols have on the air-water interface, the precipitation of the aerosols, and the temperature and moisture content that may affect them?

Study is needed to determine the behavior and biologic effects of chemicals in various environmental media and what impact they may have on the overall global ecosystem.

The improvement in quantitative risk assessment is most likely to come from a better understanding of biologic processes. Little is known about basic pharmacokinetic dynamics and the environmental mechanisms related to toxicity, other than cancer. Very little is known about actual exposure patterns, the potential of short-term biologic screening to provide early prediction of effects, and a validation of risk assessments.

Environmental Changes and Their Consequences

In order to understand environmental changes and their consequences, it is necessary to know the movement from the point of exposure to the end point of disease. There are four rough stages in environmental toxicology. They are

1. Exposure, where the ambient condition brings the organism into contact with the hazard. For example, a toxic material that may be found in the air or water.
2. The dose where the internalized quantity of the hazard creates a specific body burden that may be toxic to an organ or cell.
3. Effects seen that are markers of intermediate biologic effects that are either a step in the toxicologic process or a parallel manifestation of effect. These indicators of exposure or dose are predictors of toxicity. They may include chromosome aberrations, mutations, and cell or cell killing enzyme activity.
4. The endpoint is the ultimate toxicological effect, which may be cancer, heart disease, or other diseases or injuries.

Perturbations are complex mixtures of substances or impacts, and equally complex patterns of human activities that influence exposure. Examples include hazardous waste dumps and related groundwater problems, as well as the dynamics of the ecosystems that may be affected by these. The perturbations, which also include the introduction of new technologies or changes in resource use patterns, affect ecosystems physically and biologically. The physical changes include nutrient and energy flow, physical structure of the system, and transport and transformation of substances that have been introduced. The biological changes include direct increases in populations due to secondary effects and resulting changes in linkages and functioning of the system as a whole. The ecosystems, then, are related to exposures in humans or other animals and plants, of concerns to humans, and finally ends up with health effects that are either improper or proper. The improper health effects may be changes in target organs of the body or death to the total organism.

Recommendations for the Future

The research areas that have been previously discussed are summarized in some specific types of studies that need to be carried forward. They are

1. The dynamics of stressed ecosystems, that is, the changes that occur in ecosystems that result from people-made stresses.
2. Ecologically significant end points, that is, the stressing of the ecosystems should become a basis for potential regulatory attention.
3. Ecological markers and sentinel events, that is, a study should be done to identify early indicators of potential ecological change.
4. Total toxicity of complex mixtures, that is, research should be devoted not only to the effects of individual substances, but to the complex of common mixtures, such as solvents, agriculture runoff, and wastewater sludges, as well as airborne chemical mixtures.
5. Movement of pollutants out of sinks. Sinks are ecological areas, such as wetlands or underwater sediment and landfills, that have been used for the disposal of toxic pollutants for many years. The pollutants may move across the media. Research is needed to identify under what conditions these materials or their conversion products will move out of the sinks and into ecosystems and human environments.
6. Human exposure, that is, data should be developed on human exposure from all environmental sources, ranging from inhalation in the ambient air, in the workplace, and other indoor environments, to smoking and the ingestion of consumer products, as well as food and water, and the absorption of substances by skin contact.
7. Effects on multiple organs and physiologic systems, that is, research should be expanded to study the neurobehavioral and immune systems.
8. Plant perturbations, that is, an analysis should be made of the effect of fossil fuel combustion and the potential for global temperature increases, changes in the carbon cycles, and other possible concerns.
9. Biotechnologies, that is, a better understanding is needed for the genetic manipulation of biological organisms to determine whether potentially serious consequences may occur.
10. Linkages among ecological systems, that is, an understanding needs to be found of the transport and fate of materials between heterogenous ecosystem units.

Chapter 2

FOOD PROTECTION

BACKGROUND AND STATUS

Effective food protection is carried out on a daily basis by large numbers of personnel who attempt to make this essential component of human life safe, attractive, appetizing, nutritious, and free of disease or poison. The food and beverage industry is extremely large.

Many changes have occurred in the food industry in the last 50 years. Today we use many worldwide food sources instead of depending primarily on home-grown food or food grown nearby. Our food is mass produced and nationally distributed. Food establishments have increased in size, number, type, and complexity. The short-order food establishment has replaced, in many cases, the "greasy hamburger joints" of the past. However, problems continue to increase in the new establishments, as well as in the problem establishments of the past, which still do exist, with management being the key to the problem.

Food is contaminated by microorganisms from soil, water, air, surface, animals, insects, rodents, and people. Food is also contaminated by chemicals from soil, water, air, pesticides, herbicides, fertilizers, and radionuclides. Contamination may be introduced during processing, as demonstrated in outbreaks of botulism from vacuum-packed fish and tuna fish, or during any stage of production, transportation, storage, preparation, and serving. Raw foods are shown to contain disease-producing organisms. Salmonella is contained in 15 to 30% of raw dressed poultry. *Clostridium perfringens* is cultured from 50% of red meat in stores. *Salmonella aureus* is present in most pooled milk supplies and in roughly 20% of cheddar cheeses. Salmonella are found in 15–30% of commercial egg products, with *S. enteritidis* being a serious problem currently. *C. botulinum* type

93

E is found on raw fish coming from fresh or salt water. Shellfish are contaminated with organisms causing infectious hepatitis and cholera. More than 10,000 food items are sold in stores today. Each one has its own unique potential for causing the spread of food-borne disease.

Many people carry pathogenic organisms that can be spread through food. *S. aureus* may be found on the skin or mucous membranes of 3–50% of the population. Salmonella may be found in the feces of 0.2% of all people. *C. perfringens* is carried by 80% of the population, and 6.4% of the fecal specimens from food handlers contain enteropathogenic *E. coli*.

Eighty to 90% of total human exposure to most chemicals, including heavy metals, pesticides, and radionuclides, comes from food consumption. Food is presently contaminated heavily by chemicals. These chemicals, such as pesticides, may be deliberately applied to crops and stored commodities to kill insects. Other chemicals, such as food additives, are added to food to achieve some desirable effect, such as improved taste, color, and longevity. Drugs used in the care of animals and chemicals used in food-packaging materials may accumulate within food. Another group of chemicals enter by means of environmental pollutants or as fungal toxins. At present the most important chemicals and chemical contaminants of food for which measurements can be made include toxic metals, such as mercury, selenium, lead, arsenic, cadmium, and zinc; industrial chemicals, such as polychlorinated biphenyls, chlorinated dibenzo-*p*-dioxins, chlorinated dibenzofurans, and pesticides; fungal toxins, or mycotoxins, such as aflatoxin; and miscellaneous chemicals, such as the polynuclear aromatic hydrocarbons and nitrosamines. In addition, the level of food-borne disease due to microorganisms is increasing and will continue to increase because of the tremendous amount of food processed and sold in a variety of different types of operations.

This chapter is concerned with food-borne disease; food microbiology; plans review; physical facilities; storage, preparation, serving, and protection of food; housekeeping procedures; cleaning and sanitizing; solid waste disposal; insect and rodent control; personal health of employees; inspectional procedures; and public health laws.

SCIENTIFIC AND TECHNOLOGICAL BACKGROUND

Food Microbiology

Microorganisms are unicellular and microscopic, and perform positive or negative functions in our lives. Certain microorganisms are important in the production of vinegar (*Acetobacter*), sauerkraut (*Lactobacillus*), and bread (lactic acid). However, our major concern will be with those organisms that can cause food-borne disease or food poisoning. Microorganisms are found in soil, water, air, on animals, in body openings, and on body surfaces. These organisms can be transmitted directly to food and water by hands or indirectly by fomites, insects, and rodents, and by aerosols due to coughing and sneezing. Bacteria can exist in the vegetative or spore form. The spore is resistant to heat and chemical destruction. Aerobic bacteria, such as *Pseudomonas*, grow in the presence of oxygen. Anerobic bacteria, such as *Clostridium*, grow in the absence of air. Facultative bacteria, such as *Escherichia coli* and *Salmonella*, can grow under either condition.

Reproduction and Growth of Microorganisms

Bacterial cells reproduce by a single cell split to form two identical cells (transverse fission). This takes about 20–30 min for those bacteria that cause food-borne disease or poisonings. Bacterial growth follows four basic phases: the lag phase lasting about 2 hr; the logarithmic or growth phase, where growth is at a maximum rate; the stationary or resting phase, where bacteria die at the same rate as they are being produced (spores are produced here); the period of decline and death, or death phase, which may occur in 18–24 hr.

In order for bacteria to grow, they need an adequate temperature, moisture, presence or absence of air (depending on the organisms), and food. Psychrophiles are organisms found on uncultivated soil, lakes and streams, on meats, and in ice cream, and live best at temperatures of 0–5°C. They frequently are the cause of spoilage in refrigerated foods. Mesophiles are parasites found on plants, or may be harmless parasites on animals and humans. They live best at temperatures between 20 and 45°C. Thermophiles are organisms causing off flavors in pasteurized milk. They live best at temperatures exceeding 55°C.

Environmental Effects on Bacteria—Temperature

The temperature of the medium in which bacteria live determines the rate of growth, multiplication, and death of the organism. Bacteria can generally survive very low temperatures, even though they are dormant; high temperatures will generally kill bacteria. The consistency of the medium, the number of organisms initially present, and the species also affects the rate of bacterial death. Generally the resistance of an organism to heat is greatest at a pH of 7.0, which is favorable for growth. As the acidity or alkalinity increases, the death of organisms due to heat increases.

Radiation

Ultraviolet rays in sunlight are lethal to bacteria. Radiations of 210 nm (1 nm = 1×10^{-9} m) to 310 nm contain bactericidal power. The most effective ultraviolet wavelength against the majority of bacteria and some molds and viruses is 265 nm. Some of the shorter X-rays and gamma rays are more effective but are dangerous to people. Radiation kills bacteria logarithmically and causes the remaining cells to form mutants susceptible to attack by bacteriophages. In practical use, only air close to the irradiating ultraviolet source and surfaces close to the source are affected. Dust in the air, organic material, water, and glass interfere with the bactericidal effects. Ultraviolet light must be shielded to prevent damage to human tissue.

Pressure

Bacteria are resistant to applied pressure.

Sound Waves

Supersonic vibrations rupture bacterial cell walls and destroy bacteria. Sound waves of low intensity but high frequency are most effective.

Moisture

Moisture is needed by bacteria to secure food and dispose of waste materials. Since the osmotic pressure within the cell is usually greater than the surrounding media, water flows through the semipermeable cell membrane into the cell. By adding 10–15% sodium chloride or 50–70% sugar to the media, the water flows out of the cell, causing cell disruption and death of the bacteria. Drying the media stops the growth of bacteria and some die. As long as the product stays dry, spoilage or growth of disease-producing organisms will not occur.

Chemicals

A vast variety of chemicals will kill bacteria. These chemicals include antibiotics, alcohol, chlorine, bromine, iodine, iodophores, phenols, and quaternary ammonium chloride compounds. The effectiveness of the chemical is based on the organic load present, time, temperature, pH, specificity for organism, and mechanical action.

Detergents and Disinfectants

The detergent and/or disinfectant used in cleaning is determined by the soil load present, the type of soil, the type of water used, the cleaning operation, and the kind of disinfecting agent used for final sanitization. The degree of cleaning, which is a process of soil, tarnish, and stain removal from objects, is determined by the concentration of detergent, type of water, temperature, time, velocity, and type of cleaning procedure.

Soil varies in composition. It may contain water-soluble sugar, salts, and some organic acids, or it may contain water-insoluble mineral deposits and/or animal and vegetable fats, such as grease or carbon. Food particles may be bound to surfaces through absorption or by grease. Scale deposits may be formed by the interaction of alkaline cleaners and chemicals in the water. These chemicals include soluble iron and manganese salts, calcium and magnesium carbonate and bicarbonate, calcium sulfate, calcium chloride, magnesium sulfate, and magnesium chloride.

Adequate cleaning is enhanced through increasing the temperature of the cleaning liquid, since the rise in temperature decreases the bond between the soil and the surface, increases the solubility of materials, and increases the rate of the chemical reaction. Force or energy applied during cleaning helps to mechanically remove soil from surfaces. The cleaning procedure brings the detergent solution into intimate contact with the soil, removes the soil from the surface, disperses the soil into the solvent, and prevents redeposition of the soil on the clean surface.

Basic Detergent Mechanisms

The basic detergent mechanisms are as follows:

1. Establish intimate contact between detergent and soil through "wetting" and "penetration." Wetting reduces the surface tension of the soil and allows the detergent solution to penetrate and spread out. Penetration is the action of a liquid entering into porous materials through cracks and pinholes.
2. Displace the soil from the surface through dissolving, peptizing, and saponification. Dissolving is a chemical action of detergents that liquifies water-soluble materials or soils. Peptizing is a chemical action that brings materials into a colloidal solution. Saponification is the chemical reaction between an alkali and animal or vegetable fat to produce soap.
3. Disperse the soil in the solvent through suspension, dispsersion, or emulsification. Suspension is the action holding insoluble particles in solution. Dispersion is the action of breaking up clumps of particles and thereby suspending them. Emulsification is the action of breaking up fats and oils into small particles that are suspended in the cleaning solution.
4. Prevent redeposition of dispsersed soil through rinsibility. Rinsibility is the condition of a solution, permitting it to be flushed easily and completely from a surface.
5. Soften water through precipitation, sequestration, and chelation. Precipitation is the physical settling of particles. Sequestration is the "surrounding" or "tying up" of magnesium and calcium compounds by sequestering agents. A sequestering agent is a substance that removes a metal ion from a solution system by forming a complex ion that does not have the chemical reactions of the ion that is removed. Chelation is the same action as sequestration, except with organic materials.

Detergent Formulation

All detergents differ and should be formulated to take care of specific types of soils and surfaces. The ideal detergent should act quickly, be completely soluble, have good wetting and penetrating action, dissolve food solids easily, emulsify fats, suspend particles in solution, rinse off easily, have water-softening properties, be noncorrosive to metal surfaces, and be economical in use dilution.

Types of Detergents

Manual immersion cleaners are used in hand dishwashing or equipment-washing procedures. This cleaner is usually a blend of anionic and nonionic organic liquids or powders containing special additives for corrosion inhibition and skin protection.

Housekeeping cleaners are both liquid and powder and are used for cleaning a variety of surfaces, both by hand and machine.

Liquid all-purpose cleaners are composed of anionic and nonionic synthetic detergents that are low sudsing and will not damage or dull synthetic or wax floor finishes when used on floors.

Heavy-duty cleaners or wax scrubbers are liquid products formulated to emulsify and soften wax and synthetic floor finishes so that they may be easily removed prior to refinishing.

Low-sudsing cleaners are used in battery-operation automatic machines that both scrub and vacuum the floor. This operation is most useful on large floor areas.

Wall cleaners may be either all-purpose cleaners or heavy-duty cleaners. Occasionally, where it is necessary to remove deteriorating paint residues, a special powdered wall cleaner including trisodium phosphate may be used.

Concrete cleaners are a powder type of general cleaner similar to the powdered wall cleaner. The concrete cleaner must be more highly alkaline and specifically formulated for cleaning soiled concrete surfaces.

Machine products are either powder or liquid in form and are specially designed for cleaning dishes and utensils. These products are highly alkaline, causing skin irritation if used in manual cleaning.

Acid detergents may be made of synthetic detergents and hydrochloric acid for use in toilet bowls and urinals. They effectively remove rust and other mineral stains from toilet bowls. These strong acid detergents should never be used in bathtubs and sinks, which are generally made of porcelain. Another milder type of organic acid detergent is composed of organic acids or phosphoric acids and nonionic detergent. These products are very useful for removing mineral films from stainless steel surfaces and also from the arms of dishwashers and the milk stone found in milk vats and containers.

Detergent Evaluation

The best type of detergent evaluation is an in-use test. The manager must determine if the product is accomplishing the cleaning job required. Obviously the food service manager must take into consideration the temperature of the cleaning solution, the time it is used, the concentration, the amount of friction used, and the initial amount of soil on the utensils or equipment.

Disinfection Evaluation Methods

The recommended disinfection evaluation method is the AOAC Use Dilution Confirmation Test. The test is carried out by placing sterile, stainless-steel ring carriers in a broth culture of the test organisms and drying the rings for 20–60 min. The contaminated carriers are immersed in a tub of use solution of the germicide for 10 min at 20°C. The carriers are immersed in tubes of appropriate nutrient media and incubated for 48 hr. There must not be any growth in at least 10 separate tubes from ring carriers. In practice, the USDA often requires perfect results in 60 or more ring carriers.

Definitions

Detergent disinfectants are disinfectants with cleaning ability. They are also called germicidal detergents, detergent germicides, disinfectant detergents, and germicidal cleaners.

A disinfectant is a product that kills all vegetative bacteria but does not kill spores. Germicide and bactericide are synonymous with disinfectant.

Sanitizer is a product that kills some, but not all, bacteria. It usually reduces bacterial counts to a generally acceptable level for public health standards.

Sterilization is a process that kills all living organisms, including spores. Sterilization may be achieved by the use of saturated steam at 121–123°C for 30 min, through dry heat at 160°C for 1 hr, or by the use of certain chemicals such as ethylene oxide gas for 2 hr.

A bacteriostat is a product that prevents bacteria from multiplying without actually killing the bacteria. Bacteriostats are ineffective on hard surfaces.

An antiseptic is an antibacterial agent used on the skin.

A bactericide is a substance that kills bacteria.

Sanitizing Agents

Sanitizing agents include various forms of chlorine, iodophors, quaternary ammonium compounds, and heat. Chlorine may be available as calcium hypochlorite at 70% available chlorine, or sodium hypochlorite at 2–15% available chlorine. Hypochlorites are effective sanitizers at a minimum concentration of 50 ppm of free available chlorine and if applied for 1 min if the surfaces to be sanitized are clean. Organic material, low temperatures, and high pH influence the bactericidal effectiveness. The temperature for the sanitizer should be at least 75°F.

Chlorine may also be available as chloramine-T at 25% available chlorine, or dichlorocyanuric and tricyanuric acids at 70–90% available chlorine. These organic chlorine products having a slower bactericidal action are used at a minimum concentration of 200 ppm for 1 min. The temperature for the sanitizer should be at least 75°F.

Iodophores are soluble complexes of iodine usually combined with nonionic surface-active agents. The iodophores have a rapid bactericidal action in an acid pH range in cold or hot water. They are less affected by organic matter than hypochlorites, are nontoxic in ordinary use, noncorrosive, and nonirritating to the skin. The yellow or amber color of the solution is proportional to its concentration. Iodophores are used at a minimum concentration of 12.5 ppm for 1 min. Quaternary ammonium compounds (QAC) are effective sanitizers at a minimum concentration of 200 ppm for 1 min. They are more stable in the presence of organic matter than chlorine or iodine compounds. They are noncorrosive and nonirritating to the skin. Hot water can be used as a sanitizer if the temperature in the final rinse at the entrance of the manifold is at least 170°F for 30 sec or at least 170°F for 30 sec when equipment is immersed in hot water.

Field Tests for Sanitizers

The concentration of chlorine in rinse water can be determined by use of indicator test paper impregnated with starch iodine or by the use of an orthotolidine colorimetric comparison unit. Quaternary compounds can be tested by certain test papers that use color comparison. Iodophore concentrations can be determined by comparing the color of the solution to a color comparison kit. Hot-water sanitizing can be determined by bacteriological sampling with Rodac contact plates or swab samples. An average plate count per utensil surface should not exceed 100 colonies.

Terminology

To better understand food-borne disease, it is necessary to have some knowledge of the following terminology.

- *Active immunity.* Antibodies produced in the person through contact with disease.
- *Carrier.* An individual harboring specific infectious agents in the absence of discernible clinical disease who serves as a potential source or reservoir of infection for other humans.
- *Chain of infection.* The spread of disease from the reservoir by means of a vehicle or vector to the host.
- *Contamination.* The presence of a pathogenic organism in or on a body surface or an inanimate object.
- *Communicable disease.* An illness due to an infectious agent or its toxic products

being transmitted directly or indirectly to a well person from an infected person or animal, or transmitted through an intermediate animal host, vector, or inanimate environment.

- *Communicable period.* The time or times during which the etiologic agent may be transferred from an infected person or animal to humans.
- *Endemic.* The regular occurrence of a fairly constant number of human cases of a disease within an area.
- *Enterotoxin.* A toxin produced by staphylococci and arising in the intestine.
- *Epidemic.* The occurrence in a community or region of a group of illnesses of a similar nature, clearly in excess of normal expectancy, and derived from a common or propagated source.
- *Epidemiology.* The study of the causes, transmission, and incidence of diseases in communities or other population groups.
- *Etiological agent.* The pathogenic organism or chemical causing a specific disease in a living body.
- *Exotoxin.* Toxic substance produced by bacteria found outside of the bacterial cell.
- *Fomite.* An inanimate object not supporting bacterial growth but serving to transmit pathogenic organisms from human to human.
- *Host.* The living body, human or animal, that provides food and shelter for the disease organisms.
- *Incidence.* The number of cases of disease occurring during a prescribed time period in relation to the unit of population in which they occur.
- *Incubation period.* The time interval between the infection of a susceptible person or animal and the appearance of signs or symptoms of the disease.
- *Infection.* The entry and development or multiplication of a particular pathogen in the body of humans or animals.
- *Parasite.* An organism living on the tissues and waste products of the host within the body of the host. All disease-causing organisms are parasites.
- *Passive immunity.* Antibodies produced in another host and then injected into the diseased person.
- *Report of disease.* An official report, usually by a doctor, of the occurrence of a communicable or other disease of humans or animals to departments of health or agriculture or both, as locally required. Reports should include those diseases requiring epidemiological investigation or initiation of special control measures.
- *Reservoir of infection.* Humans, animals, plants, soil, or inanimate organic matter in which an infectious agent lives and multiplies and then is transmitted to humans, who themselves are the most frequent reservoir of infectious pathogenic agents.
- *Resistance.* The sum total of body mechanisms placing barriers to the progress of invasion of pathogenic organisms.
- *Vector.* A living insect or animal (not human) that transmits infections and diseases from one person or animal to another.
- *Vehicle.* Water, food, milk, or any other substance or article serving as an intermediate means by which the pathogenic agent is transported from a reservoir and introduced into a susceptible host through ingestion, inhalation, inoculation, or by deposit on the skin or mucous membrane.

PROBLEMS IN FOOD PROTECTION

Although our scientific technology is so advanced that we can put humans on the moon and give them protected life-support systems, including food, our food programs on Earth, in many cases, are the same as they were 50 years ago. The excellent programs are scientifically oriented and conducted by trained environmentalists; in our poor programs, investigators inspect food establishments to license them; in our nonexistent programs, the public is deprived of any type of protection from food-borne disease. Technological developments in food processing are occurring so rapidly that their public health implications are not being properly evaluated. Industry is investing more than 100 million dollars a year in these new processes. New recently discovered production hazards include botulism and typhoid organisms in commercial products, and salmonella in liquid and powdered eggs and egg products, in instant nonfat dry milk, and in drugs and diagnostic reagents. Governmental agencies have been unable to adequately supervise industry because of insufficient resources, limited authority, lack of knowledge, inadequate workforce, preoccupation with legislative response, and inadequate channels of communication between the agency and industry. Public health agencies can be satisfied no longer with expending most of their efforts in supervising the food-service industry; they must now concern themselves with the source of the raw food and the means of processing it. Because of widespread distribution of food, the potential for large outbreaks of disease has increased immensely. There is an increasing concern over long-term health effects of chemical residues, pesticides, additives, and radioactive fallout. Heavy metals, such as mercury, are found in the fish food chain consumed by humans.

Source of Exposure

There are between seven and 10 million cases of acute gastroenteritis in the United States each year. However, since less than 1% of the severe gastroenteritis cases are investigated thoroughly enough to determine the causative agents and routes of transmission, the true scope of this disease is not known.

Reporting is a major problem, since patients are either indifferent or reluctant to cause trouble, physicians usually only report fatal gastroenteritis, and industry is afraid to lose money. Investigations are hampered by nonreporting, delayed or inadequate reporting, inadequate professional staffs and laboratories, inability to obtain food samples rapidly or total absence of food samples, and lack of cooperation on the part of the institutions or industry. About 40% of the states report no outbreaks, and about 40% more report one or two outbreaks per year. California reports 30–50% of all food-borne outbreaks reported annually in the United States. Ninety-three percent of all reported cases of disease due to water, milk, and food, which involves 10,000 people annually, are caused by food products.

Food-Borne Infections

Food-borne infection is direct infection in which the organisms are ingested into the body through food or water and then continue to multiply. In some infections the organisms die off after the clinical disease is over. In other infections, such as the salmonella group, the individuals may continue to be carriers for varying periods of time. They pass off live organisms in their feces.

Streptococcus. Streptococcal infections are usually due to *Streptococcus fecalis* or *Streptococcus pyogenes.* The source of these organisms are the feces, nose, or throat. *Streptococcus pyogenes* may also come from infected wounds or dirty bandages. Almost any foods may become infected, but turkey, other meats, and milk have been most frequently incriminated. Prior to pasteurization of milk, and even on occasion subsequent to pasteurization, individuals with beta-hemolytic streptococci have infected the udders of cows or the milk itself. The incubation period is 1–3 days. The reservoir of infection is people. It is communicable for 10–21 days.

Salmonella. Salmonellosis is a costly, highly communicable disease in the United States. An estimated two million human cases occur annually. Since the diseases are primarily food or water borne, they are a potential threat to every person in the country. Although the true magnitude of the problem is not known, it has been determined that the highest incidents of reported cases occurs among the very young and elderly. The interrelationships between people, animals, and fomites make salmonellosis a difficult problem to resolve.

Salmonella are divided into three groups based on their host preferences. Group 1 is primarily adapted to people and includes *Salmonella typhi*, which causes typhoid fever, and *S. paratyphi* A, which causes paratyphoid fever. The organisms are transmitted by milk, shellfish, other food products, or water contaminated with feces from human cases or carriers. Group 2 is primarily adapted to animal hosts and includes several important pathogens of domestic animals, such as *S. cholerae-suis*, and serotypes of *S. enteritidis*. These organisms, especially serotype Dublin, cause gastroenteritis in humans. Children may be severely affected by these infections. Group 3 includes organisms adapted to either animals or humans showing no distinct preference.

In recent years, there have been numerous outbreaks of *S. typhimurium*, *S. newport*, *S. montevideo*, *S. pullorum*, *S. derby*, *S. infantis*, and *S. enteritidis*. *S. derby* was a particularly serious problem in many hospitals in the eastern United States during the 1950s and early 1960s. The problem became so severe that maternity wards and nurseries in certain hospitals were forced to close. An outbreak of *S. infantis* occurred in a major institution in the East, resulting in over 700 cases within a 30-day period. The organisms are usually spread through meat, poultry, eggs and egg products contaminated directly or indirectly through contact with human or animal carriers. In the 10 years from 1977 to 1987, there has been a fivefold increase of *S. enteritidis* in New England and the Middle Atlantic region.

A cycle of infection includes feed and fertilizer, livestock, transportation, processing of food, processed foods, and animals or humans. Processed foods, in addition to those already mentioned, that are most commonly contaminated with salmonella include poultry salads, meat pies, pressed beef, sausage, cold cuts, smoked fish, reheated meats, gravies, dried egg products, frozen eggs, synthetic creams, custard, cream puffs, prepared fish dishes, and shellfish, especially oysters.

Outbreaks of salmonellosis have also been traced to dirty cutting boards and other utensils and equipment. The incubation period is 6–72 hr. The reservoir of infection includes poultry, rodents, turtles, cats, dogs, and people. Its communicability varies from days to weeks.

Shigella. Bacillary dysentery or shigellosis is caused by different members of the genus *Shigella*. The disease is spread through moist-prepared foods, milk, other dairy products, or water contaminated directly or indirectly with feces from actual cases or from carriers. The incubation period is 1–7 days, commonly 4 days. The reservoir of infection is people. It is communicable for about 4 weeks.

Cholera vibrio. Cholera is a severe gastrointestinal disturbance caused by the *Cholera vibrio*. The vibrio enters the body through food and water. The incubation period is from a few hours to 5 days, usually 2–3 days. Communicability usually lasts for a few days, but a carrier state could last for months. The reservoir of infection is people.

Listeriosis. Listeriosis is caused by the organism *Listeria monocytogens*. It is most frequently isolated from people, although the reservoir of infection can also be wild or domestic mammals and fowl. The disease is found at the extremes of age, during pregnancy, and in immunocompromised people. It may cause septicemia, meningo encephalitis, nausea, vomiting, delirium, coma, and death. The disease may be spread by people, food, or be a nosocomial infection. The incubation period is unknown, but is probably a few days to 3 weeks. Fetuses and newborn infants are highly susceptible, with the mother shedding the organism in urine for 7–10 days after delivery.

Viral Hepatitis. Hepatitis virus A is spread through human feces to food, shellfish, and water, and is in turn consumed by people. The disease may also be transmitted by means of fomites. The incubation period is 15–50 days, usually 28–30 days. The reservoir of infection is people. Most cases are noncommunicable after the first week of jaundice. Viral hepatitis is endemic worldwide. The HBV infection is common in certain high-risk groups, such as parenteral drug abusers, homosexual men, people in hemodialysis centers, and certain health professions where there is routine exposure to blood or serous fluids. The incubation period is usually 45–180 days, although symptoms may occur in as little as 2 weeks. The reservoir of infection is people. People may be infective for years.

Brucellosis. Brucellosis, also known as undulant fever or Bang's disease, is caused by *Brucella abortus*, *B. melitensis*, and *B. suis*. The organisms are transmitted through raw contaminated milk, dairy products made from raw milk, and contact with the tissues or discharges of infected animals. The incubation period is highly variable, usually 5–30 days. Reservoir of infection is animals. There is no evidence of communicability from person to person.

Diphtheria. Diphtheria is caused by *Corynebacterium diphtheriae*. The organisms are transmitted from humans through milk or directly from human to human through discharges from the nose or throat. The incubation period is 2–5 days. The reservoir of infection is people. Communicability is typically 2–4 weeks.

Tuberculosis. Tuberculosis of the extra pulmonary type is caused by *Mycobacterium tuberculosis*. The organisms, both of the human and bovine type, are spread through raw contaminated milk or dairy products, and contact with infected animals and humans. The incubation period from infection to the primary lesion is about 4–12 weeks. The reservoir of infection is people and diseased cattle. Communicability lasts as long as there are infectious tubercle bacilli in the sputum.

Tularemia. Tularemia is caused by *Pasturella tularensis*. The organisms are spread through the meat of wild rabbits, squirrels, other animals, and bites of infected ticks and deer flies. The incubation period is 2–10 days, usually 3 days. The reservoir of infection is wild animals and various hard ticks. The disease is not directly transmitted from person to person.

Diarrhea Caused by Escherichia coli. Strains of *Escherichia coli* that cause diarrhea are of at least three different types. They are invasive, enterotoxigenic (toxin producing), and enteropathogenic. The invasive strains cause a disease that is found in the colon and that produces a fever and occasionally bloody diarrhea. The pathologic changes are similar to those seen in shigellosis. The enterotoxigenic strains act more like *Vibrio cholerae* in producing profuse, watery diarrhea without blood or mucus, abdominal cramping, vomiting, and dehydration. The enteropathogenic strains belong to the group that is associated with outbreaks of acute diarrheal disease in newborn nurseries. The diseases are typically food borne or water borne, especially in areas where there is poor protection for the food and water, and endemic diarrhea is present. Traveler's diarrhea is most usually due to an enterotoxigenic *E. coli*. The disease is spread by the fecal-oral route. Contaminated hands, as well as improper protection of food and water, may lead to the outbreak of the disease. The incubation period is 12–72 hr. The reservoir of the infection is infected people. The disease is readily transmitted from person to person or from water, food, or fomites to people.

Campylobacter Enteritis. Diarrhea caused by *Campylobacter* is an acute enteric disease of varying severity. It is caused by *Campylobacter jejuni* and *Campylobacter coli*. The organisms are an important cause of diarrheal disease in all parts of the world and in all age groups. *Campylobacter* may be responsible for more enteritis than either salmonella or shigella. Most outbreaks have been associated with foods, unpasteurized milk, and unchlorinated water. These organisms are an important cause of traveler's diarrhea. The incubation period is 3–5 days, with a range of 1–10 days. The reservoirs of infection are farm animals, cats, dogs, rodents, and birds, including poultry. The disease may be transmitted for several days to several weeks. Chronic carrier states may occur in animals and poultry, and they may become the primary source of infection.

Bacterial Food Poisoning

Bacterial food poisoning or food intoxication is due to the consumption of food containing a toxin created by bacterial growth in the food.

Staphylococcus. Outbreaks of staphylococcus food poisoning, probably the most frequent kind of food poisoning, usually affects large groups of people at picnics, church suppers, in hospitals, cafeterias, and other mass feeding operations. The organisms, which come from humans, are present in boils, carbuncles, pimples, hang nails, postnasal drip after colds, and wound infections. Humans then infect such foods as creams, hams, potato salad, chicken salad, ham salad, egg salad, meat and meat products, poultry, turkeys, custards, eggnogs, cream pies, eclairs, casseroles, and warmed-over foods. The organisms grow best at 50–120°F for a minimum of 5 hr. The bacterial growth does not alter the appearance or flavor of the food and does not produce off odors. The outbreak of the food poisoning occurs when individuals consume the enterotoxin produced by the staphylococcus. The enterotoxin causes inflammation and irritation of the stomach and intestine, resulting in vomiting and diarrhea. Since individuals vary in their susceptibility to the toxin, some become quite ill and others are not affected at all.

In July 1976 in Oakland County, Michigan, 87 senior citizens who were receiving hot meals on a special program fell ill with staphylococcus food poisoning. The beef base used in the preparation of the gravy was found to be the cause of the outbreak. Staphylococcal intoxication is due to exo-enterotoxins from *Staphylococcus aureus*, which may be present in the nose, on the skin, and in lesions of infected people and animals. The incubation period is 1–8 hr, usually

2–4 hr. The reservoir of infection is people or animals. The organisms are readily transmitted to foods to cause an intoxication.

Clostridium perfringens. *C. perfringens* food poisoning, which has increased sharply in the last few years, is caused by the consumption of large quantities of organisms in food. Although no free toxin has been found by researchers, the consumption of small quantities of *C. perfringens* does not result in illness. This indicates that the organisms do not grow in the individual and, therefore, must cause intoxication rather than food infection. Most reported outbreaks of this food poisoning have been associated with mass feeding, such as banquets at schools or hospitals and dining rooms of college residence halls.

The organisms are present everywhere, but come primarily from soil, human or animal intestinal tracts, fecal material, and sewage. Although the vegetative forms are killed by cooking the food, spore forms are not. The organisms, which are circulated through the air or by the hands of food-service personnel, are found on meat, particularly large roast beefs and pork loins, on turkeys, and in gravies, dressings, or prepared dishes containing meat, poultry, fish, vegetables, or macaroni products. The organisms require 13–14 amino acids, 5–6 vitamins, temperatures of 60–125°F (the optimum temperature is 110–117°F), and anaerobic conditions for growth. Heating of foods such as gravies, use of warmers, and slow cooling in ambient air of large pieces of meat causes the air to escape and provides excellent incubation temperatures for rapid growth of bacteria and toxin production. The incubation period is 6–24 hr, usually 10–12 hr. The reservoir of infection is soil, the gastrointestinal tract of healthy people, and animals.

Clostridium welchii. *Clostridium welchii* type A food poisoning, similar to *C. perfringens* food poisoning, is usually caused by infected food handlers. Boiled, braised, steamed, stewed, or inadequately roasted meat allowed to cool slowly and then served warmed or cold the following day is the vehicle in the spread of this organism. (See *C. perfringens* for incubation period and reservoir of infection.)

Clostridium botulinum. *Clostridium botulinum* produces in food the most deadly toxin known to humans. A small amount of the pure toxin could kill thousands of people. In the last 20 years, 399 reported cases resulted in 122 deaths. The organism is found throughout the world in the spore form in soil. The spore is resistant to heat, chemicals, and physical stress. Since home-canned, preserved, or processed foods may not be adequately cooked or processed, and since the spore is so resistant to environmental changes, most outbreaks of botulism have been traced to home-cooked, rather than commercially processed, foods. However, outbreaks of botulism, as well as food containing the toxin have been traced to commercially prepared foods. As the spores vegetate, bacteria grow and multiply, and produce the toxin in anaerobic conditions during storage. Boiling the food for a few minutes will destroy the toxin.

There are six known types of *C. botulinum*. Type A, associated with human illness, is the common cause of botulism in the United States. Type B, also associated with human illness, is most frequently found in soils in the world. Type C is associated with outbreaks of botulism in cattle, mink, waterfowl, and other animals. Type D, responsible for forage poisoning of cattle, is most commonly found in Africa. Type E, associated with human illness, is usually found in outbreaks associated with fish and fish products. Type F, associated with human illness, has only recently been isolated and is relatively rare. Type A and B toxins are found mainly in canned vegetables and fruits, although beef, pork, fish and fish products, milk and milk products, and condiments are also incriminated.

The key factors in the growth of *C. botulinum* and toxin production from the disrupted cells are pH, availability of oxygen, salt content, time, and temperature. A pH close to neutral favors growth, whereas a pH of 4.5, as is found in tomatoes, pears, and red cabbage, inhibits growth. Types A and B grow best at 95°F, but the temperature can range from 50–118°F. Type E grows best at 86°F, but can grow and produce toxin at 38–113°F. Although a food mass is aerobic, the conditions next to the bacterial cell may be anaerobic. Smoked fish can develop anaerobic conditions in the visceral cavity and under the skin. The interior of sausage can become anerobic. While some foods become foul and rancid due to growth of the organism, others show only minor changes in odor or appearance. In some foods no change occurs at all, despite the fact that the food is lethal. Vacuum-packed smoked fish is an example of a food causing serious outbreaks in recent years, yet showing no physical changes. The incubation period is 12–36 hr, when neurologic symptoms usually occur, although it may last several days. The shorter the time, the more severe the disease and the higher the death rate. The reservoir of infection is soil marine sediment and the intestinal tract of animals and fish. Botulism is not communicable from people.

Vibrio parahaemolyticus. *Vibrio parahaemolyticus* is associated almost exclusively with seafood and is found in nearly all seafood products. The number of recorded cases in Japan range from 10,000 to 14,000 annually. Several outbreaks occurred in the United States when steamed crabs were eaten. The organisms can live in salt water separate from the host. The incubation period is usually between 12 and 24 hr, but it can range from 4 to 96 hr. The reservoir of infection is in marine areas during the cold season, where the bacteria are found free in water or in fish and shellfish in the warm season. It is not spread by people.

Bacillus cereus, Food Poisoning. *Bacillus cereus* food poisoning is a gastrointestinal disorder in which there may be a sudden onset of nausea and vomiting, and in some cases diarrhea. The symptoms typically last about 24 hr. The infectious agent is *Bacillus cereus*, which is an aerobic spore former. The spore, when becoming a vegetated cell, produces two enterotoxins that are heat stable. These cause vomiting. One other enterotoxin is heat labile. This causes diarrhea. The incubation period is 1–6 hr where vomiting occurs, and 6–16 hr where diarrhea is most prominent. It is not communicable from person to person.

Chemical Poisoning

Chemicals accidentally introduced into food or leached into food from a variety of containers cause rapid illness and at times death. The incubation period is less than 30 min. The cause of the outbreak varies with the chemical and its origin.

Antimony. Antimony is leached from chipped gray enamelware by acidic foods. A typical antimony poisoning is caused by storage of a lemon punch in a large, chipped gray enamel pot.

Arsenic. Arsenic is found in ant, roach, or rodent baits, insecticidal fruit sprays, and herbicides. These poisons accidentally contaminate foods. Since arsenic is an accumulative poison, small doses over extended periods of time lead to poisoning. Arsenic is also a fairly frequent contaminant of drinking water.

Cadmium. Cadmium, present in plating materials of containers and trays, is leached into food by acidic substances, such as fruit juices. Cadmium might also seep from industrial operations into a water supply.

Chlorinated Hydrocarbons. Chlorinated hydrocarbons are synthetic chemical pesticides, such as DDT, lindane, and endrin. Since these pesticides have been used extensively and are fat soluble, they can create long-term hazards. Excess quantities of these poisons cause nervous system disorders. The main modes of entry into the environment in an agricultural setting are through the air, soil, or water. This may occur through air spraying of chemicals, ground spraying, exposure from industrial sources, and spreading of sludge on soil. The chemicals may enter the groundwater supply, directly contaminate the food supply, or be bioconcentrated in the food chain.

Copper. Copper poisoning is caused by the leaching of copper from food contact surface, such as carbonated water-machine tubes, into food or drink.

Cyanide. Cyanide, found in silver polishes, can accidentally contaminate foods.

Lead. Lead is present in lead arsenate, a pesticide used on apples. Lead may also be leached by acidic beverages from improperly glazed pottery vessels and utensils. Automobile radiators used as condensers for illegally distilled whiskey add lead to the final product.

Mercury. Mercury, commonly found in many industrial processes, commercial products, and homes, is an insidious chronic poison, producing symptoms that resemble emotional and psychological disorders. Mercury poisoning also appears in an acute form. Methyl mercury, the principle form of mercury found in food, penetrates the placenta, causing birth defects. It is also found in mother's milk. Fungicides containing mercury are used on grain consumed by farm animals, who are ultimately consumed by humans, leading to severe mercury poisoning and eventual death in humans. Fish and shellfish that have consumed mercury and are later used as food also cause blindness, paralysis, and death in humans.

Organic Phosphate Compounds. Organic phosphates, such as DDVP, parathion, and TEPP, are used as insecticides, fungicides, and herbicides. Although this group of poisons are most dangerous when inhaled, they also constitute a hazard when ingested with food.

Polychlorinated Biphenyls (PCBs). Polychlorinated biphenyls are found in wildlife in much of the world. PCBs are also found in fish; animal feed, which leads to contaminated meat, milk, and eggs; paints that have leached into feed; heat exchange fluids; and cardboard cartons from recycled paper. The potential for food poisoning will increase with the increased availability of this family of over 200 chemicals resembling DDT that are stored in the fatty tissues. PCB poisoning has occurred in Japan. In one reported case in 1968, 1000 people became ill, resulting in five deaths from ingesting rice oil contaminated with PCBs.

Zinc. Zinc poisoning is caused by zinc leaching into acidic foods stored in galvanized containers.

Poisonous Plants and Animals

Certain plants and animals are poisonous: castor beans contain a toxin called ricin; ergotism is caused by a parasitic fungus of rye (*Claviceps purpurea*); favism is due to the bean, *Vicia fava*; poisonous mushrooms are often fatal; shellfish poisoning is due to eating shellfish that have consumed a plankton called *Gonyaulax*; and cyanide is produced in green or sunburned potatoes and wild celery.

Aflatoxins. Aflatoxins are formed by molds on foods at harvest time, in storage, or in conditions of water damage. Aflatoxins are found in peanuts, brazil nuts, pecans, copra, corn, and cottonseed. Aflatoxins, known as very potent carcinogens for certain animal species, can be produced whenever proper temperature, time, humidity, and the proper strain of *Aspergillus flavus* are present. It is thought that some aflatoxins in Africa might be associated with liver cancer.

Fungi. Fungi may invade plants, causing them to undergo metabolic changes that produce toxic substances. The phytoalexins are examples of this. Genetic manipulation also produces new plant varieties having a stress mechanism that causes problems to the human being. It is known that compounds inhibiting the proteolytic activity of certain enzymes are found throughout the plant kingdom, particularly among the legumes. One of the better known inhibitor reactions is the inhibition of trypsin. Protease inhibitors may also be found in peanuts, oats, chickpeas, field beans, buckwheat, barley, sweet potatoes, rice, lyntol, lima beans, navy beans, garden peas, white potatoes, wheat, and corn. Trypsin is essential for adequate digestion and utilization of food. The presence of the inhibitors decreases the digestion and absorption of the proteins, causing reduced growth and decreased efficiency in utilizing food. Since protein is the most costly part of the diet, the trypsin inhibitors create an economic as well as public health effect.

Lectins. Certain plants contain substances that agglutinate red blood cells. These substances are called phytohemagglutinins or lectins. They are present in seeds and to a lesser extent in leaves, barks, and roots. They produce important environmental toxins, such as ricin, from the castor bean.

Hepatoxins. Certain substances, known as hepatoxins, that are toxic to the liver are found in plants. These plants are poisonous to livestock and humans. The seeds of these plants, which are of the genus *Senecio*, contaminate wheat and corn harvested from the same land.

Goitrogens. Goitrogens, natural products found in plant foods eaten by humans and animals, cause hypothyroidism, which is an enlargement of the thyroid gland. Consumption of additional iodine controls this problem. The plants in this group of the genus *Brassica* include cabbage, turnip, mustard greens, radish, and horseradish.

Saponins. Saponins, comprising at least 400 species and 80 families of plants, produce the toxin known as glycoside that hemolyze red blood cells. Saponins are found in soybeans and alfalfa, and are of particular concern, since soybeans and alfalfa are highly nutritious plants that are essential to our society.

Allergic Reactions

Individuals may have peculiar allergic reactions to any kind of food, ranging from simple headaches to unusual central nervous system disorders, including convulsions and death. Infants and children often complain of abdominal distress and sometimes of genitourinary tract problems. The cardiovascular system and skin can be involved in an allergic reaction.

Mycotoxins. Mycotoxins, produced by fungi, produce acute and chronic effects. Many humans and animals suffer with these fungal metabolites. Cases that come to the attention of public health officials are those in which large amounts of the toxin have been consumed and result in serious illness or death. Mycotoxins are secondary metabolites of molds that may be toxic, carcinogenic, or hepatoxic.

Species of *Aspergillius*, *Penicillium*, *Rhizopus*, and *Streptomyces* produce aflatoxins. Species of *Fusarium* produce trichothecenes. Species of *Penicillium* and *Aspergillus* produce penicillic acid.

There are three major classes of molds that invade agricultural products. They are field fungi, storage fungi, and advanced decay fungi. The physical factors that bring about mycotoxin production include moisture, temperature, mechanical injury, blending of grains, hot spots, and time. There must be a variety of conditions present that are favorable for the mycotoxin in order for the mycotoxins to be produced. Aflatoxins can grow on peanuts, cotton seed, seed oil, corn, legumes, dried fruits, wines, and dairy products. Zearalenone is found almost totally on corn. Ochratoxin is found on food grains such as wheat and barley. Citrinin is a co-contaminate with ochratoxins. Patulin is produced on apple rot and is found in apple juice. Penicillic acid is found on tobacco and storage grain, such as wheat, corn, peanuts, and cotton seed. Trichothecenes are found on a variety of different foods. Aflatoxins not only cause toxic effects, but also may be mutagenic and carcinogenic. When two or more toxins are combined, there may be a variety of toxic effects on experimental animals. Mycotoxins cause toxic effects in the liver, digestive tract, urinary system, the skin, hematopoietic system, reproductive organs, and nervous systems.

Parasitic Infections

Most parasitic infections are due to the ingestion of parasites in food or water.

Amoebic Dysentery. Amoebic dysentery, or amebiasis, is caused by *Entamoeba histolytica*. Water or food is contaminated with sewage or human feces containing the amoeba. Disease rates are higher in areas of poor sanitation, mental institutions, and among homosexuals. It may cluster in households or institutions where sanitation is good. The incubation period varies from a few days to several months or years. Commonly it is 2–4 weeks. The reservoir of infection is a person who is usually a chronically ill or asymptomatic cyst passer. Communicability may continue for years.

Trichinosis. Trichinosis, which is caused by *Trichinella spiralis*, has an incubation period of about 9 days, although it varies from 2 to 28 days. In heavy infections, the incubation period may be 24 hr. The live larvae are found in raw or improperly cooked pork, pork products, whale, seal, bear, or walrus meat.

Tapeworms. Tapeworms are caused by the ingestion of live larvae in raw or insufficiently cooked beef, fish, or pork. The beef tapeworm is *Taeniasis saginata*; the fish tapeworm is *Diphyllobothrium latum*; the pork tapeworm is *Taenia solium*. The beef tapeworm causes intestinal infections in people, whereas the pork tapeworm may cause intestinal infections and somatic infections by the larva. The larval disease, which is called cysticercus, is a tissue infection that may affect vital organs, and therefore, may cause fatality to occur. The beef tapeworm causes the intestinal infection, whereas the pork tapeworm may cause both diseases to occur. The fish tapeworm generally causes few symptoms. In some individuals, a vitamin B_{12} deficiency anemia may occur along with diarrhea. The incubation period for the pork tapeworm is 8–12 weeks. The incubation period for the fish tapeworm is 3–6 weeks. The reservoir of infection for beef and pork tapeworms is people. The intermediate host for the beef tapeworm is cattle, whereas the intermediate host for the pork tapeworm is pigs. The reservoir of infection for fish tapeworms is people and other hosts, include dogs, bears, and fish-eating mammals. The beef tapeworm may be communicable for as long as 30 years. The fish tapeworm is not communicable from person to person.

Giardiasis. Giardiasis is a protozoan infection, usually of the upper small intestine, which is caused by *Giardia lamblia*. The symptoms of the disease, when they are there, include chronic diarrhea, abdominal cramps, fatigue, and weight loss. The disease is transmitted by contaminated water and less frequently by contaminated food. Person to person contact may also occur and may also contribute to the spread of the disease when the hands of a contaminated person transfers the cysts from the feces of the individual to the mouth of another individual. This may occur in a variety of institutions, including day-care centers. The incubation period is 5–25 days or longer, with a possible 7- to 10-day period being most frequent. The reservoir of infection is people. The period of communicability lasts during the entire period of the infection.

Angiostrongyliasis. Angiostrongyliasis is a disease of the central nervous system that is caused by a nematode, *Angiostrongylus cantonensis*. The disease is spread through the ingestion of raw or improperly cooked snails, slugs, prawns, fish, and land crabs. The incubation period is usually 1–3 weeks, although it may be longer or shorter. The reservoir of infection is the rat. The disease is not transmitted from person to person.

Toxoplasmosis. Toxoplasmosis is a protozoal disease that may be spread from the mother to the child through the placenta, if she is infected, but may also be acquired by eating raw or undercooked infected pork or mutton. The disease may also be spread through water or dust contaminated by cat feces. The infectious agent is *Toxoplasma gondii*. The incubation period may be 10–23 days when consuming contaminated food. The reservoir of infection includes rats, pigs, cattle, sheep, goats, chickens, birds, and cats. The disease is not transmitted from person to person.

Scombroid Poisoning. Scombroid poisoning is caused by eating scrombroid fish or eating fish of the family Mahi-mahi, which is also called dolphin fish. The poisoning is due to a histamine-like substance produced by several species of *Proteus* bacteria or other bacteria. The histamine is produced from histidine in the flesh of the fish. The individual person has headaches, dizziness, nausea, vomiting, a peppery taste, a burning throat, facial swelling and flushing, stomach pain, and itching of the skin. The disease may also be caused by tuna or mackerel. The cause of the disease is inadequate refrigeration of scromboid fish. The incubation time is 10 min to an 1 hr. The disease is not transmitted from person to person, and the reservoirs of infection are the fish that have been mentioned.

SPECIAL FOOD PROBLEMS

Church dinners are usually prepared by volunteers in their homes and then transported at varying times and temperatures to the place of consumption of the food. Time, temperature, personal health, methods of preparation, and initial ingredients are uncontrolled. This has led to outbreaks of food-borne disease and intoxication.

Although fairs and special events are limited timewise, they provide an environment where tens of thousands of people congregate and eat. The food operations should be similar to a permanent food operation but usually are not. Problems include lack of hot and cold running water, inadequate refrigeration, unwholesome food, poor handling of food, flies and other insects, lack of screening, and improper waste disposal. These food operations must be licensed and inspected by the health department.

Vendors that sell food to the public from vehicles of any type must meet the same criteria as permanent food establishments. Problems related to vendors include lack of hot and cold running water, improper waste and liquid waste disposal, lack of refrigeration, improper cleaning and sanitizing, and the inability of the health department to make inspections because of the high degree of mobility of the vendors. Sanitary conditions deteriorate very quickly in this type of operation. All vendors must be licensed.

Food stores must adhere to all of the regulations of other preparation and serving establishments. Problems usually include unwholesome food or damaged and rusted cans, improper refrigeration, stacking above freezer line, inadequate cleaning of equipment and utensils, flies, roaches, rodents, and inadequate solid-waste disposal.

The remnants of consumed food create solid-waste problems. If garbage is improperly wrapped or stored, it becomes a fly and rodent attraction. If it is contaminated by highly infectious individuals, it becomes an infection problem. When garbage is removed to a landfill, it becomes a part of the overall landfill solid-waste disposal problem. When burned it creates air pollutants.

The individual infected with food-borne disease organisms may become a carrier and transmit the infection to others. Although the initial spread of disease may be due to a food-borne organism, the secondary occurrences may well be traced to other environmental factors. An example of this would be the serious and prolonged outbreaks of *Salmonella derby* that occurred in U.S. hospitals in the 1960s. The initial outbreak was thought to be caused by contaminated eggs. The secondary outbreaks were caused by the carrier spreading the organism in a variety of ways.

Individuals who become ill from food-borne disease require treatment either at home or in hospitals. The actual cost to society of food-borne illness cannot be calculated. However, if there is as little as a minimum of one million cases of food-borne disease each year in the United States, the cost of food-borne illness is probably many millions of dollars because of lost work and necessary treatment.

POTENTIAL FOR INTERVENTION

The potential for intervention in food problems related to food protection and disease outbreaks consists of the techniques of isolation, substitution, shielding, treatment, and prevention. In the area of isolation, all prepared, stored, or served food must be kept from sources of hazardous materials, especially chemicals. The food also must be isolated from insects and rodents, and from contaminating sewage and water. In the area of substitution, mechanical processes should be substituted for hand operations in the preparation of food when possible. In the area of shielding, food on a buffet table must be protected from the customers. Treatment is used for food-handlers with food-borne diseases. Prevention is the major technique utilized for intervention in disease processes. Prevention includes everything from refrigeration and freezing, adequate cooking, and proper storage, to removal of sick food-service workers from food preparation and service.

The potential for intervention generally is very good; specifically, care should be given to the selection of food and special techniques used for the handling, processing, and serving of specific types of hazardous foods.

RESOURCES

Scientific and technological resources in the area of food include the Institute of Food Technology, the Canning Trade Inc., the International Association of Environmental Milk and Food Sanitarians, the National Environmental Health Association, the American Public Health Association, various land grant colleges such as the University of Iowa and Purdue University, schools of public health such as the University of Michigan, the University of California, and the University of Minnesota. Civic associations concerned with food are basically in the area of nutrition and provision of food for individuals who are starving or who have inadequate food supplies. The governmental organizations involved in food include the Federal Food and Drug Administration, various local and state health departments, and many of the state departments of agriculture throughout the country. The U.S. Department of Agriculture is another fine resource. County extension agencies are an immediate source of help. The U.S. Senate Agricultural Committee is an important source for new legislation.

STANDARDS, PRACTICES, AND TECHNIQUES

Plans Review

A successful food-service operation is dependent on good facilities, proper arrangement and utilization of equipment, good-quality raw food, and well-trained personnel. State health departments and, where designated, local health departments are responsible for reviewing the plans of all new food establishments prior to construction. Plans and specifications, including architectural, structural, mechanical, and plumbing drawings, must be submitted to the health department, which gives preliminary approval for construction and makes an inspection prior to licensing. The inspection includes evaluating material and grouting of floors; material, finish, and color of walls and ceilings; presence of insect- and rodent-proofing devices; quantity and quality of lighting in various areas; amount and type of ventilation; type of toilet facilities for males and females; type of water supply; hand-washing facilities; construction of utensils and equipment; type of utensil washing and sanitizing; type of utensil storage; type of food-storage areas; type and quality of food refrigeration; type of locker or dressing rooms; type of sewage disposal; and type of solid-waste disposal. The food service operation is licensed and permitted to serve food after specifications set forth by the health department are met.

Food Service Facilities

Physical Plant

Floors in food preparation, serving, and storage areas should be composed of smooth, nonabsorbent, easily cleaned materials to aid in cleaning and to prevent absorption of grease, organic material, and odors. When floor drains are used, they should be screened to prevent harborage of flies and roaches, and to enable easy cleaning. Nonrefrigerated, dry storage areas need not have nonabsorbent floors.

Walls and ceilings should have light-colored, smooth, easily cleaned surfaces to facilitate good food preparation and food-service habits and to provide more adequate distribution of light. A minimum of 70 footcandles of light are needed on all food preparation surfaces and 20 footcandles on all other surfaces to provide an environment conducive to the elimination of dirt, grease, insects, and rodents, and to prevent accidents.

Plumbing and Water Supply

There must be an adequate quantity of bacteriologically and chemically safe hot and cold water under pressure from an approved source, since water can easily contaminate food directly or indirectly through equipment, utensils, and hands. Ice must be prepared from water from an approved source, and must be transported and stored in a sanitary manner, since ice is used in food and drink, and has been shown to contain numerous bacteriological contaminants.

Plumbing and drainage systems carrying human wastes have been incriminated in outbreaks of typhoid fever, paratyphoid fever, dysentery, and other gastrointestinal diseases. Transmission of these diseases usually occurs through cross-connections, overhead leakage, stoppage in drainage systems, and submerged inlets. Backflow to the potable water supply system may occur directly in coffee urns, dishwashing machines, and double-jacketed kettles. Back-siphonage is possible where submerged inlets occur as in drinking fountains, flushometer valves, flush tanks, garbage-can washers, hose outlets, ice makers, lavatories, slop sinks, steam tables, and vegetable peelers. The backup of sewage is possible in enclosed equipment, such as ice makers, refrigerators, steam tables, and walk-in freezers. Sewage should be disposed of in a public sewage system or in a properly operated on-site sewage disposal system to prevent the spread of enteric disease and mosquito breeding.

Separate sanitary toilet facilities with self-closing doors are needed to prevent the spread of enteric disease. The hand-washing facilities must have hot and cold running water, soap, and disposable towels or air dryers to prevent the spread of disease through contaminated hands. (In-depth discussions on water supply, plumbing, and sewage disposal are found in Volume II.)

Hot water has a corrosive effect on pipes. Hard-water deposits are seven times greater at 180°F than at 140°F. Water softeners should be used where hardness exceeds 125 ppm. Since hot water is the sanitizing agent, in machine dishwashing, the force and water pressure, volume of water, and temperature are very important. A water pressure of 20 lb/in.^{2} at the entrance to the rinse manifold is optimal. Water should be flowing at a rate of 9 gal/min in the rinse section. The gas or electric company should be called upon to determine the size of water heater necessary for any given establishment.

Ventilation Systems

Adequate ventilation must be provided for all rooms to reduce condensation, minimize soiling of walls and ceilings, and remove excessive heat and objectionable odors. All hoods should be equipped with noncombustible 2-in. commercial grease filters that are easy to install, remove, and clean. The surface of hoods should have sealed joints and seams, be accessible, and be easily cleaned. The function of the hood is to collect vapors, mists, particulate matter, fumes, smoke, steam, or heat. The velocity of the air capturing these items is called the capture velocity. The canopy hood is an overhead hood completely covering the equipment it is designed to serve. The filter in the hood should be a minimum of 2.5 ft above an exposed cooking flame, 4.5 ft above charcoal fires, and 3.5 ft above other exposed fires. Hoods should be so designed as to not interfere with normal combustion and/or exhaust of combustion products from commercial cooking equipment, other processes, and heat equipment.

To determine the total quantity of air exhausted from a hood, the following formula should be used:

$$Q = VPD$$

where Q is the total quantity of air to be exhausted in ft^3/min, V is capture velocity in ft/min, P is the perimeter of the open sides of the hood in feet, and D is the distance between the cooking surface and the face of the hood in feet.

Make-up air, composed of outside air equal to 100% of the quantity of the air exhausted, must be introduced into the building in such a way as to not interfere with the exhaust system.

Storage of Food

Wholesome food can be readily contaminated if stored improperly. Food must be protected against unsanitary conditions, insects and rodents, poisons, cleaners, other chemicals, and bacterial growth. Temperature, time of storage, initial contamination and moisture content are also vital factors to be considered in food storage.

Dry Storage

Unsanitary conditions result from dirt, dust, contamination by overhanging sewage pipes, sneezing, coughing, unclean utensils and work surfaces, and unnecessary handling.

Food-product insects can easily contaminate a large amount of grains, flour, and rice. These supplies should be stored in a clean, dry area, on racks at least 6 in. off the floor, and 1 ft from walls, and should be rotated at least once every 2 weeks during the summer months and once every 3–4 weeks during the winter months. The keeping quality of the food is improved because of adequate air circulation and because it is possible to prevent insects and rodent harborage under or behind the food. Mice and rats contaminate large quantities of food through defecation and urination on the food and through nibbling. This hazard can be detected by using a blacklight to determine if urine is present and a flashlight to determine if fecal material is present.

All poisons, cleaning materials, and other nonfood items must always be clearly labeled, stored in their original containers, and kept apart from all foods.

Refrigerated Food

Bacterial growth can be controlled by proper cooking and by storing food below 45°F or above 140°F in shallow containers. The temperature at which food is stored is probably the most important of all environmental conditions, since it affects the rate of decomposition and the growth of microorganisms. The time of storage is important in preventing undesirable changes in taste, odor, and quality of the raw products, and therefore, it is essential that the initial contamination of raw food be minimized. Chemicals added to food limit food spoilage, but add unusual tastes and odors if the food is stored for long periods of time. The presence or absence of air determine the amount of food decomposition due to different types of bacteria. Moisture, which aids in the growth of bacteria, has a considerable effect on food spoilage.

Refrigeration of food has been practiced for many centuries. Food has been placed in cool caves, wells, springs, and running streams. Ice has been used for at least 200 years. Today, the modern refrigerator and freezer is a major means of storage and preservation of food.

The rate of refrigeration is influenced by the heat transfer properties of the food, the volume of the food to be refrigerated, the kind of containers used, the heat conductivity of the containers, the agitation of the food, and the temperature difference between the food and the refrigerator unit. (See Table 2.1 for refrigerated storage temperatures and shelf-life for cold perishable foods.)

Potentially hazardous ingredients for salads, sandwiches, filled pastry products, and reconstituted food need to be chilled to 40°F or below.

Frozen Food

The quality of frozen food is determined by the raw products, the method of preparation, the speed of freezing, the temperature and time involved in distribution and storage, and the method of packaging. The frozen food package must be strong, flexible, and prevent entrance or escape of liquid from the package. Desiccation of frozen foods is a physical change taking place when moisture escapes. Crystallization is also a physical change, usually occurring when frozen food is permitted to defrost and is then refrozen. Bacteria grow readily in the moisture film collected at the surface of the frozen foods when the temperature rises above 32°F. The general practice of defrosting frozen foods at room temperature is considered to be poor, since the outer surface becomes a fine bacterial medium while the inner core is still defrosting. Frozen foods should be defrosted in the refrigerator at temperatures not exceeding 40°F or should be cooked immediately. Frozen poultry, fish, shellfish, and frozen leftovers are particularly

Table 2.1.   Refrigerated Storage Temperatures and Shelf-Life

Product	Temperature (°F)	Maximum Storage Period
Meats		
Bacon	28–30	15 days
Beef (dried)	62–40	6 months
Brined meats	31–32	6 months
Beef (fresh)	30–32	1 days
Fish (frozen)	5–10	6 months
Fish (iced)	30–32	15 days
Hams and loins	28–30	21 days
Lamb	28–30	14 days
Livers	20–22	6 months
Oysters (shell)	32–38	15 days
Oysters (tub)	32–38	10 days
Pork (fresh)	30–32	15 days
Pork (smoked)	28–30	15 days
Poultry (fresh)	28–30	10 days
Poultry (frozen)	0–5	10 months
Sausage (franks)	35–40	2 days
Sausage (fresh)	31–27	15 days
Sausage (smoked)	32–40	6 months
Veal	28–30	15 days
Miscellaneous		
Butter	35	6 months
Cheese (American)	32–34	15 months
Cheese (Swiss)	38–42	60 days
Cream (40%)	5–10	4 months
Eggs (frozen)	0–5	18 months
Eggs (fresh)	38–45	2 months
Milk	35–40	5 days
Oleo	34–36	90 days

prone to contamination and bacterial growth, and may easily lead to outbreaks of food infection or food poisoning if not properly defrosted and adequately cooked. Frozen-food temperatures should be below 0°F. Fishery products should be at −10°F.

Protecting Wholesomeness and Detecting Spoilage of Foods

All raw food products and processed food should be clean, wholesome, free from spoilage, free from chemical contaminants, and free from organisms that cause disease or intoxication. Meat, milk, shellfish, and similar products should come from approved sources. All leftovers should be discarded, having been handled extensively and probably having been kept at improper temperatures. All food-processing and storage areas should be free of birds, animals, insects, and rodents. Frozen foods should be defrosted in such a way as to prevent rapid growth of bacteria.

Visual Inspection of Foods

Decomposing meat may have an off odor, be slimy, and have an off color. Beef spoils from the surface inward; pork spoils from the bone outward. Decomposing fish have an off odor, sunken eyes, and gray or greenish gills. An indentation will remain on the flesh of the decomposing fish, which is easily pulled away from the bones. Decomposing poultry will be sticky or slimy, have off odors, and exhibit darkening of wing tips. Canned foods can be swollen, dented, rusted, without labels, and have off odors and off color. With the exception of large pieces of beef, all of the previously mentioned foods should be discarded. Large pieces of beef can be trimmed by the butcher and the trimmings discarded. Ground hamburger and ground pork decompose rapidly; discoloration and off odors result.

Thawing of Frozen Foods

To protect the wholesomeness of frozen foods and to prevent potential outbreaks of disease, the following thawing procedures should be followed: Cook frozen vegetables, small cuts of meat, chicken, fish, and prepared foods in the frozen state; thaw frozen fruits, juices, large cuts of meat, poultry, and shellfish in the refrigerator at 40°F.

Potentially Hazardous Foods

Fresh meats may contain *C. perfringens*. Since growth of the bacteria and development of the toxin occur best at room temperatures, refrigerate meat immediately after cooking and until used.

Fecal streptococci and *Staphylococcus aureus* are also found on meat. Partially cooked hams may contain fecal streptococci and staphylococci. Smoked hams that are cooked to an internal temperature of 150°F are not sterile and may contain organisms. Since cold cuts, including hot dogs, are handled by personnel, there is a potential hazard of staphylococcal food poisoning or salmonella food infection.

Salmonella, the primary contaminant of poultry, causes many outbreaks of salmonellosis; *C. perfringens* and fecal streptococci may also be contaminants.

Fish and shellfish coming from sewage-polluted waters may contain enteric organisms. *Clostridium botulinum* Type E is found on fish. Shellfish have been incriminated in outbreaks of infectious hepatitis and cholera.

Canned foods, such as tuna fish and mushrooms, and prepared frozen foods, such as some pizza, have been shown to contain botulism toxin.

Chicken salad, tuna salad, ham salad, potato salad, custard-filled pastries, dairy products, smoked fishes, egg salad, and other mayonnaise or cream-based salads have led to substantial outbreaks of food poisoning and food infection. The lack of proper refrigeration, improper handling of the food, and saving of leftovers are the contributing factors.

Food contaminated with water and smoke during fires, defrosted frozen foods, and unrefrigerated perishables are considered unwholesome and should be discarded.

Preparation and Serving of Food

Since food becomes contaminated easily during preparation or serving, and since organisms already present in the food can cause disease and poisoning, the following rules of food protection should be followed:

1. Use only good quality, wholesome food.
2. Keep the food clean and free of insects and rodents.
3. Clean preparation and serving areas carefully.
4. Use only well-constructed, easily cleaned equipment.
5. Keep equipment very clean.
6. Handle food as little as possible.
7. Use good personal hygiene at all times.
8. Refrigerate perishables as quickly as possible.
9. Cook foods long enough to kill organisms.
10. Keep food below 45°F or above 140°F in serving areas.
11. Provide proper sneezeguards in cafeterias and smorgasbords.
12. Use indicating thermometers accurate to +2 or –2°F to determine the temperature of the refrigerator and steam table.

Be particularly careful of ground and chopped foods, foods made of several raw materials, rich or nutritious foods, foods with high moisture content, and foods that require considerable handling during preparation, such as salads and ground meats.

Utilize the proper precleaned equipment and tools for the job. Use forks, knives, tongs, spoons, or scoops wherever possible, instead of hands. Remove small quantities of prechilled perishable ingredients from the refrigerator immediately prior to use. Wash all raw fruit and vegetables thoroughly before cooking or serving. Cook all stuffed poultry and meats to an internal temperature of at least 165°F. Cook all pork products to an internal temperature of at least 165°F. Place all custard, cream filling, and puddings below 45°F.

When food is served, the server must be careful not to place his or her fingers on the eating surface of the dishware or silverware. Plates of food should not be stacked on each other, since the undersurface of plates could contaminate the food beneath it.

When food is served in smorgasbords, buffets, and cafeterias, it should be kept behind protective guards below the speaking level of people. The food-holding equipment should be made of nonabsorbent, smooth, easily cleaned, corrosion-resistant material. Glass used must have a safety edge. Since hot or steam tables and cold tables are not intended to alter the temperature of food, but rather maintain it at the proper temperature, the food has to be preheated to 140°F internal temperatures or precooled to 45°F internal temperature. Hot tables should be maintained at 160°F in order to achieve the 140°F needed in the food. Where ice is used for cooling, the ice should be $^3/_4$ in. in diameter or less, should be clean, and should come from an approved source. Food containers should be a maximum of 6 in. deep, and food should never be stacked above the ice level. Ice storage units need drains, which have an air gap, in order to prevent backflow of sewage. Tongs, forks, spoons, picks, spatulas, and scoops should be used by food-service workers, if possible. If the customer uses these implements, they should be changed frequently. Ice cream scoops, dippers, and spoons should be stored either in a cold running water well, which is frequently cleaned, or in a clean, dry manner. Sugar should be kept in closed containers or preferably in single-service packets. The remnants of all food served to the customer should be discarded.

Where food is transported in hot and cold food trucks to various parts of an institution, special care must be taken to preheat and precool the unit. The food containers and the truck must be carefully cleaned immediately upon completion of the serving process. The set-up trays must be covered to avoid contamination by dust, aerosols, and microorganisms.

Single-service articles are an excellent method of serving food, since the articles are thrown away after each use and a premeasured portion of food can be set up at the processing plant or kitchen. Contamination by hands and improperly washed utensils is avoided. Caution must be used in the storage of single-service materials to prevent contamination by dust, dirt, sewage, insects, and rodents. Items should be discarded in lined containers with self-closing lids.

Design and Installation of Food-Service Equipment

The design of food-service equipment should be based on good research, sound engineering, good environmental practices, and the practical knowledge of experts from industry and the field of public health. The standard for equipment design is now basically set by the National Sanitation Foundation (NSF) in Ann Arbor, Michigan. This nonprofit organization develops standards based on the aforementioned criteria.

Food-service equipment should contain the following design features:

1. easily disassembled, cleaned, and maintained equipment with as few parts as possible
2. smooth, nonabsorbent, nontoxic, odorless, and easily cleaned food-contact surfaces
3. food-contact surfaces containing no toxic materials, such as cadmium, lead, or copper
4. nontoxic, nonabsorbent, easily cleaned gaskets, packing, and sealing materials that are unaffected by food or cleaning products
5. easily cleaned splash zone areas
6. food product surfaces that will not chip, crack, or rust

Existing NSF standards or criteria should be reviewed at least every 3 years to ensure that the food-service equipment keeps pace with the public health and industry advances. The competent food-service manager should obtain copies of NSF standards and review them carefully before purchasing new equipment.

The installation of equipment is almost as important as its design and construction. If equipment is installed into the wall but is not flush with the wall, roaches may easily hide behind it. Overhead equipment such as hoods must be within easy reach, enabling food-service personnel to remove the filters regularly and to clean the hoods and filters frequently. Large equipment should be mobile or installed on legs to facilitate cleaning behind and underneath. Drains should be located in areas in the floor where spillage is likely or where equipment is scrubbed. Hand-washing facilities and service sinks should be conveniently located to ensure their use when needed. Clean metal storage shelves must be provided for the storage of cleaning equipment and utensils. See Figure 2.1 for a proper layout of kitchen equipment.

Housekeeping and Cleaning in Food-Service Facilities

In a satisfactory environment in a food-service facility, soil is removed, bacteria is destroyed, and an aesthetically pleasing picture is provided through good physical facilities, proper color, adequate light, and well-managed cleaning practices. Soils are divided into four basic types: (1) fresh soil found immediately after use of equipment or facilities; (2) a thin film of soil caused by ineffective cleaning within which microbes live; (3) built-up deposits of soil caused by consistent ineffective cleaning and composed of soap films, minerals, food materials, and grease; (4) dried deposits of soil caused by drying of heavy, crusty deposits or baking of organic material onto dishes or trays by improper dishwashing.

Fresh soils and thin films of soil on floors and walls are removed by the use of a good detergent and spraying technique, and plenty of physical action. Built-up deposits are best removed by the use of detergent spraying techniques plus mechanical scrubbing machines and special heavy-duty detergents. Dried deposits on dishes are first soaked in a good detergent and hot water, and then scrubbed with plastic or metal pads. Dried deposits on baking trays are best removed by cleaning with live steam. The best approach to cleaning, however, is prevention of the accumulation of deposits and soil through good daily cleaning. (Live steam is dangerous! Be careful of its use!)

General Daily Cleaning

At the conclusion of each work day, an assigned food-service worker should wash all kitchen, storage, and serving area floors in the following manner:

1. Starting at the far end of the room, spray an area of the floor measuring approximately 100 ft^2.

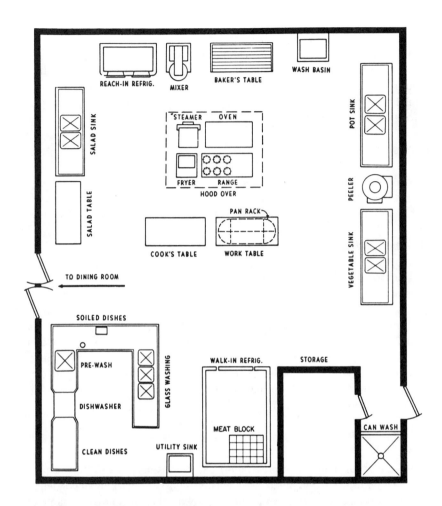

Figure 2.1.   Layout of kitchen equipment.

2. Spray an adjacent area of approximately the same size.
3. Pick up the detergent with a very clean mop and rinse in a clean pail of water.
4. Repeat steps 1, 2, and 3 until the entire floor is washed.
5. Clean floor drains.

All dry-food storage areas should be swept with a treated mop daily to decrease contamination by dust and dirt, and also to determine if an insect or rodent infestation exists. All outside areas, especially where solid waste is stored, must be checked frequently during the day and swept and washed daily. This prevents an unsightly mess and also helps destroy potential insect and rodent harborage.

Bathroom floors should be cleaned at least twice a day using the same procedure outlined for kitchen floors. In addition, the trash in the bathrooms should be emptied a minimum of twice daily, the sinks washed, and the soap and single-service towels refilled. Toilet bowls and urinals must be cleaned thoroughly. Deodorants are not a substitute for good cleaning procedures. Under no condition should food be stored in restrooms. The food-service attendant must thoroughly scrub his/her hands for at least 2 min and rinse frequently under hot running water before returning to other assignments.

Periodic Cleaning

Walls and ceilings should be washed down with a good detergent periodically, but not less than once a month. Lighting fixtures should be taken apart and cleaned and light bulbs replaced at least once every 2 months or when the bulbs burn out.

Cleaning Schedules

The establishment of definite work assignments in writing, and the development of a cleaning manual and schedule is essential to good cleaning practices. A food-service worker must be taught to disassemble and clean pieces of equipment and to utilize cleaning equipment and materials. A good food-service manager supervises this part of the food-service operation closely. The food-service manager should inspect his or her own establishment on a weekly basis using a flashlight to see behind and under equipment. A self-inspection sheet can be obtained from the health department or industry, or may be found in the chapter entitled Institutional Environment.

Cleaning and Sanitizing Equipment and Utensils

Good cleaning consists of prescraping and prerinsing with cold or warm water as soon after usage of equipment and utensils as possible; soaking in hot detergent solution, without overloading or improper stacking; power scrubbing manually or mechanically; power rinsing; sanitizing; and air drying.

All equipment, including counters, tables, carts, display cases, steam tables, shelves, storage racks, salad and vegetable bins, drain boards, meat blocks, stoves, stove hoods, coffee urns, meat and vegetable choppers, meat tenderizers, food mixers, griddles, cutting boards, ice-cream counter freezers, steam-jacketed kettles, vegetable peelers, and garbage cans, must be cleaned thoroughly each day. In all cases, empty the contents of the equipment and rinse and scrape thoroughly. For large, immovable equipment, spray all portions with a good QAC compound, scrub with clean brushes, and rinse with QAC. Scrape meat blocks and cutting boards clean and wash. Turn meat blocks regularly. Where food spillage occurs in the oven, sprinkle salt on food and heat the oven to 500°F. When the spillage is carbonized, cool oven and scrape food deposit. Ice-cream counter freezers must be emptied each night and the contents discarded. Rinse the unit thoroughly with water while still assembled, add dishwashing detergent, and operate for 1 min, rinse and drain all detergent, disassemble all parts, scrub thoroughly with a brush and detergent, rinse and reassemble, and use a final rinse of chlorine. Garbage cans should contain plastic liners, which are removed and replaced frequently. The cans should be scrubbed mechanically with a garbage-can washer or with a brush at least once every 3 days.

Hand Dishwashing

Prescrape and prerinse all dishes and utensils, and place in the first compartment of a three-compartment sink containing detergent and hot water at 95°F or above. Scrub dishes well with plastic brushes. Do not use wash cloths or sponges, since they spread microorganisms. Rinse dishes thoroughly in the second sink at a temperature of 110–120°F. In the third sink, use a chemical sanitizer or hot water at 170°F for 1 min. Remove and allow to air dry. An accurate thermometer must be kept in the sink or be readily available. In all cases, predetermine the amount of water to be used to ensure that sufficient detergent and/or sanitizer is added.

Machine Dishwashing

Prescrape and prerinse dishes and utensils. Prewash heavily soiled items carefully. Place objects in racks so that all parts are exposed to the spray. Wash at 140°F for 40 sec. Rinse at 180°F at a pressure of 15–25 lb/in.^{2} for 30 sec. Some variations in time and temperature will occur based on the type of dishwashing machine used. Suitable thermometers and pressure gauges should be attached to each machine. The accuracy of these gauges should be checked during an inspection.

A good mechanical dishwashing operation depends on the following: (1) selection of the proper-size dishwashing machine; (2) proper-size hot-water boiler and booster heater; (3) proper layout of equipment and utilization of workforce; (4) adequate training for personnel; (5) proper supervisory control; (6) adequate storage of clean dishes and utensils; (7) discarding of chipped and cracked dishes, and rusted or corroded utensils; (8) adequate ventilation of dishwashing area to reduce strain and maximize personnel efficiency.

Soiled dishes are caused by improper scraping, prerinsing or prewashing, inadequate detergent, improper wash-water temperature, inadequate time for wash and rinse, and improperly cleaned dishwasher. Check all operations carefully and make corrections where needed. Unclog wash and rinse nozzles. Films are caused by water hardness: use adequate amounts of detergent and water softeners. Greasy films are due to low pH, insufficient detergent, low water temperature, and dirty dishwasher: check alkalinity, detergent, temperature, and wash and rinse nozzles. Streaking is due to high pH and improperly cleaned dishwasher: reduce alkalinity and clean dishwasher. Cooked-on egg or other proteins are caused by improper prerinsing and prewashing, and too high a washing temperature. Spotting is due to rinse-water hardness, too high or low a rinse temperature, inadequate time between rinsing and storage: soften water, check temperatures, and allow adequate time for air drying.

Dishwashers should be cleaned at the end of each work day and descaled at least once a week. Daily cleaning consists of removing, soaking, and scrubbing scrap traps, the suction strainer, and wash and spray arms. Return to the machine and run the machine without using a load of dishes. Once a week, after disassembling and scrubbing, reassemble, add 7 fl. oz of phosphoric acid or a 2% solution of acetic acid to the wash water, and operate the machine for 1 hr; drain off the solution, add baking soda, or 2 cups of detergent, and run the machine for 15 min; drain the machine and rinse it several times.

Solid-Waste Disposal

Solid wastes containing food are unsightly, odorous, and attract insects and rodents. Unless a garbage can is in constant use in an establishment, it should have a tight-fitting lid. At night all cans should be tightly closed. The premises should be effectively screened for flies, and self-closing mechanisms should be used for all exterior and bathroom doors. Garbage grinders should only be used when the liquid waste goes to a public sewer system.

Solid waste must be stored inside or outside the premises in containers with tight-fitting lids. These containers should not absorb odors or leak water. The 10- or 20-gal cans should always have a plastic liner. Cans should be scrubbed at least every 3 days. All solid waste should be removed daily, or at the most every 3 days. Problems result from carelessness in dumping of waste, improper supervision by management, failure to empty cans at night, dirty cans, lack of a drain or clogged drain where cans are washed, failure to clean large metal storage boxes, failure to replace rusted or crushed cans and lids, and failure to use plastic bags. A complete discussion on solid-waste disposal may be found in the chapter on solid and hazardous waste in Volume II.

Insect and Rodent Control

Flies, roaches, and rodents spread disease by contaminating food. Food-product insects are a nuisance and a contaminant. Flies are present wherever there is highly organic material. They breed in floor drains or in other out-of-the-way areas. The best control for flies is removal of organic material, cleanliness, screening, and the use of safe pesticides. Roaches enter the premises through cracks; along pipes; in packages; through doors, windows, and other openings; in clothing; or through broken sewer lines. Cleanliness, sealing of roach entrances, and a good, safe poisoning program is required. Mice and rats enter establishments through holes in the walls, open doors and windows, and ruptured sewer lines. Control occurs through removal of harborage and food, poisoning, and rodent proofing. Food product insects infest dried fruit, cereal, grain, flour, rice, candy, and nuts. Control occurs through initial fumigation, rotation of stock, and destruction of highly infested foods. A detailed discussion on insects, rodents, and the use of pesticides occurs in another part of this book.

Personal Health of Employees

An employee's health, and cleanliness of person and clothing prevents the spread of many food-borne diseases. Hands are one of the greatest causes of the spread of disease. Hands must be scrupulously washed when starting work, after using the bathroom, after cleaning, and before preparing or serving food. Hands should be kept away from the head, mouth, nose, and other body parts. A 2-min hand and forearm scrub, including several soapings and rinsings, should be a routine practice for the worker. Food-service attendants should never smoke while preparing or serving food, should have clean and covered hair, and should have clean clothing changes daily. A food-service worker should not work if ill with a cold, sore throat, dripping nose, infected sores, or diarrhea. All workers with infections should be seen by a physician and a culture should be taken. All individuals with diarrhea should have a stool specimen taken. If the specimen is positive for an enteric organism causing a food-borne illness, the worker should not be permitted to return to work until three consecutive negative samples are taken.

MODES OF SURVEILLANCE AND EVALUATION

Flow Techniques and Inspectional Procedures

The department of health, or other designated departments at the state or local level, issue licenses for the operation of a food-service establishment based on laws approved by the appropriate legislative bodies. The health department formulates rules and regulations to be followed by the establishments, and conducts special and periodic evaluations on site to determine compliance and whether licenses should be issued or renewed. All new food service establishments must submit plans for approval by the health department prior to construction and be evaluated on site prior to opening.

The environment health practitioner inspects the food-service establishment using the flow technique, whereby the flow of food from the point of delivery to the point of disposal is followed (see Figure 2.2) and problems are recorded. Simultaneously, cleanliness and the condition of floors, walls, ceilings, lighting, ventilation, hand-washing facilities, and bathrooms are checked, as well as plumbing problems, water supply, sewage disposal, housekeeping procedures, and insects and rodents inside and outside of the premises. The environmentalist

Preparation Delivery	Storage	Preparation	Equipment	Serving	Cleaning and Storage	Waste Disposal
Raw foods	Refrigerator	Baking	Ovens	Milk containers	Removal of soiled utensils and dishes	Garbage disposal units
Finished products	Freezer	Frying	Stoves	Steam tables	Equipment cleaning	Garbage cans
	Dry storage	Cooking	Deepfryers	Cold tables	Dishwasher operation	Garbage rooms
		Salad making	Steam kettles	Counter freezers	Dishwashing techniques	Dempster dumpers
		Mixing	Slicers	Serving utensils	Use of detergents	Garbage can cleaning
		Grinding	Grinders	Dishware	Sanitizing	Cleaning of exterior premises
		Chopping	Toasters	Silverware	Storage of clean utensils, pots, pans, and dishware	
		Separating	Coffee urns	Single service utensils	Storage of single service articles	
		Combining	Pots and pans	Ice cream dippers		
		Food handling	Cutting boards	Ice handling		
			Meat blocks	Food handling		
			Griddles	Automatic vending equipment		
			Can openers			

Figure 2.2. Flow chart for food establishment inspection by type of operation. The environmental health worker checks construction and cleanliness of equipment, temperatures, and time, where applicable, during inspection.

should always carry a flashlight, clipboard, inspections sheets, pads, sanitizing testing equipment, light meter, and applicable rules and regulations of the health department.

If possible, the food-service manager or supervisor should accompany the environmental health practitioner on the on-site evaluation; otherwise, at the conclusion of the on-site evaluation, the environmental health practitioner should ask the food-service manager to go back through the establishment to see the major problems noted. The environmental health practitioner should complete the on-site evaluation form from his or her notes, and determine the length of time needed to correct each problem. If conditions are bad, another evaluation should be made within 1 week. If the health of the public is potentially endangered, conditions should be corrected at once. The environmental health practitioner should teach the food-service manager to set up an adequate cleaning schedule and should make recommendations on the carrying out of necessary corrections. It is important to discuss the public health reasons for compliance of the establishment with certain rules and regulations. Additional on-site evaluations are made as needed. When the food-service establishment will not comply, further action such as warning letters, administrative hearings, license suspension, license revocation, and court hearings are necessary. (See Figure 2.3 for an example of an inspection sheet.)

Food-Borne Disease Investigations

When the health department is notified that a possible food-borne disease outbreak has occurred, it must move quickly to obtain the necessary information from persons who ingested the suspected food. Surveys of the food preparation, storage, and serving operations must be made, and samples taken and processed rapidly. Unfortunately, many outbreaks of food-borne disease occur at large gatherings, such as church dinners, reunions, and picnics. This type of food operation complicates the investigation procedures.

Epidemiological Study Techniques

Environmental health personnel and epidemiologists should complete detailed questionnaires for all persons who may have eaten the food or drinks suspected of having caused the food-borne disease. Once a pattern develops concerning the kinds of foods involved and the places in which the food was consumed, additional environmentalists should be detailed to make complete studies of all food operations. These studies must include a food history. The food, its source, method of preparation, storage, and refrigeration must be determined. As part of this study, remnants of suspected foods should be collected aseptically in sterile containers or in their original containers and sent, refrigerated, immediately to the laboratory for analysis. It is necessary to properly identify the origin of the sample, the date and time of collection, the name of the environmentalist, and a brief

FOOD SERVICE
INSPECTION REPORT _____
                    HEALTH JURISDICTION

KEY: S = Satisfactory
      U = Unsatisfactory
      R = Repeat Violation
      N = Not Apply

Name of Establishment	Owner or Operator's Name	Address

Person Interviewed &	Date	Time	am	Current	Yes ( )
Title			pm	License	No ( )

Rating   *Rating*

Facilities

1. Floors—clean, good repair
2. Walls—clean, painted, good repair
3. Ceiling—clean, painted, good repair
4. Lighting—fixtures clean, good repair, adequate
5. Ventilation—adequate, proper design
6. Filters—duct work clean, adequate
7. Bathrooms—convenient, self-closing doors, clean, free from odors, adequate supplies
8. Handwashing Facilities—adequate, good repair, hot & cold running water, convenient, clean soap, single use towels or dryers
9. Plumbing—sewage & water supply, properly constructed, no cross-connections
10. Hot and cold running water
11. Screened doors & windows or other fly proofing devices

Food Protection

12. Ice—approved source
13. Approved source
14. Wholesomeness
15. Refrigeration of perishables below 45°F; Freezing below 0°F
16. Hazardous cooked foods above 140 °F
17. Storage of foods
18. Protection of food from contamination
19. Preparation of food
20. Serving of food

Equipment and Utensils

21. Food Service Equipment—clean & good repair
22. Materials for cleaning & sanitizing
23. Equipment for cleaning—dishwashers, sinks, etc.
24. Cleaning Operation—temperatures, procedures, etc.
25. Sanitization—hot water or chemical
26. Cleanliness—equipment
27. Cleanliness—utensils, dishes, etc.
28. Storage of utensils, silverware, dishes, pots, pans, etc.
29. Construction of equipment, utensils, etc. (including thermometers)
30. Cleaning procedures for food contact surfaces
31. Cleanliness of food contact surfaces

Personal Health

32. Employee free from boils, diarrhea, etc.
33. Cleanliness of hands, garments, etc.

Solid Waste

34. Garbage storage & disposal
35. Rubbish storage & disposal
36. Outer premises

Insects and Rodents

37. Flies, roaches, rats, mice

Sampling Type

38. Sample number

Item	Recommendations	Date of Compliance

Figure 2.3.   Inspection sheet.

statement about the symptoms of the patients and the suspected organisms or chemicals. Water samples should also be taken. All sewage systems and plumbing should be evaluated. The presence of insects and rodents must be determined. It is essential as part of this study to determine if any food handlers were ill within a period of 6–8 weeks prior to the onset of the disease. If an individual food handler had diarrhea or vomiting, a fecal sample should be taken. Food handlers should also be checked for any boils, carbuncles, and respiratory infections that could cause an outbreak of staphylococcus food poisoning.

Medical diagnosis of patients combined with good epidemiological data may determine the causative agent in the outbreak and assist the environmentalist in preventing further outbreaks of this type. If commercially prepared food is the cause of the food-borne disease, it is necessary to obtain data identifying the source of the food, when and where it was purchased, and any identifying codes or marks on the containers. This information should be immediately forwarded to the Food and Drug Administration. If the outbreak of disease is traced to a hospital or college cafeteria, intensive inspection, coupled with changes in procedures and good food-handler training, will help to prevent further outbreaks. If the outbreak of disease is traced to a church dinner or picnic, the environmentalist should provide information and suggest the presentation of a food-handling course to the group involved.

To gather data properly, an Individual Food-Borne Outbreak Report, a summary of Food-Borne Disease Reports, a Study Form for Food Preparation Establishments, and a History of Food Handlers Report should be used. These forms are obtained from local or state health departments (Figures 2.4 and 2.5).

HAZARD ANALYSIS/CRITICAL CONTROL POINT INSPECTION

Hazard analysis/critical control point inspection (HACCP) is an in-depth inspection process used to resolve or prevent disease outbreaks by identifying the foods at greatest risk and the critical control points. These foods may be critical because they are naturally contaminated foods, large volume preparation, multistep preparation, and temperature changes in the foods. The critical control points are those few steps in a process that are most important to bacterial contamination, survival, or growth. They also are an operation or part of an operation where actual or potential risks are usually found and where preventive or control measures can be exercised that eliminate, prevent, or minimize a hazard that has occurred prior to this point. The HACCP analysis helps determine the critical control points to prevent disease. The establishments are chosen for the program by use of a system similar to triage in mass casualties. The program may be very valuable in certain instances. Contact the Food and Drug Administration for further information.

FOOD PROTECTION CONTROLS

Until recently, the major controls used for food protection were control of temperature and time, and the proper preparation of food in a clean manner. Although these measures are still essential, there are new concerns due to the influx of chemicals into our food supply. Since specific controls do not exist at present, they are mentioned in a succeeding chapter under research needs.

Streptococcus infection control measures include pasteurization of milk and other dairy products, exclusion of persons with known streptococcal infections, and antibiotic treatment of known carriers or contacts.

1. Where did the outbreak occur?	2. Date of outbreak: (Date of onset 1st case)
State _____ (1,2) City or Town _____ County _____	_____ (3-8)

3. Indicate actual (a) or estimated (e) numbers	4. History of Exposed Persons	5. Incubation period (hours):
Persons exposed _____ (9-11)	No. histories obtained _____ (18-20)	Shortest ____ (40-42) Longest ____ (43-45)
Persons ill _____ (12-14)	No. persons with symptoms _____ (21-23)	Approx. for majority _____ (46-48)
Hospitalized _____ (15-16)	Nausea ____ (24-26) Diarrhea ____ (33-35)	6. Duration of Illness (hours)
Fatal cases _____ (17)	Vomiting ____ (27-29) Fever ____ (36-38)	Shortest ____ (49-51) Longest ____ (52-54)
	Cramps ____ (30-32) Other, specify ____ (39)	Approx. for majority _____ (55-57)

7. Food-specific attack rates: (58)

Food Items Served	Number of persons who ATE specified food				Number who did NOT eat specified food			
	III	Not Ill	Total	Percent III	III	Not Ill	Total	Percent III
_____								
_____								
_____								
_____								
_____								
_____								
_____								
_____								
_____								
_____								

8. Vehicle responsible (food item incriminated by epidemiological evidence): (59,60)

9. Manner in which incriminated food was marketed: (Check all applicable)	10. Place of Preparation of Contaminated Item: (65)	11. Place where eaten: (66)
(a) Food Industry (61) (c) Not wrapped .... □ 1 (63)	Restaurant .... □ 1	Restaurant ...... □ 1
Raw .... □ 1 Ordinary Wrapping ..... □ 2	Delicatessen .... □ 2	Delicatessen ..... □ 2
Processed .... □ 2 Canned .... □ 3	Cafeteria .... □ 3	Cafeteria .... □ 3
Home Produced Canned–Vacuum Sealed .. □ 4	Private Home .... □ 4	Private Home .... □ 4
Raw .... □ 3 Other (specify) .... □ 5	Caterer .... □ 5	Picnic .... □ 5
Processed ...... □ 4	Institution:	Institution:
(b) Vending Machine ... □ 1 (62) (d) Room Temperature ..... □ 1 (64)	School .... □ 6	School .... □ 6
Refrigerated .... □ 2	Church .... □ 7	Church .... □ 7
Frozen .... □ 3	Camp .... □ 8	Camp .... □ 8
Heated .... □ 4	Other, specify ...... □ 9	Other, specify .... □ 9
If a commercial product, indicate brand name and lot number		

DEPARTMENT OF HEALTH, EDUCATION, AND WELFARE
PUBLIC HEALTH SERVICE
CENTER FOR DISEASE CONTROL
BUREAU OF EPIDEMIOLOGY
ATLANTA, GEORGIA 30333

Figure 2.4. Investigation of a food-borne outbreak.

Salmonella controls include pasteurization of milk and other dairy products, control and certification of shellfish areas, water treatment and chlorination, elimination of flies from human feces, and prohibition of food handling by individuals with a history of typhoid or paratyphoid fever.

Salmonellosis control measures include the use of good personal hygiene, removal of sick food handlers from duty, obtaining cultures from food handlers with diarrhea infections, and restricting them from work until three consecutive fecal samples are negative; storage of susceptible foods at temperatures below 40°F; avoidance of the danger zone for salmonella growth, which occurs between 60°F and 120°F (serve all hot foods above 140°F but precontrol to 165°F); never stuff chicken or turkey the night before cooking; preparing small quantities of such things as egg salad, potato salad, chicken salad, or turkey salad; analyzing

12. Food specimens examined: (67)

Specify by "X" whether food examined was original (eaten at time of outbreak) or check-up (prepared in similar manner but not involved in outbreak)

Item	Orig.	Check up	Findings Qualitative	Quantitative
Example: beef	X		C. perfringens, Hobbs type 10	2×10^6/gm

13. Environmental specimens examined: (68)

Item	Findings
Example: meat grinder	C. perfringens, Hobbs Type 10

14. Specimens from patients examined (stool, vomitus, etc.): (69)

Item	No. Persons	Findings
Example: stool	11	C. perfringens, Hobbs Type 10

15. Specimens from food handlers (stool, lesions, etc.): (70)

Item	Findings
Example: lesion	C. perfringens, Hobbs type 10

16. Factors contributing to outbreak (check all applicable):

	Yes	No
1. Improper storage or holding temperature	☐ 1	☐ 2 (71)
2. Inadequate cooking	☐ 1	☐ 2 (72)
3. Contaminated equipment or working surfaces	☐ 1	☐ 2 (73)
4. Food obtained from unsafe source	☐ 1	☐ 2 (74)
5. Poor personal hygiene of food handler	☐ 1	☐ 2 (75)
6. Other, specify	☐ 1	☐ 2 (76)

17. Etiology: (77, 78)

Pathogen _____
Chemical _____
Other _____

Suspected .... ☐ 1 (79)
Confirmed .... ☐ 2
Unknown .... ☐ 3

18. Remarks: Briefly describe aspects of the investigation not covered above, such as unusual age or sex distribution; unusual circumstances leading to contamination of food, water; epidemic curve; etc. (Attach additional page if necessary)

Name of reporting agency: (80)

Investigating official: | Date of investigation:

NOTE: Epidemic and Laboratory Assistance for the investigation of a foodborne outbreak is available upon request by the State Health Department to the Center for Disease Control, Atlanta, Georgia 30333.

To improve national surveillance, please send a copy of this report to:
Center for Disease Control
Attn: Enteric Diseases Section, Bacterial Diseases Branch
Bureau of Epidemiology
Atlanta, Georgia 30333
Submitted copies should include as much information as possible, but the completion of every item is not required.

Figure 2.5.   Laboratory findings including negative results.

feed for salmonella contamination and maintaining disease-free animals; reporting immediately all suspected food infections to public health authorities and retaining samples of food; and making complete epidemiological studies.

Shigellosis control measures include strict personal hygiene, elimination of carriers as food handlers, refrigeration of moist foods, cooking of foods at 165°F prior to serving, and elimination of flies.

Cholera control measures include filtration and chlorination of drinking water supply, growing of shellfish in certified sewage-free areas only, removal of known cases from food-handling operations, and vaccination when an outbreak of cholera occurs.

Hepatitis can be prevented by observing the same controls as stated under cholera.

Brucellosis, diphtheria, and bovine tuberculosis control measures include pasteurization of milk and dairy products and isolation of carriers. In addition, for brucellosis and bovine tuberculosis, sick animals should be eliminated.

Tularemia control measures include the use of protective gloves when handling wild animals and the elimination of arthropods.

Staphylococcus food poisoning control measures include removing food handlers who have nasal discharges or skin infections, cooking foods thoroughly, refrigerating foods immediately, using custard pastries within 2–4 hr after preparation and discarding the remainder, and washing hands thoroughly and frequently.

Clostridium welchii or *C. perfringens* control measures include the elimination of known carriers from food handling, cooking meat thoroughly before consumption, cooling meat rapidly in the refrigerator immediately after cooking, using drippings from meat for gravies only on the day of cooking, discarding all leftovers, and preparing chicken, turkey, or beef pot pies only from freshly cooked pieces of meat.

Botulism can be prevented by proper cooking of foods. If home-processed foods are used, they should be boiled for a minimum of 15 min before eating. Discard any canned or bottled foods that show signs of spoilage including off odors, off tastes, gas or foams, and off colors. If botulism is suspected, have a physician administer the proper antitoxin immediately.

Vibrio parahaemolyticus can be killed by thorough cooking.

Where substantial mold damage appears, food should not be consumed because of the potential for mycotoxins.

Amoebic dysentery control measures include protection of water supplies from human excreta, elimination of carriers from food preparation, proper sewage disposal, and filtration and chlorination of water supplies.

Beef and pork tapeworm control measures include proper inspections by trained veterinarians and thorough cooking of beef and pork. For fish tapeworms, cook fish thoroughly and avoid all raw smoked fish.

Trichinosis control measures include cooking garbage fed to pigs to an internal temperature of 137°F, eliminating rats from pig farms, and cooking all pork to an internal temperature of 165°F.

Chemical poisoning control measures include eliminating utensils and containers that may leach the chemical into solution and discontinuing the use of dangerous pesticides; protecting all food and food contact surfaces when using pesticides; and storing all chemicals in their original containers away from food storage, preparation, and serving areas.

Intoxication from poisonous plants or animals can be prevented by not eating these plants or animals.

Food-Service Training Programs

The key to a safe, sanitary food program is food-service supervisors and personnel properly trained in good techniques of purchasing, storing, preparing, transporting, and serving wholesome food. Scheduling of cleaning operations, cleaning and sanitizing techniques, and proper use of equipment and materials must be understood and the necessary materials must be made available. Although all food-service personnel are important, the supervisor is the key. The supervisor schedules operations, teaches personnel, orders supplies, and inspects the establishment daily. Assistance is available from local and state health departments, who are happy to conduct necessary training programs. The supervisor should attend special training sessions held by the health department, and should be tested and certified on an annual basis by the health department. Renewal of the establishment's license should be contingent on this certification.

An educational program directed at homemakers and individuals who prepare food for mass feeding operations, such as church dinners, is useful in controlling the numerous outbreaks of disease that may occur in the home or at special functions.

SUMMARY

Food protection is an essential part of the environmental health program, considering the vast variety of food produced, the enormous potential for disease, and the short- and long-range problems of chemicals entering our food supply. Food-borne disease outbreaks probably account for millions of illnesses each year. The various long-range problems related to food-borne disease cannot even be estimated.

Food protection consists of proper plan review, the development and use of adequate facilities in a correct manner; the procurement and use of correct equipment; necessary cleanliness; proper storage, preparation, and serving of food; and adequate disposal of food remnants. The cleaning and sanitizing techniques utilized are an essential facet of food protection. Important factors are also the health and potential for spreading disease of humans. A variety of inspectional techniques are utilized on a regular and special basis, depending on the nature of the food establishment and the types of problems likely to occur. Proper techniques and food-borne disease investigations eliminate outbreaks and prevent future outbreaks of disease. All states and many countries and localities have food protection programs.

RESEARCH NEEDS

To avoid undo repetition, research needs will be discussed in the chapter on food technology.

Chapter 3

FOOD TECHNOLOGY

BACKGROUND AND STATUS

Food is a perishable product consisting of proteins, carbohydrates, fats, vitamins, minerals, water, and fibers. It is used by an organism in sustaining growth, repairing tissues, maintaining vital processes, and furnishing energy. Great pleasure is derived from the proper preparation of food and from its consumption. Food is altered favorably or unfavorably by microorganisms, enzymes, insects, and environmental changes. Microbes help produce sauerkraut, bread, and cheese; ripen olives; and ferment milk. The action of the yeast *Saccharomyces cerevisiae* produces carbon dioxide in bread, making it porous and causing it to rise. The starches and proteins in bread are split by the action of the yeast, becoming more digestible, and characteristic flavors and aromas are produced. However, if certain spore-forming bacilli, principally *Bacillus mesentericus*, are introduced with the flour or yeast, discoloration and unpleasant odor results and the bread becomes unappetizing.

In this chapter, contamination, spoilage, and disease potential of certain types of foods will be discussed, including milk and milk products, poultry, eggs, meats and meat products, and fish and shellfish. The nature of enzymes, microbiological spoilage, factors affecting spoilage, techniques of preservation of foods, chemical preservatives, additives, pesticides, fertilizers, and antibiotics will also be discussed. The production, processing, testing, and environmental problems associated with each food group will be addressed. Applied on-site evaluation techniques and inspection forms will be introduced as needed. The foods were selected as the most frequent contributors to food-borne diseases, intoxication, or the kinds of food spoilage that the environmental health practitioner is likely to encounter.

During the early 1970s, famine with drought and starvation occurred in the Sahara Desert of Africa and in the Indian subcontinent. The United Nations estimated that at least 460 million people, excluding the communist countries of Asia, were suffering from serious food deficiency in 1970. It is estimated that 10–20 million young children are severely afflicted with the protein calorie malnutrition diseases of kwashiorkor and marasmus. Most of these children die without treatment. Another 200 million suffer from all other forms of malnutrition. The food crisis resulted from a decline in world food production and an increase in population. In the developing countries, the annual population growth averages nearly 2.5%. At this level the population doubles every 28 years. If this trend in population growth continues, there should be an estimated 1.2 billion people in India by the 21st century. In the 1980s starvation is occurring in many Third World countries. The situation is not anticipated to be improved in this century.

To combat hunger and death, and to improve the economy of many countries, more land is being cultivated and more environmental resources and man-made chemicals are being utilized. Water, which is a prime resource for increasing food output, is plentiful in some areas and essentially lacking in others. The changing of water patterns to increase productivity have also brought disease. In Egypt, for example, the Aswan Dam stimulated a large outbreak of schistosomiasis. Fertilizers when used properly are excellent for increasing crop yield, but when improperly used they become contaminants and potential cancer-causing substances.

Plant disease, insects, pests, and weeds caused an estimated 30% loss in potential food production throughout the world, resulting in a sharply increased use of pesticides. Unfortunately, some pesticides are inherently dangerous or are dangerous when misused. In the Canete Valley of Peru, the cotton crop was heavily treated with DDT and other hydrocarbons. The sprays killed not only the pests, but also their predators. The pests eventually developed resistance to DDT, whereas the predators did not, resulting in lost cotton crops. Numerous times pesticides contaminated raw food products and were concentrated as they were processed through the food or became a part of the runoff problem, resulting in varying levels of water pollution.

Additives are added to food to enhance flavor, taste, and keeping quality. Unfortunately, unintentional or intentional additives may be the cause of cancer or other related diseases. There is serious concern over the use of sodium in processed foods, since the sodium may produce adverse effects in hypertensive individuals. Other contaminants found in raw food before processing and remaining in the food until consumption include mercury, lead, cadmium, zinc, copper, manganese, selenium, and arsenic. In addition, a variety of pesticides, such as chlorinated hydrocarbons, polychlorinated biphenyls, and a vast group of chemicals are entering our food supply.

Animal feed may induce a toxic response in humans or animals. Unfortunately, there is a lack of research in this area, although scientists know the toxicants are found in animal tissue. Milk and dairy products are of particular importance, not only to the young but also to vegetarians. Milk and dairy products contain estrogens, nitrates, antibiotics, pesticides, radionuclides, and mycotoxins. Eggs may cause allergic disorders in children and adults due to a natural allergic response or to estrogens and antibiotics found in egg yolks. Aflatoxins were found in Iowa corn in 1990.

There has been a rapid increase in the use of frozen prepared foods and vacuum-packed foods. Improper processing of these foods has resulted in outbreaks of botulism. The situation in food technology can only worsen in the coming years, unless extreme methods are taken to control the many unwanted substances found in raw food products and unless research is conducted in the many unknown areas of contamination. Our understanding of acute responses of food poisoning is relatively good; our understanding of chronic responses, after years of consuming various types of products, is still very poor.

SCIENTIFIC AND TECHNOLOGICAL BACKGROUND

Chemistry of Foods

Enzymes produce desirable and undesirable changes in foods. As an example, enzymes may tenderize meat or cause off colors, off odors, and off tastes. An enzyme is an organic catalyst that speeds up a chemical reaction without being altered itself. It has the following characteristics: it acts most rapidly at body temperature, it has a high degree of specificity for a given test, it is produced by living cells including microorganisms, it continues to react after harvest of plants or slaughter of animals, it is retarded by low temperatures, it is destroyed by boiling, and it may react either within the cell (endozyme) in which it was produced or outside of the cell (exozyme) in the tissue.

The inactivation of the enzyme phosphomonoesterase by heat is a reliable means of measuring the efficiency of pasteurization of milk and ice cream in the phosphatase test. The enzyme is totally destroyed by proper pasteurization. However, false positives can occur due to certain bacteria present in milk, including streptococcus and aerobacter.

Food Additives and Preservatives

A food additive is any substance or mixture of substances other than basic foodstuffs added during the production, processing, storage, or packaging of foods. Additives are used to preserve, emulsify, flavor, color, and increase nutritive value. Intentional food additives include preservatives, antioxidants, sequestrants, surfactants, stabilizers, bleaching and maturing agents, buffers, acids, alkalines, colors, special sweeteners, nutrient supplements, flavoring compounds, and natural flavoring material. Unintentional food additives or contaminants include radionuclides, insect parts and excreta, insecticides and herbicides, fertilizer residue, other chemicals, dirt, microorganisms, and any other unintentional materials.

Antioxidants preserve freshness in meats by preventing rancidity of fats and preserve appearance and taste of fruits by preventing discoloration. Mold inhibitors, such as calcium or sodium propionate, keep bread fresh for longer periods. Emulsifiers, such as gum arabic, are used to maintain the consistency of French dressing. Stabilizers and thickeners, such as lecithin, are used in ice cream. Monosodium glutamate, nonnutritive sweeteners, and other agents enhance the flavor of food. Additives that add nutritive value to foods include iodized salt to prevent goiter; vitamin D in milk to prevent rickets; thiamine, niacin, riboflavin, or iron in bread and cereals to provide individuals with sources of these essential vitamins and minerals; and vitamin A in margarine to help prevent malnutrition. The major restrictions on additives are that they do not cause short-term or chronic poisoning or cancer, do not make decomposed food appear fresh, or do not cause any other unfavorable side reactions. For example, in curing meats, nitrites must not exceed 120 ppm combined with an activator of ascorbates or erythorbates at 550 ppm.

In addition to the nitrates or nitrites that are purposely added to food, considerable amounts of these substances are introduced into the food chain through the use of nitrogen fertilizers. Certain vegetables accumulate nitrogen compounds based on the amount of nitrate and molybdenum in the soil, the light intensity, and existing drought conditions. Further high concentrations exist from the accidental introduction of large amounts of nitrates or nitrites in the foods. When tests are conducted to meet the nitrate-nitrite standards, they should not only include levels added, but levels that already exist. Nitrates may also be introduced by groundwater usage.

Sodium chloride is added to diets in varying degrees in individuals. There is a suspected relationship between sodium chloride and human hypertension. The amount of salt added as an additive to food may well be a serious problem to the individual who has potential or existing hypertension.

Phosphates are introduced into food through the processing of poultry, in the soft-drink industry, and in the production of modified starches. Phosphates may also be introduced through the use of various fertilizers. Excessive daily intake of phosphorous causes a premature cessation of bone growth in children, subsequently affecting final adult height.

Additional knowledge is needed concerning the effects of the intentional addition of food additives to our processed foods. Cyclamates have been removed from the market. It is difficult to predict which additives previously thought to be safe will be evaluated and found potentially hazardous to humans. Even the widely used saccharin is now under suspicion. The Food and Drug Administration puts out a Food Additive Status List, which states the status of food additives that may be harmful to people.

Nitrates, Nitrites, and Nitrosamines

Nitrates and nitrites are used to preserve meats, including bacon, ham, hot dogs, pastrami, smoked fish, cured poultry, etc. Nitrite curing is more rapid than nitrate curing and, therefore, is used more often. Nitrites are added to the cured meat as a preservative. They are also effective microbial inhibitors. They retard the development of rancidity, enhance the flavor of certain meats, and maintain the bright red color of meats. The nitrite in the cured meat does not actually add color, but fixes the color pigment, myoglobin. Nitrites are not themselves carcinogens but rather are precursors. Nitrosamines are produced from nitrites and amines, which may be found in various places, including the decomposition of proteins. Amines are found in food, drugs, and pesticides. Nitrosation is the reaction between nitrous acid and secondary or tertiary amines. Nitrous acid comes from the nitrite added to the meat, and amines come from the meat protein. Over a hundred nitrosamines have been isolated and identified, with approximately 75% of these being carcinogens. Nitrosamines can cause cancer in animals after a single dose. Nitrosamines may also be found in imported and domestic beer.

Sulfites Used in Food

Sulfites have been used as preservatives in wine and other beverages, as well as in foods. Sulfiting agents retard the oxidation process of uncooked food, including vegetables, and extend the time of visual appeal. Fruits, green vegetables, especially lettuce, peeled or cut potatoes and apples, shrimp, seafood, and seafood salads are commonly sprayed or dipped into sulfiting agents. A sulfiting agent refers to sulfur dioxide, and several forms of inorganic sulfite that liberate sulfur dioxide under various conditions of use. Currently sulfur dioxide, potassium and sodium metabisulfite, potassium and sodium bisulfite, and sodium sulfite are considered to be safe for use in foods by the FDA. Sulfiting agents are added to foods because they control enzymatic and nonenzymatic browning, are antimicrobial, and act as an antioxidant, as well as a reducing agent. Sulfiting agents are used in a wide variety of food products.

Unfortunately it has been discovered that some individuals, especially asthmatics, may have a potentially severe adverse reaction to sulfites. The sulfites may induce an attack of asthma. Further, in some rare instances other types of hypersensitivity have occurred. Therefore, it is necessary that where sulfiting agents are permitted to be used, especially on salad bars, individuals be advised of this by the placement of a placard at the salad table.

Aspartame

Aspartame is marketed either as Equal® or Nutrasweet®. Aspartame is low in calories, has no odor or aftertaste, but because of its sweetness acts as a sugar substitute, and therefore, is useful in the diet of diabetics, heart patients, overweight patients, etc. There are some questions concerning the potential health effects of aspartame. Aspartame degrades into the diketopiperazine at certain temperatures. Diketopiperazine is suspected of causing cancer in rats. The metabolism of aspartame, which is a peptide, proceeds in the same way as the metabolism of proteins. After metabolism, aspartame components become aspartic acid, phenylalanine, and methanol.

In some women of child-bearing age, the natural blood levels of phenylalanine fluctuate wildly. Therefore, small doses of aspartame might contribute to much higher levels of the phenylalanine of the individual. Some researchers have indicated that certain people had difficulty in metabolizing phenylalanine. High levels of phenylalanine in mothers apparently cause mental retardation in their babies. At lower levels, phenylalanine could affect the development of the fetus brain, thereby potentially decreasing IQ. Because aspartame is so useful in the diet of so many people, it is essential that further research be done in the area of the potential health effects of aspartame on people. If these health effects are proven to occur, certain restrictions or advisories should be printed on the aspartame containers.

F D and C Dyes

F D and C dyes in themselves are not carcinogenic, however, when the dyes are metabolized there is a possibility, depending on which dye is utilized, that a carcinogenic substance may be produced. As a result of this, at least 13 F D and C dyes have been banned from food use in the United States. The gastrointestinal tract is an important metabolic organ. It contains mostly anaerobic bacteria, which aid in metabolizing F D and C dyes with the possible production of azo-reduction, which is the most important metabolic reaction of azo dyes that can contribute to carcinogenesis.

Anabolic Hormones

Anabolic hormones are used in cattle to increase weight. There is a potentially serious concern about the adverse affects to children who have been exposed to these hormones. In Puerto Rico, a study was done on young children that indicated premature development of the breasts, high incidence of ovarian cysts, very early development of vaginal bleeding, and uterus enlargement due to exposure to anabolic hormones.

PROBLEMS OF FOOD TECHNOLOGY

Contamination

Animal products contain microorganisms that are part of their normal flora, are present because of a disease process, or have been added at any point during the killing, processing, storage, preparation, and serving of the product. The organisms may be added by humans or from the environment. About 8% of fresh eggs contain microorganisms; dirty eggs are covered with organisms that easily penetrate the shell. The organism may be salmonella, shigella, and *Cholera vibrio.* Meat contains pseudomonas, proteus, *E. coli,* clostridium, and many other organisms. Although milk of a cow is sterile in a healthy udder, it frequently becomes contaminated by streptococci normally present in the milk ducts. Diseased cattle may be infected with pathogenic staphylococcus, streptococcus, tuberculosis, or brucellosis organisms.

Plant products, such as lettuce, cabbage, carrots, fruits, and vegetables, are contaminated with microorganisms from the air and soil, and by workers handling food.

Those bacteria causing food-borne disease or food-borne intoxication are of particular concern. The spores of *Clostridium botulinum* require temperatures well above the boiling point for destruction. Although intoxication is infrequent, it is of such a serious nature that particular care must be taken in preparing high-risk foods. This is especially true in home canning. Some molds will produce toxic materials known as mycotoxins. The best known of these are the aflatoxins, which have been found on peanuts, rye, wheat, millet, jellies, and jams.

Food Spoilage

The composition of food is important. Proteins are especially susceptible to spoilage by spore-forming gram-negative rods, such as pseudonomas and proteus, and by molds. Carbohydrates are particularly affected by yeasts, molds, streptococcus, and micrococcus. Fats undergo hydrolytic decomposition and become rancid. The pH of acidic foods, such as fruits, is low enough to prevent most bacterial spoilage; however, yeasts and molds grow well on these products. Nonacidic foods are subject to bacterial spoilage. Foods with a moisture content exceeding 10% will support the growth of microorganisms. Sugar and salt concentrations create an osmotic pressure that disrupts organisms: 5–15% salt will inhibit bacteria: 65–70% sugar inhibits molds. The presence or absence of oxygen determines the kinds of microorganisms that may or may not grow.

Food spoilage is due to two principal causes: chemical, including the enzymes just discussed, and biological, caused by bacteria, molds, and yeasts. Spoilage in most fruits is caused by molds and yeasts. Damage to the surface of the fruit increases the rate of deterioration. Spoilage of vegetables is due to bacteria and molds. Spoilage of properly refrigerated meat is usually confined to the surface. This is the reason why in the "aging" of beef, the surface is cut away and the interior portion, which is now more tender, is used. However, putrefactive decomposition occurs rapidly in ground meat and ground fish, because the bacteria are distributed throughout the food mass. Fish fillets become slimy and proteolysis occurs within several days because of the enormous number of microorganisms present initially and the handling techniques used. Shucked shellfish contains large numbers of bacteria that can lead to putrefactive decomposition if the shellfish are not refrigerated immediately. Milk, since it is drawn from the cow at temperatures favorable to rapid bacterial growth, must be immediately cooled to prevent bacterial decomposition.

Canning Spoilage

Underprocessing of canned foods results in microbial spoilage in the presence of high heat-resistant spores. This spoilage by microorganisms is a potential problem, since the raw food products and the processed foods may be contaminated by air, water, hands, equipment, and added ingredients. Flat, sour spoilage is characterized by the production of acid without the production of gas. A slightly disagreeable odor and change of color may be present in the canned food. A swollen can of food may be caused by gas-forming thermophilic organisms. Swollen cans of meat are most likely caused by organisms producing putrefaction. Anaerobic bacterial growth in canned foods is of particular concern, since *Clostridium botulinum* may be causing the problem, rather than other members of the anaerobic family. Some characteristics of anaerobic activity are offensive odors, black sediment or residue, and reduction of oxygen. Spoilage in cans may also be due to a reaction between the food and the metal in the can, producing hydrogen gas; overfilling the can at too low a temperature; freezing the liquid portion of the food; inadequate removal of oxygen from the can before sealing; or sulfides present in the foods.

Sources of Exposure

Extraneous Materials. Insect parts and rodent hairs are unwanted components of raw agricultural materials and processed food. Although large amounts of these materials are unacceptable, little is known about the relationship of the quantity of material to microbiological contamination. More research is needed in this area.

Pesticides and Fertilizer. Raw food products are contaminated by pesticides, herbicides, and fertilizers. After washing, a residue of the chemical may be present. The Pesticide Chemical Act of 1954, as updated by The Federal Insecticide, Fungicide, and Rodenticide Act of 1988, prohibits interstate shipping of raw agricultural foodstuffs containing a residue of a pesticide unless it is safe or if the residue is within the tolerance level established by the Food and Drug Administration as safe. A further discussion on pesticides appears later in this book.

Radioactive Fallout. Radioactive material mixed with Earth, soil, and rock are spread over a large area when a nuclear weapon is exploded or when a nuclear accident occurs. The fallout causes harm to people, animals, crops, food, water, structures, and fields. Radioactive strontium and iodine are most dangerous. Radioactive strontium is chemically similar to calcium; it enters the bones and may cause cancer. It has a half-life of 28 years. Radioactive iodine, which has a half-life of 8 days, may cause cancer of the thyroid gland. The major danger to people is that strontium-90 may be taken in with grass by cows and then passed through milk to humans. Milk sheds are sampled regularly for levels of strontium-90.

Antibiotics in Food. Such antibiotics as penicillin, streptomycin, bacitracin, Chloromycetin, Terramycin, and Aureomycin are used in the preservation of fish and meat, primarily to retard spoilage. Antibiotics and sulfa drugs are commonly used in treatment of bovine mastitis in cows. Antibiotics have also been used as food additives in uncooked ground beef and pork products to preserve keeping quality. Until recently, the Food and Drug Administration permitted the addition of chlorotetracycline and oxytetracycline to poultry water to extend shelf life. The major question for consumers is whether the antibiotics are toxic if ingested over extended periods of time. Another serious question is whether substantial usage of antibiotics in foods causes the development of resistant organisms that may be harmful to humans. Other concerns include the potential for human allergic reactions to the antibiotics.

Heavy Metals and the Food Chain. The heavy metals are considered to be atimony, arsenic, cadmium, chromium, lead, mercury, nickel, silver, thallium, and uranium. All individuals are exposed to heavy metals from a variety of sources, including natural sources, land pollution sources, water pollution sources, and aerial sources. The natural sources are due to chemical weathering and volcanic activities. In freshwater systems, chemical weathering of igneous and metamorphic rocks, as well as soil, causes these trace metals to enter surface waters. The decomposition of plants and animals also adds trace metals to the surface waters. Dust from volcanic activities, smoke from forest fires, aerosols, and particulates on the surface of oceans add additional amounts of trace metals.

Mining operations contribute higher levels of heavy metals, because the material brought from below the surface of the ground is now exposed to the air. Additional heavy metals are added by the corrosion of water pipes and the manufacturing of consumer products. Wastewater treatment by means of the activated sludge process usually removes less than 50% of the heavy metals present in the liquid. Sewage sludge, when placed on top of the ground or sprayed on the ground, becomes a source of cadmium, lead, mercury, and iron. Industrial sludge also adds a significant amount of heavy metals. Storm water runoff from urban areas contribute lead, cadmium, chromium, and zinc. The salt used on streets may increase the mobilization of metal ions, such as cadmium, mercury, and lead. Leachate from sanitary landfills can increase the levels of lead and mercury. Agricultural runoff, which includes soil erosion as well as animal and plant residues, fertilizers, herbicides, fungicides, and other pesticides, adds to the heavy metal problem.

Aerial sources of heavy metals come from the smelting of metallic ores. Other aerial sources come from fossil fuels and the addition of metals to fuels as anti-knock agents. These enter the body via these routes:

1. through respiratory surfaces
2. through adsorption from water onto body surfaces
3. from ingested food particles or water through the digestive system

There are two main routes of entry into plants:

1. through the plants surface above the ground
2. through the root system

Terrestrial animals take up heavy metals through food and water. Further, a significant amount of absorption may occur through the lungs in some situations. Heavy metals can be bioconcentrated, as there is a progression upward through the food chain, and therefore, there may be a biomagnification of the heavy metals in larger animals.

Environmental Problems in Milk Processing

Environmental problems on dairy farms vary with the size of the operation and the age of the structures and equipment. The older, smaller operations may have problems with poor ventilation, poor lighting, and improper construction of the physical facility. Common to all farms are the increased problems of waste handling, flies, excessive use of pesticides, improper storage of pesticides, inadequate and improper cleaning of pipelines, milking equipment, and bulk storage tanks. A safe water supply is essential.

The basic problems in milk plants include cleaning of equipment and the constant use of properly operated, accurate instruments.

Environmental Problems in Poultry Processing

Basic problems in poultry plants include cleanliness of walls, floors, drains, equipment, conveyor belts, and interior and exterior surroundings. Blood is frequently splattered and mixed with dirt and feathers. This becomes very difficult to remove without constant, well-organized, efficient cleaning procedures. The amount of solid waste produced causes a serious removal problem. Sewage systems are frequently overloaded and the BOD (biological oxygen demand) is enormous. Rats frequently invade chicken-slaughtering plants because of the availability of food, blood, and other waste materials. Flies and roaches may become a serious problem. Water used in the chilling process is readily contaminated by birds, thereby continuing the cycle of contamination and bacterial growth. Improper supervision of cleaning, inadequate cleaning schedules, and a poor understanding of cleaning techniques increase problems. Quaternary ammonium chloride compounds cannot be used as sanitizers in poultry drinking water.

Environmental Problems in Egg Processing

Environmental problems include all those previously related to poultry. Further, because of the nature of the egg, bacteria may be introduced and grow rapidly due to improper handling, unhealthy workers, unclean equipment, utensils, facilities, contaminated uniforms, insects, rodents, and general debris. The water supply may be unsafe, and there may be an inadequate quantity of hot and cold running water under pressure for use in all preparation areas. The solid-waste problem is enormous because of the large quantity of shells and other materials that must be removed. A serious problem occurs when individuals, equipment, or air move from the raw eggs storage and breaking areas to the pasteurizing and packaging egg-processing areas. Improper removal of detergents and use of chemicals readily contaminates eggs. The use and storage of insecticides must be carefully controlled, since the eggs may readily absorb odors, as well as a variety of chemical contaminants.

Organisms penetrate the eggshell and grow quickly within the egg itself. Since eggs are used raw in eggnog and partially cooked in scrambled eggs, the chance of spreading food-borne disease is high. In addition, eggs are frequently used in specialized therapeutic diets for high-risk populations, such as the very young, sick, chronically ill, and aged. The feed ingredients are of concern since they may be contaminated. Clean, fresh eggs properly handled after production are not a public health problem. However, where the eggs are dirty and/or cracked, a potential condition arises in which pathogenic organisms could be introduced.

Environmental Problems in Meat Processing

The environmental problems relating to meat encompass all areas of stockyards, slaughterhouses, meat-processing plants, retail meat-cutting operations, meat storage, and sales. Concern exists about the removal of large quantities of solid waste, intense fly problems, and the presence of rodents while the cattle are in stockyards. Disease may be spread from animal to animal and from human to animal.

At the slaughtering plant, organisms are added during many parts of the killing process. All equipment, appurtenances, and parts of the physical structure, such as walls, ceilings, and floors, contribute contaminants because of improper cleaning techniques. Adequate quantities of potable hot and cold running water may not be available for cleaning of equipment, physical structure, and for hand washing. Cutting instruments and food-contact surface may be improperly cleaned and sanitized. Meats may contain such pathogenic organisms as brucella, salmonella, streptococcus, mycobacterium, pseudomonas, and any other organisms that may inadvertently be introduced. Pork is hazardous since it contains *Trichinella spiralis*. Refrigeration must be constantly monitored, since it is so essential to the control of bacterial growth. Employee health, personal cleanliness, and uniforms are also of concern.

At the retail market, all utensils, saws, and packaging equipment are readily contaminated and frequently improperly cleaned. Adequate quantities of hot and cold running water may not be available for equipment and hand washing. Employee health and personal hygiene practices, which are so essential, since the meat is frequently handled, may be poor. The walls, ceilings, cutting tables, and other pieces of equipment are frequently improperly cleaned. Meat may be overstocked in the display cases and not rotated properly. Once again refrigerator temperatures are of considerable concern. The number of microorganisms on meat varies from as few as 100 per gram of beef to several million per gram of beef, depending on slaughtering and processing techniques and the ultimate condition of the meat. Obviously, any chopped or ground meat may have bacterial counts in the millions.

Environmental Problems in Fish and Shellfish Processing

Fish and shellfish are potential vehicles for many organisms that cause disease in people. Fish acquire organisms from polluted water, harvesting equipment, processing equipment, or infected human beings. Many shellfish, such as oysters, clams, and mussels, are consumed raw, posing a serious public-health hazard. Infectious hepatitis, salmonellosis, typhoid fever, and cholera have been spread by shellfish. In recent years, outbreaks of botulism type E were caused by fish. The heat processing of fish may not create a high enough temperature to kill organisms. Coagulase-positive staphylococci are frequently isolated from crab and shrimp meat. Fish meal is of particular concern, since it is produced in the United States or abroad, and it may be contaminated with innumerable organisms plus the feces from rodents and birds. Shigella from imported shrimp have been involved in outbreaks of dysentery. Lately concern has increased about fish containing mercury and other chemicals found in polluted waters.

The major environmental problems in the harvesting of shellfish are due to the water in which the shellfish are cleaned, the harvesting boats, lack of proper prewashing of the shellfish to remove mud, the bacteriological quality of the washing water, and the disposal of human excretion.

The major environmental problems at the shucking and packing plant are cross-contamination from the shucking room, which is a dirty room, to the packing room, which must be kept immaculately clean; the presence of flies, which spread enteric organisms; improper plumbing, with the resultant possible spread of disease-causing organisms; inadequate hand-washing facilities; improper personal hygiene of workers, including inadequate hand cleaning; inadequate cleaning and bactericidal treatment; inadequate refrigeration of shell stock and shucked shellfish; and contamination of single-service containers.

Since employees can easily cut their fingers or hands with sharp knives or shells, it is essential that anyone with an infection be removed from the shellfish-processing areas. At present there is no mandatory federal standard for shellfish quality. Although a state may have strict standards, it cannot stop shellfish from coming in from states with lesser standards. This would be a violation of interstate commerce.

Impact of Food Processing on Other Environmental Problems

Food processing has an enormous impact on other environmental problem areas. Enormous quantities of solid waste are produced that must be processed in some manner, discharged to a receiving stream, burned, or removed to a landfill. The materials cause water pollution, air pollution, and soil pollution. For an in-depth discussion of these concerns, see Volume II.

Food Quality

Establishing standards or guidelines for food is a complex problem. Although microbiological standards have been established for water, milk, and dairy products, attempting to establish microbiological standards for solid foods is quite difficult. Virtually all foods, unless prepared in a sterile manner, will contain bacteria. The poorer the practices, the higher the level of bacteria. However, numbers of bacteria, although indicative of possible poor handling practices, may not necessarily indicate exposure to microorganisms that may cause disease. Frequently the number of microorganisms exceed 10 million per 1 g of food. Numbers of microorganisms may be increased sharply if there are roaches, flies, ants, rats, or mice present in the establishment.

Imported Foods

Imported foods may be of considerable concern to the population. The foods may come from processing plants that are run in a very unsanitary manner. Water used in processing may be contaminated. The raw food stuff may be contaminated with microorganisms, pesticides, or other chemicals. Leaking jars, dented cans, moldy food, and insect parts have been found in areas where imported foods have been stored.

Economics of Food Processing

Food production is controlled in many cases by the amount of fertilizers available and the energy that can be utilized. High-cost energy tends to decrease the amount of food produced in various parts of the world.

Inflationary forces in the United States and in other industrial countries have caused the cost of food to rise sharply. The farmers in certain areas are facing difficulties in getting adequate prices for the food they produce. The short-range effect may be to drive some of the smaller farmers out of business. The long-range effect will probably be further increases in the cost of food. The larger farmers tend, especially in the cattle-raising business, to concentrate the yields of food and therefore concentrate the byproducts, which become environmental contaminants.

POTENTIAL FOR INTERVENTION

Potential for intervention in the causation of disease or injury due to food processing varies with the type of processed food. With raw food products, a reduction in the use of pesticides under controlled conditions and the proper application of fertilizers and pesticides should help reduce the potential for disease due to these contaminants. A better understanding of the use of additives and preservatives, and the establishment of limitations based on this understanding, may be of further assistance in reducing disease potential. However, considerable research is needed in the entire area of food processing of raw foodstuffs to better understand the techniques for intervention in the disease process.

In the specific preparation of milk, meat, eggs, shellfish, and other highly perishable foods, the standards, practices, and techniques exist and the potential for intervention and disease prevention is excellent.

RESOURCES

Scientific and technical resources include the food industry, the National Environmental Health Association, the International Association of Milk, Food, and Environmental Sanitarians; the American Public Health Association; the National Sanitation Foundation; the American Dietetic Association; the University of Iowa; the University of Michigan School of Public Health; the University of Minnesota School of Public Health; the University of California School of Public Health; and many others.

Civic associations interested in food include the North American Vegetarian Society, the Nutrition Today Society, and the Society for Nutrition Education.

Governmental organizations include all state health departments and departments of agriculture, and many local health departments, the Environmental Protection Agency, the Department of Energy, and various sections of the Department of Health and Human Services, including the Center for Disease Control, the National Cancer Institute, the National Institute of Environmental Health Sciences, the National Institute of Occupational Safety and Health, the National Center for Toxological Research, and the Food and Drug Administration. In addition, the Department of Defense and the Consumer Product Safety Commission are concerned with food products. Departments of Agriculture of the various states and the federal government are important resources in a variety of areas. The United States Senate Agricultural Committee is an important source of legislation and information. The United States Fish and Wildlife Service Department of the Interior is also an important resource.

The Food and Drug Administration, Consumer Affairs and Small Business Staff, Department of Health and Human Services, 5600 Fishers Lane, Room 13-55, Rockville, MD 20857, is helpful to individuals trying to get additional information about food and food technology. The Food and Nutrition Service, Department of Agriculture, Room 512, 3101 Park Office Center Drive, Alexandria, VA 22302, is useful for those concerned with nutrition aspects related to food. The Food Safety and Inspection Service, Room 1163, South Building, Department of Agriculture, Washington, DC 20250, is useful for all areas of food and food technology. The National Agricultural Chemicals Association, 1155 15th Street, N.W., Washington, DC 20005, have experts in the areas of agricultural chemicals, including pesticides and their effects on food.

STANDARDS, PRACTICES, AND TECHNIQUES

Food Preservation Techniques

The shelf life of food can be extended by inactivating or destroying enzymes or by inhibiting or destroying microorganisms. Food preservation techniques include the use of drying, low and high temperatures, irradiation, salt and sugar, and chemical preservatives. Various chemicals are also added during processing to enhance flavor and to promote better color and odor.

In all food preservation, it is essential to start with as clean a raw material as possible and to prevent contamination from hands, soil, water, air, surfaces, processing, and storage equipment.

Drying

Drying, also called dehydration or desiccation, is one of the oldest methods of food preservation still in use today. Dates, figs, and raisins are air dried. Forced hot air is used in drying potatoes, fruits, and vegetables. Milk and eggs are sprayed into heated cylinders. Enzymes are inactivated prior to drying by blanching, which is the removal of color by boiling, or by the use of sulfur dioxide or sulfites. Blanching removes sticky substances, checks browning, and makes vegetables easier to work with. Reconstitution of the dry food is aided by clustering the particles during drying, by adding a foaming agent such as nitrogen during concentration of the food prior to drying, or by causing the food to explode into particles. Drying will kill some organisms, but spores of bacteria, yeasts, and molds will survive and can later cause outbreaks of food-borne disease, or food spoilage.

Freeze Drying

Food is first frozen and water is then evaporated from the ice crystals without melting them. A small amount of heat is applied and pressure is reduced in this process, called sublimation.

Refrigeration and Freezing

Low temperatures retard food spoilage. The shelf life depends on the kind of food. Freezing is the removal of heat from a product. The formation of ice crystals in the food is due to free water or air being trapped in the package. During the freezing process, moisture from the warm food will form a vapor on the packaging material and will freeze more rapidly than the food mass. Ice crystals break down cell structure and cause a physical deterioration in food quality. This can be avoided by rapid freezing of high-quality raw materials. Freezing does not improve the quality of food, nor does it destroy many of the pathogenic organisms that may be present.

Heating

The rate of kill of microorganisms at high temperatures varies with the species of the organisms, presence of spores, quantity of organisms, consistency, pH and type of food medium, temperature, time, moisture, quantity of food, and ease of heat conduction and convection. To avoid overcooking a given food, which could affect food quality, the temperature and time of cooking should be adequate to kill pathogenic organisms, but not necessarily thermophilic or thermoduric organisms.

Pasteurization is a heat-treatment process well below the boiling point, which will destroy all harmful microorganisms and improve the keeping quality of food. It is used in the dairy industry and also in the preservation of dried fruits, syrups, honey, and juices.

Boiling is a common method of preparing and preserving food. Boiling does not kill all organisms, since at atmospheric pressure the temperature may not reach 212°F. Therefore, adequate time must be provided for a total bacterial kill. Although heating of food to boiling will destroy botulinum toxin, the spores may survive boiling temperatures for a considerable period of time. Baking in an oven at a temperature of 350–400°F may not cause the food temperature to rise to 212°F because of moisture present in the food and because baked goods have low heat conductivity. Frying kills organisms on the surface of the food, but may not affect the organisms within the food mass.

Canning

Canning is the preservation of food by subjecting selected prepared foods to

Table 3.1. Canning Flow Diagram

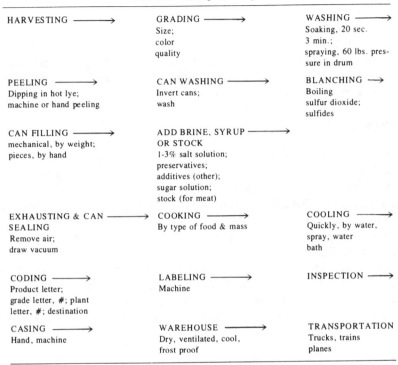

high temperatures in a permanently sealed container (Table 3.1). Containers may
be made of glass and sealed with tin, aluminum, paperboard, or plastic caps, which
are either automatically vacuum sealed or screwed on. Metal cans made from tin-
coated steel or aluminum are also used. Lacquered cans are used for highly
colored foods to prevent color loss. Can sizes are as follows: no. 1, 10.94 oz; no.
2, 20.55 oz; no. $2\frac{1}{2}$, 29.79 oz; no. 3, 35.08 oz; no. 10, 109.43 oz.

The entire canning process is important. However, certain key items should be
noted:

1. Use high-quality food from which gross dirt and microorganisms have been removed.
2. Heat-process foods as soon as the air has been exhausted.
3. Cool promptly to avoid overcooking and undesirable changes in texture and flavor
   and the germination and multiplication of spores of thermophiles.
4. Since food is commercially sterile (all bacteria, but not spores are killed), the
   resistant spores of *Clostridium botulinum* may not be destroyed in nonacid foods.

Testing. Canned products should be tested by incubating samples at 100°F to test
for swells, incubating samples at 130°F for 2 weeks to test for flat sours, inserting

thermocouples in food center to check temperature, placing spore-forming test organisms in marked cans, and taste-sampling for quality.

Problems in Canning and Bottling. The following problems may occur in canning: microbial contamination, pinholes developing from chemical action of food on container, cans collapsing, glass containers breaking because of impact, thermal shock, internal pressure, and toxins developing.

Radiation

Certain selected food, meat, fish products, vegetables, and spices have been preserved by radiation. However, there are still certain questions concerning food-product safety. Organisms in food are destroyed by direct hits, causing ionization and death. Some lethal effects are caused chemically by free radicals produced in the solvents. In irradiated water, hydroxyl ions, which are strong oxidizing agents, are formed. The hydrogen ions present are strong reducing agents. The oxidizing and reducing agents will not only kill organisms, but also produce off flavors, and changes in taste, appearance, odor, and texture. Other undesirable changes in food include a rise in the pH of meats, increases in H_2S, and destruction of thiamine, ascorbic acid, riboflavin, niacin, and other vitamins. On the beneficial side, radiation sterilizes very quickly, kills insects in grains effectively, and keeps the temperature of the food from rising significantly.

Milk

Milk is the fluid secretion of the mammary glands, practically free of colostrum, containing a minimum of 8.25% nonfat milk solids and 3.25% milk fat. Milk contains the following average constituents: water, 87.3%; fat, 3.7%; protein, consisting of casein, 2.9%; albumin, 0.5%; lactose, 4.9%; and ash, 0.7%, which includes salts of calcium, copper, iron, magnesium, manganese, potassium, and zinc. Milk contains lactose in true solution, casein in permanent suspension, and butter fat in temporary colloidal suspension. Milk is an excellent source of vitamins A and B_2, and a fair source of vitamins B_1 and E. It also contains vitamins D, K, niacin, pantothenic acid, B_6, B_{12}, folic acid, citrin, P-amino benzoic acid, biotin, choline, and inositrol. It also contains enzymes, epithelial cells, leukocytes, yeasts and molds, and extraneous foreign material and bacteria from within the udder from milking, people, processing equipment, and poor techniques.

Milk Production

Clean, safe, nutritious milk comes from healthy cows that live and are milked in an environment where clean, sanitized equipment, rapid cooling, and proper handling and storage techniques are utilized (Table 3.2 contains a diagram of the milking flow process).

Table 3.2.   Milk Production Flow Diagram

MILKING

COW ⟶
Wash udder with warm water
and then with chlorine;
strip, examine
foremilk;
rinse teat cups
sanitize teat cups

MILKING MACHINE ⟶
Prewash, sanitize;
rinse in cold water;
sanitize;
attach cups to udder;
set timer
strip udder

STRAINING ⟶
Stainless steel
strainer with pad;
material on pad
indicates level of
sanitation in
milk house

COOLING AND ⟶
STORAGE
Mechanical milk cooler;
cool below 45 °F rapidly

TRANSPORTATION ⟶
Cans cooled below 45 °F;
bulk below 45 °F in
stainless steel tank;
wash, sanitize
every 24 hrs.

MILK PLANT

MILK PROCESSING

TRANSPORTATION ⟶
Raw milk—cans; bulk

HOLDING TANK- ⟶
RAW MILK
Storage below 45 °F

BALANCE TANK →
Constant level of
liquid

BOOSTER PUMP ⟶
Moves milk

REGENERATION ⟶
Hot pasteurized milk on
positive pressure side of plate
heats up cold, raw milk on
negative pressure side

METERING PUMP
Opposite side
regenerator from
holding tank;
regulates amount
of milk passing

PASTEURIZATION

HOMOGENIZER ⟶
Suspends milk, milk
particles; diffuses
equally

Vat ⟶
145° F for 30 min.
higher milk fat or added
sweeteners—150 °F for
30 min.
*High-temperature
short-time*
161 °F for 15 sec.;
higher milk fat or added
sweetners—166 °F for
15 sec.
Ultra high temperature
by steam injection at
191 °F for 1 sec.;
194 °F for 0.5 sec. or
212 °F for 0.01 sec.

REGENERATION ⟶
Cold, raw milk on
negative pressure side
cools hot
pasteurized milk

COOLER ⟶
Quickly lowers
milk temperature

PIPES ⟶
To storage containers;
to fillers

FILLERS ⟶
Bottles; cartons; cans;
bulk containers
CONSUMER

TRANSPORTATION ⟶
Below 40 °F

RETAIL STORE →
40 °F

Vat Pasteurization. The vat pasteurizer consists of a stainless-steel jacketed vat
with a cover, agitator, inlet and outlet valves, airspace heater, and indicating,
recording, and airspace thermometers. It is essential that the milk is protected
during the entire process, that pasteurization temperatures of 145°F be maintained
for a minimum of 30 min, and that the milk be agitated.

The vat may be operated in the following manner:

1. Heat milk to pasteurization temperatures and maintain these temperatures by spraying hot water on inside of the double-jacketed vat or by surrounding the inner jacket with heated coils.
2. Partially preheat milk by means of a heater and then bring milk up to pasteurization temperatures as in 1.
3. Preheat milk to pasteurization temperature and then maintain as in 1.
4. Remove milk from vat either partially cooled by turning off heating system or at pasteurization temperature.

The cover must have overlapping edges, have all of its openings protected by raised edges, and have all condensate diverted from it.

The inlet and outlet of the vat must have leak-protector valves with the following design:

1. seat in every closed position
2. grooves at least $^3/_{16}$ in. wide and $^3/_{32}$ in. deep at center
3. stops not reversible
4. close coupled valve seats flush with inner wall of pasteurizer or so close to the wall that milk in the valve inlet is no more than 1°F colder than pasteurized milk

To ensure that all milk is pasteurized, the inlet and outlet lines must be properly sloped to allow drainage of all milk; the inlet valve must be closed during the holding and draining periods; and the outlets valves must be closed during filling, heating, and holding.

Constant agitation of the milk during pasteurization ensures that all of the milk is uniformly heated and properly pasteurized. The airspace heater is used for the same purpose for the foam above the milk. The temperature of the air must be 5°F higher to ensure proper pasteurization.

The indicating thermometer and airspace thermometer show the temperature of the milk at any given time. The recording thermometer, which is attached by wires to a chart-and-pen arm, records the temperature at which the milk is being processed. The recording thermometer, which is checked daily against the indicating thermometer, must never read higher than the indicating thermometer. Charts should indicate the preheating time, pasteurization time, precooling time, airspace temperature, indicating thermometer temperature at a given time, record of unusual occurrences, name of milk plant, operator's name, location of recorder, and date.

High-Temperature Short-Time Pasteurization. High-temperature short-time (HTST) pasteurization (Table 3.3) consists of a system including:

1. Drawing of cold raw milk at about 40°F from a constant-level tank.
2. Raw milk heated as it flows on the negative pressure side of thin stainless-steel

Table 3.3. High-Temperature Short-Time Flow

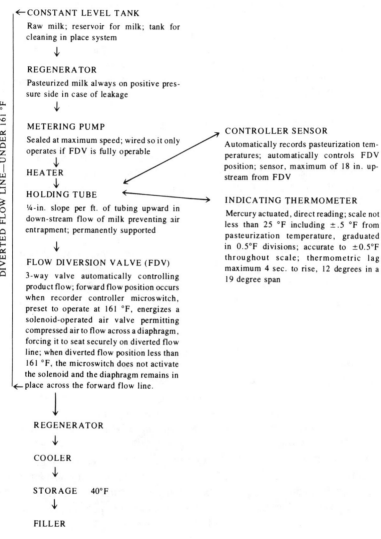

plates in the regenerator, by hot pasteurized milk flowing in the opposite direction on the positive-pressure side of the plates.

3. Raw milk still under negative pressure passing through a positive displacement timing pump that pushes it under positive pressure through the rest of the system.

4. Heating raw milk to at least 161°F.

5. Raw milk under pressure at 161°F flowing through holding tube for a minimum of 15 sec.
6. Milk flowing past indicating thermometer and recorder controller at 161°F or higher and then through the forward-flow position of the flow-diversion valve. If the temperature is less than 161°F, the milk flows through the diverted-flow position into the diverted-flow line and back to the raw, constant-level tank.
7. Pasteurized milk flowing through the regenerator and cooling.
8. Milk passing through cooler where temperature is dropped to 40°F and being stored in a vat until packaging.

The recorder chart must have a scale with a span greater than 30°F, including 12°F from the diversion temperature. The temperature must be accurate to within 1°F at diversion temperatures. The chart must not rotate more frequently than once every 12 hr. The thermometric response must be 4 sec to travel 12°F within a 19° range. The date, number of location, name and address of plant, reading of indicating thermometer at a given point, amount and name of product, product temperatures at beginning and end, time in forward-flow position, unusual occurrences, and operator's name must be recorded on the chart.

It is always important to maintain proper pressure throughout the system.

Ultra-High-Temperature Pasteurization. Ultra-high-temperature pasteurization is accomplished by steam injection or by the use of a vacuum. In the steam process, automatically controlled injected steam is used to heat the milk to the following pasteurization temperatures: 191°F for 1 sec, 194°F for 0.5 sec, 201°F for 0.1 sec, 204°F for 0.05 sec, 212°F for 0.01 sec. The product is then rapidly cooled and passed through the flow-diversion valve before storage. The reason that the flow-diversion valve, which operates the same as in the HTST process, is located after the cooler is that a safety hazard, due to flashing steam, occurs if milk is diverted at 212°F; the milk would be diluted by steam if diverted at high temperatures; the response time of the thermal limit controllers would be too slow to prevent the forward flow of raw milk.

In the vacuum process, the milk must be heated to at least 194°F. Constant temperatures must be maintained in the pasteurizing zone by a constant uninterrupted, adequate quantity of steam. The steam pressure must be 35 lb/in.^{2} and the vacuum drawn in the chamber must be 8 in. Ultra-high-temperature pasteurization is used to increase the shelf life of milk, inactivate enzymes, and improve flavor.

Dry Milk. Milk is dried by spraying in a vacuum chamber. Salmonella surviving in the dry milk are a potential hazard.

Cleaning

Proper cleaning is absolutely essential during all phases of milk production and milk processing. One technique involves disassembling all equipment, rinsing thoroughly with cold water, scrubbing all equipment with brushes using hot water and detergent, rinsing thoroughly, reassembling, and sanitizing with 200 ppm chlorine. Sanitization may also be accomplished by the use of steam for 5 min or hot water at 180°F for 5 min. Acid cleaners should be used when necessary to aid in cleaning. Cleaning in place is a technique in which the circulating system is not disassembled. Instead, a cleaning solution is circulated through the system at a velocity of at least 5 ft/sec. The solution, which is alkaline and nondepositing, is heated to 120°F and circulated for at least 15 min. The solution is then drained and the system is thoroughly rinsed. Finally, chlorine or hot water is used for sanitization.

Ice Cream and Frozen Desserts

Frozen desserts are pure, clean, frozen, or semifrozen foods prepared by freezing while stirring a pasteurized mix composed of edible fats, nonfat milk solids, sugar sweeteners, flavors, and other ingredients such as eggs, fruits, and nuts. The pasteurization process is the same as used in milk, except that in the vats the pasteurization temperature must be 155°F for 30 min instead of 145°F for 30 min, and in the HTST system the pasteurization temperature is 175°F for 25 sec instead of 161°F for 15 sec. Contamination may enter the mix when fruits and nuts are blended into the mix after pasteurization and cooling, and before freezing. Poor techniques, careless handling, or contaminated raw products can raise the bacterial count of the mix.

Counter Freezers or Soft Ice-Cream Freezers

Counter freezers or soft ice-cream freezers are a continuation of the frozen dessert process. Handling of the mix after delivery to the stores and improper cleaning of the equipment contributes to extremely high bacterial counts. Typically, the mix is delivered in plastic bags, packed in boxes, or packed in cartons similar to large milk containers. The mix is poured into the dispenser and then added to during the day as needed. At the end of the day, in many instances, the mix is drained into a bottle or jar and saved for the next day. The dispenser may or may not be rinsed or cleaned.

Proper technique consists of carrying out the following steps each evening: Drain mix at end of day to waste; rinse dispenser thoroughly with cold water; wash inside of machine with hot water and detergent for at least 5 min; rinse unit thoroughly; add to dispenser 200 ppm chlorine in water and allow to stand for 2 min; drain thoroughly and air dry. Before refilling the following morning, the dispenser should be once again sanitized with 200 ppm chlorine for 2 min and allowed to drain.

Table 3.4. Poultry Processing Flow Diagram

LIVE BIRD ⟶ Healthy; vigorous	INSPECT AND WEIGH ⟶	SHACKLE ⟶
BLEED ⟶ Thoroughly; quickly	SCALD ⟶ 135 to 140 °F; helps remove feathers	DEFEATHER ⟶
WASH ⟶ Spray	EVISCERATE ⟶	INSPECT ⟶
WASH ⟶ Inside; outside	CHILL ⟶ Quickly below 35 °F; freeze	PACKAGE Cut up; whole; fresh, chilled, frozen; in prepared meal, either frozen, canned, or fresh (chicken roll)

Poultry

Poultry are domesticated birds, such as chickens, turkeys, ducks, geese, and pigeons, that are raised for food consumption. The widespread use of poultry and poultry products, the substantial amount of handling and processing, the kinds of bacteria indigenous to the birds (salmonella), and the long history of poultry-associated food-borne diseases makes poultry processing an important area of concern in public health.

Poultry are mass produced in special housing with automated feed and water supplies. The feed ingredients may contain salmonella or other organisms transmitted to the chicks and eventually to humans. Salmonella may also be introduced into the environment of the poultry by dust, air, domestic animals, rodents, fomites, human waste material, flies, pigeons, and other birds. (See Table 3.4 for a diagram of the poultry process.)

Eggs

Eggs are produced from domesticated chickens, ducks, geese, turkeys, and other domestic fowl for human consumption. Eggs are mass produced in special units in which hens are automatically fed and provided water. (See Table 3.5 for a diagram of the egg process.) Egg pasteurization is usually performed in a high-temperature short-time pasteurizer and is similar to HTST milk pasteurization.

Meat

Fresh, raw meats refer to the regular retail cuts of beef, veal, lamb, and pork. Processed meats are meats plus other ingredients, additives, and spices that have received special treatment, such as curing, smoking, or canning.

In order to simplify the slaughtering process, only the slaughtering of cattle will be discussed. (See Table 3.6 for a diagram of the slaughtering process.) In the slaughtering plant, the bones, feet, and condemned parts will be sent to rendering

Table 3.5.   Egg Processing Flow Diagram

WHOLE SHELL EGGS

HEN ⟶	SHELL EGGS ⟶	
CANDLE ⟶	CLEAN ⟶	SHELL TREATMENT →
Classify; eliminate spoiled eggs, blood spots; separate cracked eggs	Spray wash with water at same temperature as eggs	Immerse eggs at 75 °F in oil; rotate for 10 min.
STORE ⟶	SUPERMARKET	
40 °F; well-ventilated area; clean area; relative humidity 85-90%	40 °F	

FLOW CHART FOR LIQUID OR FROZEN EGGS

HEN ⟶	SHELL EGGS ⟶	CANDLE ⟶
CLEAN ⟶	SANITIZE SHELLS ⟶	BREAK SHELLS ⟶
	Spray	Asepticly; separate white from yolk
	Do not use a quat	
	Rinse off excess sanitizer	
MIX THOROUGHLY ⟶	CLARIFY OR FILTER ⟶	COOL ⟶
		40 °F
FERMENT ⟶	PASTEURIZE ⟶	ADD ADDITIVES ⟶
To remove sugar	140 °F for 1.75 min. for whole eggs; 134 °F at pH9 for 1.75 min. for liquid egg whites; 142 °F for 1.75 min. for liquid yolk	Salt; sugar; adjust pH
PACKAGE ⟶	FREEZE	
	Rapidly to 0 °F; defrost when needed in refrigerator at 45 °F	

FLOW CHART FOR DRIED EGGS

HEN ⟶	SHELL EGG ⟶	CANDLE ⟶
CLEAN ⟶	STORE ⟶	SANITIZE ⟶
	40 °F for 72 hours before sanitizing	
BREAK SHELLS ⟶	MIX ⟶	CLARIFY ⟶
HOLDING VAT ⟶	PASTEURIZE ⟶	SPRAY DRY ⟶
40 °F for maximum of 48 hours		Under high pressure; drying temperature 340 °F
COOL ⟶	STORE	
By coils to 90 °F; final cooling to 50 °F	In sterile barrels or cartons	

plants. Unusable portions of the meat are used for pet food. However, pet-food processing must be a totally separate operation from human-food processing.

Refrigerated beef, after being cut into prime or retail cuts, may be wrapped and frozen quickly at temperatures of –10 to 0°F. Quick freezing and proper wrapping

Table 3.6.   Meat Processing Flow Chart

SLAUGHTER ⟶	ANTEMORTEM ⟶	KILL ⟶
Few animals at a time	EXAMINATION Inspectors and Veterinarians condemn or retain animals if the animal is in a coma, seriously crippled, or dead dying, has a disease or temperature of 105 °F	Stunning by sharp blow on head or electrical prod; hung by hind legs on rail; jugular vein slit
DRESS ⟶	CARCASS WRAPPING ⟶	PROCESS CARCASS ⟶
Blood completely drained; head removed; skin cut; belly opened; legs removed; skin removed; viscera removed; viscera inspected by Vet, or meat inspector (post-mortem inspection); carcass split; hide inspected	Wrapped in cloth for shaping	Chill within 1/2 hr. after slaughter by high velocity, cold air system or refrigera- tion; humidity in cooler 88-92% hold in cooler 1-3 days or until rigor mortis is over, may hold longer for aging and tenderizing temperature of 32 °F
CUT FOR RETAIL ⟶ At 32 °F; primal cuts, such as, roast, steak, stew	DISPLAY IN RETAIL MARKET 32 °F	

are essential to preserve the quality of the food. Freezing may cause slight changes in flavor or color.

Specialized Processing of Meats

Curing, the addition to meat of sodium chloride mixed with sugar and spices, was originally developed to preserve meats without the use of refrigeration. Sodium nitrites and nitrate are added to stabilize the color, since most people prefer a pink look to their meat. Sugar is added to counteract the hardening effect of salt and to improve flavor. These ingredients can be dissolved in water to form a brine solution or brine cure. The meat can either be soaked in the brine, injected with brine, pumped with brine through the arteries, or rubbed with a salt and brine solution.

The smoking of meats is primarily for the addition of tastes, although some preservation takes place. Hardwood chips and sawdust are used in the smoke houses. Humidity and temperature are carefully regulated. Long-cured meats are smoked at an internal meat temperature of about 110°F. Short-cured meats are smoked at temperatures of 140–150°F and finally at temperatures of about 200°F. The length of smoking varies from a few hours to several months.

Cased meat products include sausages, frankfurters, and luncheon meats. Sausage is usually made from chopped meats, including beef, pork, veal, lamb, and mutton. It contains fillers, such as cereal, starch, soy flour, dry milk, and added water. Sausages may be cooked or smoked. Frankfurters, a type of smoked sausage, contain any of the above products and may also contain chicken; they are always fully cooked before being sold. Luncheon meats contain a variety of meat products, other ingredients, spices, and water; they are usually precooked before being sold.

Canned meats are usually cooked after packing. The storage of meats in cans depends upon the processing. In almost all cases, canned hams must be refrigerated.

Fish

Fish may be preserved by icing, salting, smoking, freezing, or canning. The fish is harvested from the ocean and may be processed on the boat. If so, they are gutted, scaled, filleted, and iced. The fish may also be iced on the harvesting boat and then taken back to a fish-processing plant. Unfortunately, even artificially produced ice is not sterile because of handling, production, and storage, and may contaminate the fish.

In the smoking process, dehydration occurs, which helps in preservation. Smoking, however, is basically used for improving taste. The fish is first salted and then smoked with hardwood fumes in the same way that meat is smoked. Smoking may last a few hours or many days.

Fish preserved by canning are heat treated and commercially sterile. The keeping quality of the canned fish is based on the kind and amount of cooking.

Fish should be flash frozen. One of the problems is that during defrosting, a considerable amount of water is lost from the fish and the flavor is impaired. When salting, fish must first be cleaned and cut, immersed in a 20% salt brine for 30 min, and placed in water-tight containers.

Fresh fish, when inspected, will have the following characteristics: firm; no off odor; no discoloration or slime; eyes bright, full, and moist; red gills, not grey or brown; tight scales.

Shellfish Harvesting and Processing

Shellfish include all edible species of oysters, clams, and mussels. (See Table 3.7 for a diagram of the shellfish production process.)

MODES OF SURVEILLANCE AND EVALUATION

Milk Testing

Tests are performed on raw milk as it enters the plant to determine odor, appearance, off color, blood, viscosity, foreign matter, sediment, and temperature. (See Figure 3.1 for an example of a milk farm inspection report.)

Laboratory tests conducted on raw milk include

1. direct microscopic count or breed count, which measures bacteria count, and identifies the type of bacteria and the source of contamination

Table 3.7.   Shellfish Processing Flow Chart

GROWING AREA ⟶	HARVEST ⟶	TRANSPORT ⟶
Sanitary survey by Public Health Service; evaluate sources of actual or potential pollution; bacteriological study of water and shellfish; chemical analysis of water and shellfish; graded as approved, conditional approval, restricted, or prohibited	From approved area only; from conditionally approved area only under supervision of health authorities, to ensure that shellfish are purified; from restricted areas only under controlled purification and supervision of health authorities; never from prohibited areas; human body waste must not be discharged from the harvest boat into the water	Boats and truck, shellfish stored in thoroughly cleaned bins and away from bilge water; bins must be sanitized
WASH SHELLFISH ⟶	SHUCK ⟶	PACK ⟶
Wash away mud; use potable water under pressure	Separate from packing operation; not subject to flooding; fly tight screening	Clean room; surfaces washed, sanitized; impervious floors and walls; fly control measures; adequate safe, sanitary water supply; approved plumbing— avoid cross connections and submerged inlets; strict hand-washing procedures; use clean, sanitized single-service containers; label with packer's name, address, number, state, and date
REFRIGERATE ⟶	CLEAN EQUIPMENT AND SANITIZE	
Refrigerate quickly below 40 °F; freeze quickly	Washing sinks, made of impervious, non-toxic materials; storage, shucking, and packing rooms cleaned within two hr. after completing operations; sanitize with 100 ppm available chlorine	

2. methylene blue test, which gives a rough estimate of the bacterial load; decolorization in half an hour indicates a high bacterial load, whereas decolorization in 8 hr indicates a low bacterial load
3. resazurin test, which is similar to the methylene blue test
4. standard plate count
5. thermoduric bacteria test, which indicates use of unclean equipment on the farm or in transportation
6. coliform test, which indicates use of unclean equipment in production

Laboratory tests conducted on pasteurized milk include

1. standard plate count, which determines the number of living bacteria and indicates the sanitary conditions under which milk was processed
2. phosphatase test, which determines the efficiency of pasteurization
3. coliform test, which determines the presence of coliform and indicates contamination after pasteurization
4. residual bacterial count from milk processing and storage equipment and containers

County	State		Date

Farm	Location	Date _____ Time _____	Ratings N-Not Apply U-Unsatisfactory S-Satisfactory

Item		Rating	Item		Rating
1.	Cow healthy & from disease free herd.		5.	Utensils & Equipment:	
2.	Milk barn:		a.	smooth, non-absorbent non-toxic	
a.	adequate lighting		b.	good repair	
b.	adequate ventilation		c.	easily cleanable	
c.	floors, walls, ceiling clean		d.	clean	
d.	floors, walls, ceilings, good repair		e.	sanitized	
3.	Cow yard:		f.	proper storage	
a.	graded, drained, clean		6.	Milk storage:	
b.	manure stored properly & removed		a.	below 45 °F within 2 hours	
4.	Milk house:		7.	Milking:	
a.	floors, walls, ceilings, smooth & good repair		a.	cleanliness of flanks & udders	
b.	adequate lighting		8.	Fly control	
c.	adequate ventilation		9.	Storage of Pesticides	
d.	all openings screened		10.	Safe water of adequate quantity:	
e.	facilities clean		a.	hot & cold water	
			11.	Proper sewage disposal	
			12.	Adequate toilet facilities	
			13.	Adequate handwashing facilities	

Recommendations   Time for Compliance

Environmentalist _____

Figure 3.1.   Milk farm inspection report form.

5. screening tests for abnormal milk
6. disc assay for presence of antibiotics
7. radioactive material testing

Vat Pasteurization Equipment Tests

All indicating thermometers, recording thermometers, and leak-protector valves used in vat pasteurization must be checked at installation and at least quarterly afterwards to determine if they are operating properly. The indicating thermometer, which must be accurate to +0.5 or –0.5°F within the pasteurization range or +0.5 or –0.5°F within the airspace range compared to a standard thermometer, is tested by inserting both thermometers into a well-agitated can of hot water at 145°F and making a direct reading. However, the indicating thermometers must read not less than the pasteurization temperature during the entire holding period. The error allowed in the thermometer's accuracy is taken into account when determining the indicating temperatures.

All recording thermometers and pen arms must be checked at installation, quarterly, and when needed, and must be able to adjust to a wide range of fluctuating temperatures. The recorder thermometer is adjusted to read exactly the same as the previously tested indicating thermometer while immersed in a can of agitated water at 145°F. The recorder sensor bulb is then immersed in a can of boiling water for 5 min and into a can of ice water for 5 min. Finally the recorder element is placed back into the can of agitated water at 145°F. The final temperature on the recorder chart must never read higher than the indicating thermometer.

The rotating time of the recorder chart is tested at installation and at least quarterly to make sure that the elapsed time is accurate. This is done by comparing the elapsed time on the recorder chart to an accurate watch.

The recording-pen arm temperature notation is compared to the indicating thermometer while the product is in the vat. The recorder must not be higher than the indicating temperature.

Proper functioning of the leak-protector valve is determined by disconnecting the outlet pipe and carefully checking for leakage around the valve while product pressure is being exerted on the valve.

High-Temperature Short-Time Pasteurization Equipment Tests

Indicating thermometers, including airspace thermometers, must be checked upon installation and at least once every 3 months against a standard thermometer to make sure that they are within +0.5 and –0.5°F accuracy of the test thermometer. This is done by inserting the thermometer in an agitated 10-gal can of water heated to within –3°F of pasteurization temperatures and taking a direct reading.

Measure the thermometric response of the indicating thermometer on the pipeline by determining the time that it takes to raise the temperature of the indicating thermometer, which has been placed in a water bath 19°F higher than the lowest reading on the thermometer and above pasteurization temperatures. It should rise 12°F within 4 sec.

The recording thermometer should be checked against the indicating thermometer at installation and quarterly by an environmental health worker and daily by the plant operator. The recording thermometer should never read higher than the indicating thermometer. The recording thermometer's temperature accuracy is tested in the same way that vat pasteurization temperature is tested, except that 161°F is used instead of 145°F. The time accuracy of the recorder controller is measured in the same way.

Cut-in temperatures of milk flow should be determined by raising the product temperature from 3°F below pasteurization temperatures to pasteurization at a rate of 1°F per 30 sec and determining the temperature at cut in. Cut-out temperature is determined by raising the pasteurized product flow slightly and then decreasing at a rate of 1°F per 30 sec. Determine the temperature at cut out. This ensures that milk is reaching full pasteurization temperatures.

The thermometric lag of the recorder controller must not exceed 5 sec. This is determined by inserting the recording thermometer in a can of hot water that is 7°F above pasteurization and determining the time it takes the mercury to rise from 12°F below pasteurization and the cut-in mechanism by the controller.

Check the flow-diversion valve for leakage. Measure the response time of the flow-diversion valve, which should not exceed 1 sec, by determining the elapsed time between the instant of activation of the cut-out mechanism and the fully diverted flow position.

Determine the time of product flow through the holding tube by electrodes and injection of 50 ml of saturated salt solution at the starting point. Determine the time it takes to reach the second electrode. This should be a minimum of 15 sec.

Poultry Testing

High bacterial counts indicate poor handling and processing techniques and improper refrigeration. Clean birds should be held at 32–34°F for a maximum of 4 days. If slime develops, the bacterial count per square centimeter of surface will probably be in the millions. A total count of organisms can be made using a swab technique or by pressing an agar plate against the surface of the bird. A rinse technique using sterile water may also be used.

The dye, resazurin, may be used to determine the condition of a bird, with swab samples in a solution noting the length of time the sample takes to decolorize to a fluorescent pink color. A fresh bird takes greater than 8 hr, a good bird takes greater than 5 hr but less than 8, a fair bird takes greater than $3\frac{1}{2}$ hr but less than 5 hours, and a poor bird takes less than $3\frac{1}{2}$ hr.

Egg Testing

The tests used with milk pasteurization equipment are also used with egg pasteurization equipment. In addition, total counts are taken from the whole eggs, egg white, egg yolk, salted yolk, sugared yolk, and blends. Tests are taken of liquid, frozen, or dried eggs. Yeasts and molds are also evaluated. Of particular concern are salmonella bacteria. Therefore, special tests are conducted to determine the kinds and quantities of salmonella present. Bacteria grow very rapidly in broken-out egg products when the temperature rises above 60°F. Although freezing and drying decreases the number of viable bacteria, these processes do not eliminate them, and therefore, frozen and dried eggs can and have caused outbreaks of food-borne disease.

Meat Testing

Programs monitor pesticide residues, cleaning agents, and antibiotics in meat tissue. Inspections are made at many stages during meat processing by the U.S. Department of Agriculture, the Food and Drug Administration, and state and local health authorities. Labels are required to show all ingredients used in any type of processed meat. Total plate counts may be taken from fresh meats, cured meats, canned meats, and a variety of meat products. Samples are usually taken of the surface and the interior of the meat. In addition, tests are conducted for yeasts and molds. Direct microscopic examinations are also used. Generally the test for contamination is not meaningful unless there is a population of at least one million organisms per gram of meat. The pH, salt, and nitrite levels are also determined.

Shellfish Testing

Shellfish are tested for coliform on a routine basis. However, where outbreaks of specific diseases, such as infectious hepatitis, cholera, typhoid fever, and bacillary dysentery occur, specific tests are conducted for these organisms. The bacterial content of shellfish can be sharply increased and added to during any phase of harvesting, processing, and storage.

FOOD QUALITY CONTROLS

The federal government protects all food shipped in interstate commerce through the enforcement of the following acts: Pure Food and Drug Act, passed in 1906 and later updated; the Wholesome Meat Act of 1967; the Wholesome Poultry Products Act of 1968; and the Egg Products Inspection Act of 1970. Many federal agencies are involved in protecting the food supply.

All states have health departments and departments of agriculture that provide additional protection through inspection services. Counties, cities, and multiple county health departments provide for local inspection services to protect the food supply. When food processing is poor or unsanitary, necessary legal action is taken.

SUMMARY

The production of adequate quantities of food to satisfy the needs of the world will continue to be an extremely serious problem because of the sharp increase in world population and the decrease in death rates brought about by better medical care. The production of food brings with it inherent difficulties, such as the use of scarce water supplies, energy supplies, and potentially hazardous fertilizers and pesticides. Additives to enhance color and taste and to preserve food are increasingly under suspicion of having the potential to cause a variety of diseases, including cancer. Current techniques of food technology have been incriminated in outbreaks of botulism from lack of adequate controls during food processing. As the quantity of prepared frozen foods and food shipped to varying parts of the country and the world increase sharply, a set of controls must be established to prevent the spread of disease.

RESEARCH NEEDS

Toxicants that may be part of plant foods require further study. Long-term feeding studies need to be conducted in experimental animals to determine special problems, such as sensitization and allergies. Analytical methods should be developed to measure the important toxicants found in foods. These toxicants include the pressor amines, goitrogens, anti-nutritional factors, safrole-related compounds, and allergic hemagglutinins. Sampling techniques must be utilized both on raw agricultural products and prepared products. Studies of plant breeding to reduce the level of toxicants found in certain selected food plants are needed. New plant varieties should be investigated to determine if genetic manipulation has caused an increase in the production of toxicants. Research is required to determine if existing plants produce abnormal toxic metabolites under certain stress conditions. A catalogue of toxic constituents of poisonous plants should be produced. A rapid, reliable, quantitative analytical method should be developed for paralytic shellfish poisoning. A complete study of the chemical nature of various poisons must be conducted on all paralytic shellfish poisoning. The mycotoxins must be studied in the laboratory and field situation to determine why they are produced. Methods must be developed for measuring the incidence of mycotoxin contamination in raw and finished foods. Studies are needed to determine the biotransformation of mycotoxins that occurs in food animals and to determine the potential for contamination of the food for human beings. Toxicological studies are needed to determine the amount of mycotoxin to which humans are exposed. A complete study of the biological activities of nitrates, nitrites, and nitrosamines in foods and a study of the effect of these chemicals on humans is needed. Research is needed on the effects of high intakes of sodium chloride and phosphates. Special emphasis should be placed on the relationship of these chemicals to hypertension and bone growth. Because of the large numbers and quantities of compounds found in foods, ongoing studies should be made to determine the interaction of food additives with major dietary constituents, the alterations of these dietary items, and possible resulting effects. Considerable research is needed in understanding the movement of mercury through the food chain. An understanding is needed of the adverse effect of lead and on the range of adverse effects of cadmium, copper, zinc, magnesium, selenium, and arsenic found in food. Research is needed on pesticides to determine the level of food contamination and their adverse effects in humans. Additional research should be carried out on the consumption of food additives, the interaction of food additives and other residues present in food, the chemical nature of many of the additives, and the chemical changes brought about by cooking. This fundamental research is essential to determine which additives should be eliminated from food and which should remain.

At present the FDA has listed 86 chemicals, including many of the coloring substances, as potential carcinogens. The use of risk assessment in some of these determinations is an excellent technique. The first uses of risk assessment appears to have been for aflatoxins in 1978 and for PCBs in 1979. In 1982, risk-assessment techniques were used to determine carcinogenic impurities in color additives that were not covered by the Delaney clause. Finally, risk assessment was extended to substances covered by the Delaney clause. Beside direct food additives, which may be potential carcinogens, indirect food additives may come from the packaging materials and food-processing equipment, color additives used for ingestion and for external purposes, potential contaminants of food or color additives, and unavoidable environmental contaminants in food and cosmetic ingredients. In certain substances, such as saccharin, although the FDA has found the substance to be potentially carcinogenic, the benefit that the substance gives to the population at large is more important than the potential risk that has been shown in some studies with laboratory animals. The FDA also allows the continued use of methylene chloride to decaffeinate coffee, although it is banned for use in cosmetics. The use of this decaffeination agent is justified on the grounds that through assessment the risk is minimal to people. The indirect food additives, such as the packaging materials that have been mentioned may come in contact with food, as well as processing equipment. The FDA has banned outright two indirect food additives. They are flectol-H and NBOCA. The FDA has also banned certain uses of bottles made from acrylonitrile copolymers and polyvinyl chloride because of the concern of leaching of these chemicals into the food. There may be numerous other chemicals that are part of packaging materials that should be studied as potentially hazardous to people through consumption of food.

Animal drug residues have been tested since 1962. The Congress in 1962 passed the DES (diethylstilbestrol) Proviso that allows the use of carcinogenic drugs in animals providing that the residues cannot be detected in edible portions of tissue or foods derived from the living animals. These substances need to be studied further, since there may be potential risk for carcinogenic activity in humans, over time, as was shown later with the drug DES. DES became the only animal drug successfully banned, since there was human evidence of potential cancer to humans. Other research concerns include the potential hazard to the food supply by groundwater or surface water contaminated by a variety of organic chemicals that may be toxic and hazardous. Additional study should be done on food crops to determine if airborne pollutants that may be hazardous to people can be carried forward during the processing of food and therefore create potential problems. The study of sludge-related contaminants put on land, and their potential effect in the food chain, need to be further researched.

Chapter 4

INSECT CONTROL

BACKGROUND AND STATUS

Insects and rodents are responsible for numerous outbreaks of disease in humans and animals. These pests or vectors also cause extreme annoyance because of bites. They contaminate huge quantities of food, which must be destroyed because of the potential spread of disease. Contamination can be caused by vectors coming into contact with disease-causing microorganisms and carrying the organisms on their bodies or by vectors depositing feces or urine on food.

Vector-borne disease varies in prevalence from state to state and from year to year according to weather and climatic conditions. In determining the potential incidence of these diseases, it is necessary to identify and determine the prevalence of the vectors that act as primary transmitters. For instance, in an outbreak of viral encephalitis in St. Louis, it was determined that mosquitoes spread the disease to the South and through Illinois, Indiana, and other states in the country. Other types of data include the incidence of a given vector-borne disease in an area; the type of climate, including maximum and minimum temperatures, rainfall, prevailing winds, and their daily and monthly occurrence; life cycles of the vectors; seasonal fluctuation in vector populations and their requirements for life; any unusual conditions that may help spread the given vector-borne disease, such as floods, earthquakes, and other natural disasters.

Insects are extremely adaptable to the human environment. They are found in the air, on and under the soil, and in fresh or brackish water. They live on or in plants and animals, and compete fiercely with other species as parasites. Insects have caused enormous problems for many centuries. With numerous insecticides available, insects continue to plague humans, for they are capable of developing a resistance to insecticides.

Rat fleas have transmitted plague organisms to millions of people over the last 15 centuries. An estimated 25 million people or 25% of the population of Europe died during the great plagues of the 15th, 16th, and 17th centuries. Today, in the United States, humans still contract plague from rodents. Sylvatic plague, which is contracted from wild rodents, particularly groundhogs, in the western part of the United States, is seen most frequently today. In recent years, plague occurred in some of the southwestern states, including New Mexico. The potential for serious outbreaks of plague always exists as long as rats are present, Indian rat fleas are present, and the organism can be transmitted from rodent to rodent and from rodent to human. Pneumonic plague is a highly contagious, pneumonia-like disease related to bubonic plague. Fortunately, pneumonic plague has not appeared recently in this country.

The majority of the important human vector-borne diseases cannot be prevented by vaccines or by chemotherapy. Their control is based on the ability to reduce the source of vectors and the contact between vector and humans. Although in the last 25 years there has been a sharp reduction in many of the insect-borne diseases, they have not disappeared, and in some areas because of changes in environmental conditions or social conditions, the diseases are on the increase again. By 1972, malaria was eradicated from most of the world, yet in 1973 and 1974, during the period of conflict between Pakistan and India, malaria increased. Yellow fever was under control for many years, but there was a sharp increase in Nigeria in 1969. There was an increase in plague in South Vietnam from 1963 to 1973, and whenever wars have occurred in other areas, an increase in typhus and relapsing fevers followed. As a result of movement from urban to rural areas, there has been a sharp decrease in available sewage systems and an increase in overflowing sewage and mosquito problems. In California, people moving into the woodlands in the suburbs have become exposed to the vectors of California encephalitis and dog-heart worm. As artificial lakes are created to enhance the aesthetic beauty of areas, mosquito problems increase. Schistosomiasis increased with the creation of major dams, especially the Aswan Dam in Egypt. There were devastating epidemics of malaria in Pakistan after the construction of the Sukkar Varrage. In India a severe epidemic of malaria occurred after the construction of the Mettur Dam. A detailed assessment of status is difficult to make, since there are limited statistics available that can be organized into some uniform analysis.

SCIENTIFIC, TECHNOLOGICAL, AND GENERAL INFORMATION

Description of Insects

To identify insects, one must understand something about insect biology. Insects have a hard outer skin called an exoskeleton. The exoskeleton serves as a means of attachment of muscles and also protects the internal organs from injury. In most insects, the outer parts of the body wall are hardened into place. These hardened areas, or sclerites, are joined together by flexible body wall segments called intersegmental segments. The scleritis may be covered with numerous small structures, such as hairs, scales, protuberance, and spines, which are useful for identification. Insect bodies are divided into three main regions: the head, thorax, and abdomen. The class arachnid, comprised of such insects as ticks, mites, scorpions, and spiders, has one or two main segments instead of three (see Figure 4.1 for adult arthropods of public health significance).

External Structure

The following discussion is not meant to be a detailed description of all insects. For this type of information, it would be well to obtain a book on entomology. The first body region, or head, contains mouth parts, antennae, large compound eyes and simple eyes (ocelli). The mouth parts are used in chewing, sponging, or piercing-sucking. Insects that have chewing mouth parts, e.g., roaches and silver-fish, grind solid food. Insects that have sponging mouth parts, e.g., house flies, blow flies, and flesh flies, suck up liquid or readily soluble foods. They eat such solids as sugar and all liquids; they regurgitate a drop of saliva to dissolve the sugar. Insects such as mosquitoes, deer flies, sucking lice, and fleas have piercing sucking mouth parts. They easily pierce the skin of animals and humans, and suck their blood. The number of pairs of antennae vary with the type of insect, as do the eyes.

The second part of the insect, the thorax, contains the three sets of legs and may contain two sets of wings attached to the last two segments of the three-part thorax. Insects are identified at times by observing the thorax. The wings, when present, contain a reinforcing structure called veins. The arrangement and number of wing veins aids in insect identification.

The third body part, or abdomen, contains the spiracles, which are the external openings of the respiratory system and the external reproductive organs.

Internal Structure and Physiology

The digestive system of an insect consists of an alimentary canal. This canal runs from the mouth to the anus and is divided into several sections. Digestion takes place in the canal, and undigested food plus waste is removed as feces. The digestive system is an important part of the transmission of disease, since the insect can pick up an organism from one host and pass it on to another during feeding and defecation.

The circulatory system is not enclosed in blood vessels but circulates freely throughout the body cavity. In most insects, blood is colorless or greenish-yellow. Blood does not carry oxygen or expel carbon dioxide. Its major function is in the removal of waste products from body cells.

The nervous system of the insect contains a brain located in the head, a double

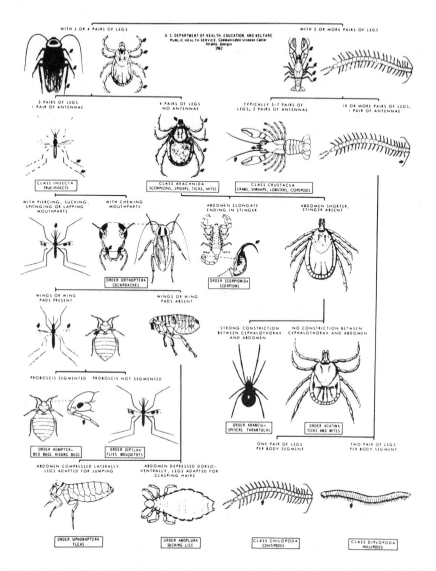

Figure 4.1.   Pictorial key to major classes and orders of adult arthropods of public health importance. *Source: Pictorial Keys to Some Arthropods and Mammals of Public Health Importance* (Washington, DC: United States Department of Health, Education, and Welfare, Public Health Service, 1964), p. 3.

nerve cord extending backward along the ventral surface of the body cavity, and a nerve center or ganglia.

Air enters the insect through spiracles into large tracheal trunks to tracheae and tracheoles.

Most insects have two sexes that mate to produce eggs. Insects that lay eggs are oviparous, and insects that deposit larvae are larviparous. Some insects have only one sex, the female, and reproduce without fertilization. Ants and bees are examples of unisexual reproducing insects. This occurs part of the time to produce workers, which are then sexually sterile.

Insect Development

The life cycle of the insect starts with the fertilization of the egg and is completed when the adult stage is reached. Some insects go through an incomplete metamorphosis. This means that their life cycle consists of egg, nymph, and adult stages. Examples of insects with incomplete metamorphosis are roaches and body lice.

Complete metamorphosis occurs in four stages, as egg, larvae, pupa, and adult. Examples of insects with complete metamorphosis are mosquitoes, flies, butterflies, and moths.

Insects molt or shed their protective exoskeleton while a new exoskeleton is being exposed. After the first day of molting, a new outer skin hardens and becomes colored.

Insect Senses and Behavior

Insects have a sense of touch, taste, smell, hearing, sight, balance, and possibly orientation. Since insect skin is hard, the sense of touch is conducted through antennae or feelers, as well as other parts of the body that are sensitive to contact and pressure. The sense of smell, located primarily in the antennae, is highly developed to aid in locating food, mates, and a place to deposit eggs. Insects vary in their degree of hearing ability. Sound waves are picked up by fine sensory hairs or special organs appearing on the side of the abdomen or the lower part of the front legs. They also hear through cuplike organs on the antennae. Visual power varies considerably among insects.

Although there is no proof that insects reason, they do have a series of instinctive, highly complex patterns that they follow. Certain stimuli cause them to behave in certain ways.

TYPES OF INSECTS OF PUBLIC HEALTH SIGNIFICANCE

The insects to be discussed include flies, fleas, lice, mites, mosquitoes, roaches, ticks, and food-product insects. It is important to understand the biology, characteristics, breeding habits, and mechanisms of spread of disease of each of these insects in order to institute adequate chemical, physical, biological, or mechanical controls.

Fleas

Fleas breed in large numbers where pets and livestock are housed. They spread readily through homes, buildings, and yards, attacking pets, livestock, poultry, and people. Fleas are the vectors of many important diseases, such as bubonic plague, murine typhus, tapeworms, salmonellosis, and tularemia.

Fleas obtain food and spread disease by biting humans and animals. Some individuals have a simple swelling as a result of the flea bite; others have severe generalized rashes. The species of fleas that commonly bite humans include the cat flea (*Ctenocephalides felis*), the dog flea (*Ctenocephalides canis*), the human flea (*Pulex irritans*), and the oriental rat flea (*Xenopsylla cheopis*).

Female fleas lay their eggs after mating on the hair of animals. The eggs drop to the floor and may become enmeshed in mats, webs, overstuffed furniture, or cellar floors. When the eggs grow to fleas, they seek a blood meal from animals or humans. In the case of rat fleas, the adult life is usually spent on the Norway or roof rat. Fleas may leave their animal host and bite humans (see Figure 4.2).

Flea Biology

Fleas are small, wingless insects that range in length from 1 to 8.5 mm. They feed through a siphon or tube after they have bitten a warm-blooded animal. The female must have a blood meal before producing eggs. Fleas cannot tolerate extremes in temperature and humidity. This is why they live close to the animals or humans that they infest. They may live on the individual's body or in the individual's shelter.

Fleas go through complete metamorphosis. Under favorable conditions, this may occur in 2–3 weeks. The eggs are laid in small batches over a long time period, which is interrupted by blood meals. The adults are usually able to feed within 1 day after emerging from the cocoon. The adult flea can live several weeks without food. This is the reason why people can take their dogs or cats along on vacation and return to still have a flea infestation.

Flies

Flies annoy humans; they bite, infest human flesh and the flesh of domestic animals, attack and destroy crops, and cause numerous, serious diseases. Although biting flies do not contain any type of toxin, they can kill the victim when attacking in the thousands. The victim may die of anaphylactic shock as the result of all the foreign protein injected into the body. Eye gnats do not bite, they damage the delicate membrane of the eye.

Many different species of flies lay eggs or larvae in the flesh of the humans and animals. The larvae invade the flesh and produce a condition known as myiasis.

Thousands of species of flies exist. The ones that will be discussed here are those most common to humans and those that cause the most disease. These flies

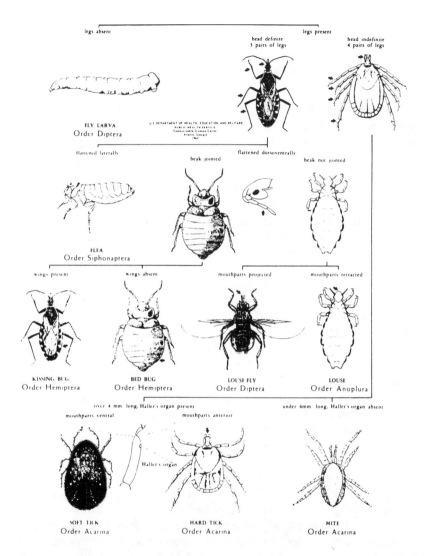

Figure 4.2.   Pictorial key to Groups of Human Ectoparasites. *Source: Pictorial Keys to Some Arthropods and Mammals of Public Health Importance* (Washington, DC: United States Department of Health, Education, and Welfare, Public Health Services, 1964), p. 4.

include the common house fly (*Musca domestica*), the stable fly (*Stomoxya calcitrans*), the face fly (*Musca autunnalis*), the blow fly or bottle fly (*Calliphora cynomyopsis, C. phaenicia, C. lucillia*, and *C. phormica*), the flesh flies (*Sarcophaga*), and the horse or deer fly (*Tavanus chrysops*). The house fly is the most important vector of microorganisms causing food-borne and infectious diseases. They are most active during daylight hours, when they move from food source to food source to obtain a balanced diet. They utilize sugar and starches for energy and to extend their life span. They utilize proteins for the production of eggs. Flies feed about three times a day. Between feedings, they rest on floors, walls, interior surfaces, ground, grass, bushes, fences, and other surfaces. During feeding on solid food, the fly regurgitates fluids to dissolve the solids, since they can only take up liquid food. During this feeding process, they regurgitate fluids containing microorganisms that cause disease and they also defecate. Flies are inactive at night. They usually rest on upper surfaces of rooms (see Figure 4.3).

Fly Biology

Flies belong to the order Diptera. This order has one or no pair of wings and halteres. Halteres are tiny knoblike structures located behind the wings and are considered to be the second pair of wings. The adult fly has three distinct body regions—the head, thorax, and abdomen. They have large compound eyes, one pair of antennae, sponging, and rasping or sucking mouth parts. The mesothorax, or middle section of the thorax, is much larger than the first and second sections. The size of the mesothorax is necessary for the powerful wing muscles used by the fly in flight.

Flies go through complete metamorphosis. The larvae feed differently and have a different habitat from the adult. The pupae are usually quiet and often are enclosed in a heavy puparium.

The house fly lives in almost all parts of the world, except the Arctic, Antarctic, and at extremely high altitudes. The normal life cycle runs from 8–20 days under average summer conditions. The female fly starts laying eggs within 4–20 days after it reaches adulthood. The eggs are deposited in batches of 75–150. The average female will lay five or six batches in cracks and crevices in the breeding medium away from direct light. The eggs hatch in about 12–24 hours during the summer months. The larvae grow quickly in the breeding medium. In warm weather the usual time for the larvae to develop through the three stages is 4–7 days. When the growth is completed, the larvae migrate from the growth medium and go into soil or under debris for the pupa stage. The pupa stage may be as short as 3 days or, at low temperatures, as long as several weeks. Following the pupa stage, the adult emerges. As an adult, the fly will mate and start the breeding process all over again. It is common to have two or more fly generations per month during warm weather. Adult flies can be kept alive for long periods of time at 50–60°F. This is why flies live through the winter and are ready to mate and start new fly populations during the warm months. Almost any kind of fresh, moist, organic matter is suitable for fly breeding. Flies can live in cereal, grain, animal manure, garbage, garbage-soaked soils, urine-soaked soils, or anywhere organic materials are present.

House flies move rapidly into new areas by flying. They travel as far as 6 mi within a 24-hr period, and possibly as far as 20 mi from their source of breeding. Flies are inactive below 45°F and are killed by temperatures slightly below 32°F.

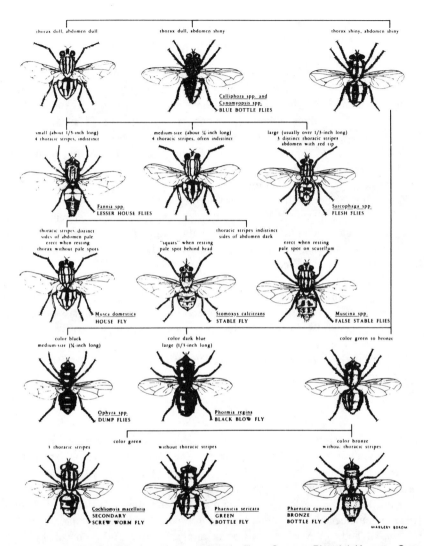

Figure 4.3.   Pictorial key to common domestic flies. *Source: Pictorial Keys to Some Arthropods and Mammals of Public Health Importance* (Washington, DC: United States Department of Health, Education, and Welfare, Public Health Services, 1964), p. 37.

Their maximum activity occurs at 90°F. There is a rapid decline of activity at higher temperatures, and death occurs at 112°F. Humidity has a definite relationship to temperature. When the humidity is high and the temperature is low or very high, flies die rapidly. Above 60°F, flies live longest when the relative humidity is 42–55%. Flies tend to move toward light and are inactive at night. This is why the ordinary flytrap is successful. Flies are sensitive to strong air currents, and therefore, are quiet on extremely windy days. During high winds, house flies may be carried as far as 100 mi. The natural enemies of flies include fungi, bacteria, protozoa, roundworms, other anthropods, amphibians, reptiles, birds, and humans.

Face flies are barely distinguishable from house flies in physical characteristics. The larvae develop in fresh animal feces and the pupa burrow into the soil. Adults hibernate in houses and barns. They suck blood and other exudates from the surfaces of mammals, although they cannot pierce the skin. They are particularly annoying because they are found around the eyes, nostrils, and lips.

Stable flies have a piercing proboscis, which sucks blood from humans or animals. They are vicious biters, usually found around stables and horses. Stable flies are suspected of transmitting disease, although no proof exists at this point.

Blow flies or bottle flies have varying degrees of iridescence in their basic coloring. Their eggs are deposited on carcasses of animals, decaying animal matter, or in garbage containing animal matter. Eggs may also be deposited on fresh meat or decaying materials. On humans, the larvae of the flies can invade living tissue and cause a condition known as myiasis.

Flesh flies are comprised of a large number of species. They are usually light grey and have three, dark longitudinal stripes on the thorax. The abdomen has a checkered pattern. Most of the species breed in the flesh of animals or in their stools. They generally do not enter buildings, and therefore, are not very significant as vectors or nuisances.

Horseflies and deer flies are large flies that produce painful bites. They create a nuisance around swimming pools, streams, and sunny portions of damp woods.

Lice

Lice bite severely, causing skin irritation and sometimes a generalized toxic reaction. They inject an irritating saliva into the skin during feeding. The itching for some people is extremely annoying. Lice have the capacity to act as vectors for numerous diseases. They are particularly a problem during times of war and disaster. The three basic kinds of human lice are the body louse (*Pediculus humanus*), the head louse (*Pediculus humanus capitis*), and the crab louse (*Phthirus pubis*). Today lice infestations occur rarely, but they still do occur. In 1974, in a school system in Indiana, a lice infestation occurred that spread throughout the school system (see Figure 4.2).

Louse Biology

Lice go through an incomplete metamorphosis. The stages are egg; first, second, and third nymph; and adult. The egg, also called a nit, attaches itself to the scalp, skin, or area around the pubic hair, as well as to underclothes. Body heat causes the eggs to hatch. The nymph then goes through its three molting stages until it becomes a sexually mature adult. The nymphal stages require 2–4 weeks if the clothing is removed from the body periodically and 8–9 days if the lice remain in contact with the body. The total life cycle may take about 18 days. Lice move quickly from person to person by means of clothing, bedding, or close contact. Infested brushes and combs are also excellent means of transmission of lice.

Mites

Mites are very small arthropods that are difficult to see with the naked eye. They do not have a distinct body segmentation. Their life cycle is very short, usually from 2 to 3 weeks. Mites increase in number very quickly under favorable conditions. They often infest food, stuffed furniture, and mattresses, and occur sometimes by the hundreds of thousands or millions. They enter premises with birds or rodents. The major problem is that they may cause several types of diseases, as well as annoyance or infestation of the body (see Figure 4.2).

Mite Biology

Since mites differ so greatly from one another, it is not possible to give a general picture of them. The mite lays eggs that hatch into larvae and then pass through two or more nymphal stages to the adult stage. When trying to identify mites, use a good entomology textbook.

Mosquitoes

Mosquitoes are small, long-legged, two-winged insects belonging to the order Diptera and the family Culicidae. The adults differ from other flies by having an elongated proboscis and scales on their wing veins and wing margin. Of the 2600 species of mosquitoes, approximately 150 are found in the United States. The ones of most concern include anopheles, aedes, culex, culiseta, mansonia, and psorophora. The discussion that follows will be concerned with the mosquitoes most hazardous or annoying to humans.

Female mosquitoes have piercing and blood-sucking organs; males do not suck blood.

There are four stages in the life cycle of the mosquito. These include the egg, larvae, pupa, and winged adult. The eggs are laid and hatched in quiet, standing water. The female lay eggs in batches of 50–200 or more, and may lay several batches of eggs. Some species glue the eggs together into a floating mass. Other species deposit eggs singly on the water, on the side of containers above the waterline, or in moist depressions. The incubation period is 3 days. It then takes 7–10 days for the larvae to mature. The larvae become adults in 4–10 days, during which time the larvae skin is shed four times. Mosquito larvae feed on minute plants and animals or fragments of organic debris.

It is difficult to determine the life span of adult mosquitoes in natural settings. However, in most of the southern species, the life span is probably only a few

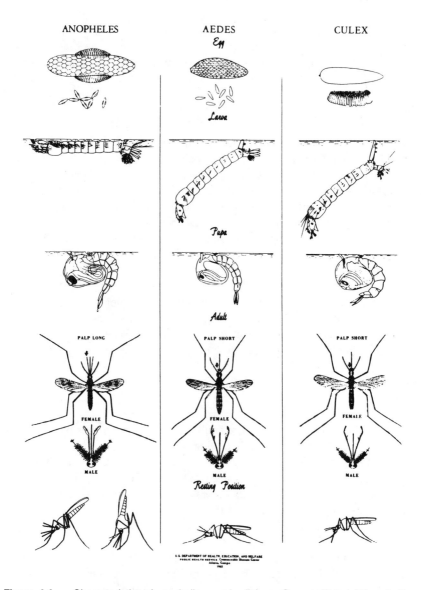

Figure 4.4.   Characteristics of anophelines and culicines. *Source*: *Pictorial Keys to Some Arthropods and Mammals of Public Health Importance* (Washington, DC: United States Department of Health, Education, and Welfare, Public Health Services, 1964).

weeks during the summer months. In the North, the female of culex, anopheles, and other species hibernate.

Mosquitoes inject saliva during feeding. It is the saliva that causes the itchy feeling (see Figure 4.4).

Mosquito Biology

Roughly equal numbers of male and female mosquitoes are produced. The male usually stays near the breeding place and mates with the female soon after reaching adulthood. Flight habits vary with the species of mosquito. Some stay within 100 yd of human habitation, and others travel as much as 10–20 mi. The preferred hosts, which include cattle, horses, other domestic animals, and humans, vary with the kind of mosquito. The female requires 2 days or more to digest a blood meal, lay eggs, and seek another blood meal. Mosquitoes may breed four or five times before they pass on disease to animals or humans. Since the spread of disease is intimately related to a given mosquito and its life cycle, each of the mosquitoes and the diseases that they transmit will be discussed.

Aedes. Of the 500 species of aedes in the world, about 40 are common to the United States. They are of greatest concern in the northern areas. All aedes lay their eggs singly, on the ground, at or above the waterline, in tree holes, or in a variety of containers. The eggs hatch after the area has been flooded. Some species can survive long dry periods. Breeding places for *Aedes* vary tremendously. They may breed in small, wet depressions left by rains or melting snows. Some breed in coastal salt marshes; others in irrigation ditches. The *Aedes* are vicious biters and bloodsuckers. Some attack during the evening hours, some during the day, and some will bite at any time.

Aedes aegypti is a vector for yellow fever. It is found in the southeastern and southern United States. It is a semidomesticated mosquito that lives very well in artificial containers in and around human habitation. Mosquitoes may breed in anything that will hold water. The life cycle can vary from 10 days to 3 weeks or more. *Aedes aegypti* are susceptible to cold and cannot survive the winter unless they are in the southern United States. The adults prefer humans to other animals, and therefore, spread disease from human to human. They usually bite around the ankles, under coat sleeves, or on the back of the neck. The adult lives about 4 months or more. Their flight range is normally from about 100 ft to 100 yd.

Anopheles. *Anopheles* mosquitoes are found throughout the United States. Most of the anophelines have spotted wings, while the culicines have clear wings. Anophelines rest in a position where their head, thorax, and abdomen are in a straight line, usually at an angle of 40–90°, while the culicines rest in a position almost parallel to the surface.

Anopheline eggs are laid singly on the water surface and float there. They are usually laid in batches of 100 or more, and hatch within 1–3 days.

The larvae are found either in fresh or brackish water. The larval stage takes from 4–5 days to several weeks, depending on the species of mosquito and the environmental conditions, especially the water temperature. The larvae eat microscopic and plant life. Most adults are active only at night. They spend the daytime resting in dark, damp shelters. Their peak of activity occurs just after dark and just before daylight. Flight range is from less than one mile to several miles.

Anopheles quadrimaculatus is the most common species of anopheles that causes malaria. It is frequently found in houses and is likely to attack humans. The mosquito is distributed throughout the southeastern United States and into the Midwest and Canada. It breeds primarily in permanent freshwater pools, ponds, and swamps that have aquatic vegetation or floating debris. It is most frequently found in shallow waters, in sunlit or densely shaded areas. The larvae can withstand temperatures below 50–55°F, although development will not be completed. Development starts to progress at temperatures of 65–70°F and is most active at temperatures of 85–90°F. It takes 8–14 days at the 85–90°F range for the larvae to complete their cycle. The female, after mating, lays eggs for 2–3 days after the first blood meal. A single mosquito may lay as many as 12 batches, totaling over 3000 eggs. The adults are inactive and rest during the daytime in cool, damp, dark shelters, such as buildings, caves, and under bridges. Feeding and other activities occur almost completely at night. They feed readily in houses on humans and on other warm-blooded animals. Usually the adults do not fly more than one-half mile from the breeding place.

Anopheles albimanus are found in southern Texas and the Florida Keys. The larvae live in freshwater ponds or brackish water. The adults enter dwellings during the night and bite people, and leave at dawn for forested areas. They have not been known to cause malaria in the United States.

Anopheles freeborni is the most important vector of malaria in the western United States. These mosquitoes enter homes and animal shelters and bite avidly at dusk and dawn. They breed in any permanent or semipermanent water that is at least partially exposed to sunlight. The larvae are also found in slightly brackish water. This species is particularly adapted to rice fields and sunny arid regions. The mosquitoes leave their hibernating places in February, obtain blood meals, and lay their eggs. As the season progresses and the area gets dryer, the mosquitoes travel longer distances, sometimes 10–12 mi to seek shelter. During the winter, they are in semihibernation.

Culex. The culex include about 300 species, of which 26 are known to inhabit the United States. Only about 12 of these species are common or pests or vectors of disease. The culex mosquitoes breed in quiet waters, artificial containers, or large bodies of permanent water, and often breed in areas where there are large quantities of sewage. The eggs are deposited in rafts of 100 or more. They float on the water surface until they hatch in 2–3 days. The adult females are usually inactive during the day, but bite during the night.

Roaches

Most of the 55 species of roaches known in the United States live outdoors and are not a problem to humans. Only five species are usually found indoors in the

United States. They mechanically carry dirt and organisms that may spread disease, and they destroy fabrics and book bindings. Their odors, which are quite offensive, ruin food. Roaches are broad, flattened, dark or light brown or black insects. They usually run at night to seek food and hide out during the daytime. When roaches are seen during the day, they are generally present in very large numbers. The females lay eggs in capsules called oothecae.

Roaches have been around for about 400 million years. Their fossil remains, which are similar to today's roaches, are abundant in the strata of certain types of soils. Roaches are present everywhere. It is important that moisture be present where they exist. They feed upon cereals, baked goods, small amounts of grease, glue, starch, wallpaper binding, fecal materials, dead animals, etc. During the feeding process, they regurgitate a brown liquid that may contaminate the food. They commonly rest in crevices, behind moldings of doors and window frames, in areas where pieces of equipment are joined together, under food preparation tables, in cabinets and closets, in the asbestos covering of hot-water lines, in cardboard boxes, in and under debris, and in almost any other place where a tiny crack may exist. They are also found in the backs of radios, television sets, paper bags from the supermarket, and in the pockets or cuffs of pants or other places in clothing (see Figure 4.5).

Roach Biology

The roach passes through incomplete metamorphosis. The stages include the egg, nymph, and adult. Roaches have an oval, flattened shape with the head hidden. They feed on almost anything. The eggs are enclosed in capsules called oothecae that are either deposited or carried at the end of the abdomen.

American Roach. The American roach, *Periplaneta americana*, is widely distributed throughout the world in mild climates, although it is a native of tropical and subtropical climates. The adult is reddish brown to dark brown. It usually forages for food on the first floor of buildings. Although it does not normally fly, it has the ability to do so. The female carries the egg capsule for a day or two, and then glues it to some object in a protected area. In 2 or 3 months, about 12 nymphs will hatch from each capsule. Adult females live for at least a year. During this period of time, they lay many capsules. The nymph, which is the same color as the roach, passes through 13 molts as it matures during the course of a year. It can grow very slowly where conditions are unfavorable. The length of the roach ranges from $1^1/_2$ to 2 in.

Brown-Banded Roach. The brown-banded roach, *Supelle supellectilium*, is found throughout the world, although its main habitat is the tropical areas. The adult is about $^1/_2$ in. in length, light brown, and has mottled wings. The female is reddish brown and lighter. The eggs are laid and hatched in the same way as the American

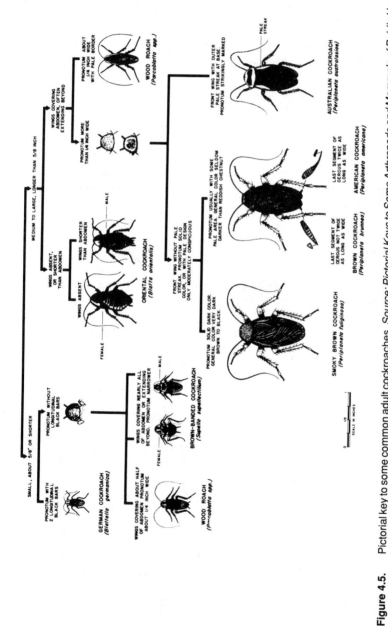

Figure 4.5.   Pictorial key to some common adult cockroaches. *Source: Pictorial Keys to Some Arthropods and Mammals of Public Health Importance* (Washington, DC: United States Department of Health, Education, and Welfare, Public Health Services, 1964), p. 18.

roach. The color of the nymph is the same as the adult. It matures in 4–6 months. It lives in almost any part of buildings.

German Roach. The German roach, *Blattella germanica*, has a very wide range. It is most commonly found indoors and is the roach that readily becomes resistant to insecticides. The adult is light brown with longitudinal stripes on the back. The female carries oothecae for about a month and then drops it a day or so before the eggs are ready to hatch. There are about 30 nymphs in each capsule. The adult is about $^6/_{10}$ in. in size. The color of the nymph is the same as the adult. It undergoes six molts and matures in about 4–6 months. It lives in kitchens, bathrooms, and other areas.

Oriental Roach. The oriental roach, *Blatta orientalis*, is spread throughout the world. The adult is black to dark brown, with vestigial wings in the female. It is quite sluggish. It usually forages in the basement or first floor of buildings. The eggs are laid in an egg capsule, which is glued to some object in a protected place. About 12 nymphs hatch from each capsule in about 2–3 months. The roach is 1 to $1^1/_4$ in. in length, with the female being a little longer than the male. The color of the nymph is the same as the adult. It matures in about a year.

Ticks

Ticks belong to the class Arachnida. They are not true insects, but are important vectors of significant arthropod-borne disease (see Figure 4.2).

Tick Biology

Ticks belong to the same class as mites, spiders, and scorpions. They have three characteristics that distinguish them from insects: the head, thorax, and abdomen are all fused together in one body region; no antennae; and four pairs of legs in the nymph and adult stages instead of three. Ticks are divided into two groups, soft and hard.

The tick has a piercing organ that has barbs, allowing the tick to anchor itself to the host. It also has cutting organs to allow the piercing organ to enter the skin for a blood meal. The leathery body covering is capable of being greatly distended with blood. The female tick is somewhat flattened and tapered toward the anterior end before it takes its blood meal. Ticks are adaptable for blood sucking because they have a powerful, pumping pharynx. The blood is pumped backward into the esophagus and then into the stomach.

Ticks metamorphize completely. The female hard tick will feed once and lay large batches of up to 18,000 or more eggs. Most of the soft ticks take several blood meals and then deposit 20–50 eggs in a batch. The eggs hatch in 2 weeks to several months, depending on the environmental conditions. The larvae have difficulty attaching themselves to a host, and therefore, may go through long periods of starvation. After a blood meal, the engorged larvae usually will drop to the soil, shed its skin, and enter the nymph stage. The nymph may wait for long periods for a suitable host. After it has its blood meal, the nymph drops from the host, molts, and becomes the adult. The life cycle may take from less than a year to 2–3 years or longer. Ticks survive if they can get another blood meal. During this period the sexual organs are formed, and both the female and male tick become bloodsuckers and feed several times before copulation. The male hard tick will have a blood meal, usually copulate with one or more females, and then dies. The female drops to the ground and starts to develop eggs over a period of several days. Once the eggs are deposited, the hard tick female dies. Unfed hard ticks can survive frigid winter weather if they are sheltered. Engorged ticks and eggs are less resistant. Ticks are most active during the spring and summer. They usually die quickly if the area is dry. Soft ticks can exist in much dryer situations.

Food-Product Insects

Stored food-product insects become a serious problem in the kitchen, in the warehouse, pantries, or any other place where dried foods are stored. These pests eat or contaminate the food, making it unfit for human consumption. They are also annoying, since they leave the infested food and crawl or fly around the area. These food product insects include the confused flour beetle, saw-tooth grain beetle, flour grain beetle, larder beetle, black carpet beetle, cabinet beetle, grain weevil, rice weevil, Indian meal moth, Mediterranean flour moth, and bean and pea weevils. These insects may forage on dry foods, furs, skins, wool, silk, cheese, and almost any other product made from dried meats, cheese, or other organic materials.

ENVIRONMENTAL AND HEALTH PROBLEMS CREATED BY INSECTS

Destruction of Food and Material

Insects infest cattle, causing weight loss, annoyance, and eventually a reduction in the supply of acceptable food. Further, they infest stored foods and contaminate them with feces, body parts, or live insects. Common stored food-product insects include grain and flour beetles; *Dermes pidae* beetles, which feed on a variety of plant and animal products, including birds, skins, live meat products, woolen and silk materials, cheese, and cereal grain products; grain weevils, which feed on stored whole grain; and bean and pea weevils, which infest dry beans and dry peas.

Transmission of Disease

Disease is transmitted from human to human, from human through vector to human, from human through vector to animal to human, or from human through vector through animal through vector to human. It is very important to understand these various types of disease transmission, since once the life cycle of the disease organism is known, it becomes much simpler to break the chain of infection and to stop the outbreak of the vector-borne disease.

Disease is transmitted mechanically, for example, flies pick up microorganisms of dysentery, typhoid fever, or cholera on their feet or body hair and transmit them to humans or their food; or biologically through various parts of the life cycle of the arthropod. In developmental biological transmission, the parasite grows or multiplies in the insect. For example, the viruses causing yellow fever or encephalitis increase in number within the body of the mosquito. In cyclical biological transmission, the parasite goes through an essential part of its life cycle in the arthropod, without increasing in number. For example, in filariasis, the mosquito sucks up a number of microfilaria into its gut and eventually these organisms enter the thoracic muscles, where they pass through several stages of their life cycle and then move to the proboscis of the mosquito. When the mosquito feeds, it infects the host. In cyclodevelopmental transmission, both an essential cycle occurs in the arthropod host and the parasites increase in number. An example of this would be the parasite *Plasmodium vivax*, which causes malaria.

In some cases, humans are accidentally infected with a disease that ordinarily would be an animal disease. This is called a zoonoses. An example of this is jungle yellow fever, where jungle mammals such as monkeys are normal hosts and wild mosquitoes transmit the disease from monkey to monkey. From time to time, the insect will bite a human, who then develops the disease. Another example of this is viral encephalitis transmitted by mosquitoes, such as St. Louis encephalitis, where the normal hosts would be wild birds.

Numerous tick- and mite-borne diseases are spread by eggs from the infected parents to the next generation. This is congenital transmission of the pathogen. The eggs are laid either on humans or animals, and when they hatch and become insects they transmit the disease to the host. Examples of this would be tick-borne typhus, tick-borne encephalitis, tularemia, relapsing fever, and scrub typhus.

People can also be poisoned by the bites and stings of insects. The victim may suffer an insignificant blister, anaphylactic shock, or death. Anaphylactic shock can result from an individual reaction to the protein or a large amount of protein injected into the body by the arthropods. Some insects, such as the black widow spider, cause death by their extremely poisonous toxin.

An infestation occurs when living insects burrow into the skin of the host. Lice or scabies burrow under the skin causing infestation. Table 4.1 lists some of the arthropods that affect human comfort.

Table 4.1 Some Arthropods Affecting Human Comfort

Common name	Scientific name of arthropod	Effect on man
Mites		
Chigger	*Trombicula alfreddugesi*	Intense itching; dermatitis
Rat mite	*Ornithonyssus bacoti*	Intense itching; dermatitis
Grain itch mite	*Pyemotes ventricosus*	Dermatitis and fever
Scabies mite	*Sarcoptes scabiei*	Burrows in skin, causing dermatitis
Ticks		
Hard ticks	*Amblyomma spp.*	Painful bite
	Dermacentor variabilis,	
	D. andersoni and	Tick paralysis; usually fatal
	Ixodes spp.	if ticks not removed
Soft ticks	*Ornithodoros spp.*	Some species very venomous
Spiders		
Black widow spider	*Latrodectus mactans*	Local swelling; intense pain and occasionally death
Brown spider	*Loxosceles reclusa*	Extensive loss of affected body tissue
Scorpions	Order: Scorpionida	Painful sting; sometimes death
Centipedes	Class: Chilopoda	Painful bite
Mayflies	Order: Ephemerida	Asthmatic symptoms from inhaling fragments
Caddis flies	Order: Trichopera	Asthmatic symptoms from inhaling hairs and scales
Lice	*Pediculus humanus and Pthirus pubis*	Intense irritation, reddish papules
Bugs		
Bed bugs	*Cimex lectularius*	Blood suckers; irritating to some
Kissing bugs	Family: Reduviidae	Painful bite: local inflammation
Beetles		
Blister beetles	Family: Meloidae	Severe blisters on skin from crushed beetles
Rove beetles	Family: Staphylinidae	Delayed blistering effect
Caterpillars	Order: Lepidoptera	Rash on contact with hairs or spines
Bees, wasps, and ants	Order: Hymenoptera	Painful sting; local swelling
Fleas	Order: Siphonaptera	Marked dermatitis frequent
Flies		
Punkies or biting midges	*Culicoides* spp.	Nodular swelling, inflamed
Black flies	*Simulium* spp.	Bleeding punctures; pain, swelling
Sand flies	*Phlebotomus* spp.	Stinging bite; itching; whitish wheal
Mosquitoes	Family: Culicidae	Swelling; itching
Horse flies	*Tabanus* spp.	Painful bite
Deer flies	*Chrysops* spp.	Painful bite
Stable flies	*Stomoxys calcitrans*	Painful bite

Source: *Introduction to Arthropods of Public Health Importance,* H. D. Pratt and K. S. Littig, DHEW Publication No. (HSM) 72–8139 (1972).

Insect-Borne Diseases

Bubonic Plague

Bubonic plague is a highly fatal infectious disease, with toxemia, high fever, reduced blood pressure, rapid and irregular pulse, mental confusion, and prostration. Bubonic plague can become pneumonic plague when the disease spreads to the lungs. The incubation period in bubonic plague is 2–6 days. The incubation period in pneumonic plague is 3–4 days. The etiological agent is *Pasteurella pestis*. The insect involved is the oriental rat flea, *Xenopsylla cheopis*. The reservoir of infection is wild or domesticated rats, and possibly other rodents.

Colorado Tick Fever

Colorado tick fever is an acute fever lasting 2–3 days, with occasional encephalitis occurring. The incubation period is 4–5 days. The etiological agent is a virus. The insect involved is an infected tick, most frequently *Dermacentor andersoni*. The reservoir of infection is small mammals.

Dengue Fever

Dengue fever causes a sudden onset of fever, intense headache, joint and muscle pain, a rash, and prolonged fatigue, as well as depression. The incubation period is typically 3–15 days, but most often 5–6 days. The etiological agent is a virus of the group B Togaviruses. Various aedes, including *Aedes aegypti* are responsible for causing this disease. The reservoir of infection is people, mosquitoes, and monkeys.

Encephalitis

Arthropod-borne viral encephalitis includes eastern equine encephalitis, western equine encephalitis, California encephalitis, and St. Louis encephalitis. The symptoms of encephalitis include a short period inflammation, which involves parts of the brain, spinal cord, and meninges. Symptoms also include headaches, high fever, stupor, and disorientation. Fatality rates range from 5 to 60%. The incubation period is 5–15 days. The etiological agents vary with the type of the disease. Eastern and western equine encephalitis are caused by a virus in group A Togaviruses, whereas St. Louis encephalitis is caused by a virus in the group B Togaviruses. Various mosquitoes, including *Culex tarsalis, Culiseta melanura,* and other mosquitoes, transmit the disease. The reservoir of infection is suspected to be birds, rodents, bats, and reptiles. Surviving mosquito eggs or adults may keep the virus alive over the winter.

Lyme Disease

The early symptoms of Lyme disease includes a rash that starts out flat or raised and may develop into blistering or scabbing in the center. The lesions may have a bluish discoloration. Other symptoms include fatigue, headache, neck stiffness, jaw discomfort, pain or stiffness in muscles or joints, slight fever, swelling of glands, or conjunctivitis. If the symptoms of the disease are untreated, later symptoms occur, which include complications of the heart, nervous system, or joints. The incubation period ranges from 3 to 32 days after a tick bite. Typically a red ring-like lesion develops at the site of the tick bite. The etiological agent, which is a spirochete, is called *Dorrelia durgdorferi*. A tiny tick found in northeastern United States, and as far south as Virginia, called *Lxodes dammini*, spreads the disease. In the south and southwest, *Lxodes scaturlaria* spreads the disease. On the west coast, *Lxodes pacificus* spreads the disease. Dogs, cats, horses, cows, and goats are domestic animals that may carry the ticks and the disease. Also, wild animals may carry the ticks and the disease. These are the reservoirs of infection.

Malaria

The symptoms of malaria include fever, chills, sweating, headache, shock, renal failure, acute encephalitis, and coma. In nontreated cases, there is a fatality rate of greater than 10%. The incubation period varies with the type of plasmodium causing the disease. *Plasmodium falciparum* has an incubation period of 12 days. *Plasmodium vivax* has an incubation period of 14 days. *Plasmodium ovale* has an incubation period of 14 days. *Plasmodium malaria* has an incubation period of 30 days. In certain strains of *Plasmodium vivax*, the incubation period is 8–10 months. The etiological agents have just been described. The anopheles mosquito transmits the disease. The reservoir of disease is people.

Relapsing Fever

The symptoms of relapsing fever include periods of normal temperature and then elevated temperature. In untreated cases, the fatality rate is usually 2–10%, although it may exceed 50% in the epidemic louse-borne disease form. The incubation period is 5–15 days but usually 8 days. The etiological agent is *Borrelia recurrentis*. The disease is transmitted either through lice or ticks. The reservoir of infection for the louse-borne disease is people. The reservoir of infection for the tick-borne disease is wild rodents or through transovarian transmission.

Rickettsial Pox

The symptoms of the disease include skin lesions, chills, fever, rash, headache, muscular pain, and malaise. The incubation period is probably 10–24 days. The etiological agent is *Rickettsia akari*. The rodent mite, *Liponyssoides sanguineus*, is the insect that spreads the disease through its bite. The reservoir of infection is the infected house mouse.

Rocky Mountain Spotted Fever

Rocky mountain spotted fever is also known as tick fever or tick-borne typhus fever. The symptoms of the disease include a sudden onset of a persistent fever, headache, and chills. A rash appears on extremities and spreads to most of the body. There is a 20% fatality rate in untreated cases. The incubation period is 3–10 days. The etiological agent is *Rickettsia rickettsii*. The disease is spread by the bite of an infected tick or by contamination of the skin with the crushed tissues or feces of the tick. The reservoir of infection is the tick, which may transmit the disease either through transovarian or transstadial passage within the tick. The tick may also be transferred from animals to people.

Scabies

The symptom of the disease is an intense itching, especially at night. The incubation period ranges from several days to several weeks. The etiological agent is *Sarcoptes scabiei*. The insect involved is the mite. The means of spread is through direct contact or by indirect contact by means of underclothing. The reservoir of infection is people.

Scrub Typhus

The symptoms are skin ulcers followed by fever, headache, profuse sweating, eruptions on various parts of the body, cough, and pneumonitis. The fatality rate ranges from 1 to 40%. The incubation period is from 6 to 21 days, but usually 10–12 days. The etiological agent is *Rickettsia tsutsuganushi*. A mite causes the disease. The reservoir of infection is the infected larval stages of mite.

Tularemia

Tularemia, which is also known as deerfly fever or rabbit fever, causes ulcers to form at the site of the innoculation by the insect. Headache, chills, and a rapid rise of temperature occurs. There is about a 5% fatality rate in untreated cases. The incubation period is 1–10 days, but usually 3 days. The etiological agent is *Francisella tularensis* (*Pasteurella tularensis*). Although the disease may be transmitted by handling or ingesting insufficiently cooked rabbit or hare meat, or by the drinking of contaminated water or inhalation of dust from contaminated soil, it is also spread by a variety of insects. They include the deerfly; the mosquito aedes; the wood tick, *Dermacentor andersoni*; the dog tick, *D. variabilis*; and the lone star tick, *Amblyomma americanum*. The reservoir of infection includes many wild animals, especially rabbits and beavers.

Typhus Fever — Epidemic (Louise-Borne)

Epidemic typhus fever causes symptoms of headaches, chills, prostration, fever, pain, and toxemia. In untreated cases, the rate of fatality varies from 10 to 40%. The incubation period is from 1 to 2 weeks, but usually 10 days. The etiological agent is *Rickettsia prowazeki*. The most common insect spreading the disease is the body louse, *Pediculus humanus*. The reservoir of infection is people.

Typhus Fever — Flea-Borne

Flea-borne typhus fever, also known as murine typhus or epidemic typhus fever, causes the same symptoms as the louse-borne epidemic typhus fever but they are milder. If the individual is untreated, the death rate is about 2%. The incubation period is 1–2 weeks but usually 12 days. The etiological agent is *Rickettsia typhi*. The insect involved is the flea, primarily the rat flea, which is *Xenopsylla cheopis*. Rats are the reservoir of infection, but cats and other wild or domestic animals may spread the disease.

Yellow Fever

The symptoms of the disease are fever, headache, backache, prostration, nausea, vomiting, and jaundice. The fatality rate in areas where yellow fever is endemic is less than 5%. In other areas, it may be as high as 50%. The incubation period is 3–6 days. The etiological agent is a virus of the B Togaviruses. The disease is spread by the bite of the infected *Aedes aegypti* mosquitoes.

Fleas

There are many important diseases transmitted by fleas beside bubonic plague. Murine typhus fever is a disease caused by *Rickettsia typhi*. The organism is transmitted by fleas to humans. The important vector in the United States is the oriental rat flea. The disease is spread from rodent to rodent and occasionally to humans. Infection occurs when the site of the flea bite is contaminated by the flea's feces. Although murine typhus fever has largely disappeared, it still could be the cause of a serious epidemic.

Fleas spread tapeworm infestations. They act as an intermediate host for several tapeworms that usually infest dogs and cats or rodents, and occasionally infest humans. The dwarf tapeworm (*Hymenolepis mana*) frequently infests children.

Salmonella enteritidis, which often causes outbreaks of food-borne disease, may be transmitted from rats and mice to humans via fleas. The flea contaminates food with its feces or it contaminates humans directly.

Flies

Flies transmit disease mechanically or biologically. Since flies breed and feed in human feces, they are capable of carrying organisms mechanically to food or to people on their mouthpart, in their vomitus, on their hairs, which are on their body and legs, on the sticky pads of their feet, or in their feces. Mechanically transmitted diseases include typhoid fever, paratyphoid fever, cholera, bacterial dysentery, amoebic dysentery, pinworm, roundworm, whipworm, hookworm, tapeworm, and salmonella other than those that are the causative agents of typhoid and paratyphoid fever. Biologically transmitted diseases include African sleeping sickness, onchocerciasis (blinding filariasis), loiasis (African eyeworm disease), bartonellosis, and sandfly fever.

Lice

Lice are responsible for the spread of epidemic typhus fever, trench fever, and relapsing fever. Typhus fever has resulted in tens of thousands of deaths and hundreds of thousands of illnesses. Typhus fever is caused by a rickettsia. The rickettsia goes through part of its life cycle within the louse. It may change from the organism, causing murine typhus fever, which is rather mild, or to the mutant causing epidemic typhus fever, which is extremely serious. The rickettsia multiplies rapidly within the midgut of the louse after it has ingested blood with the organism *R. prowazeki*. The epithelial cells of the midgut become so loaded with rickettsiae that they rupture and are released into the contents of the gut and feces of the insect. The insect may die as the result of this. The disease is transmitted through bites or feces deposited on human skin.

Trench fever caused over a million cases of illness in soldiers on the western front in World War I. In World War II, the organism and disease once again appeared on the German-Russian front. The organism responsible for this disease is *Rickettsia quintana*.

Relapsing fevers are a group of closely related diseases in which the patient's temperature rises and falls at regular intervals. Louse-borne relapsing fever caused epidemics in parts of Europe, Africa, and Asia. When the louse bites the patient with relapsing fever, it takes in about 1 mg of blood which may contain many spirochetes. The spirochetes leave the gut of the louse and reappear in great numbers in the blood of the louse in about 6 days. The louse becomes infected for life. The disease is not transferred to human beings by bites or feces, but only by crushing the louse or damaging it in some way so that the blood contaminates the skin or mucous membrane of the human being.

Mites

Scabies or mange-like conditions are produced primarily by the mange or itch mites. Sometimes the mites cause only mild infections, but often they can cause serious infections, serious skin irritations, and severe allergic reactions. Outbreaks of scabies have occurred, especially during wartime and other emergency periods when many people are together in close quarters. The mites burrow under the skin and leave open sores to become the source of the secondary infection. The victim becomes very pale and loses considerable sleep. The individual may have intense itching, redness, or rashes. The primary means of transmission of the scabies mite is through physical contact with an infected person. Itch and mange mites are also found on domestic animals.

Chiggers transmit the rickettsia causing scrub typhus, a disease that has afflicted Americans, especially during wartime while in the Orient. Scrub typhus occurs most frequently where there are tall grass fields and neglected coconut plantations. The rickettsia are transmitted from infected victims through the egg to the larvae, which then feeds on rodents and humans. In past outbreaks of scrub typhus, mortality rates varied from 3 to 50%, depending on the virulence of the strain of rickettsia. Chigger bites are thought to transmit hemorrhagic fever. The disease, which is probably caused by a virus, affected American troops in Korea. The disease is also found in Siberia and Manchuria. It is fatal in about 5% of the cases.

The house-mouse mite transmits *Rickettsia akari* from the house mouse to humans. This causes outbreaks of rickettsial pox. Another possible disease transmitted by mites is encephalitis, which is usually found in fowl or wild birds. A number of different mites, especially the chiggers and rodent- and bird-mites, may cause skin irritation or dermatitis. These mites are found in tropical or subtropical areas, including the subtropical areas of the United States. They are most abundant in wooded areas, swamps, roadsides, or any place inhabited by birds or wild rodents. Occasionally mites have been found to invade the respiratory tract of laboratory animals such as dogs, monkeys, and birds. Mites may serve as the intermediate host for tapeworms. The tapeworm is transmitted from one host to another. Of course, the possibility always exists that humans are the inadvertent host.

Mosquitoes-Aedes

Aedes aegypti is the principal mosquito vector of yellow fever. The virus lives in the victim's blood for 3–4 days at the onset of illness. During this time, a mosquito can bite the infected human and start the process of spreading yellow fever once again. The mosquito becomes infectious after a period of 10–14 days and can harbor the virus for life. There is no known means of infecting human beings other than by the mosquito bite, which transmits the filterable virus to the person. Destroying the breeding places of *Aedes aegypti* is the principal means of controlling the spread of disease.

Other problem-causing aedes are discussed briefly in the following section. *Aedes atlanticus-tormentor-infirmatus* are vicious biters and have been known to drive cattle from woodlands. The virus of California encephalitis was isolated from these mosquitoes. *Aedes canavensis* are serious pests in woodlands and also transmit the virus of California encephalitis. *Aedes cantator* live in salt marshes and migrate to shore town and resorts, and can be a nuisance on the east coast. *Aedes cinerus* occasionally pester the northern states. *Aedes borsalis* is a severe pest for humans and cattle throughout the arid and semiarid parts of the western United States. It is a vicious biter and attacks during the day or night. *Aedes nigromis* inhabits the western plains from Minnesota to Washington and south to Texas and Mexico. This mosquito is a serious pest that bites severely in the daytime. *Aedes punctor* and other related species are important pests in woodlands and recreational areas in northeastern United States and mountains of the West. *Aedes sollicitans* live in salt marshes and are severe pests along the Atlantic and Gulf Coasts and are also found in the Midwest. They are fierce biters that migrate just before dark. *Aedes seencerii* are found on the prairies of Minnesota, North Dakota, and Montana, and in several northwestern states. They are fierce biters that attack during the day. *Aedes sticticus* are found mostly in the northern states. They are usually quite abundant after a flood. They are severe biters during the evenings and also bite in cloudy or shady areas during the day. The *Aedes stimulans* are found in the northern states. They bite during the daytime. *Aedes taeniorhynchus* are found on coastal plains along the Pacific Coast. They are most abundant in salt marshes and are severe pests and fierce biters, primarily at night. *Aedes triseriatus* are fierce biters found in tree holes, old tires, tin cans, and other artificial containers. *Aedes trivittatus* are fierce biters and annoying pests found in the northern United States. They usually rest in grass and vegetation during the day. *Aedes vexans* are found in rain pools or flood waters in the northern states from New England to the Pacific Coast. The adults travel 5–10 mi from breeding places. They viciously bite at dusk or after dark. There are many other species of *Aedes* too numerous to discuss in this volume; the principal species have been covered.

Anopheles

The major vectors of malaria include *Anopheles albimanus* in the Caribbean region, *Anopheles freeborni* in the western part of the United States, and *Anopheles quadrimaculatus* in most parts of the world and east of the Rocky Mountains.

Humans are the only important reservoir of human malaria. The disease is transmitted when the anopheles ingest human blood containing the plasmodium in the gametocyte stage. The parasite develops into sporozoites in 8–35 days. The sporozoites concentrate in the salivary glands of the mosquitoes and are injected into humans as the mosquito takes its blood meals. The gametocyte appears in the blood of a host within 3–14 days after the onset of the symptoms. Malaria is also transmitted from person to person through blood transfusions or contaminated syringes.

The incubation period varies, depending on the organism, from 12 days to 10 months. The infection may be transmitted as long as the gametocytes are present in the patient's blood. The disease is carried by a patient for as little as 1 year or as long as a lifetime depending on the organism. The *Plasmodium vivax* is usually carried from 1 to 3 years. *Plasmodium falciparum, Plasmodium ovale*, and *Plasmodium malariae* are also malaria parasites.

Other troublesome anopheles include *Anopheles punctipennis*, which is a vicious biter out of doors and apparently does not enter homes readily. It breeds in rain barrels, hog wallows, grassy bogs, swamps, and margins of streams. *Anopheles walkeri* breeds in sunny marshes along the edges of lakes and in saw grass. It readily bites humans and is a good laboratory vector of malaria.

Culex

Culex pipiens pipiens, the northern house mosquito, lives in the northern United States; *Culex pipiens quinquexasciatus*, the southern house mosquito, lives in the southern United States. Both of these mosquitoes are severe pests. They are extremely annoying, bite fiercely, and feed continuously on humans. Both of these mosquitoes breed in huge quantities in rain barrels, tanks, tin cans, storm-sewer catch basins, poorly drained street gutters, polluted ground pools, cesspools, open septic tanks, effluent drains from sewage disposal plants, and in any other highly unsanitary condition. The mosquitoes do not migrate far from their area of harborage. They are capable of transmitting innumerable diseases.

Culex tarsalis are naturally infected with the viruses of St. Louis and Western encephalitis, and have the capability to transmit either of these diseases. The mosquito bites birds and then transmits the virus to other birds, horses, or humans. It is the most important vector of encephalitis in humans and horses in the western United States. The insect is active soon after dusk and enters buildings for its blood meal. It is widely distributed west of the Mississippi River, and also in southern Canada and northern Mexico. The larvae develop in a variety of aquatic settings. They develop in arid and semiarid regions, wherever water is trapped, in effluent from cesspools and other organic materials, and in artificial containers. The females deposit 100–150 eggs in a raft. Hatching occurs within 18 hr. Adults are active from dusk to dawn and remain at rest in secluded areas during the day. They are found on porches, shaded sides of buildings, under bridges, or in other protected areas. The majority of these mosquitoes rest in grass and shrubs or along the banks of streams. They fly up to 11 mi, even though they normally stay within 1 mi of their breeding place.

Encephalitis, which is caused by an arthropod-borne virus, is a group of acute, inflammatory diseases of short duration that involve parts of the brain, spinal cord, and meninges. The disease is not transmitted directly from human to human, but from mosquito to bird to mosquito to human.

Culiseta

There are 10 species of culiseta in the United States, of which five are fairly widespread. Although they are relatively unimportant as pests, two of the species are naturally infected with encephalitis viruses. It is not known, however, whether they spread encephalitis to humans. These two species are *Culiseta inornata*, which are frequently found in cold water, and *Culiseta melanura*, which are found in most of the eastern United States from the Gulf states to Canada.

Mansonia

Of the three species of Mansonia found in the United States, one is widespread and common. *Mansonia perturbans* is a serious biter and pest found in the southern and eastern states, and also in the Great Plains, Rocky Mountain, and Pacific Coast states. In Georgia, this mosquito was found to be naturally infected with the virus of Eastern encephalitis. It breeds in marshes, ponds, and lakes that have a thick growth of aquatic vegetation. The larvae develop slowly. The difficulty with this mosquito is that it cannot be killed in its larvae stage by an ordinary surface larvacide, since its breathing equipment is inserted into plants through which it breathes. The females bite during the daytime in shady humid places, but they are most active in the evening and the early part of the night. They enter houses and bite viciously.

Psorophora

Thirteen species of *Psorophora* are found in the United States, 10 of which are rather widely distributed in the southern and eastern states. Their breeding habits are similar to the *Aedes*. *Psorophora ciliata* which breeds in temporary pools, is a vicious biter and a serious nuisance in the south, midwest, and eastern United States. It attacks during the day or evening. *Psorophora confinnis* is a fierce biter that attacks in the day or night. Their numbers are often so great that they kill livestock. They are principally found through the southern United States and in Nebraska, Iowa, New York, Massachusetts, and southern California. They breed in temporary rain pools, irrigation waters, or seepage pits. They are a great problem in rice fields. *Psorophora cyanescens* is a severe pest that attacks in the day or night. It is particularly abundant in Oklahoma, Arkansas, Alabama, Mississippi, Louisiana, and Texas. *Psorophora ferox* is a vicious, painful biter found in wooded areas of the south and east.

Roaches

Although there are no records of outbreaks of disease caused by roaches, it is possible that they act as vectors. Roaches cross areas containing sewage or fecal material. They also cross food and could mechanically transmit enteric organisms from their breeding places to food. It is also possible that they transmit organisms through their feces and through the small amount of material regurgitated while they are feeding. Roaches have been shown to carry salmonella in their gut.

Ticks

Ticks can transmit pathogenic organisms mechanically or biologically. Their mouthparts become contaminated when feeding on an infected animal and may in turn contaminate another animal or human while feeding. This is mechanical transmission. Further, ticks serve as reservoirs of viruses, rickettsiae, spirochetes, bacteria, and protozoa. These organisms are transmitted from the infected adults through the eggs to the larval, nymphal, and adult stages. This is called transovarial or transstadial transmission. This mechanism is of considerable concern because the tick-borne disease can survive adverse weather conditions, since the organisms are being transmitted through the eggs.

Tularemia or rabbit fever is caused by the bacillus *Pasteurella tularensis*. This bacterial disease infects numerous hosts, including rabbits, rodents, and humans. Ticks transmit this disease organism to the genera *Dermacentor* and *Haenaphysalis*. In addition, the deer fly, which is also a blood-sucking arthropod, may be an important vector of the spread of tularemia. Ticks transmit the disease organisms to their own eggs, keeping the organism alive from generation to generation.

There are two peaks of tularemia in the United States, occurring during the fall-winter season and also during the spring-summer season when the ticks and deer flies are most prevalent.

Protozoal diseases spread by ticks include cattle tick fever, Texas cattle fever, and anapromosis. The diseases are spread primarily from cattle to cattle. Rickettsial diseases include Rocky Mountain spotted fever and Q fever. Rocky Mountain spotted fever is caused by the organisms *Rickettsia rickettsia*. The wood tick, *Dermacentor andersoni*, and the rabbit tick, *Haenaphysalis leporistalustris,* spread the disease from animal to animal and from animal to human in the western part of the country. The American dog tick, *Dermacentor variabilis*, is the most important vector in the eastern part of the country. The Lone Star tick, *Amblyomma americanum*, is probably a vector of Rocky Mountain spotted fever in Texas, Oklahoma, and Arkansas. These *Dermacentor* ticks easily spread spotted fever, because the larvae and nymphs feed on infected animals and the adults attack humans or other large animals. Spotted fever has been reported in 46 of the 50 states in the United States. The western strain of the disease is much more virulent than the eastern strain.

Rocky Mountain spotted fever resembles typhus fever, but it is a more severe infection. The mortality rate may be as high as 25%. The rickettsiae invade the endothelial cells and smooth muscle of the arterioles, which may cause obstruction and destruction of the vessels. Although the disease is not transmitted directly from human to human, the tick remains infected for life and continues to transmit the disease to humans for periods of up to 18 months.

Q fever is a rickettsial disease caused by the organism *Coxiella burnetii*. Outbreaks of the disease often occur in stockyard workers. The wood tick and Lone Star tick, among others, spread rickettsiae to humans. The disease is also airborne or is passed through raw milk from infected cows. The tissues and feces of ticks become massively infected. It is suspected that the most common means of transmission occurs when the organisms are inhaled in dust or droplet form. About 1% of untreated victims die of Q fever. Relapsing fevers, which are tick or louse borne, are caused by spirochetes. The diseases occur on all continents, with the possible exception of Australia. The spirochetes causing relapsing fever most frequently in the United States belong to the genus *Borrelia*. The louse-borne type of the disease is usually due to *Borrelia recurrentis*. The tick-borne disease is spread through soft ticks in the genus Ornithodoros found in 13 western states. The spirochetes are transmitted through the offspring of the ticks or through their rodent hosts, which then serve as reservoirs for the disease.

The viral disease spread by ticks is Colorado tick fever. New cases occur in the Rocky Mountain areas of the west. The wood tick, or *Dermacentor andersoni*, carries the disease. Although dog ticks are shown to carry the virus, they have never been incriminated in any outbreak of the disease. The reservoirs of infection include ground squirrels, porcupines, and chipmunks.

Ticks, in addition to causing the above-mentioned diseases, may also cause a paralysis of considerable concern to both physicians and veterinarians.

Food-Product Insects

Stored food-product insects are a nuisance and contaminate food in large quantities. They must be destroyed and removed from the food supply. Their role as agents in the spread of food-borne diseases has not been proven.

Impact of Insects on Other Environmental and Health Problems

Pesticides are the major means of control of pests today. Unfortunately, the World Health Organization estimates that approximately 500,000 cases world-wide of pesticide intoxication occur annually and that about 1% of the victims die as a direct result of contact with the pesticide. Pesticides also cause resistance in a disease-bearing vector, thereby exacerbating disease production.

Economics

It is difficult to make a good economic analysis of a pest control program. Public health specialists are unable to evaluate the associated costs and benefits of large or small programs, and economists fail to deal adequately with the benefits or values of health vs illness, the value of human life, or even the value of a reduction in the number of pests. There also is uncertainty about the technical feasibility of various types of controls. Even the standard techniques used, such as source reduction, larvaciding, and adulticiding, vary tremendously with region and country. Nowhere has a major effort been made in specific data collection to justify or to establish cost-benefit ratios.

POTENTIAL FOR INTERVENTION

The potential for intervention is excellent in the general area of insect control and for specific pests. The techniques of intervention include removal of adult harborage, adulticiding, larvaciding, and biological control. (For additional information, see the section on controls.)

Intervention by isolation techniques is difficult to accomplish, since insects move rapidly through a given area. Substitution is another inadequate technique. Shielding can be accomplished by use of proper-sized screens or blow-down air devices. Treatment is used for disease carriers but not effectively for the insects. Prevention is the strongest potential intervention technique in the spread of particular insects.

RESOURCES

Some scientific and technical resources available include Tulane University School of Public Health, Colorado State University, various land grant colleges such as Pennsylvania State University and Purdue University, the American Public Health Association, the National Environmental Health Association, and the Entomological Society of America.

The civil and professional associations in pest control include the American Mosquito Control Association, the International Association of Game-Fish and Conservation Commissioners. Pest control operators associations and chemical manufacturers associations are other important resources.

Government resources available include state health departments, many local health departments, the U.S. Department of Agriculture, the National Bureau of Standards, the Department of Defense, the Environmental Protection Agency, the Food and Drug Administration, the National Science Foundation, and Department of Health and Human Services, and county extension agents.

STANDARDS, PRACTICES, AND TECHNIQUES

Mosquitoes are by far the largest group of insects controlled by humans. The standards, practices, and techniques utilized vary tremendously from area to area. Generally, surveys are made to determine the extent and location of the mosquito populations and then extensive larvaciding or adulticiding programs are put into effect. In the event of an outbreak of disease, extensive larvaciding and adulticiding programs are again carried on in those areas where mosquitoes are most prevalent. In many areas there are unorganized mosquito control programs. These programs are put into effect when citizens demand action because of the annoyance due to mosquitoes. Generally public works employees or contracted commercial firms will do the actual control work. The work consists of the use of larvacides and fogging units, or mist units for adulticiding. Control may be done in a haphazard way or it may be carried out weekly on a regular basis where specific high-volume mosquito areas exist. Where organized mosquito programs are in existence, the community is taught first how to reduce mosquitoes through environmental controls and adequate prevention; then proper engineering approaches are used, as well as larvacides and adulticides. Although the individuals carrying out this work must be licensed by their states, and adhere to the principles of the Environmental Protection Agency, there still is a substantial need to establish proper standards, practices, and techniques for insect control.

MODES OF SURVEILLANCE AND EVALUATION

Inspections and Surveys

Fly Surveys

The major value of a fly survey is to determine those areas in which the flies are in greatest concentration, to determine the kinds of flies in order to use effective controls, and to determine the potential spread of disease. There are several different types of fly surveys. They include fly traps, fly grills, and reconnaissance surveys. The selection of the sample area should be based on a clear understanding of what information is needed, the kinds of problems anticipated, the areas where the greatest number of flies are found, and the areas where the greatest refuse and other solid-waste problems exist.

Fly traps consist of baited traps, fly paper, or fly cones. The fly traps have the advantage of capturing a cross-section of the kinds of flies in a given area. They also provide an approximate count of the various species. The baited fly trap may be lined with animal feces, sugar, fish heads, or some other decaying matter. The flypaper strip gives a rapid count of flies, but this only will represent a small portion of the fly population. Flies are drawn to fly cones by the attractant, and then as they attempt to escape, they move upward toward a light that is in the fly cage.

Fly grills are the most widely used means of surveying for flies. The grill depends upon the natural tendency of flies to rest on edges. The grill is placed over a natural attractant, such as garbage, manure, or other decaying materials. The number of flies landing on the grill during a 30-sec period is tabulated. It is possible to physically count the number of flies and make some judgement of the different kinds of flies on the grill. If there is a huge number of flies, the grill should be subdivided and a count made on the subdivided portion. A minimum of

10 counts should be made in each subdivision sample, and the five highest counts should be recorded.

It is possible to make a rapid survey by driving through an area or walking through an area in which you would expect to find large numbers of flies. A rough estimate can be determined by checking the usual fly-breeding areas and making a physical count.

In an attempt to determine if a fly survey or a fly control program is effective, it is wise to make a fly survey initially in the usual breeding areas and then return after the breeding areas have been eliminated and resurvey. Obviously, chemical control procedures should be used prior to the cleanup procedures so that the flies do not move on to other areas.

Mosquito Surveys

Surveys are necessary for the planning, operation, and evaluation of mosquito control programs. The basic survey is used to determine the species of mosquitoes, sources, breeding places, locations, quantities, flight distance, larval habits, adult resting places, and any other information that may be needed in the actual mosquito control program. It is also necessary to continually evaluate the ongoing program. This is done by conducting surveys on a periodic basis during each season of the year, especially the mosquito-breeding season, to determine the effectiveness of chemicals and other controls. The surveys consist of making field evaluations of the quantity of mosquitoes, the types of mosquitoes, the kinds of mosquito breeding areas and potential mosquito breeding areas, the numbers and types of larvae, and the development of accurate, comprehensive maps that indicate where all the larval and adult problems may be found.

A larval mosquito study may be carried out by an environmentalist who goes to areas where there are artificial breeding places and standing bodies of water and removes water containing larvae by means of a white enamel dipper, which is about 4 in. in diameter. The dipper has an extended handle so that it may be lowered into the body of water. Some environmentalists prefer to use white enamel pans that are about 14 in. long, 9 in. wide, and 2 in. deep. The pan is used to sweep an area of water until the pan is half full. The water is then examined for larvae. Another technique is to examine cisterns, artificial containers, or other areas with a flashlight. A large bulb pipette or siphon is used to draw out water with larvae into a pan or onto a slide, and then observations are made. The environmentalist should always, after observing the habits of the larvae in the water, record the number of dips he or she makes and the count of larvae. A rough estimate can be made over time of larvae increases or decreases by going to the same area and using the same dipping procedures.

Adult mosquito surveys can be carried out by setting up bait traps, window traps, carbon dioxide traps, insect nets, light traps, or checking the daytime resting places. Bait traps are put in areas where mosquitoes are expected to be found. The mosquitoes are attracted to a bait animal that is placed in a trap. When they enter the trap, they cannot bite the animal and they cannot be released. The mosquitoes are taken alive and evaluated. Bait animals have been such animals as horses, calves, mules, donkeys, and sheep. Window traps use the same principle as the animal bait trap, except a human being is the attractant. The trap is set on the outside of the window. The mosquitoes are attracted to the building by the presence of the human being and then they enter the trap and cannot leave. Insect nets are used for collecting mosquitoes in grass and other vegetation. Light traps are traps that contain a light that acts as an attractant to the mosquitoes. Once the mosquitoes enter, they are incapable of leaving the trap area. An evaluation can be made of the adult resting places. The type of resting place depends on the kind of mosquito. This material has been previously covered.

Once all the information is gathered and a determination is made of the kinds of mosquitoes that are present in an area, comprehensive maps are drawn, and the control program is established and put into action.

Food-Product Insect Surveys

The easiest means of determining the presence of food-product insects is searching through dry-food products and looking for live insects. The insects may also be present along the seam of the food bag. At times the insects can be seen crawling near the food or flying about. In any of these cases, a food-product insect problem exists and preventive measures should be taken.

CONTROLS

Before DDT was used for mosquito problems, the principal control techniques consisted of identifying the species of mosquito and its breeding habits, and then designing a method that would eliminate the larva, either by means of engineering techniques or by means of chemical larvacides. Engineering techniques consist of improving drainage areas, adjusting the slope and depth of streams, and using methods and techniques for the removal of any obstructions hampering the flow of water, which in turn increases the potential for mosquito breeding. Biological control is the technique utilizing parasites, predators, and pathogens of insects to eliminate them. It also includes the use of sterilization techniques, genetic manipulation, management of the habitat, and biologically produced chemical compounds. In recent years there has been considerable interest in this area; however, with the exception of *Aedes aegypti* and *Anopheles gambaie*, disease-causing insects are indigenous to the occupied areas, and the classical biological control approach is largely eliminated. Biological control is basically the fostering and manipulation of the natural enemies of insects. Fish are an excellent biological control for mosquito larvae. There are over 1000 species of freshwater fish that consume mosquito larva and pupa. However, the one that is most frequently used is the minnow *Gambusia affinis*. This minnow is a surface feeder and inhabits permanent waters. It has been used in the control of anopheline mosquito larvae to reduce malaria, and *Aedes aegypti* to reduce yellow fever. The technique of building insects out by removing harborage along with chemical control is now called integrated pest management.

Insect growth regulators are used as a means of control. Attractants will lure the insect to trap or poison. Repellents are used on the skin and body as a means of control to prevent the insect from alighting and biting, or otherwise contaminating the individual. Since certain pesticides are being banned as a result of new research, it is essential to reevaluate the chemicals being used very frequently. Read the labels for proper usage.

Specific Scientific and Technical Controls

Flea Control

In flea control, infested animals and their habitats are treated. Animals are treated by shaking a dust containing 10% methoxychlor or 5% carbaryl. These chemicals have a slight toxicity and a long residual period. Particular attention should be paid to the back, the neck, and the top of the head. The powder should be rubbed thoroughly into the hair of the animal. The animal's quarters or the house that is infested with fleas should be treated with either carbaryl, methoxychlor, or ronnel. Each of these chemicals are slightly toxic and have long residual periods. If there is concern about a rodent flea infestation, it is important to dust the rat run with any of the previously mentioned chemicals. The dust will then get onto the rats or mice and will kill the fleas that are present. It is essential that the fleas be killed before a rat-control program starts. Otherwise, the rats will die and the fleas will leave the dead animals and possibly attack humans. In addition, infested premises may be treated with 2% malathion or 0.5% diazinon. In barns and yards, you may use 2.5% diazinon or 2.5% ronnel, malathion as a 4% dust, or carbaryl as a 5% dust.

Fly Control

The four basic fly-control techniques include proper solid-waste disposal, chemical control, mechanical and physical control, and biological control.

Since house flies breed in garbage, decaying matter, feces, dead animals, and any kind of organic material, it is essential that this solid waste be removed and stored properly. The storage should take place in plastic bags within metal cans with tight-fitting lids. The refuse should also be stored, depending on the kind, in concrete containers (for animal manure), in large metal containers with tightly closed lids, or in refuse rooms where the doors and windows are flyproof. The maximum size of the can should be not greater than 30-gal capacity, since it becomes difficult to remove heavier cans. Where garbage, heavy refuse, or ashes are stored, the maximum size of the cans should be 20 gal. Collection should take place during the summer at least twice a week, since flies go from the eggs to the pupa stage within 5 days and then move into the ground. By removing the cans at least once every 4 days, the fly eggs, larvae, or pupa are removed and disposed of. During the winter, the solid waste should be removed at least once a week. In food operations and in other situations where there are large quantities of organic waste material, this material should be removed on a daily basis. Solid-waste disposal should occur in either an incinerator or a properly operated sanitary landfill. Garbage that is used for hog feeding must be cooked to an internal temperature of 137°F before it is served to the hogs. One other technique utilized for disposal is through the home garbage grinder.

Animal feed and feces become breeding areas for flies. Animal manure should be adequately stored, and dog and cat feces should be picked up and removed in order to prevent fly breeding. It is also essential to check the animal feed periodically to make sure flies are not breeding in it. In a home situation, dog food or cat food left outside becomes an attractant

Large fly populations can breed in high weeds. It is therefore necessary that weeds be controlled either chemically or physically. The type of chemical chosen for control of flies is determined by the effect that is desired. Residual sprays are put on surfaces on which flies will alight. These sprays should last for considerable periods of time. The chemicals include 2–5% malathion mixed with 6–12% sugar emulsion or a 1% diazinon emulsion. Space sprays are used in areas where quick knockdown of flies is desired, and where the aim is to kill the adult fly in large quantities. The space spray should contain 0.1% synergized pyrethrum plus 1% malathion. Fly cords should be impregnated with 2.5% diazinon. The cord should be made of strong cotton, $^3/_{32}$ in. in diameter and strung at a rate of 30 linear ft of cord per 100 ft^2 of floor area. Cords are accepted because flies will rest on them absorbing the insecticide. A larvacidal treatment should consist of 2.5% diazinon emulsion, 1% malathion emulsion, 2.5% ronnel emulsion, or 2% DDVP emulsion. Dry baits can consist of diazinon, malathion, ronnel, or DDVP plus sugar. For wet bait add water.

It is important to realize that insects become quite tolerant and resistant to insecticides. As a result of this, the insecticides that are now recommended and probably the safest to use for humans at this time may not be those recommended for future use. The environmental health specialist should contact the Center for Disease Control in Atlanta, Georgia or the local county extension agent if he or she has any questions concerning the problems of fly control and chemical safety.

It is known that flies not only are resistant to certain insecticides, but apparently develop an inherited resistance over several generations. As a result of this, flies are able to either absorb a lower rate of insecticide, store the insecticide without being killed, excrete the insecticide, detoxify it, or use an alternate mode of accomplishing its body functions when a chemical blocks its normal route.

There are several techniques used in applying insecticides. They include space spraying for the adults; mist spraying through large blowers or insecticide bombs for adults; fog generators, which produce an aerosol or smoke for adults; and residual spraying with hand sprayers for adults and larvae.

Flies can also be controlled chemically by treating the animal by the use of fly repellents or by the use of fly attractants to draw flies into traps or onto poison.

One of the most effective physical controls is fly screens, which must be 16-mesh in size. The screens have to fit tightly to the windows or door frames. Fly traps, which have been discussed under survey techniques, can be utilized for some fly control. Flies can be electrocuted if they cross an electrically charged field or alight on a screen that is electrically charged. Air shields can be installed so that air blows down and out from the doorway, thereby preventing flies from entering a building.

The biological control of flies is based on the use of sterile flies and on the dissemination of organisms that are pathogenic to the insect. These techniques are basically in experimental stages and are not generally used. Predatory animals will consume a certain quantity of flies; however, it is necessary to be cautious that the predators do not themselves become pests.

The most important means for control of flies is through the elimination of food and garbage. Therefore, it is extremely important that all exterior areas and all interior areas are thoroughly cleaned and all solid waste is removed.

Within the home, flies are best chemically controlled by the use of synergized pyrethrum plus malathion, diazinon, or baygon. Each of these chemicals must be used very carefully. DDVP strips may be used to kill flies. These resin strips will last for about 3 months. One strip should be used per 1000 ft^3 of space. Care must be taken in the use of DDVP strips around food or food services and where individuals have respiratory difficulties.

Louse Control

Body louse control is accomplished by washing the clothes of the individual in very hot water, drying, and then ironing. Wool clothing, which is especially good for louse habitation, should be thoroughly dry cleaned and then pressed. Clothing may also be put in a plastic bag and allowed to remain for a period of 30 days. This will kill the existing lice as well as the eggs. The individual can be treated with a 1% malathion powder, which is effective against eggs and adult lice.

Head lice are controlled by cutting the hair very short and then washing it thoroughly. It is essential to kill both the lice and the eggs. The best technique for head louse control is to shampoo, dry the hair thoroughly, tilt the head back, cover the eyes, apply a 25% benzyl benzoate emulsion, which is worked into the hair and scalp, and combed through the hair. Shampoo the hair again after 24 hr, dry, comb, and brush it to remove the dead lice and loosen the eggs.

In crab louse control, the pubic hair should be washed very thoroughly and the hair shaved off. A vaseline ointment containing pyrethrum or a 1% lindane cream or lotion may be used.

It is essential to complete louse control by eliminating all the bedding and clothes used by the lice-infested individuals. These garments and materials should be washed very thoroughly or dry cleaned. It is also essential that hair is washed frequently and thoroughly.

Mite Control

Mites are controlled through the use of chemicals or proper cleanup. In trying to control mites within a house, or even within areas where animals live, pyrethrum bombs can be used as knockdown agents. Obviously the pyrethrum has to be synergized. As soon as the mites are knocked down, they must be vacuumed up and destroyed. For residual treatment use 1% malathion. This insecticide should be applied onto the top of foundations, around the plate and ends of joists, baseboards, edges of floor areas, and around windows and doors. On the outside, mites can be controlled with 0.5–1% malathion at a rate of 2.5 gal of 40% emulsifiable concentrate per acre.

Human infestation can be controlled by use of 68% benzyl benzoate or 12% benzocaine. The chemicals should be left on the body for 24 hr before washing. The second treatment is needed in 10–14 days. Scabies may also be treated with eurax, which is a salve containing 10% N-ethyl-O-crotontoluide in a vanishing cream. Two applications should be given at 24-hr intervals.

Mites are controlled through elimination of rats, house mice, birds, and so forth. With chigger mites, the environment should be modified to permit plenty of sunlight and air to circulate freely.

Larval and Adult Mosquito Control

Mosquito larvae are controlled mechanically, biologically, or chemically. Mechanical control involves emptying and removing all temporary containers, such as old tires and cans. If possible, depressions in the ground should be smoothed out. High weeds, which trap water, must be cut on a regular basis. It is also important to ensure that all sanitary landfills or any other type of landfill has a final slope of 0.1–0.5 ft/100 ft for drainage purposes. Clean, straighten, and drain all ditches so that water runs freely. Any growth in bodies of water, particularly around the shoreline, must be removed so that the water flows readily rather than sits stagnant. Of particular importance is the closing or sealing of all seepage pits, septic tanks, and other areas where sewage may be trapped in a stagnant situation. Impounded waters can be a source of mosquito larvae breeding. This can be remedied by cleaning the major vegetation off the shoreline, filling, and removing the water from low places. The water level should be so controlled that a reduction in level of about $1/_{10}$ ft per week will occur during the mosquito-breeding season. Additional research should be done in the areas of mechanical mosquito larvae control, since there are huge numbers of mosquitoes in marshy areas and in areas where certain types of food, such as rice, are grown. It would be detrimental, at present, to change the ecology of these areas. On irrigated lands, the mosquito problems can be considerable. They generally find habitation in stored water, the irrigation conveyance and distribution system, the irrigated land, and in drainage systems. The key to control is to use only the amount of water necessary and to have all systems freely moving so that stagnation of water cannot occur. The biological technique for eliminating larvae would be to stock the lakes and other impounded areas with top-feeding minnows of the gambusia variety. In residential areas, it is important to change the water in the bird baths, stock garden ponds with goldfish, and remove all containers holding stagnant water.

Chemical control is accomplished by applying #2 fuel oil, #2 diesel oil, or kerosene on the surface of the water. The petroleum is toxic to the eggs, larvae, and pupa. Pyrethrum larvacides have been used for many years. Methoxychlor may act as a systemic poison and as a contact poison, which penetrates the body wall or respiratory tract. In addition, baytex and malathion have been used effectively in mosquito larvae control.

At dark, mosquitoes are best controlled by use of screens, repellents, and by space spraying. Screens should be 16×16 or 14×14 mesh to the inch to keep mosquitoes out. In areas where very small mosquitoes exist, such as the *Aedes aegypti*, it is necessary to have a screen with a mesh of 16×20 or 16×23. Bed nets are useful in temporary camp areas where mosquitoes are quite prevalent. Bed nets are made of cotton or nylon cloth with 23–26 meshes/in.

There are several types of mosquito repellents on the market today. They include bimethyl thalate, Indalone, and biethyl toluamide. The repellents are applied to the neck, face, hands, and arms. They will prevent mosquito bites for a period of from 2 to 12 hr. This protection depends on the person, the species of mosquito attacking, and the abundance of mosquitoes available.

Space spraying for adult mosquitoes is accomplished by the use of aerosols, fogging, misting, dusting, or by airplane application. The chemicals generally used include carbaryl and malathion. Care must be taken to avoid large concentrations of pesticides in any type of space spraying in residential areas. Where residential areas are space sprayed, personnel must ensure that the trucks have blinking red or orange lights as signals and that children are not driving their bicycles in and out of the mists. Serious accidents have been caused when automobile drivers were blinded by the fog or mist and drove into children on bicycles.

Residual spraying is another technique for the destruction of adult mosquitoes, where 2.5% malathion is used. The chemicals are applied to surfaces, and they continue to kill mosquitoes for a period of 10 to as much as 32 weeks, depending on the type of insecticide and mosquito. Another valuable technique is the use of strips of DDVP in enclosed basins. These strips continue to give off vapona insecticide over a period of time and will kill the mosquitoes within the catch basins.

It is important to realize that mosquito-borne outbreaks of disease have not been eliminated from this country or other countries. In the years to come, all health departments should develop adequate, comprehensive mosquito-control programs in conjunction with the general public and the federal agencies.

Roach Control

Roaches are controlled by keeping the premises extremely clean and by discarding paper bags, food cartons, and other materials from food stores and warehouses. Roaches are controlled chemically by the following insecticides: diazinon, 2% spray or 2% dust; malathion, 3% spray or 4% dust; baygon 1% spray or 2% dust. Borax is also used to control roaches by placing the powder in out of the way areas, such as wall openings, under cabinets, under bathtubs, in attics, and in other hard to get at places during construction. Roaches that repeatedly cross these areas will be killed by the chemical.

Tick Control

An individual is protected against ticks by keeping clothing buttoned, trouser legs tucked into socks, and by wearing the type of clothing that will prevent ticks from penetrating and getting to exposed body parts. Individuals should always inspect their clothing and their bodies after they have been through a tick-infested area. The ticks should be removed immediately and destroyed. If ticks become attached, the simplest technique for removal is a slow straight pull that will not break off the mouth part and leave it in the wound. A drop of chloroform, ether, vaseline, or fingernail polish rubbed over the tick and the area will help the removal. Tick repellents can also be used. However, there is no one repellent that appears to be perfect for all ticks. Clothing may be treated with indalone, dimethyl tareate, or benzyl benzoate. Dogs and cats may be dusted with 0.75–1% rotenone or 3–5% malathion.

Ticks inside of buildings are controlled by the use of baygon or 0.5% diazinon. Sprays should contain 0.2–0.5% DDVP as a fumigant to drive the ticks out from behind baseboards and out of the cracks in walls. Tick-infested areas on the outside may be treated with a 4% malathion dust or DDVP. The control program must go on for many months so that the infected eggs can be destroyed as they become adults. Another technique for the removal or destruction of ticks is to remove the host. Where dogs are infested with ticks in a certain area, the area and the dog should be carefully treated, and then the dog should be kept away from the usual sleeping quarters, which prevents the ticks from getting a blood meal.

Food-Product Insect Control

Control of food-product insects include the following: (1) locate the source of infestation by examining the foods and areas where infestation is most likely; (2) clean all cabinets and storage areas thoroughly by vacuuming and washing with hot water; and (3) keep all shelving dry and spray with appropriate insecticides. These insecticides may include diazinon of 0.5% concentration or 1% baygon; (4) after drying and spraying, put only fresh food back into the area; and (5) make sure that foods are rotated frequently. During the hot months, rotate the dry foods at least once every 2 weeks. During the cold months, the foods should be rotated at least once every month.

Public Health Laws

A typical public health law gives the state board of health the power to develop rules and regulations to control nuisances that are dangerous to public health, to control fly and mosquito breeding places and to control the spread of rodents. These provisions are typically found in the powers and duties of the state board of health. In addition, the food codes that come under the general public health laws have specific provisions that food establishments shall be protected by all reasonable means against flies, other insects, and rodents. Further, housing codes stipulate that dwelling units, school units, or any other type of dwelling or shelter must be free of insects and rodents. A health code may also contain a provision making it unlawful for any individual to maintain any lands, places, buildings, structures, vessels, or watercraft that are infested by insects and rodents. It may also stipulate that any person owning, leasing, occupying, possessing, or having charge of any land, place, building, structure, stacks, or quantities of wood, hay, corn, wheat, other grains, vessels, or watercraft must eliminate infestations of insects or rodents or be prosecuted under the pest eradication sections of the Health Code.

SUMMARY

Vector-borne disease will continue to be a major deterrent to human settlement and agricultural development in certain areas of the country and the world. The insects discussed were of public health importance, including fleas, flies, mites, mosquitoes, roaches, ticks, and food-product insects. These insects destroy food material and cause disease. The potential for intervention is excellent, providing that humans are willing to use a comprehensive program of habitat removal, biological control, chemical control, and good public health education. The pesticides used in chemical control, while effective, have caused potential hazards to humans, either through the short-range problem of intoxication or the long-range potential hazard of carcinogenesis or mutagenesis. The destruction of the vector should be supplemented by chemotherapy and vaccines to limit disease potential. In working toward a good pest-control program, the potential hazards to the environment should be considered. The use of pesticides for all purposes should be coordinated to reduce the environmental degradation while increasing the health potential for society.

RESEARCH NEEDS

Additional study is needed to determine the best technique to be used for the control of arthropod vectors of disease or other public health pests, and the significance of this control on the ecology and behavior of the various ecosystems. A reporting system where information from health departments of cities, counties, and states is sent to the Center for Disease Control to help quickly identify the existing levels of insect or other pest problems, is needed. Additional study is needed to determine the problems related to environmental manipulation. Tolerance levels for biting nuisances should be established. It is necessary to have a combined study by agricultural experts, medical experts, and environmental experts to determine the problems of poison persistence and resistance as it relates to pesticide management.

Chapter 5

RODENT CONTROL

BACKGROUND AND STATUS

Rodents are members of the order Rodentia. These mammals have teeth and jaws adapted for gnawing. The order Rodentia includes squirrels, beavers, mice, rats, lemmings, porcupines, and chinchillas. For the purpose of this chapter, three rodents — the Norway rat, the roof rat, and house mouse — will be discussed (Figure 5.1).

These rodents, known as murine rodents (they are actually of the subfamily Murinae), are important because they are found everywhere that humans are found. Murine rodents live in buildings, destroy food and property, endanger health, and compete with humans for existence. They are found in all parts of the country and are dangerous whether they are out of doors or within the home.

In rural areas rodents pose the greatest problems by destroying and contaminating food. They inhabit improperly managed solid-waste disposal areas and attics, or other parts of the home, depending on the species.

In urban areas, rodent infestations occur in luxurious suburbs where mice inhabit houses and rats inhabit lawns, undersides of doghouses, or any area where construction has taken place or sewers have been disrupted. The Norway rat typically lives in sewers or along creek or river banks and migrates to areas where food may be found.

Rat infestations become extremely significant within the inner city. To simply count rat bites or determine the amount of food loss or property damage is inappropriate. Human misery increases as the rat continues to spread. The rat is an important part of the inner city community. It is a symptom of the breakdown of the community, not only from an environmental standpoint, but also from a social and economic standpoint.

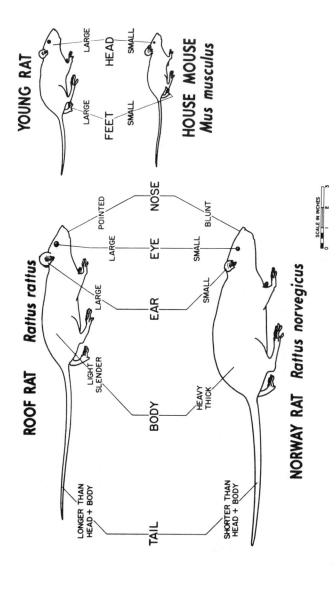

Figure 5.1.   Field identification of domestic rodents. *Source: Pictorial Keys to Some Arthropods and Mammals of Public Health Importance* (Atlanta, GA: U.S. Department of Health, Education, and Welfare, Public Health Service, Communicable Disease Center, 1964), p. 55.

Rat control within the city is an extremely complex problem that must be resolved by involving not only the various municipal and state agencies, but also by involving the community itself. Behavioral patterns must be altered, lifestyles changed, social and economic factors improved, and serious concern with the rat problem shown if the rodent problem, including disease and community degradation, is to be eliminated.

More than 60 million people live in 300 cities where rats are a major environmental problem in the United States. The Urban Rat Control Program was started in 1969 to eliminate the conditions that allowed rats to grow and multiply. The program emphasized the control and improvement of both the physical and social environments. As a result of this program, intensive efforts in rodent control realized considerable solid-waste removal and some reduction in rat populations. However, although the federal government by 1979 had spent many millions of dollars in attempting to remove or reduce rat populations, the results were insubstantial. The program was temporarily effective, but as soon as it was removed from an area, the rat problem generally returned. However, the City of Philadelphia and some other communities are maintaining effective rodent control efforts as of 1991.

Rodents are developing immunity to anticoagulants in both the United States and Europe. As their resistance increases, so too will their numbers.

SCIENTIFIC, TECHNOLOGICAL, AND GENERAL INFORMATION

Characteristics and Behavior of Murine Rodents

Murine rodents are capable of almost continual reproduction and produce large quantities of newborn mice and rats. Some of the young are killed or eaten. The newborn rodent is born blind, but gains sight in about 12–14 days. For about 21 days it depends on its mother for food. If forced to, it can start to live by itself within 3–4 weeks of birth. Sexual activity and fighting occurs in about 2–3 months. The young learn through experience and imitation. At 3 months, they are very active and completely independent (see Table 5.1 for detailed description of murine rodents).

Daily life patterns rarely vary, unless there is a sharp increase in the rodent population or a decrease in the food supply. When food is abundant, rodents are most active during the first half of the night; when food is in short supply, they are active at any time. Generally, however, if rodents are seen during the day, there are large numbers present. Rodents tend to use regular paths and runways. They live as close to their food source as possible and travel to it with the least amount of exposure. This is why runways tend to be along the edges of walls or through covered areas.

Rodents have an avoidance instinct to alien objects that aids survival. Rodents will sense and avoid unusual changes within the environment. They may avoid

Table 5.1. Description of Murine Rodents

Characteristics	Norway Rat (*Rattus norvegicus*)	Roof Rat (*Rattus rattus*)	House Mouse (*Mus musculus*)
Weight	16 ozs.	8-12 ozs.	½-¾ ozs.
Total length including the tail	12¾-18 in.	8¾-17¾ in.	6-7½ in.
Head and body	Blunt muzzle; heavy, thick body	Pointed muzzle; slender body	Small muzzles and body
Tail	Shorter than head and body	Longer than head and body	Equal to or longer than head and body
Ears	Small, close set	Large, prominent	Large, prominent
Hind foot	1½ in. or longer	less than 1½ in. in length	¾ in. in length
Teeth	Strong, well-developed; single pair of insicors in upper and lower jaw	Same	Same
Color	Great range	Black, tawny, gray	Gray, brown, white
Distribution	US and Southern Canada	Southern gulf states and Pacific coast	US and Southern Canada
Environmental Distribution	Share human habitat	Less dependent on humans; live away from human habitation	Least dependent on humans; live anywhere
Gestation	22 days	22 days	19 days
Mating	The female can mate again within 48 hrs after giving birth	Same	Same
Litter size	4-6 litters of 8-12 young each	4-6 litters of 6-8 young each	8 litters of 5-6 young each
Number weaned	20 per female annually	Same	30-35 per female annually
Length of life	One year	Same	Same
Harborage	Outdoors—in buildings, under foundations, in waste disposal areas; indoors—between floors and walls, under solid waste, in any concealed area	Outdoors—trees and dense vines; indoors—attics, between walls, enclosed spaces	Nest anywhere
Range	100–150 ft	100–150 ft	10–30 ft
Droppings	To ¾ in.; capsule-shaped	To ½ in.; spindle-shaped	To ⅛ in. rod-shaped
Food	Omnivorous—garbage, meat, fish cereal	Omnivorous—usually vegetables, fruits, cereals	Omnivorous—usually cereals
Daily Food requirements	¾-1 oz. dry food; ½-1 oz. Water	½-1 oz. dry food; 1 oz. water	Nibbles; $1/10$ oz. dry food; $3/10$ oz. water

new food for several days. When poisoning it is wise to prebait the area with unpoisoned bait for several days before poison is added. This is necessary when the poison kills swiftly; when the poison is an anticoagulant and death is slow, it is possible to start poison baiting immediately. In environments where there are constantly new strange objects, rodents are less prone to avoidance. Rodents will also avoid areas containing several dead rodents, around poisoned bait, out of a sense of danger.

Rodents have tremendous versatility in climbing, jumping, reaching, and swimming. Mice and roof rats are excellent climbers; Norway rats vary from fair to good. Rodents can climb the vertical walls of most brick buildings as long as they can get a toenail hold. They can also climb or cross-wires easily. Rats can reach as much as 13 in. along smooth vertical walls. They can make a standing high jump of about 2 ft and with a running start, they can jump a little over 3 ft. Mice can high jump more than 2 ft with a running start. Rats can jump outward 8 ft horizontally when jumping or dropping at least 15 ft downward. With a running start, the distance is increased. Rats can swim as much as a half mile in open water. If they are placed in a tank of water, they will repeatedly dive to the bottom to look for exit pipes. This is why they survive so well in sewer lines and use them as a transportation route.

As mentioned, rodents live close to food sources if they can maintain their safety. Rats usually live within 100–150 ft of their food source, while mice live within 10–30 ft. Rodents will build their nests in any relatively quiet hiding place. Outside, they burrow into the ground. Norway rats burrow 5–6 ft; roof rats will only burrow in the absence of Norway rats; house mice will burrow extremely well if there is no other harborage present. Rodents also nest in piled-up trash, lumber piles, discarded appliances, around stables, under animal houses, in old garages and sheds, and on the banks of streams, creeks, and rivers. Indoors, rodents nest between double walls, under floors, above ceilings, in attics and basements, and in any closed-in spaces behind counters, equipment, or stairwells. Their nests are usually bowl-shaped, and are about 8 in. in diameter for rats and 5 in. in diameter for mice.

Rodent Senses

Rodents have the senses of touch, vision, smell, taste, and balance. In addition to the normal sense of touch, rats have highly sensitive whiskers with a complex nerve net at the base of each. Their bodies are covered by guard hairs, facilitating night travel.

Rodents are apparently color blind. Their sense of vision is underdeveloped, although they are able to detect objects as far away as 45 ft.

Rodents have a keen sense of smell. They are attracted to the entrance of new bait boxes by the urine of other rodents, if the odors are strong enough to travel some distance. Human odor is commonplace, since rodents live with humans constantly.

The sense of taste is not as well developed in rodents as it is in humans. They probably cannot taste poisonous material in baits with the possible exception of red squill, which has a very bitter taste. Bait shyness is developed in rodents because the rodents become sick, rather than because they are able to taste the poison within the bait. In the case of anticoagulants, the poisoning is done so slowly, over a period of 7–10 days, that the rodent does not connect sickness with the anticoagulant.

Rodents have an excellent sense of balance. They usually land on their feet if tossed into the air. Rats and mice can sometimes fall two to three stories without injury. Mice have fallen as much as four stories without injury. This sense of balance enables rodents to cross from one building to another over wires and cables, or cross from a tree into the open window of a house.

Rodent Signs

Signs of rodent habitation apart from the sight of live or dead rodents include sounds, droppings, runways, tracks, rub marks, and burrows. The most frequently seen outdoor rodent sign is a burrow or gnawed material. The most frequently seen indoor rodent sign is rodent droppings. Obviously, the most positive proof of an infestation would be to sight live or dead rodents. As mentioned, rodents travel at night; rodents seen during the day suggest a sizable infestation. A technique used in determining whether rodents are present in a building at night is going into the building and suddenly turning on the lights or flashing a large-beam flashlight. Either the rodents will be seen or the noise of their flight will be heard.

Sounds of rodents are very unique. You can hear them running, gnawing, or scratching within double walls or floors. You may also hear squeaking noises from young rodents in the nest or the noises of adults fighting.

Droppings are a sure sign of the presence of rats and mice. The Norway rat's droppings are $3/_4$ in. long by $1/_4$ in. in diameter. The roof rat's droppings are generally smaller than those of the Norway rat. The house mouse droppings are usually about $1/_8$ in. long. The age of the dropping can be determined by color and texture. If the droppings are fresh they are usually black, soft, glistening, and appear moist. If they are old, they are brittle, graying, and dry (see Figure 5.2).

Since rodents follow the same path frequently, they establish runways, tracks, and rub marks. The runways are easily identified because the area is smooth from the constant movement of the rodents. The area is also relatively free of dust, in contrast to surrounding areas. Tracks are detected by shining a light obliquely with the ground. The tracks are marks in dusty or muddy areas. Dark smears or rub marks are found along the walls, pipes, or rafters from the grease on the coats of the rodents. Generally the runways are along walls, under boards, behind stored objects, and in, around, or under accumulated solid waste. Tracing the runways, harborage, food and water supplies, and the methods of entry of the rodents into the building facilitates proper control measures.

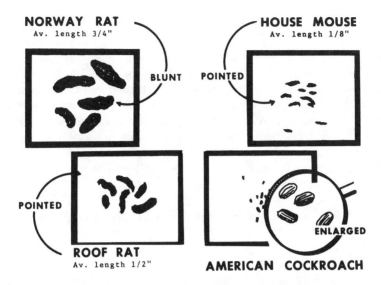

NORWAY RAT
Av. length 3/4"

BLUNT

HOUSE MOUSE
Av. length 1/8"

POINTED

POINTED

ROOF RAT
Av. length 1/2"

ENLARGED

AMERICAN COCKROACH

Figure 5.2.   Recognizing rat and mouse signs. *Source: Control of Domestic Rats and Mice* (Atlanta, GA: U.S. Department of Health, Education, and Welfare, 1972), p. 7.

Rodent Population Characteristics

A rodent population behaves like any other society. It goes through the characteristic stages of birth, growth, maturity, and death. Each population area has the capacity to support rodent populations by the environment. This includes the amount of food, harborage, living space, and population in the area. The population size, affected by reproduction patterns, tends to increase in the spring and fall. Mortality rates counterbalance population increases. Female rats and mice live longer than males; however, accurate mortality data is difficult to obtain. Population density is also affected by movement. Rats and mice generally spend their lifetimes in a limited area. If within the home area there is a problem of sharply changing temperature or population increase, portions of the population will move. Rodents frequently migrate to additional food sources. This behavior is important in a successful rodent control project, since if the food source or harborage is removed prior to poisoning, the rodents will scatter and sharply increase the rodent population of other areas. The major food source for rodents is garbage, stored food products, crops, and grains. Harborage is a limited problem since rodents are so adaptable. Rodents are predators and will, in the absence of a food source, feed upon each other; they are, of course, the victims of predators as well. Where the population of rodents is high, more are killed by predatory animals. Cats, with the exception of Siamese cats, are poor rodent predators. Although cats are good mouse catchers, they generally play with the mice rather than kill large quantities of them. Dogs, while good rat catchers, are not used for rodent control work since they are often mutilated by the rodents. The presence of domestic animals also provides a source of food for the rodents since owners leave pet food and water out for their pets, providing an excellent fresh food source for the rodents on a continuing basis. Other rodent predators are snakes and birds. Rodents will compete fiercely for food and water, nesting sites, harborage, and females in heat. When Norway and roof rats compete, the fiercer Norway rats generally are successful. Norway rats will compete with each other — many times over females. This competition among members of the same species is associated closely with their social organization. The aggressive animals live closest to the food source, whereas the least aggressive animals live furthest away from the food source and are in most danger of attack by predators.

RODENT PROBLEMS AND THEIR EFFECT ON HUMANS

Rodents cause psychological, sociological, and economical human suffering and a variety of diseases. It is estimated that each rat damages between one and ten dollars worth of food and other materials each year by gnawing or feeding. They contaminate 5–10 times more food and materials than they utilize. Many fires of unknown origin are attributed to rats gnawing through the insulation of electrical wires. They also gnaw through walls, floors, and doors to either obtain food or to grind down their teeth, which grow at a rate of 5 in. a year. Americans lose an estimated 500 million to one billion dollars annually from rodent damage.

It is difficult to properly determine the number of rat bites occurring each year, since many are unreported. Many years ago the author, while involved in a community rodent control program, conducted a survey of several city blocks within the inner city. At least 1500 rat bites had occurred during a year in the target rat control area. It was difficult to get good statistics, since many of the individuals interviewed failed to report the rat bites to the health department or other medical authorities. An estimated 45,000 individuals are bitten by rats each year. Apparently these figures are low. In order to get a proper figure, it would be necessary to conduct extensive surveys of many inner city areas throughout the country.

Rats bite helpless infants and defenseless aged or invalid adults. Rat bites are disfiguring; they may become infected and in rare cases lead to death. The psychological problem associated with rat bites is enormous. The individual is extremely frightened by the occurrence and may bear emotional scars for years.

During the course of one of the rat-control projects participated in by the author, a mother came running out of the door screaming that her child was being bitten by a rat. When the health department extermination crew ran into the building, they found a large Norway rat clinging to a 6- or 7-year-old girl. The rat had to be beaten into unconsciousness before it would let go of the child. Rats that bite once will probably bite again. That is why it is so important to report rat bites and then to exterminate the premises where the bite has occurred.

Sixty million Americans live in dilapidated and dirty inner city sections within the nation's larger cities. To these individuals the rat is the uninvited king that bites, destroys, stalks its prey, makes irritating noises, defecates, urinates, and spreads disease. It is a symbol of urban decay. Empty lots piled 3 and 4 ft high with garbage, trash, and debris or houses with the plaster falling down, the electrical wires showing, the floor boards ripped up, the staircases collapsing, and overpowering odors are home to rats.

Today the sewer systems of our major cities are infested with rats. Sewers provide easy entrance and exit to areas, food, liquid, shelter, and a constant temperature. The food found in sewers includes food wastes, undigested food in fecal material, ground garbage from garbage disposal units, and food that is washed or thrown into sewers. Rats enter sewers by burrowing into the ground, and then through a broken pipe into the sewer, or they swim through storm drains. Combined sewers have a far greater rate of rat infestation than the modern sanitary sewer, since the opportunity for entrance is increased. During the summer months, when the level of liquid decreases within the sewer, the rats may take to the exterior areas and search for food in yards and other areas. Rats generally will not leave the sewer area unless the competition for space and food becomes acute. Then they move to a new section of the sewer or leave the system altogether. Obviously, since rats are constantly in contact with all manner of infectious organisms, the opportunity for spreading disease is multiplied. When rodent problems occur in new neighborhoods, they are frequently traced to broken or improperly connected sewer lines.

Diseases

Rats and mice may be the reservoir of infection for many human diseases, including murine typhus fever (see insect chapter), plague (see insect chapter), rat bite fever, rickettsial pox (see insect chapter), salmonellosis (see food chapter), trichinosis (see food chapter), and Weil's disease.

Rat Bite Fever

Rat bite fever or Haverhill fever is transmitted through the teeth and gums of a rat infected with the organisms *Streptobacillus moniliformis*. This disease, which may then also be spread by milk, is highly infectious and causes acute febrile symptoms. If untreated the fatality rate is 7–10%. The incubation period is 5–15 days, usually 8 days. The reservoir is infected rats.

Weil's Disease

Weil's disease or leptospirosis is caused by the organisms *Leptospira incetrohaemorrahagiae*. The individual becomes infected by direct or indirect contact with infected rodent urine. The organism enters the body through the mucous membranes or through minute cuts or abrasions of the skin. The disease ranges from mild to severe, but is seldom fatal. The incubation period is about 10 days, with a range of 4–19 days. The reservoir of infection is rats or farm and pet animals.

POTENTIAL FOR INTERVENTION

Isolation, substitution, shielding, treatment, and prevention techniques are all forms of intervention in controlling rat problems. The potential of each technique varies from excellent to poor, depending on the individual's involved, the nature of the housing areas, the concern of citizen's organizations, and government. Isolation is ineffective since it is almost impossible to isolate rodents. Substitution is another unacceptable procedure. Proper rodent-proofing techniques in all areas are used in the shielding technique. Treatment is the application of rodent-specific pesticides. Commercial applicators use the anticoagulants and other chemicals, whereas government programs use red squill, raticate, and anticoagulants. Prevention is by far the most effective technique for reducing rodent problems. Prevention is the removal of all food and harborage, and proper maintenance of property to eliminate rodent entry.

RESOURCES

Scientific and technical or industry resources include, among others, the National Pest Control Association, the area schools of agriculture, the National Environmental Health Association, and the American Public Health Association.

Civic and professional resources consist of the community civic associations responsible for neighborhood beautification, neighborhood housing improvement, and specific rodent control programs. Pest control operators associations and chemical manufacturers associations are important resources.

Governmental resources include local and state health departments, the federal urban rat-control program, and county extension agents. The City of Philadelphia Department of Public Health Division of Environmental Health is an important resource. They prepare a report annually of their activities, entitled Philadelphia Environmental Improvement Program.

STANDARDS, PRACTICES, AND TECHNIQUES

Rodent Poisoning and Trapping

The safest rodent poisons include the anticoagulants, such as warfarin, pival, fumarin, diphacinone, and red squill, (a safe single-action poison). Although many other rodent poisons are used for rat and mouse control, they are too toxic to be used in the home. Some of the more toxic poisons include zinc phosphide and ANTU. Rodent poisoning is discussed in detail in the chapter on pesticides.

Rodent trapping is only fairly successful for rats, since rats drag traps, if necessary chew off a leg caught in a trap, or even cause a trap to malfunction.

Mouse trapping is successful if single mouse traps, rather than four-unit traps, are used if the trap is placed perpendicular to the wall and within the runway and if fresh bait is used on a routine basis. The bait should be changed each day or every other day. Mice are attracted by peanut butter, bacon, nuts, and other fresh foods. The bait should always be tied to the trigger so that the mouse will not steal it. Another technique of baiting is to take cornmeal or some other tasty food and make a tiny path up to the trigger and cover the trigger with the food.

Glue boards, which are similar to fly paper but sturdier, have been used but are not very successful. The rodent sticks while crossing the glue board and must then be killed and disposed of.

Food and Harborage Removal

The potential for rodent infestation is directly related to the amount of food and harborage present. Rodents find food in exposed garbage, pet food, or stored food, such as grains, fruits, meats, and vegetables. Rodents also find food sources in gardens, on farms, in horse manure, and any other type of organic material. Although they prefer fresh food, rodents will eat anything.

Most harborage is provided by poorly stored rubbish and trash, lumber, abandoned vehicles, discarded appliances, old shacks and sheds, high weeds, double walls, false ceilings, drawers, attics of houses, and any other area where they may be able to conceal themselves and nest. Frequently, rodents are found in cellars or closets within the home behind piles of dirty, discarded clothing.

An effective rat-control program requires removal of available food supplies and shelter for the rodents. As many rats as possible must first be exterminated to avoid scattering during a comprehensive cleanup program. Reduced food supply decreases the rodent population because of competition for food and harborage. The cleanup technique is extremely important, not only for the reduction and final elimination of most of the rodents, but also for the elimination of flies, roaches, mosquito-breeding areas, and for the reduction of potential disease.

All refuse should be stored in 30-gal metal cans with tight-fitting lids. Plastic is unacceptable because rodents chew it and also plastic cans tend to tip easily. Heavy duty rubber cans may be used providing the lids screw on tightly. A good refuse container is rust-resistant, watertight, tightly covered, easy to clean and handle, and of heavy-duty construction. Fifty-five gallon drums, such as oil drums, should never be used as a refuse storage container, since the lids do not fit properly on the container and when full they are too heavy to lift.

Remove pet food immediately after mealtime, making certain to dispose of all spilled pet food as well. Pet food has been a vital source of rat problems in numerous instances.

All dry food products should be stacked on pallets 6 in. above the floor and 12 in. from walls. Rotate food at least once every 2 weeks during the summer and once a month during the winter. Clean behind and under dry food on a daily basis. All spilled foods should be removed immediately. It is important to use a bright light or flashlight when sweeping storage areas to locate all spilled food and to determine if rodent signs are present.

Materials such as lumber and pipes should be stored in columns above the ground. Since high weeds contribute not only to rodent problems but also to fly and mosquito breeding, it is essential that they be trimmed regularly and destroyed.

Abandoned automobiles are rodent harborages and should be removed and disposed of promptly in keeping with air pollution or solid-waste ordinances. Rickety old sheds are excellent rodent shelters. Materials usually found in these sheds should be removed and such sheds either refurbished or torn down. Numerous rodent problems are traced to old sheds, which provide excellent harborage for rats.

Solid-Waste Disposal and Storage

Solid waste should be collected at least twice a week during the summer and at least once a week during the winter. All business areas should have solid waste collected on a daily basis. Many of the large metal containers found behind businesses are open invitations for rats to come, eat, and live. If these containers are improperly covered and maintained, food is spilled, and units grow odorous and attract flies, roaches, and rodents. It is essential that these containers are perfectly maintained, that the surrounding areas are policed several times a day, and that the lids are placed firmly on the units. Compactor trucks are best for collection of garbage and other types of solid waste. Compactors cannot be used for the collection of large items, such as refrigerators and lumber. Compactor trucks prevent blowing and spilling of materials, are leakproof, and are easy to load and unload. Solid waste taken to properly operated sanitary landfills sharply reduces potential rodent problems. Properly operated incinerators will destroy solid-waste materials; however, when improperly operated, partially burnt organic material is removed from the incinerator and becomes an excellent rodent attractant.

Garbage may be ground up and sent through the sewers to sewage treatment plants. This technique is fine for the home but provides enormous quantities of fresh garbage for the rodents living in the sewers. Composting can be utilized; however, if the compost heap is improperly maintained, it also becomes a rodent attractant and supplies rodent food. Further information on solid-waste disposal will be found in the chapter on solid and hazardous waste found in Volume II.

Rodent Proofing

Rodent proofing is the elimination of all holes $\frac{1}{2}$ in. or larger for rats and $\frac{1}{4}$ in. or larger for mice to prevent these rodents from entering an establishment. Rodent proof all doors, windows, gratings, vents, pipe openings, and foundation walls within easy access of rats. Techniques include cuffing and channeling, screening, use of metal guards, concrete, and curtain walls. Channeling and cuffing is the application of metal bent at right angles to the bottom and sides of doors. Kickplates are also used on doors. The door must be lowered so that the opening beneath it is less than $1/_2$ in. from the floor. Vents and windows must be screened against rodents and flies. Existing fire-control ordinances or rules and regulations should be checked to avoid violation. The screening should be 17-gauge galvanized hardware cloth for protection against rats, and 19-gauge ($4 \times 4^1/_2$ in. mesh) for protection against mice. Where sheet metal is used, it must be 24-gauge galvanized sheet metal.

Metal guards are protection against rats entering a building by coming across wires and pipes. The metal guard is conical in shape, with the cone facing outward. The openings around pipes or conduits have to be covered with either sheet metal patches or with concrete bricking mortar. The difficulty with the use of concrete is that if it is in an area, especially around radiators, where expansion and contraction takes place, the concrete will expand and contract at a different rate than the wall and may easily crack and crumble, thereby creating a new rat opening. Curtain walls, which are L-shaped walls, are made of concrete. They are poured along the foundation walls to keep rats from burrowing under the foundation. The curtain has to be at least 2 ft below ground level, 1 ft in length, and 4 in. in height. (See Figure 5.3 for these types of rodent proofing.) Rodent proofing also calls for the elimination of dead spaces, such as double walls, double floors, or enclosed areas of stairways, whenever possible. If elimination is impossible, then these areas should be checked frequently for rodent infestation and measures taken as are needed for control. Once a building has been rodent proofed, adequate rat and mouse control work should take place within the building before it is occupied.

Continued Maintenance

Continued maintenance is a necessary part of any rodent control activity. Once food and harborage are removed, the last rodent is killed, and rodent proofing is complete, many assume the job is finished. It is not, since there are innumerable ways in which rodents may still invade a premises. New pipes may be put into a building. New construction may take place. Doors or windows may be left open. Walls may crumble. Screens may be broken. Garbage and trash may start to accumulate again, and new harborage may be created. It is essential that the individuals responsible, whether they be tenants, landlords, or homeowners, thoroughly survey the premises at least once every 2 weeks to determine if a rodent problem is redeveloping. Flashlights should always be used indoors to check for rodent signs. Continued maintenance is the key to any private or public rodent control program. Without it the conditions will not only reoccur, but the rat problem will be substantially increased, since it takes a period of time before a rat population will find some sense of equilibrium within its environment.

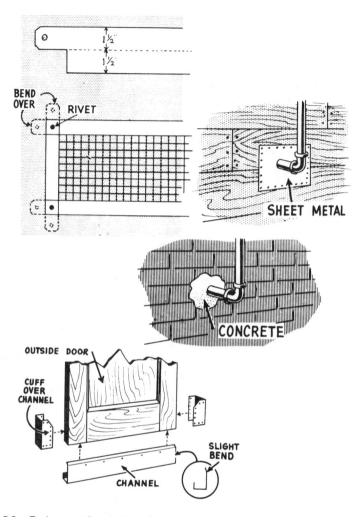

Figure 5.3. Rodent-proofing devices. *Source: Control of Domestic Rats and Mice* (Atlanta, GA: U.S. Department of Health, Education, and Welfare Public Health Service, Communicable Disease Center, 1964), p. 22.

MODES OF SURVEILLANCE AND EVALUATION

Rodent surveys are conducted by professional personnel, technical personnel, members of the community, or businessmen who are trained to look for rat signs. There are four basic types of rodent surveys. The first one is a survey in a single building, business, or industry where rodents are suspected. The surveyor looks for previously discussed rodent signs, for food and water sources, for improper solid waste storage and removal, for rodent openings, and for any other potential rodent harborage. This information is noted on a survey form and the exact place where each of the problems occur is marked on a rough map on the back of the form. It is wise to use a system of symbols with explanations to avoid cluttering the map.

The second type of survey is one in which the environmental health practitioner or technician makes a study of an area and determines the types of problems in a given city block. This study includes such information as the address of a building, information concerning the building (i.e., residential, business, vacant lot, etc.), potential rodent food and harborage, active rodent signs, potential rodent entries, and so forth. A map of the area is then drawn. It is also wise, at this time, to make a comprehensive housing survey to determine the rodent problem within houses and the types of housing problems needing correction.

A third type of survey is one in which a professional health educator or environmentalist goes from house to house with a questionnaire to determine from residents the types of rodent problems encountered. Again this information could be put on a checksheet. The information should be indicated as a questionnaire survey, in contrast to a physical survey. Together the two surveys are of considerable value in a rodent-control program.

A fourth type of survey is one in which the environmentalist goes out to a vacant area, farmland, park, or other outdoors area away from homes to determine where rodent conditions exist. A detailed map is drawn of the area to indicate precisely where rodent problems are found and the extent of the problems.

It is wise to utilize a simple complaint form to record rodent or other problems on an individual basis.

RODENT CONTROL

Many of the techniques of rodent control include poisoning, trapping, removal of food sources, removal of harborage, storage of solid waste, alternate methods of disposal of solid waste, and continued maintenance. Additional control measures are used in large communities, single homes, sewers, and outside areas.

Community Rodent Control

Although rodent problems are not unique to any part of the urban environment, those that are of greatest concern exist within the inner city. Slum tenements and poorly kept houses, coupled with debris, litter, and other solid waste, contribute to the large rat populations. Where single-family, three-story, or four-story dwellings once existed, now dwellings have been divided into six- to eight-family apartments. The population density of a given dwelling can rise from 4 or 5 to 50 or 60. Obviously, bathroom accommodations are lacking and there is insufficient space for adequate living. Many of these dwellings have numerous rat entrances, usually bad sanitary conditions, and severe rat infestations. The human population is generally highly transient. There seems to be a lack of cohesiveness between members of the same building.

Even where civic organizations and block councils exist, only a few residents may have participated in electing the officers and boards of directors. At times, individuals may simply appoint themselves as officials. Generally these associations start out to improve their neighborhood by keeping taprooms out, obtaining stoplights or stop signs, improving recreational facilities and housing conditions, and possibly by getting involved in some health matters. The typical single-block organization consists of a group of people who live near to each other on a given street. These individual block organizations range from totally inactive to highly active. Usually within each block one finds one or more civic-minded people who are willing to work together on a project to improve the neighborhood. In many cases, the larger civic association does not truly represent the wishes of the individual block groups, since the more vocal and better-educated individuals tend to come to civic association meetings, whereas the less vocal tend to stay home and be concerned with individual problems. Unfortunately, if you only use these organizations, rat-control projects will be unsuccessful, since the people who need the most help are generally those who do not attend the civic association, home and school association, or health and welfare meetings. To stimulate this group of individuals to work together, it is important to go into each individual block and work on a house-to-house basis to attempt to set up small meetings in the homes. A good approach is to knock on doors, introduce yourself officially as a member of the health department (make sure you always show your identification card), and explain the severity of the rat problems in the area, as well as in other areas. The environmentalist should also give some indication as to the nature of the programs that can be brought into the area to help improve it. These individuals within the blocks are then encouraged to come to a mass meeting at a local school to further discuss the rodent control problems and the kind of things that might be done to improve the environment. Each individual should be encouraged to bring friends, neighbors, and relatives to the meeting. Announcements should be distributed the day prior to the meeting and on the day of the meeting. Sound equipment should be used on a truck to announce the community meeting for rodent control at the local school. Explanations of the rodent problem given over the loudspeaker should be simple. The meeting place and time should be spelled out clearly.

It is essential at this mass meeting to have representatives from all the civic associations and the various official agencies to discuss informally the needs of the community as they see them, to listen to citizen problems, and then to explain some proposed techniques for conducting a community rodent-control program. It is important that the individuals at this meeting vote to participate in each of the phases of the rat program. During such a meeting each of the citizen complaints should be recorded and submitted to the appropriate official agency for correction, where possible. In one such meeting in which the author was involved, the community's major concern was the fact that the young people were constantly getting into trouble with the police. There was an apparent language barrier. When this problem was identified, the police department sent community relations people who spoke the language of the community into the area and worked with them to improve community-police relations. This helped not only to satisfy a need of the community, but it also indicated that the official agencies were interested in the public and that it would be a good idea to work together on community rodent control.

Another technique used to organize blocks is to pull together members of the community by the use of sound equipment at the time that rodent poison is placed by the health department. Here the members of any given block have an opportunity to meet face to face. This is an important time to hold an election of officers for that given block. Once officers are elected, the health educator in conjunction with the larger community organization should assist the block in developing its own block organization.

Another technique is to ask individuals, when you are going door to door prior to the first mass meeting, about the individual who they turn to when they have problems. Frequently, key people in a block will be identified. These individuals are natural leaders and could be utilized as such in a block organization. There are innumerable block organization techniques. The key is not necessarily which technique to use, but rather to help organize a block, stimulate the people to work together, and try to maintain the organization so that a ready supply of willing workforce exists for continued maintenance to keep the program from falling apart. These block organizations may be used later in helping bring satellite health clinics into an area to help increase immunizations, and so forth.

An important person in a block organization can be the health aide. This individual is a member of the community hired by the health department who receives specific health training. As a member of the community and therefore recognized by the community, and as a member of the health team, this individual can greatly assist in developing a comprehensive rodent-control program, as well as other comprehensive health programs.

Presurveys

Three types of presurveys may be used. First is the horseback type of survey, where the environmentalist rides up and down alleyways or walks the city blocks to get some approximate idea of the kinds of existing housing conditions, the amount and type of solid waste, and the kind of rat signs readily visible. This survey gives a rough idea of the needs of the community.

The second kind of survey, the opinion attitude survey, should contain several basic points. (1) What is the extent of the community organization, including the size and scope of the group, potential leaders, and desire to organize? (2) What are the needs and attitudes of the people toward elimination of roaches, rats, and poor housing? (3) What is the extent of the pest infestation, including mice, rats, and roaches? (4) What action has the citizen taken against roaches, mice, and rats? (5) How many complaints has the individual made about pests, and to whom have they been made? (6) How many rat bites have occurred in the person's house in the last year? How many have been reported to the health department? How many have been reported to others? (7) Has the individual heard about rat bites in the neighborhood within the last year? If so how many?

The information in this opinion attitude survey, when compiled, gives the environmentalist some understanding of the community's self-concept and whether the individuals are anxious to help correct the rodent situation.

The technical survey consist of a professional evaluation of rodent signs inside and outside of the property, of potential rodent entrances, of the availability of food and harborage, and of the rat-proofing problem. The data should be compiled and combined with the information in the opinion attitude survey to make a good evaluation of the existing rodent problems within the community.

The survey data should be kept and used as a means of comparison after completion of the program. This will give the environmentalist a means of comparing the results of the actions taken and also the results of continued maintenance.

In addition, it is well to add the material from the files of the health department relating to rodent complaints, solid-waste complaints, and rat bites that were reported to the health department within the areas under study. It is recommended that a comprehensive map with proper coding be drawn to indicate the degree of severity of rodent infestations and also the other environmental problems associated with rodent infestation.

Area Selection

The area selected for a community rodent-control program should be based on the extent of community organization, the willingness of groups to form and work together, the number of complaints for a 3-year period relating to rats and solid waste, the number of rat bites for the past 7 years, and the various surveys that have been conducted. These proposed areas should be stated clearly on a map and in writing. The number of houses involved, the severity of infestation, and the extent of cleanup should then be presented to a committee composed of regulatory agencies and civic leaders. The ultimate decision of where to work should be made jointly by these representatives of the communities and the official agencies.

Planning

In order to plan an effective community rodent-control program, it is necessary to establish a working committee structure. The steering committee can be composed of representatives of the health department, housing department, police sanitation unit, solid-waste department, and community relations department. The functions of a steering committee are to plan details of operation, resolve day-to-day problems, supply general knowledge of existing community resources, determine on a departmental basis the availability of workforce, act as liaisons between the project and the individual official departments or agencies, and help evaluate and redirect the program when needed.

A second committee, called a citizens advisory committee, should be established and can be composed of representatives of the health and welfare council, civic associations, block councils, block leaders, unofficial agencies, a local religious leader, a local school principal, a representative of the landlords, a business person, and others as needed. The functions of this committee are to review the suggested programs developed jointly by official agencies and citizens and make recommendations to the health director or environmental health director who is the coordinator of the rodent-control program regarding changes or additions. These recommendations may include selecting certain city blocks for intensive rat-control activities, techniques to assist in stimulating community interest, and practical methods of carrying out effective community rodent-control programs. The steering committee and citizens advisory committee should define the problem clearly in writing and establish goals, as well as develop the program and help in its implementation. The written definition of the problem is based on the previously mentioned surveys. The goals should be of two types: (1) long range, to establish and maintain dwelling units, yard, building, and other areas that are free of rats; to help provide a community with a spirit of self-help and pride, leading to neighborhood betterment; to improve the total dwelling unit and neighborhood; to stimulate interest to undertake similar programs in other infested areas. (2) immediate goals, to remove garbage, trash and other solid waste from yards, homes, buildings; to remove rodent harborage; to poison and otherwise destroy the rats in the neighborhood; to remove abandoned automobiles; to cut down weeds in vacant areas; to rodent-proof homes; to train community leaders in the various phases of rat control; and to encourage block organizations to promote successful programs and continued maintenance.

The Program

The community program consists of a series of meetings and actual rat control activities. The first meetings, which are for stimulation and interest, have already been discussed. The major meeting called at a school or other similar facility has also been discussed. It might be added that, during the course of this major meeting, the responsibilities of government and each of its branches and the responsibility of each of the groups of tenants, landlords, businesses, and industry should be explained. The next meeting called should be a training meeting. Personal safety, safety of the child and family, removal of dead rodents, and the covering up of rodent odors are taught at this time. It is important to note that only certified environmentalists or pest-control operators can place rodent poisons. The training meeting is also used to present a total picture of how to rid an area of rats. Therefore, further discussions, demonstrations, and visual aids in the removal and storage of solid-waste material, in rodent control through rodent proofing, and in the necessity for constantly policing potential rodent problems and continually maintaining a rodent-free environment are conducted. At times, the general meeting may be combined with the training meeting. The individuals within the project will decide which is better for their community.

During a discussion of cleanup techniques, it is important to mention the team approach, the use of teenagers, and the use of volunteers. Everyone can help, from a small child to an aged person. All that is necessary is to remove the solid waste to the curb in the best manner possible so that the solid-waste agency workers can pick up the material and remove it. Additional meetings are held as needed during the course of the project to explain various aspects of the project. It is important to establish at the first meeting a timetable for the project and then to reinforce the dates of the timetable at each meeting.

In addition to meetings, other techniques used to stimulate community help are handouts, and especially, mobile sound equipment, since the spoken message, especially in the early evening, is received by all, including the working people within the family, who are then given the opportunity to participate in the rodent control program.

The second phase of the program consists of the actual implementation of activities. Implementation of activities are as follows. (1) First poisoning — red squill used preferably and applied by the health department. (2) Second clean up. The latter should be carried out by the community and any volunteers such as school children and Y-teen groups. All solid waste, including junk, garbage, refuse, and debris, have to be removed from basements, cellars, and vacant lots. The community supplies the workforce to bring this material to the curb. The city supplies the necessary personnel to pick up the material and truck it to the landfills, recycling center, or incinerators. (3) Additional poisonings, using anti-coagulants. (4) Total cleanup carried out by the community on a special cleanup day. (5) The city should go into a direct enforcement program to get the individuals who do not comply with the wishes of the community to remove all their solid waste and destroy all their rats. Inspections should be made of all of the areas. Owners of houses in violation should be cited and if necessary brought to court. Vacant houses should be cleaned and sealed, and the cost should be assessed on the owners. Anti-litter, -garbage, and -refuse storage ordinances should be enforced strictly. All vacant lots and alleyways should be cleaned by the community, and the waste should be removed by the solid-waste agency. (6) The program should be evaluated by the environmentalist by making careful postsurveys of all areas. A comparison should be made between the presurvey and the postsurvey to determine the effectiveness of the program. (7) The community, with the assistance of the environmentalist, should now stimulate their block organizations to work on continued maintenance and rodent proofing, where rodent proofing is possible. Each individual should attempt to rodent proof his or her own structure.

The schedule for the program is generally as follows: first week—first poisoning; second week — first cleanup; third week — second poisoning; fourth week — second cleanup; fifth week — additional poisonings; sixth week — survey; seventh week forward — continued maintenance.

The area-baiting concept is extremely important since rat pressures are reduced initially and during cleanup, rats will not scatter. Eventually, as more and more solid waste is removed, the rats will die from lack of food or will be poisoned by the additional rat poison available to them.

Individual Rodent Control

When an individual is faced with rodent problems, the premises should be surveyed to determine where the rats are coming from and appropriate measures of poisoning, cleanup, and rodent proofing should be taken. It is important that the individual contact the health department and inform them of the problem. In this way, the health department can determine if this is more than a local situation. An individual should never use rodent poisons other than anticoagulants. If the situation is beyond the individual's control, a trained exterminator should be called to carry out the necessary rodent-control functions.

In businesses, trained exterminators should be utilized for rodent control. However, owners should be alert to the use of highly toxic substances. Rat problems are the responsibility of the individual business owner.

Sewers

Rodents can be controlled in sewers by use of a poison, usually red squill, raticate, or an anticoagulant in paraffin blocks placed along the edges of the sewers. The rats consume the poison and usually die in the sewer.

Outside Areas

It is important to use adequate rodent poisoning. If the area is completely away from homes, carbon monoxide or chloropicrin can be pumped into the burrows to destroy the rats. Other rodenticides, such as red squill and anticoagulants, may be used.

Government Programs

Local and state governments have been involved in rodent control through the enforcement of nuisance laws, food laws, and other types of public health laws. The federal government became involved in rat control in 1967, when Congress authorized 40 million dollars for a 2-year period for rat prevention. In the ensuing years, additional millions of dollars were allocated to the Department of Health and Human Services for rat-control projects in various communities. Since 1969, the Partnership for Health Amendment (PL 90-174) has initiated new rodent-control programs and strengthened existing ones in cities with serious rat problems. Since 1969, programs have been initiated in New York, Atlanta, Baltimore, Charlotte, Chicago, Cleveland, Milwaukee, Nashville, Norfolk, Philadelphia, Pittsburgh, St. Louis, and Washington, DC These rodent-control programs generally consist of three phases: (1) preparatory phase, during which the concept was sold to community officials and mass media on how to control rats; (2) the attack phase, consisting of surveys, education, cleanup, poisoning, and code enforcement; and (3) follow-up phase, which included continued maintenance. As part of these federal efforts, plans were drawn up for comprehensive rodent-control programs. These included citizen participation, community information, and education; effective local administration; adequate municipal service, including garbage, trash, and junk pickup; enforcement of codes and ordinances; removal of dilapidated buildings; systematic poisoning; training and employment of local residents in rodent-control efforts.

In 1979, the federally funded rat-control projects were once again renewed. In the 1980s, rodent control became solely a state and local effort. Philadelphia and New Orleans used some funds from Preventive Health Block Grants for rodent control. Ultimately local citizens working with local government must solve the problem.

SUMMARY

Rodents live close to humans and can readily cause disease or other health-related problems. Although at present the level of disease caused by rodents is low, the potential is always there. Rodents not only destroy property and cause fires and disease, they also bite helpless individuals and are the symbol of urban decay. Rodent control can be carried out if a given community is willing to work together to remove all the food and harborage, apply the necessary rodent poisons, and carry out proper rodent proofing. Rodents will continue to be a serious problem in our society because of a lack of adequate control programs.

RESEARCH NEEDS

It is necessary to develop a better understanding of the resistance that rats are starting to acquire toward anticoagulants. The mechanism of resistance must be understood in order to prevent the elimination of this relatively safe type of rodent poison. New rat-control methods are needed. Research should be conducted for the development of single-dose chemosterilants and repellents. It is also necessary to determine how to motivate citizens to keep their homes and neighborhoods rat free.

Chapter 6

PESTICIDES

BACKGROUND AND STATUS

Pests cause a reduction in size, yield, storage, and market quality of food and spread diseases. To control pests, a series of chemicals, called pesticides, have been developed. Pesticides include acaricides or miticides, used against mites; algaecides, used against algae; attractants, used to attract insects, birds, and other animals; chemosterilants, used to interfere with reproduction; defoliants, used to remove leaves form plants prior to harvest or to eliminate unwanted plants; fungicides, used against fungi; herbicides, used against weeds; insecticides, used against insects; molluscides, used against slugs and snails; ovicides, used against insect eggs; repellents, used to drive animals or insects away; rodenticides, used against rats, mice, and other rodents.

Pesticides are used as aerosols, sprays, dust, in granular form, or as baits. They may be effective on contact, be taken up by the plant, enter the lungs or trachea of animals, or be eaten. The quantity of the pesticide used, the type of pesticide, and how it is used are very important in the control of troublesome pests.

Pesticides produce useful and harmful effects, depending on the type and quantity used and the method of application. About half of all pesticides are used in farming. Roughly 5% are used by governmental agencies and the balance by residential and industrial users. Currently, we are using approximately one billion pounds of pesticides annually.

Pesticides are categorized by their lifetime of effectiveness as follows: nonpersistent, lasting several days to about 12 weeks; moderately persistent, lasting 1–18 months; persistent, which include most of the chlorinated hydrocarbons such as DDT, aldrin, and dieldrin, lasting many months to 20 years; permanent, including mercury, lead, and arsenic, lasting indefinitely. PCBs or polychlorinated biphenyls used in asphalt, ink, and paper behave very much like the persistent pesticides and require close control to avoid contamination of the environment.

Pesticides that degrade or deteriorate rapidly are also of great concern because of their extreme toxicity and because they are nonselective in their action on animals, humans, and pests. Organophosphates would be an example of this type of pesticide.

A pesticide moves through an ecosystem in numerous ways. It is introduced by surface application, spraying, or other techniques and may stay in the air or be washed down by rain. The concentrations of the pesticide continue to increase in the soil over time, and where leaching occurs, the pesticide can move into surface or underground water supplies. Some pesticides become tightly bound to soil particles, polluting the surface waters when the surface particles are washed into them by the force of heavy rains. Some pesticides are ingested by minute, aquatic organisms and scavengers and become concentrated as they move up through the food chain. It is known that oysters, for instance, will concentrate DDT in their tissues 70,000 times greater than amounts found in the surrounding waters. Fish also concentrate pesticides as part of the food chain. Eventually, the pesticides may reach humans and, at least in the case of DDT, are stored in the fatty tissues.

The major pesticide laws in effect in the United States were totally rewritten in 1972 and updated in 1978 and in 1988 as amendments to the 1947 Federal Insecticide, Fungicide, and Rodenticide Act (FIFRA). This law forbids anyone, including the federal government, from using a pesticide contrary to label instructions and gives the Environmental Protection Agency (EPA) the authority to restrict the use of pesticides to trained persons. The law applies to interstate and intrastate use and sale of the product. It provides screening procedures for new pesticides suspected of causing cancer, birth defects, or mutations. Based on this law, the EPA has taken action against the use of Kepone, DDT, Aldrin, dieldrin, heptachlor, chlordane, mirex, and mercury-based pesticides. At present, the only exception to these actions is made by the EPA if the agency feels that the benefits outweigh the potential adverse effects and no alternatives are available, if a significant health problem occurs without its use, or if an emergency exists.

Under present EPA orders, all individuals, including public health workers, who are involved in the use of pesticides, must take comprehensive examinations and be registered in the use of pesticides by category of employment. Since the use of pesticides is increasing, the dangers will increase and the current status in succeeding years will deteriorate unless further action is taken.

Current Issues

When the U.S. EPA was founded in 1970, the FIFRA authority was transferred from the U.S. Department of Agriculture to the Environmental Protection Agency. In 1972 the Federal Environmental Pesticide Control Act passed by Congress was an amendment to the original Pesticide Act mentioned earlier. It provided for direct controls on the use of pesticides, for classification of certain pesticides into a restricted category, for registration of the manufacturing plants, and for a national monitoring program for pesticide residues. Environmental effects and risks were added to the pesticide registration process.

The 1972 FIFRA amendments required a review of all of the registered products being used. The review was to be completed by 1975. Unfortunately because of the large amount of data that was needed and the large number of products to be assessed, the General Accounting Office (GAO) determined it would take until the year 2024 for this work to be done. As a result of this, the entire reregistration process simply broke down.

From the very beginning, the EPA had problems with the pesticide regulation program and with the process of implementing the 1972 amendments to FIFRA. However, this act still did not resolve the many problems related to pesticide regulation.

The issues were registration, tolerances for pesticide residues, federal preemption of state tolerances, reregistration, inert ingredients of pesticide formulations, regulatory options, and pesticides in groundwater.

Under FIFRA the regulation of pesticides was done through the registration of the individual pesticide products. The products were not permitted to present unreasonable adverse effects to people or the environment if the pesticide was used on a food crop or animal feed. The regulations also required that a maximum acceptable level of pesticide residues remaining on a treated crop be determined by the EPA, and be monitored and enforced by the Food and Drug Administration. The data required for registration of pesticide products included health and environmental data, environmental fate, carcinogenicity, chemistry of the product, toxicity to fishlife, and mutagenicity.

The EPA established tolerances for pesticide residues in foods. Tolerances determined the maximum amount of pesticide residue that could be permitted in food or animal feed without being considered to be an adulteration of the product. There were several kinds of tolerances of any given pesticide, since the foods may have been utilized as raw products or as a mix in preparing a finished food product. This was quite confusing.

The federal government, under the 1988 law, preempted the tolerances established by state law. The problem was that the EPA standards were currently considered to be a floor and not a ceiling for standards for any given pesticide. Therefore, the states could have more stringent standards. However, if this occurred, it could have interfered or hindered the flow of products through interstate commerce and hindered the marketing of pesticide products, pest treatment services, and the treated commodities. The states argued that the data that had been used to determine the allowable amount of pesticides in the food or feed product may not have been accurate, and over a long term it was the right of the states to protect their citizens.

The FIFRA amendments authorized the EPA to conduct a "generic" review of the active ingredient's safety. The EPA had identified some 600 active ingredients considered to be commercially important among the 1500 active ingredients officially registered with them. There were an estimated 40,000 studies about these pesticides in the EPA's files. Only two pesticide active ingredients had been reregistered under this process until this point.

The EPA also investigates inert ingredients. An *inert* ingredient is that part of the pesticide formulation that is not intended to have any pesticidal activity. It is used either to dilute the pesticide or to propel it or deliver it in some manner. Unfortunately, some of the *inert* ingredients have potentially adverse effects on people. Vinyl chloride gas is a human carcinogen that has been used as an aerosol propellant. Of the approximately 1200 compounds used as *inert* ingredients in pesticides, the EPA has determined that 55 are known toxics that may cause animal cancer or nerve damage. Fifty-one compounds are structurally related to the known toxic compounds. Nine hundred compounds are of unknown toxicity.

The EPA can revoke the registration for a pesticide if the information indicates that the product presents an unreasonable risk to human health or the environment. This process is called *deregistration*. If the chemical is cancelled, the stockholders, owners, and other individuals can demand payment of the EPA for their losses.

Pesticides and Groundwater

Pesticides can reach the groundwater supply through misuse or mismanagement related to waste disposal, spills, leaching, etc. There are four main issues related to pesticides in groundwater. They are

1. Pesticides need to be detected in groundwater.
2. The EPA needs to determine at what level pesticide residues in groundwater should trigger action.
3. If a pesticide is detected in groundwater as a result of normal use and if the groundwater pesticide limits will be exceeded, the EPA needs to decide what is an appropriate remedy.
4. The EPA needs to decide if a pesticide should be immediately prohibited if it reaches substantial levels.

The EPA estimated from a 1986 survey that 30 states had found wells contaminated with one or more of 60 different pesticides.

Other Issues

Additional issues in the use of pesticides include food contamination, potential indoor air pollution, preparation of professional pesticide applicators, and pesticides related to fish and wildlife. Pesticides in foods or on foods have become a major issue today. The scare related to the purposeful contamination of two grapes in 1990 brought an entire food supply to a total halt and created untold problems for the country providing the food, the agencies evaluating the food, and the public, who in some way end up paying for the food through taxes or through the disposal of the rotted fruit. There are some questions concerning the amount of pesticides that are introduced into the food chain and the concentration of these pesticides in the finished product because of the use of raw products that have been exposed to the chemicals. The consumer must be protected, but the food supply must not be destroyed.

A variety of pesticides are used on lawns and on pets. The individuals applying these pesticides are untrained in the storage, mixing, application, and disposal of the pesticides. Pesticides may be stored in garages or other enclosed areas, and therefore, constitute a potential indoor air pollution problem.

Professional pesticide applicators need to be certified and recertified by their state agencies. Many states have more stringent standards than the minimum standards established by the U.S. EPA. It is therefore necessary to determine the standards of a given state and then decide how best to train the applicators and test them.

Pesticides may have a long-range effect on fish and wildlife, depending on the type of accumulation that may occur. It has been known for years that DDT causes the thinning of egg shells, which in turn prevents the successful hatching of the chick in a variety of birds. Even a pesticide that is as valuable as diazinon has been found to be harmful to fish and wildlife and, therefore, can no longer be used on golf courses and sod farms.

SCIENTIFIC, TECHNOLOGICAL, AND GENERAL INFORMATION

Types of Pesticides of Public Health Importance

The chapter will basically be concerned with the types of pesticides used to control insects, ticks, mites, spiders, rodents, and plants of public health significance. Pesticides should have certain qualities to be acceptable for use. They should be specifically toxic to harmful insects, etc., harmless to humans, inexpensive and easily used, rapidly degradable to nontoxic substances, nonflammable, noncorrosive, nonexplosive, and nonstaining.

Insecticides may be used as stomach poisons, contact poisons that penetrate the body wall, fumigants that enter through the insect's breathing pores, desiccants that scratch or break the body wall or absorb into the waxy protective outer coating. Pesticides may also be listed as larvicides that kill larvae, ovicides that kill the insect's eggs, or adulticides that kill the adults.

Inorganic Insecticides and Petroleum Compounds

Prior to 1945, numerous inorganic pesticides were used widely. These included the arsenical, Paris Green, used against the potato beetle; hydrogen cyanide, used against red scale; lead arsenate, used against the gypsy moth; and sodium arsenite, an insecticide, and a weed killer. All arsenicals are now banned. Compounds of copper, zinc, and chromium were also used as pesticides. Chlorine and sulfur made extremely toxic compounds and were used along with salts of arsenic, lead, mercury, and selenium. Unfortunately, many of these compounds were quite toxic to humans. Also, some of the insects developed resistance to certain inorganic pesticides.

Petroleum oils, such as kerosene, diesel oil, and #2 fuel oil, were used as mosquito larvicides. These oils, which are still in use, have certain toxic properties, because they penetrate the tracheae of larvae and pupa of mosquitoes and anesthetize them. A fraction of these oils mechanically interfere with the breathing process of insects, causing suffocation. Sulfur acts as a repellent against chiggers. Borax is still used in buildings for roach control with varying results.

Botanicals

Probably the earliest pesticides used were pyrethrum, which is extracted from the flowers of *Chrysanthemum cinerarifolium*, rotenone, which is derived from Peruvian cuve, and red squill, which is derived from the inner bulb of the plant *Urginea maritima* belonging to the lily family. These chemicals are highly specific to the pests and have a very low toxicity to humans. Today pyrethrum is used primarily in combination with other insecticides. Pyrethrum continues to be a quick insect knockdown agent. Synthetic compounds similar to pyrethrum, called allethrin and resmethrin, have been developed and are utilized in the same manner as natural pyrethrum. Rotenone is used to kill fish without leaving toxic byproducts for human beings. Natives of some tropical countries crush and throw plants such as Derris and cuve into the water, and the chemicals present in the plants paralyze the fish. Rotenone is also utilized for killing of fleas and other ectoparasites on domestic pets. Red squill in its fortified state is used effectively in killing Norway rats. Because of the difficulty of obtaining these pesticides from abroad during World War II, and the military need for chemicals to kill disease-producing insects, the United States developed a series of organic pesticides in the early 1940s. In addition, such insecticides as DDT were recognized in Switzerland in 1939. Benzene hexachloride was recognized as an insecticide in 1940 in France and England.

Chlorinated Hydrocarbons

The chlorinated hydrocarbon insecticides are combinations of chlorine, hydrogen, and carbon, They act primarily as central nervous system poisons. The insect goes through a series of convulsions and finally dies. The first major chlorinated hydrocarbon was DDT. The technical name is dichlorodiphenyl trichloroethane. DDT has been highly useful for the control of mosquitoes, flies, fleas, lice, ticks, and mites, reducing considerably the level of malaria, plague, typhus fever, yellow fever, encephalitis, and so forth.

Although DDT has been banned for general use in the United States, with the exception of a serious uncontrollable emergency, it is still an effective chemical for the control of mosquitoes, which may cause malaria or other diseases. DDT enters the ecosystem and is stored in animal fat. However, in public health a decision must be frequently made as to the relative importance of one hazard vs another. Therefore, DDT is still used abroad in the interior of homes in areas where malaria is prevalent. The DDT, if applied carefully, will leave a residue on the structures that will last from 6 to 12 months. It also will not readily escape into the environment. The dosage in the residual spray should be 100–200 mg/ft^2.

The chlorinated hydrocarbons are divided into the DDT group and the lindane-chlordane group. The DDT group contains two benzene rings; the chlordane group contains either a *benzene hexachloride* or two modified benzene rings.

Methoxychlor and DDD are part of the DDT group. Methoxychlor is safer than DDT, since it is less toxic to mammals. It is utilized in many household sprays and aerosols. It is readily metabolized and eliminated in the urine of vertebrates. It also does not remain within the environment for more than a short period of time. It is used to kill mosquito larvae and control flies and insects that attack livestock; it is also used in agricultural areas. As a larvicide, methoxychlor is applied at a rate of 0.05–0.20 lb/acre. DDD is now banned.

Benzene hexachloride or BHC was widely used in public health and in agricultural programs. It has a musty odor and a short residual life. The gamma isomer of BHC has significant insecticidal activity. BHC is used currently abroad as a residual spray at a dosage rate of 25 or 50 mg/ft^2 for the control of mosquitoes causing malaria or other diseases. It will have a residual effect for about 3 months.

Lindane is the pure gamma isomer of BHC. It is highly effective as a stomach poison and as a residual contact insecticide. It has been used for control of lice, ticks, and other insects. Lindane vaporizers, used to control flies in food establishments, are dangerous and should never be used. In Southeast Asia, lindane was used in irrigation waters to control rice stem-borers. Unfortunately, the chemical killed the fish that were the protein sources for the local population. In overseas areas where the vector of plague, the flea *Xenopsylla cheopis,* still exists and where DDT is not effective, it is recommended that a 1% lindane solution be applied to ensure adequate control of this rodent flea. Lindane may also be used by professional personnel for the control of fleas on household pets, provided the pets are not under 2 months of age. Lindane in a 1% emulsion may be used for treating infested household sites within the house and in the yard. One percent lindane powders are used to control body lice. The brown dog tick can be controlled with a 0.5% lindane spray, with spot treatments on baseboards, floors, wall crevices, and areas where the animal sleeps. Lindane can still be used for dogs. It is banned for any type of fumigation.

Chlordane is dissolvable in many solvents but not in water. It is used to produce oil solutions, emulsifiable concentrates, wettable powders, and dusts. It acts as a stomach poison, contact insecticide, and fumigant. It is effective in spot control of ants, American roaches, silver fish, and has also been used extensively on soil insects, particularly termites. Chlordane has been extremely effective over the last several years, although German roaches have built up a resistance to it. It is probably the least toxic of the chlordane series, which includes chlordane, heptachlor, aldrin, dieldrin, endrin, isodrin, and toxaphene. In July 1975, the EPA suspended the use of chlordane because it is suspected of causing cancer in animals and it readily contaminates the environment.

Heptachlor has been used effectively for mosquito larvicide control. Heptachlor production was also suspended by the EPA as a suspected link to cancer in animals and because it is highly toxic to humans. It remains for long periods of time in the environment and readily contaminates the water, soil, and air. Aldrin and dieldrin, considered to be effective chemicals for insect control, were found to be highly toxic when misused, killing fish, birds, mammals, and even human beings, and is also suspected of carcinogenic activity; these too have been suspended from use since 1974. However, dieldrin still is used overseas at a rate of 25 or 50 mg/ft^2 as a residual spray in those areas where malaria or other mosquito-borne diseases are prevalent.

Endrin is one of the most poisonous of the chlorinated hydrocarbons. It is highly toxic, persists in the environment, and is a hazard to animals and humans. It should be used only when necessary and under strict supervision. It is not recommended for general use. It is now banned for general use.

Isodrin and toxaphene are both highly toxic chemicals, persisting for long periods of time in the environment. They are not recommended for use as insecticides. They are now banned for general use.

Chlordecone, better known as kepone, is a very effective chlorinated hydrocarbon when used as insect bait. It may last as long as 1 year without being altered. When used in proper dosages, that is, 0.125% peanut butter bait, it is effective against both roaches and ants, and produces a high kill. However, kepone, when used improperly or produced improperly, can be very toxic to humans. Therefore, the production of the chemical should be closely monitored, the disposal of waste in the chemical process controlled, and the bait carefully handled. Kepone can be placed in a paraffin bait and still be very effective for the control of American roaches. Kepone has been banned.

Organophosphates

The organophosphates are derived from phosphoric acid to inhibit the enzyme cholinesterase. In many cases, they have replaced the chlorinated hydrocarbons, because they are effective against insects that have become resistant to the chlorinated hydrocarbons; they are biodegradable; they do not contaminate the environment for long periods; and they have less long-lasting effects on organisms that are not meant to be treated with these chemicals. However, they vary tremendously in toxicity. The organic phosphorous insecticides include TEPP, chlorpyrifos, dichlorvos, phosdrin, and parathion, which are highly toxic; bayer 29493, baytex, dimethoate, fenthion, DDVP, and diazinon, which are moderately toxic; abate, gardona, dipterex, malathion, and ronnel, which are slightly toxic.

TEPP is used in greenhouses and on fruits and vegetables. It is highly toxic when mishandled and causes severe poisoning. It contaminates the environment for short periods of time.

Phosdrin and parathion are highly toxic insecticides that are fatal to humans if only one drop is placed in the eye. They should only be used by highly experienced, licensed, commercial operators. They are used as larvicides for mosquitoes at a rate of 0.1 lb/acre in rural areas away from children and animals.

DDVP, dichlorvos, and vapona are all the same compound. This pesticide is useful as a fumigant because it is highly volatile. It is quite toxic, but breaks down quickly. It is generally used in fly control as a spray or fog or in impregnated strips. In strip form it is effective for mosquito control for 2.5–3.5 months if used at a rate of one strip per 1000 ft^3. It may also be used in catch basins with one strip suspended 12 in. below the catch-basin cover, per basin. Dichlorvos is used in a sugar solution as a bait for fly control. It is mixed with water and used as an outdoor space spray for flies or mixed with water and used as a fly larvicide. Dichlorvos, since it has only a short residual life, is most effective rapidly and presents a short residual hazard to the environment. However, vapona strips should not be suspended over food, for a drop of the chemical could collect at the bottom of the strip and fall into the food, creating the potential chemical food poisoning. It would not be wise to hang vapona strips in areas where individuals suffer from upper respiratory ailments, since the chemical is discharged slowly over a long period of time and could become either an irritation or hazardous to the individual.

Diazinon is utilized in fly and roach control and other insect control problems related to vegetables and fruits. The residual period in the environment is fairly short, varying from 1 week to at most 2 months. Diazinon should not be used in poultry farms, since it is toxic to birds. In fly control, diazinon is mixed with petroleum compounds or water and used as a space spray or mixed with water and used as a larvicide. In roach control, diazinon is effective in reducing or eliminating all roaches, including German roaches. Only the German roach appears to be developing some resistance in some areas of the country to the compound at this time. Solutions usually contain the following concentrations of diazinon: spray, 0.5%; dust, 1%. However, pest-control operators are permitted to use 1% spray and 2–5% dust. Diazinon cannot be used on golf courses or sod farms.

Dipterex or trichlorfon is used in sugar and water as a bait for flies. In areas where garbage, organic materials, and manure are controlled, the fly bait works effectively and rapidly. Dipterex is also used as a bait with some success in roach control.

Abate is used at a rate of 0.05–0.1 lb/acre for larvicidal control of mosquitoes. Chlorpyrifous is used in mosquito control for ground-applied outdoor space, spraying at a rate of 0.0125 lb/acre. It is also used as a mosquito larvicide at a rate of 0.05–0.125 lb/acre. It is effective in control of roaches at a 0.5% concentration. When painted on a surface over which German roaches crawl, it has a strong residual effect, causing a kill of 90% of roaches for up to 1 year. The chemical persists in the environment from several days to as long as 1 year. Dimetholate is used for outdoor space spraying for flies and also for larvicidal control outdoors for flies. Fenthion must be used by trained mosquito-control personnel only. It is used in ground-applied outdoor space spraying for mosquitoes at a rate of 0.001–0.1 lb/acre. It is used as a larvicide at a level of 0.05–0.1 lb/acre, but must be carefully handled by trained personnel. Also, it is effective in roach control, but must only be used by pest-control operators as a spray at a rate of 2.0% concentration.

Gardona is a relatively safe, nonsystemic, broad-spectrum organophosphate used for fly and tick control. However, it is highly toxic to bees. It persists in the environment for only short periods of time. Naled is moderately toxic to animals. It has a very short residual period in the environment. It is used as an outdoor ground-applied space spray at a rate of 0.02–0.1 mg/f^{t2} for mosquito control. It is also used as an outdoor space spray for flies in liquid form, and as a bait in a sugar solution. Ronnel or korlan is used to control flies in an agricultural area. Its toxicity is slight and its residual period in the environment is very short. It is hazardous to livestock or dairy food.

Carbamates

The carbamates are derived from carbonic acid. Most of the carbonates are contact insecticides. They lower the cholinesterase level and act as nerve poisons, similar to the organic phosphorous compounds. Several of them produce a rapid knockdown, as produced by pyrethrum. The carbamates include sevin, which is also called carbaryl and dimetitan; baygon, which also called propoxur; and landrin.

Carbaryl is widely used in public health and agriculture. It is one of the safer insecticides for animals, but is highly toxic to bees. It is formulated only as a solid, which is then used as a wettable powder, slurry, or dust. It remains for a relatively short period of time in the environment. Carbaryl dusts are used in a 2–5% concentration to kill fleas on dogs and cats older than 4 weeks. It is also used in the United Sates to kill the oriental rat flea in murine typhus control programs. The sprays and dusts have been used in adult mosquito control. Carbaryl, when used as an outdoor space spray, is concentrated at a level of 0.2–1.0 lb/acre for the control of mosquitoes.

Dimetitan is highly toxic when ingested and moderately toxic when absorbed through the skin. It is impregnated into plastic bands and suspended near the ceilings of farm buildings for use in fly control.

Propoxur (baygon) acts as a stomach poison and contact poison in roach control and also in tick control. The spray has a long-lasting residual contact. This insecticide also differs from others in that it has a flushing or irritating action that forces insects out of hiding areas, and it has a rapid knockdown action. Baygon is used to control mosquitoes, flies, sandflies, ants, other insects, and the resistant brown dog tick. It has some toxicity for animals. Apparently, it lasts for short periods of time in the environment as a contaminant. Baygon is used in a dosage of 100–200 mg/ft^2 for residual spraying in mosquito control. For roach control, it is used either as a 1% spray or a 2% bait.

Landrin has been tested by the World Health Organization as a residual spray in anopheles mosquito control.

Fumigants

Fumigants are gases that kill body cells and tissues after penetrating the body wall and respiratory tract of insects. They are purchased in either solid, liquid, or gaseous form. Fumigants are of limited use in public health work due to special hazards. They are flammable, toxic, highly reactive, costly, tend to corrode metals or damage dyes in fabrics, and lack chemical stability. The fumigants include hydrogen cyanide, which is extremely hazardous to animals and humans; methyl bromide, which has little or no warning odor; carbon disulfide, which is highly flammable and explosive: chloropicrin (tear gas), which is highly irritating; ethylene dibromide, which is now banned, desorbs very slowly from certain products; ethylene oxide, which is highly flammable and explosive; phosphine, which may be a fire hazard; and sulfuryl fluoride, which is not recommended for food fumigation.

The fumigants are important because they do provide a means of destroying large quantities of insects that infest food and may also be utilized in homes where there are severe insect infestations. It is extremely important that all fumigants be handled very carefully and that they be applied only by trained, licensed pest-control operators who understand the nature of the chemicals and their hazards.

Desiccants or Sorptive Dusts

Certain desiccants, which in effect damage the outer waterproof layer of the arthopod's exoskeleton by either absorbing the fatty or waxy material or by abrasion, are used in insect control. These desiccants include finely powdered silica gels and silica aerosols.

Other Types of Insecticides

Attractants are materials used to lure insects into traps or to make poison baits more inviting. Attractants include such things as sugar, peanut butter, and fish. Sex hormones have also been tried. Some chemical attractants include methyl eugenol, ethyl acetate, and octyl butrate.

Repellents are substances that produce a reaction in insects that makes them avoid animals or humans. A good repellent works for several hours; is nontoxic, nonirritating, and nonallergenic; has a pleasant odor; is harmless to clothing and accessories; is effective against many insects; and is stable in sunlight. Some repellents include oil of citronella, sulfur, dimethyl phthalate, and indalone.

Piperonyl butoxide is a compound that acts as a synergist when added to insecticides. It is a low hazard compound in itself, is not known to create environmental problems, and is most effective when used in combination with insecticides that require a booster to do an effective killing job.

Red Squill

Red squill is an extremely effective rodenticide against Norway rats. It may be used fresh with live bait or water bait. Fortified red squill should be used, since the effectiveness of this natural chemical, which comes from the inner bulb scales *Urginea maritima*, varies in effectiveness with production techniques and the time of storage of the pesticide. Red squill kills very rapidly, and although a single rat may develop shyness to bait, rats in general will continue to come back and feed upon it for periods of time if the bait is mixed properly. In the past, it had always been recommended that red squill be mixed with fresh baits. This is an effective means of control of rats for a short period of time. However, if the red squill is mixed with cracked corn and rolled oats and bound with peanut oil, the peanut oil will not only help preserve the bait for weeks under all types of conditions, but also will act as an attractant to the rats, who will come back and feed upon the red squill bait. The poison is of considerable usefulness in public health, since it is one of the least hazardous rodenticides to humans and domestic animals. Red squill causes animals to regurgitate; since Norway rats cannot regurgitate, the poison produces cardiac arrest, convulsions, and respiratory failure. It is used at a ratio of 3.5–10.0% in baits.

Anticoagulants

The anticoagulant poisons include fumarin; diphacinone; indandione; PMP, also known as valone; warfarin; warfarin plus, also known as sulfoquinoxalin; and chlorophacinone. Each of these poisons can be used in liquid or dry bait. They provide excellent control of Norway rats, roof rats, and also mice. Since the anticoagulant poison depends on an accumulative action, it is necessary for rodents to feed upon poison bait for a period of several days. Generally, it takes from 1–2 weeks to get an effective kill of the rodents present. Some resistance has been noted to warfarin in Europe and also in some parts of the United States. However, the anticoagulants are still extremely useful and safe. It would take a large dose of anticoagulant poison bait to cause any harmful effects in humans or animals. The chances of getting such large doses are apparently insignificant.

Fumarin is odorous, nonflammable, and soluble in water and oil. It is highly toxic to rats and mice. It does not deteriorate in baits. It does present a slight problem of secondary poisoning to cats, dogs, and to the individual applying the poison. Diphacinone is highly toxic to rats, cats, dogs, and rabbits. Again, the quantity of poison necessary to harm cats, dogs, and rabbits tends to protect them. It is only hazardous to wildlife and fish if they feed on it continuously for a period of days. No deterioration is found in the bait. There is a slight possibility of hazard to the individual who is using the poison.

Indandione is an odorless compound that is also known as pival. It is soluble in water and oil, is moderately toxic to dogs, and is hazardous to fish and wildlife if eaten continuously over a period of time. It is a slight hazard to the person applying the poison.

PMP is odorless, insoluble in water but soluble in oil, does not deteriorate in baits, has a slight chance of causing secondary poisoning to cats and dogs, and creates a slight problem for the applicator.

Warfarin is relatively insoluble in water and should not be added to baits that contain much vitamin K. It is highly toxic to cats, moderately toxic to dogs, and relatively nontoxic to humans. There is no deterioration in the baits. There is a slight hazard to the applicator. Warfarin plus is warfarin containing hydroxycoumarin and sulfaquinoxalin. This compound is available in ready-to-use baits. Sulfaquinoxalin inhibits vitamin K-producing bacteria in rodents and therefore increases the effectiveness of the warfarin. Smaller amounts are needed for control. Chlorophacinone in laboratory tests has shown good results against Norway rats.

Other Rodenticides

There are a variety of other rodenticides available. They will each be discussed. ANTU, or alpha naphthyl thiourea, is a compound that causes death by inhibiting the clotting of blood and causing internal hemorrhaging. It is very toxic to Norway rats, but less effective on other species. There is a medium degree of hazard in its use, since there is no known antidote for it. Tartar emetic is the best substance used when the compound is ingested accidentally. ANTU should not be used more than once a year, because the rat population will refuse the bait. The poison has also been used as a 20–25% tracking powder. ANTU is quite toxic to dogs, cats, and hogs. It is ineffective against roof rats.

Phosphorus bait is a fast-acting poison effective on Norway rats, roof rats, and roaches. It is highly toxic to humans, especially children. Phosphorus baits should only be used when absolutely necessary, in the absence of children, away from food, and under the strict supervision of pest-control operators.

Zinc phosphide is usually used to kill rats and mice. It is generally prepared as a 1% bait with meat or diced fruit. Tartar emetic can be added to this product to make it less hazardous to humans, however, it should never be placed in any area where children, dogs, or cats could consume it. It is an extremely hazardous poison and therefore should be used with the greatest of care.

Sodium fluoroacetate, or 1080, is an extremely effective poison against rats and mice. However, the poison is extremely hazardous to people, and therefore should only be used with the greatest of care and by highly skilled, licensed pest-control operators only. It causes death by paralyzing the heart and central nervous system. The degree of hazard of this chemical is so great that it is recommended that it either never or rarely be used. The residue of the poisons must be destroyed by burning in an open field away from any possible human or animal activity. The operator must be extremely cautious in the destruction of this residue. This chemical has now been banned.

Fluoracetamide, or 1081, is very effective for rat control in sewers, in either dry or watered baits. It should not be used for any other purpose. It is extremely toxic to both humans and animals. It should be applied only by trained pest-control operators and should either never or rarely be used. The remains of the poison and the dead rodents must be burned in an open field or buried so that they cannot be dug up. Extreme caution must be utilized if 1081 is used in rodent control.

Norbormide, better known as raticate, is a dicarboximide. It is highly toxic for rats; slightly toxic for mice. It is extremely stable in all baits and environments.

Thallium sulfate is used as a slow-acting rat and mouse poison. It produces a variety of neurological, circulatory, and gastrointestinal symptoms. Because of the danger in the use of thallium sulfate, it has been banned since 1972 by the EPA and therefore should not be used.

Arsenic trioxide, sodium arsenate, and sodium arsenite are odorless powders used for mouse and rat control. They are fast acting, but tolerance to the chemical can develop. The arsenic compounds are extremely toxic to humans. They should only be used where absolutely necessary, away from children and animals, and should be used only by trained pest-control operators. Arsenicals are now banned.

Strychnine and strychnine sulfate are odorless compounds used in mouse baits. They may also be toxic to rats and are highly toxic to humans. It is extremely important that strychnine compounds not be used unless absolutely necessary and then away from humans, pets, and other animals, and used only by professionally trained pest-control operators. These chemicals have now been banned.

Zinc phosphide is effective against rats and mice. Since it is extremely toxic to all animals, including humans, it should only be used where absolutely essential and should be kept away from all animals and humans. It should be only applied by trained pest-control operators.

As can be seen, many of the rodenticides are extremely dangerous to humans or other animals. It is essential that rodenticides be selected with great care and that they be used by trained professional people. It is preferable, whenever possible, to utilize either anticoagulants, red squill, or raticate, rather than the other types of rodenticides, since the chance of harming nontarget animals and humans is reduced.

Herbicides

There are over 100 different chemicals that act effectively as herbicides. They affect plants either through contact, as a systemic poison, or as a soil sterilant. Herbicides are important in public health work because they are used for the control of weeds and therefore will reduce insect and rodent harborage, as well as decrease the amount of pollen present in the air. Contact herbicides kill plants through direct contact. They may be selective or nonselective and kill all plants. Systemic herbicides are also either selective or nonselective, and therefore pose a problem to plants other than those that one would want to destroy. Soil sterilants unfortunately may remain in the environment for long periods of time, and therefore pose a problem to the environment. Inorganic herbicides are derived from the inorganic acids in which hydrogen is replaced by a metal. They produce a burning effect when coming into contact with the plants. Examples are calcium arsenate, sodium chlorate, and sodium borate. The metal organic compounds include those that have a metal ion complex combined with an organic portion of the molecule. The herbicides are usually used to control large areas of weeds, such as on railroads and highway right-of-ways. An example is disodium methane arsenate.

A third group of herbicides are the carboxyl aromatic herbicides. This group has a carboxyl group and an aromatic group. They work as contact, systemic, and soil sterilants. They are categorized by five basic types. (1) Phenoxy herbicides, which are systemic in nature and usually last 30–60 days. They are only slightly toxic to humans and animals. Examples are 2,4-D (cautionary statement now added for grazing animals) and sesone. (2) Phenolactic acid is used for aquatic weed control and weed control in right-of-ways. (3) Benzoic acid compounds have a longer soil resistance and low toxicity to mammals. Examples are benzac and trysben. (4) Phthalic acid compounds act to prevent weed germination. They are persistent for about 30 days in the soil and are relatively nontoxic to mammals. Examples are dacthal and endothall. (5) Phthalamic acid compounds also prevent weed germination. They are relatively safe to humans and other warm-blooded animals. An example is alanap.

Other herbicides include aliphatic acid herbicides, which contain a carboxyl group and are temporary soil sterilants. Examples are dowpon and trichloracetic acid. Substituted phenol herbicides are used for contact killing and are applied by sprays in such areas as railroads and highway right-of-ways. These include dinoseb and pentachlorophenol. Their toxicity to mammals varies from moderate to very toxic. The nitrile herbicides are used for killing the seeds of broadleaf weeds. They are usually used in agricultural weed control.

Biological Controls

A series of techniques is being tried on an experimental basis to sterilize male insects and set them loose in areas where females are present. The aim is to reduce the fertility of the insects and therefore utilize a biological control. Predators have also been utilized to destroy insects. The only problem is that the predator may become a pest itself. In rodent control, biological control occurs when various environmental forces are utilized in the destruction of the rats by eliminating food supplies and harborage. The rat population density decreases. If a rat population is not tampered with by humans, it usually expands beyond the environment's ability to support the number of rats. The overcrowding results in disease and competition, and therefore an increased mortality rate and decreased natality rate. Eventually, the population drops to a size that the environment can support.

Other Types of Pesticides

In addition to the pesticides mentioned, fungicides and bactericides are used to prevent plant disease caused by fungi and bacteria. The nematicides are used to control nematodes, which attack plants. Molluscicides are used to control molluscs, which affect fishing areas or plant areas. Piscicides (fish killers) are used to treat public waters. Their objective is to remove rough or trash fish from restocking lakes or game fish lakes. Avicides (bird killers) are used to control birds and pigeons in areas where they are troublesome and damage crops.

PROBLEMS CREATED BY PESTICIDES

Pesticides in the Environment

Pesticides may enter the environment by means of the air route, the food route, and the soil route. The environment is contaminated through the indiscriminate, uncontrolled, unmonitored, and excessive use of pesticides by all types of people, including the owners of households.

The Air Route

Pesticides enter the air by means of aerial spraying, the use of mists and fog machines, and the application of pesticides by individuals using pressure containers. The pesticides, depending on the size of the particles and the volume discharged, the velocity of the air current, the temperature of the air, and other factors, may stay within a given area or may contaminate areas other than those intended. It is essential that great care be taken in the application of pesticides by the air route. The drift and weather conditions must be considered carefully. The human hazards are caused by inhalation, skin absorption, and ingestion of the pesticide. Pesticides may be transported over long distances if they attach to dust particles in the air. Further, they may be mixed in with other chemicals produced by a variety of air-polluting situations, causing secondary chemicals to form, which are in themselves very hazardous.

The Water Route

Pesticides enter surface waters by being washed from the surface of the soil or from plants, houses, and agricultural areas. Some pesticides percolate down into underground water supplies through a water flow, providing access to pesticides injected into the soil purposely for the control of insects, or through rain or snow, which washes the pesticides into the soil and slowly helps them percolate into the underground water. The use of pesticides must be carefully controlled and bodies of water should be regulated. Lakes and other bodies of water should be studied carefully before pesticides are used for either mosquito control or water-weed control, since the pesticides may end up causing more harm than good. It should be recognized that pesticides and fertilizers are used extensively by homeowners and farmers. All of these chemicals have a tendency to be washed into bodies of water through surface drainage and through storm sewer pipes.

The Food Route

Pesticides, from time to time, have caused disastrous consequences when stored in the same vehicles in which food was being transported. It is essential that this be forbidden, since any breakage or leakage would cause chemical food poisoning. Food treated with pesticides must contain the minimum quantity for effectiveness on insects and must not cause harm to humans.

Soil

The persistence of pesticides in the soil creates a situation in which not only is the soil contaminated, but additionally, the air may become contaminated by soil particles, or water may become contaminated by runoff. The chemicals used should not be hazardous and should degrade rapidly within the soil. Those chemicals that are taken up by plants in the soil and are hazardous to humans should not be utilized.

The persistence of a pesticide in the soil depends partially on how it is transferred to the soil. Is it done through leaching, erosion, evaporation, or through an uptake of plants. The persistence also depends on how the pesticide is degraded. Erosion is still another factor. The algae, fungi, and bacteria found in the soil may use the organic chemicals present as a source of energy, and therefore may reduce some of the amounts of pesticides found in the soil. Chemical reactions may destroy some of the activity of pesticides, while enhancing the activity of other pesticides. Diazinon is broken down in acid conditions, however, the opposite is true for malathion.

Household Use

Homeowners are often the ones most apt to contaminate their immediate environments or to provide opportunities for accidental poisoning of their children, themselves, and others around them. Many homeowners have no concept of the proper use of pesticides. They fail to read and understand the labels, use the pesticides under hazardous conditions, and generally store pesticides in areas where children play or can gain access, or in which fires may occur. Pesticides are misused in gardens, in homes, and in the care of household plants.

Effects of Pesticides on Humans

It is difficult to fully evaluate the human risks of chronic pesticide problems, since very few studies have been made in this area and because of the complex nature of the problems involved. Further, variables such as age, sex, race, socio-economic status, diet, state of health, length and state of exposure, and pesticide concentration level all profoundly affect the human response to pesticides. Although there are cases of acute pesticide poisoning, one should not extrapolate these results to chronic low-level exposures. The individual is not only exposed to pesticides in the environment, but also to dusts and a variety of environmental conditions that can alter the human response to any specific pesticide.

It is known that the organochlorine compounds, such as aldrin and dieldrin, can increase the excitability of the nervous system and may damage the liver. However, it is difficult to establish a correct diagnosis, since the symptoms vary. Certain compounds can penetrate the unbroken skin. Lindane is believed to cause hematological disorders.

The organophosphate insecticides inhibit cholinesterase enzymes. There is a great variation in acute toxicity from one compound to another. Organophosphates penetrate the skin easily. Carbamate pesticides also inhibit cholinesterase, but since the enzyme deactivates rapidly within the human body, it is difficult to measure the exposure based on this deactivation.

DDT is an organochlorine compound that becomes stored in the fat tissues of animals and humans. It also has considerable effects on fish and wildlife.

The difficulties involved in trying to gather adequate information on the human effects of pesticides contribute to the considerable uncertainty in determining which pesticides are safe and which are harmful. It is impossible to do a general epidemiological study, since all individuals have had some pesticide exposure; therefore, a control group cannot be selected. More often studies are made of either acute poisoning due to pesticides among occupational groups, children, or other individuals who have accidentally poisoned themselves. From these data, we can generally determine the kinds of effects that humans have in the event of acute poisoning. However, even in acute poisoning, inadequate amounts of data are available, and it is difficult to make careful judgements about the absolute danger of the given pesticide, unless the pesticide is taken into the body in large amounts. Chronic poisoning becomes an even more complex and confusing issue.

The routes of entry of the pesticide include absorption through the intestines due to ingestion, absorption through the lungs due to inhalation of airborne pesticides, and penetration through the intact skin or absorption directly into the bloodstream through broken skin. The route of entry depends very much on the group of individuals studied and the use of the pesticide. Absorption through the intestine occurs when the residues that remain on food are ingested. This is probably the major route through which pesticides enter the body. In addition, accidental poisoning of children generally occurs through ingestion. Inhalation occurs when bug bombs or aerosol sprays are used to control roaches and other pests in the homes or when individuals inhale particles from fogs and mists that are used to control mosquitoes in exterior areas. Most skin contamination occurs in occupational environments. Pesticides may also be hazardous to humans because of the possibility of fires or explosions. This subject has been previously discussed.

Considerable study is being conducted on the possibility of pesticides causing cancer. In the future, many of our existing pesticides may be banned because they are carcinogenic in animals. These studies have to be conducted with great care to ensure that false conclusions are not drawn. It will be important not to ban pesticides that are valuable to humans and, at the same time, to protect humans from a new burden of additional carcinogenic agents.

Resistance to Pesticides

Pests, particularly insects, develop resistance to pesticides. Some insects are less susceptible to certain insecticides, and some are affected but not killed by insecticides. Over a period of time, resistant insects survive exposure to insecticides and reproduce new generations of increasingly resistant insects. Generally, two forms of resistance develop in insects: physiological and behavioral. In physiological resistance, the insect develops an immunity to the poison. The exoskeleton becomes less permeable to the insecticide, the insecticide is detoxified into less toxic chemicals, or the insecticide may be stored harmlessly in the body tissues or be excreted. Behavioral resistance is the ability of the insect to avoid lethal contact with the insecticide because it has developed protective habits or behavioral patterns. This includes such things as mosquitoes changing resting places and the avoidance of baits by flies. Resistance of insects to insecticides is increasing. In fact, cross-resistance has been known to occur. Studies have shown that insects resistant to certain types of chlorinated hydrocarbons may also develop resistance to the organophosphates. Examples of these insects would be the housefly, and the mosquitoes *Culex tarsalis* and *Aedes nigromaculis*. It is necessary from time to time to conduct surveys to determine if insects are developing resistance to a given insecticide, and if so, the insecticide should be changed.

Economics

The major question in environmental economics as it relates to pesticides is whether the damage costs resulting from blighted crops, poor health, and higher death rates are greater or less than the potential benefits from increased crop yield, reduction of disease, and so forth. It is difficult to obtain a true picture of the cost-benefit ratio in the use of pesticides generally. Each pesticide needs to be studied and judged on an individual basis.

POTENTIAL FOR INTERVENTION

Intervention strategies include the use of the techniques of isolation, substitution, shielding, treatment, and prevention. The potential for intervention varies from poor to excellent, based on the understanding of the short-range and long-range problems associated with the particular pesticide; the preparation, storage, use, and disposal of the pesticide; and the training and ability of the pesticide applicator. Proper storage of pesticides to avoid fire, explosions, and contamination of individuals and their food are techniques of isolation. Substitution is the use of a less hazardous pesticide. Shielding is the use of safety glasses and protective clothing. Treatment is the technique used on an individual who becomes contaminated with a given pesticide. Prevention is the overall process of keeping pesticides out of the eyes, off the skin, away from the lungs, and away from clothing. Good housekeeping is part of the overall prevention technique.

RESOURCES

Scientific, technical, and industry resources include the Entomological Society of America, the Pest Control Operators Association, the American Public Health Association, the National Environmental Health Association, various land grant colleges and universities, and chemical manufacturers associations.

Civic resources include the National Audubon Society, the Wildlife Society, and the Environmental Defense Fund.

Governmental resources include state and local health departments, the Department of Agriculture, the National Bureau of Standards, Department of Defense, the Environmental Protection Agency, Department of Health and Human Services, the National Science Foundation, and so forth.

The area of pesticides continues to change so rapidly that what is used today may be banned tomorrow, and what does not exist today may exist tomorrow. Public health workers or environmental health workers should obtain a document from the U.S. EPA entitled, "Suspended, Cancelled and Restricted Pesticides." A second source is entitled *Handbook of Pest Control* by Arnold Mallis. The third set of references would be the current rules, regulations, and laws of the Environmental Protection Agency. A fourth resource is the Library of Congress, and a fifth resource is the United States Senate Agricultural Committee.

The National Pesticide Telecommunications Network may be reached by phone by calling 800-858-7378, which is a toll-free call. The Pesticide Producers' Association, 1200 17th Street, N.W., Washington, DC 20036, is another resource. The Synthetic Organic Chemical Manufacturers' Association, 1330 Connecticut Avenue N.W., Washington, DC 20036, is also another useful resource.

STANDARDS, PRACTICES, AND TECHNIQUES

Application of Pesticides

Pesticides are applied to standing water by the use of spray equipment. Oil solutions or emulsions may be utilized as well. The adult mosquito is controlled by the use of aerosol bombs, which are successful in the destruction of flying insects such as flies and mosquitoes. Fogging and misting techniques are used in large community programs as the major method of controlling the adult insect, especially the mosquito. Fog and mist are produced through specialized equipment mounted on driven vehicles and blown out into the open air from the equipment. Dusting is used largely in agricultural areas. The dust is usually applied by airplanes. Residual spraying is carried out by applying the spray to surfaces upon which the insects will alight or rest, usually from hand held equipment. Fumigation is the technique of releasing large quantities of aerosols quickly into an area. In catch basins, vapona strips are hung. This insecticide is a fumigant. It is slowly released from the solid material.

The type of equipment utilized varies from the hand sprayer, which may be 1–3 gal in capacity, to the aerosol bomb; to the compressed air sprayer, which is usually 3–4 gal in capacity; to a variety of power equipment; to aircraft application.

Pesticide Labels and Names

All pesticides must be registered for use. The registration can be revoked by the EPA if the pesticide is determined to be a hazard to the community. Pesticide labels and names are very important. The label on a pesticide container specifies the name of the manufacturer; the name of the product; the active chemical ingredients and percentages of concentration; the type of chemical, whether it is an insecticide, rodenticide, and so forth; recommendations for specific uses; directions for use; precautions in storage; and precautions during use by personnel.

Since many chemicals can be sold under the same brand or trade name, it is essential to identify the actual, active chemical ingredients. The generic or common name of the ingredients must appear on the label. The label must state specifically what pests are controlled and what rate of application of insecticide is needed to control the pests affected by the particular pesticide. Since the application of the pesticide is extremely important, instructions should be listed clearly and carefully regarding correct and safe methods of application. All labels should specify the necessary precautions to be taken by personnel. Insecticide labels must say, *Keep away from children.* Where highly toxic products are used, skull and crossbones are required on the label; plus the word *POISON* in red letters. The label should also contain the word *Warning*, instructions for handling, the antidote, and the statement, *Call a physician immediately.* Where the pesticide is not as toxic as stated above, the words *warning* or *caution* should be placed on the label, and it should be stated that a pesticide is a hazard and should be kept out of the hands of children. Common names of the pesticides should be used when possible.

Pesticide Formulations

To use pesticides properly, it is necessary to have some understanding of the form in which they come and also dilution specifications to achieve the proper formulation effective for a given pesticide in a given situation. Solvents dissolve insecticides, dispersing them evenly through the solution. The solvent is a carrier as well as a dilutant, depending on the amount used. Solvents may include volatile liquids, such as xylene, which evaporates after use and leaves a residual deposit, or nonvolatile or semivolatile liquids, such as petroleum distillates, which leave the surface coated with the solution of the toxicant. The emulsifiers are surface-active agents that allow liquids to be mixed within liquids. The emulsifier forms a thin film around each minute droplet of oil, thereby keeping the oil from coalescing or separating into oil and water. Wetting agents or spreading agents allow the pesticide to penetrate a surface. Adhesives are used to improve the quality of the deposits of the pesticide. An example of such a material would be gelatin or gum. Perfumes and masking agents are utilized to cover the unpleasant odors of certain pesticides. An example would be oil of wintergreen. Synergists are materials that help a pesticide work more effectively. Examples of synergists are piperonyl butoxide, sesamex, sulfoxide, and propylisome. Carriers and dilu-tants for dust are used to help deliver the insecticide more readily and in an inexpensive manner. Examples of these carriers would be aptapulvite, bentonite, calcite, diatonmite, and talc.

An insecticide is used in various forms. The technical grade insecticide is the insecticide in concentrated form plus inert material. It is basically the purest chemical form. The insecticidal dusts are prepared by blending toxic ingredients, such as diazinon, into an inert carrier, such as talc. Wettable powders are the toxic ingredients plus the inert dust and a wetting agent such as sodium lauryl sulfate. Water is added and the material is agitated to form a suspension. The material must be continually agitated so that the chemical will not settle out. Wettable powders are used frequently in out-buildings. Insecticidal solutions may be pur-chased or prepared either as a concentrate or as a finished spray. The concentrates are diluted with oil to make the usable solution, which becomes a finished spray. The solution is not soluble in water. Emulsifiable concentrates are concentrated solutions to which an emulsifier has been added. When the concentrate is mixed with water and the emulsifying agent, the insecticide spreads throughout the water. A finished emulsion is made by diluting a concentrate with sufficient water to form the concentration needed for use. Aerosol bombs typically contain 0.1–0.6% pyrethrums; 1–2% of malathion, or baygon; 10–12% petroleum distillates, and 85% propellant, such as petroleum distillates.

It is essential to have the correct dosage of pesticides for the proper type of pest control. Excess pesticides may cause unsightly residues, ruination of edible crops, toxicity to crops and animals, excess costs, and a specific hazard to animals and humans. When preparing pesticides for use, it is necessary to determine the safe time intervals for application; the amount of residue acceptable by the Food and Drug Administration; whether a combination of pesticides are safe or compatible; if products for animals are treated, the toxicity of the pesticide to be used, the manner in which the equipment was cleaned prior to pesticide use, the rate of application of the pesticide, the size of the area, and the safety precautions to be taken by the operators. Always follow the instructions on pesticide containers to determine the dilution of the pesticide to the proper concentration.

Pesticide Application and Equipment

Pesticides may be applied as solids, liquids, or gases. Solids are usually applied as dusts by means of hand dusting, shaker cans, bellows, vault dusters, or power dusters. Solids may be applied as granules, pellets, gelatinous capsules, or poisonous baits. Liquids may be applied by pouring, painting, spraying, ejecting, or through aerosol bombs. The method of application is based on whether the pest will be killed, safety of the operation, the expense, durability of the pesticide, and potential problems. Hand sprayers range from something as old fashioned as a flit gun, used for bedbug control, to various types of compressed air or other sprayers. The sprayers are regulated for the type of spray desired, the quantity desired, and the most desirable pressure for delivery of the spray. It is essential that all parts of the sprayer, including the hose and all fittings, be checked carefully to make sure that they are thoroughly clean and do not leak. A wand is a slender metal tube that is placed on the sprayer and connected to a nozzle that extends the length of the range of the sprayer. The nozzle is the most important part of the sprayer. It has to be fitted so that it will effectively deliver the kind of spray and the concentration desired.

Dusters are utilized in areas where rodent ectoparasites are found, in areas where a fire hazard results from the use of oils, and in areas where oil and water are not used advisably.

Powered sprayers or foggers are generally mounted on some type of mechanical equipment. The power unit distributes a large amount of spray at a high pressure. It is essential that this type of spraying be carefully controlled to avoid annoying residents of a home and also to avoid creating an accident hazard. Children have a habit of traveling in and out of fogging machine units, thereby creating accident hazards to the spraying crews, to automobiles, and to themselves. Wind direction and wind velocity must be taken into account whenever power fogging or spraying is utilized. The equipment should be operated to spray perpendicular to the wind and should be directed as close to the ground as possible. Mists and fogs are generally utilized to quickly knock down high concentrations of adult insects. They are economical and can be utilized to cover large residential areas in short periods of time. The mechanical fog generator breaks the insecticide into fine particles and then blows it into the air at high speed. Other types of fog generators are available that operate on either a pulsation principle, thermal principle, or steam principle.

Insecticides may also be injected into the soil to control a variety of pests, especially termites. Fumigation equipment is used to kill insects and rodents in large areas where people are not present. These would include railroad boxcars, the hulls of ships, certain aircraft arriving from overseas, in areas where arthopod-borne diseases are prevalent, in food storage warehouses, and in food-processing plants. Great care must be taken in fumigation.

Insecticides may also be applied by airplanes or helicopters. The insecticide must not drift into residential areas and affect animals or humans. Weather conditions, time of day, and so forth must be taken into account when aerial dispersion of insecticides is used.

Storage of Pesticides

Pesticides must be stored with great care, since they may be highly flammable, explosive, or toxic. Most insecticides, fungicides, and rodenticides can be stored in the same room. However, herbicides are quite volatile, and special precautions should be taken with their storage in order that they do not contaminate the storage area or escape to the outside, thereby damaging plants.

All storage areas should maintain the following precautions: Storage areas should be locked and located away from food, animal feed, plant seed, or water; pesticides should be stored in a dry, well-ventilated place, as directed on the labels; storage areas should be clearly marked as a pesticide storage area; a sign should be posted listing the types of pesticides stored and the hazards therein; pesticides should be kept in their original containers; containers should be checked periodically for leaks, tears, and spills; an inventory list should be kept in order that outdated materials are eliminated and shortages are accounted for.

In the event of a fire, it is extremely important that the firefighters and the general public be protected from the fumes, residues, or washings due to the fire. Fires may originate because petroleum distillates are present; aerosol containers become overheated and explode; other flammable or explosive solvents are present; finely divided dust or powders explode; chlorates, which are flammable or explosive, are present; ammonium nitrate fertilizers are stored; and/or calcium hypochlorites are present, and may cause spontaneous ignition and explosion if contaminated by organic substances.

The hazards that exist are due to the presence of organophosphates, carbamates, and chlorinated hydrocarbons, which are highly toxic. Further, fumes from solvents, the presence of gases, or any combination of substances may be toxic. Care must be taken that runoff water from firefighting, which may contain highly toxic pesticides in quantity, does not enter any of the environmental pathways.

In the event of a pesticide fire, a qualified physician should be available, hospitals should be alerted as to the type of potential hazard, and the firefighters should be protected from poisoning by use of proper protective clothing and special self-contained breathing equipment. Great care must be given to the cleanup after the fire to make sure the firemen have removed all traces of the pesticide from their clothing and boots. All of the areas where the pesticides are present must be cleaned with the utmost of care, and the resulting contaminated water must be trapped and treated.

Transportation of Pesticides

Transporting toxic chemicals can be very hazardous, with the possibility of accidents or leakage. It is essential that all regulations of the Interstate Commerce Commission and the Department of Transportation concerning proper identification of vehicles carrying hazardous chemicals be followed. The individuals operating these vehicles must be fully aware of procedures to follow in the event of an accident or emergency. Volatile pesticides should never be within the section of the vehicle holding passengers or the driver.

The transporting vehicles should be properly built in order that powders within paper bags will be protected from rain and will not be punctured or torn. A vehicle should be able to be easily cleaned. Where pesticides are in liquid form, they must be in tightly closed original containers. Glass containers are not recommended. If they must be utilized, they should be packed and transported in such a way as to avoid breakage. Since pesticides are affected by high and low temperatures, they should be removed from trucks as soon as possible after delivery and stored in safe, locked facilities.

In the event of an accident involving the vehicle transporting pesticides, the drivers should immediately avail themselves of protective clothing and respirators, and should inform the fire department, police, department, and health department of the accident. It is essential that fire, police, and health personnel understand the nature of the toxic material and the hazards therein. It is also extremely important that the public be kept as far away from the accident site as possible.

Disposal of Pesticides

Since pesticides in concentrated form or even in diluted form may constitute extremely serious hazards, unused pesticides, as well as the empty containers, must be disposed of in a safe manner. In all cases it is necessary to follow the laws, rules, and regulations stipulated by states, the federal government, and local legal bodies. Some types of disposal include ground disposal and incineration. Ground disposal may be very dangerous if the pesticide will contaminate either surface or groundwater supplies. Where this does not occur, it is possible to bury the pesticide in containers that will not deteriorate. This must be a deep burial in order that the pesticide will not be dug up inadvertently. Burial should not occur in normal landfills (see the chapter, Solid and Hazardous Waste).

Many pesticides are destroyed during incineration. Incineration is acceptable if the resultant fumes or resultant waste escaping into the air is scrubbed out of the gases, concentrated, and either reincinerated or finally buried in containers that will not deteriorate. Pesticide containers should be handled separately and should be buried only in areas in which the water percolating through the ground will not carry the pesticide into water supplies.

Ethylene dibromide (EDB), which has now been banned, is a perfect example of a chemical that is now considered to be very harmful. Leaking drums of EDB in Missouri showed how bad the disposal process had become.

MODES OF SURVEILLANCE AND EVALUATION

Surveillance and evaluation consists of the epidemiological approach, the major incident approach, analysis, and monitoring. The epidemiological approach is used to study the general population and its exposure to various pesticides over long time periods. It is used on specific populations, such as workers exposed to pesticides during production, use, and storage, or individuals in communities exposed to large amounts of pesticides, especially individuals living in small farm communities. The epidemiological approach is a technique evaluating mortality and morbidity records of these individuals.

The major incident approach is used in situations such as the contamination of the city of Hopewell, Virginia and its surroundings by the pesticide kepone. This particular incident caused widespread kepone poisoning. At least 70 victims were identified and 20 of these victims were hospitalized with untreatable ailments, including apparent brain and liver damage, sterility, slurred speech, loss of memory, and eye twitching. The National Cancer Institute reports that kepone causes cancer in test animals. Kepone was found in shellfish 60 mi down the James River from Hopewell, Virginia. This type of incident is dangerous to a given community. It does provide, however, large quantities of data for further analysis and evaluation. It is hoped that these data can be utilized in preventing other types of major incidents.

Analysis and monitoring provide information on the methods by which chemical escape into the environment, the levels at which they are harmful, the types of controls currently utilized, and effective control techniques. Analysis consists of identification of the chemical, substance, and chemical entities, and the quantitative measurement of amounts present. It also provides toxicological evaluation of the product, its isomers, byproducts, secondary products, and unreacted intermediate products. During analysis, instruments capable of detecting concentrations in the range of 0.01 to 100 ppm should be used. For some chemicals it is necessary to detect even smaller quantities. Monitoring tracks specific chemicals through the environment. To monitor properly, it is necessary to consider a wide range of concentrations and potential toxicities and to understand the limit of detection of a given analytical method and the behavior of the chemical in the environment. How, when, and what to monitor, and when to stop monitoring, are complex decisions that must be made by competent professionals.

Chemical structure, reactivity, basic physical and chemical properties, proper analytical methods and monitoring strategies, gathering of reliable analytical data, storage of the data, and retrieval and use of the data are essential to the proper techniques of surveillance and evaluation.

CONTROL OF PESTICIDES

Safety

Pesticides must be used with great care, since they are hazardous to humans. Most victims of pesticide poisoning are either workers in the occupational preparation or occupational use area, individuals who inadvertently have been affected by improper use of pesticides, or children who have eaten the pesticide. Individuals may become chronically ill by being exposed for long periods of time or acutely ill by being exposed to large quantities over a short period. Pesticides follow the respiratory route, are absorbed through the skin, or are ingested with food.

The solvents used for dilution may also be toxic. Poisoning of children under 5 years of age by pesticides is a serious public health hazard. It is essential that these materials be stored out of the reach of children. Pesticides should not be stored in pantries, under sinks, or in garages where children can reach them.

Safety rules include proper reading and understanding of labels; proper preparing and applying of the insecticide; proper storage of the insecticide in the original containers with the original labels; mixing of pesticides in well-ventilated areas; mixing and applying flammable pesticides in such a way that they are not near fires, defective wiring, smoking, or hot areas; avoiding eating, drinking, or smoking where pesticides are used; wearing of appropriate clothing and headgear; avoiding contamination of food and water of humans and animals; avoiding inhalation of sprays and dusts; keeping equipment in good operation condition; avoiding the storage of partially used pesticides; proper disposal of pesticide containers; proper transportation of pesticides; proper storage of pesticides; and understanding of first aid measures where needed.

When accidental poisoning occurs, speed is the most essential concern. Proper treatment must be given at once and the individual taken to a hospital immediately. Poison information centers are available and usually can be reached by the telephone operator or local hospital. The poison information center or physician will supply information on immediate first aid, depending on the poison taken. It is essential that the label be read to determine the chemicals present and the antidotes and techniques used to counteract the poison. In no case should an individual be made to vomit if he or she is unconscious, in a coma, or convulsing, or after consuming petroleum products and corrosive poisons. If eyes have been affected, they should be washed immediately with cold water for at least 5 min. Any delay may result in permanent injury. If the individual is in a poisonous atmosphere, remove him or her as soon as possible to fresh air. It is best to reiterate that speed is absolutely essential in any type of pesticide poisoning. It is also essential to determine as soon as possible what chemicals are used and to get proper help from medical authorities.

Laws and Regulatory Agencies

The Federal Insecticide, Fungicide and Rodenticide Act of 1947 was superseded by the Federal Environmental Pesticide Control Act of 1972. The law forbids anyone, including the government, from using pesticides in a manner contrary to the label instructions and gave the EPA the authority to restrict use of certain pesticides to trained personnel in approved programs. It also extended control to all pesticides sold intrastate and interstate. In June 1975, regulations were published establishing a screening procedure. If a new pesticide was chemically suspected of causing cancer, heart defects, or mutations, it had to undergo testing before being declared safe. Any suspected pesticides that were already registered could retain their registration until a pending test proved that they were dangerous. These regulations also required that data be developed to determine when farm workers may reenter fields. On October 11, 1974, the EPA suspended the registration of aldrin and dieldrin because new evidence showed that they were an imminent hazard. Animal experiments indicated that they were carcinogenic. The courts held that the burden of proof of the safety of a pesticide rested on the registrant and not with the government. The court also accepted animal test results to indicate the cancer risk to humans. In July, the EPA issued a notice of intent to suspend the use of heptachlor and chlordane based on animal experiments indicating that these pesticides may cause cancer. Pesticides containing myrex and phenyl mercury compounds had also been cancelled. Although the EPA has the right to lift a ban in the event of a public health or other national emergency, generally they did not do so. There has been only one major exception involving the use of DDT that occurred in 1974 in the forests of the northwest. DDT has been banned since 1972.

Under the 1972 act, all pesticides distributed, sold, offered for sale, held for sale, shipped, or delivered were required to be registered with the EPA. Registration consisted of filing a statement with the EPA Administrator giving the name and address of the applicant or any other name appearing on the label; the name of the pesticide; a complete copy of the labeling of the pesticide, including directions for use; upon request, a full description of tests made of the complete formula; and a request for the type of classification, whether it be for general use, restricted use, or otherwise. The EPA Administrator then approves the registration of the pesticide. Another important part of the law stated that the pesticide must be used only by certified applicators. Therefore, pest-control operators have to obtain certification. Federal certification is handled by the EPA; state certification was submitted by the governor of each state to the EPA for approval. All operators after certification have to earn continuing certification hours credit to retain their license.

Registration of any pesticide is cancelled if the registrant does not reapply within a 5-year period. In addition the EPA can cancel the registration if the pesticide is found to be hazardous in any way. The EPA has the right to stop the sale and use, and to order removal and seizure of any pesticide that it deemed hazardous or being utilized in a hazardous manner.

The 1972 law (Section 171.3, categorization of commercial applicators of pesticides category number 8) stipulated that all state, federal, or other governmental employees using or supervising the use of restricted pesticides in public health programs have to be certified. Therefore, under the 1972 law, all pesticide operators and all public health workers involved in the control of pests of public health importance had to be certified by October 21, 1976, by the EPA and also by the state in which they were operating. The law was again updated in 1978 and in 1988.

The 1972 law was updated in 1975 to require impact statements and to require the EPA Administrator to submit actions concerning health matters related to pesticides to a scientific advisory panel. The comments of the advisory committee and the administrator have been published in the *Federal Registry* (see Current Issues for Update on 1988 FIFRA Act Amendments in this chapter).

SUMMARY

Pesticides move through the various ecosystems in the environment in a variety of ways. When pesticides are ingested or otherwise carried by the target species, they will stay in the environment. They may be recycled rapidly or further concentrated through bioconcentration as the pesticides move through the food chain. Most of the large volume of pesticides utilized do not reach their intended areas and therefore become contaminants. Pesticides are introduced into the environment by spraying or by surface application. The storage in body fat of a pesticide is based on its chemical nature, physical state, means of application, and atmospheric conditions. The persistence of pesticides in the air is influenced by both gravitational fallout and the washout caused by rain. They build up in the soil to concentrations that affect the various ecosystems. They may contaminate soil water, air, and various organisms. Pesticides can cause direct problems to humans through airborne contamination, drinking water, or contamination of food. The pesticide problem is increasing, despite the fact that the Departments of Health and Human Services, and the Environmental Protection Agency, are working diligently to remove dangerous pesticides from the market. However, without pesticides our society would be in great trouble, because insects and rodents would consume valuable food supplies and cause a variety of serious diseases.

It is obvious from the preceding material and concerns under discussion in this chapter that the problem of various pests is unresolved and will probably continue as long as we exist. It is also clear that pesticides are hazardous and must be used with the greatest of care. Public health officials and pest-control operators must be constantly on the alert to ensure that pesticide usage is proper. With new research, effective pesticides may be eliminated from the market because of their potential hazards as toxins, fire and explosion hazards, and because they are potential carcinogens.

RESEARCH NEEDS

Research is needed on all common pesticides used for the control of pests of public health significance. It is necessary to determine if the various chemicals can cause a resistance in the pest that might reduce or eliminate the effectiveness of large groups of pesticides. Chlorinated hydrocarbons must be studied to determine their reactions with food constituents and their potential harmful effect on humans. Several pesticides have been shown to be altered by sunlight. It is important to study photochemical alterations to determine if the efficiency of the pesticide is reduced in the natural environment. What are the issues relating pesticides binding to macromolecular components of plant and animal tissues used as food? Little is known about the chemistry of these bound residues. Organophosphate pesticides may be affecting commercial sprayers, since they have a significantly higher anxiety score on standardized tests than the control groups. This area also needs further study. Laboratory animals have difficulty in learning and memory because of exposure to pesticides. Chemical agents, such as the anticholinesterase pesticides in high doses, acutely affect the nervous system. Studies should be made over long periods of time at low dose levels to determine if they cause irreparable damage.

Under The Federal Food, Drug and Cosmetic Act, the EPA, in cooperation with the U.S. Food and Drug Administration, sets allowable limits for pesticide residues in food. These limits, which are called *tolerance levels,* are supposed to protect human health while allowing for the production of an adequate, wholesome, economical food supply. Tolerance limits need to be reevaluated for many existing chemicals

Techniques of genetic engineering, as they relate to the production of genetically engineered microorganisms under the FIFRA Act, need to be studied further. Although many pesticides have been approved in the past, they need to be further studied to determine if more stringent standards should be applied to their use. Studies should be conducted on dioxins, which are byproducts of the production of pesticides. Studies should be conducted of the gas stream to determine the chemicals present and their potentials for causing serious health effects when pesticides are destroyed through incineration. Additional research is necessary in reducing the health risk of pesticides and also in reducing the risk to the environment. Evaluations of indoor air should be conducted to determine the level of pesticides present in the indoor air of the home and also the potential hazards. All new and previously registered pesticides should be screened for their potential to contaminate groundwater. Continuing studies are needed to determine the potential health effects of some 55 inert ingredients found in pesticide formulations. The EPA has invalid or fraudulent health data on 36 pesticides including 35 used on food. Additional studies in this area are necessary.

Chapter 7

THE INDOOR ENVIRONMENT

BACKGROUND AND STATUS

The indoor environment in its broadest sense includes housing, the immediate external environment of the structure, and other facilities that we may work in.

Each of us understands to some extent the importance of our home environment. However, the physical environment for millions of less fortunate families is conducive to disease, behavioral problems, accidents, and social problems. Many times during the course of history, attempts have been made to create more attractive housing and to eliminate health-related environmental hazards. This chapter is concerned with the origins of housing problems, the types of health problems related to housing, neighborhoods and their effects on housing, the types of housing, the environment within dwellings, the structural soundness of dwellings, various techniques for improving housing conditions, indoor air pollution, and injury control.

In 1626, the Plymouth colony passed a law stating that new houses must be roofed with board. In 1648, wooden or plastered chimneys were prohibited on new houses in New Amsterdam, and chimneys on existing houses had to be inspected regularly. In Charlestown in 1740, it was declared that all buildings must be made of brooker stone because of a disastrous fire that had occurred. The reason that these communities began passing housing laws was the frequent fires that destroyed many lives, houses, possessions, and food. During the 1600s, some Pennsylvanians were living in caves along the water banks. This was declared illegal in 1687. Most of the early housing codes were primarily concerned with the fireproof nature or size of the structure and not with the basic indoor environment, which was very poor. Outdoor privies were used. In 1657, New Amsterdam passed a law that rubbish and filth could not be thrown into the streets or canals.

In the 1800s during the Industrial Revolution, millions of immigrants arrived in the United States unable to support themselves and had to seek housing accommodations with relatives or had to move into housing in poor condition. The cities, particularly New York, grew extremely overcrowded, and housing became the source of numerous outbreaks of disease. Only the fire department had any authority, until 1867, when the Tenement Housing Act of New York City was passed. This was the first comprehensive rule of its kind in the country. It stated that each room used for sleeping or occupied by an individual had to have adequate ventilation by means of a transom window, if windows to the outside were not available. The roof, stairs, and banisters had to be properly maintained. Al least one water closet or privy had to be provided for every 20 persons. All houses had be cleaned to the satisfaction of the board of health, and all cases of infectious disease had to be reported. Authority was given to condemn houses unfit for human habitation. Additional laws were passed over the years to modify and strengthen the initial act.

In 1901, a new, improved Tenement House Act was passed, stipulating the amount of required space per person, the fireproofing of the structure, and the required number of bathrooms, room sizes, and so forth. Housing legislation was also enacted in Philadelphia, Chicago, and other cities. In 1892, the federal government passed a resolution that authorized an investigation of slum conditions in cities of more than 200,000 people. Very little money was allocated for the study, which limited the actual investigation. Although housing laws continued to be modified, enforcement of the codes was poor until the 1920s, when the state of housing in the United States grew extremely unsatisfactory, both within the cities and in the farm areas. By the 1930s, President Roosevelt reported to the country that one third of the nation was ill fed, ill housed, and ill clothed. The first federal housing law was passed in 1934, establishing a better system for residential mortgages through government insurance. The Federal Housing Administration (FHA) was created to carry out the objectives of the law. Many other governmental agencies became involved in the next few years in mortgage lending and in an attempt to clear slums and to increase urban renewal. However, it was not until after World War II that substantial housing acts were passed. The first one of these was the Federal Housing Act of 1949.

Industrialization, coupled with a substantial move from farms to cities, helped to create the severe neighborhood and housing problems within the last 100 years. Industries established themselves wherever they were most profitable. If they contaminated the land, water, or air, it really was not of much importance. Obviously, since people had no access to the modern automobile, they lived close to industry or within convenient public transportation distance. As new groups moved into areas to fill job needs created by industrial expansion, neighborhoods and houses were filled by a highly transient population. The nature of the population, the overcrowding that occurred, the nature of the industrial pollutants, and the cheaply constructed housing used as dwellings by workers were the factors contributing to the severe housing problems of the 20th century.

The 1940 census indicated for the first time the type of housing conditions found in cities. New Haven, Connecticut decided to assign penalty scores to four items of the census. The resultant maps showed the concentration of substandard and presumably slum conditions. On the basis of these findings, the city made a sample environmental survey of the dwelling units. This survey indicated the condition of the houses and the apparent condition of the neighborhoods. Points were given to such things as toilets, baths, means of exit, lighting, heat, rooms without windows, deterioration, number of people per room, and the sleeping area in square feet per person. This survey indicated where slum clearance and housing improvement was needed. It was evident that any rebuilding program or slum clearance program had to be carried out in relation to the overall city plan. Revision and enforcement of housing regulations and control of blight in nonslum areas were necessary parts of the new thoughts on housing. Problems common to most areas included overcrowding due to large families, poor quality of dwellings available for minorities, and substandard properties in relation to the amount of rent paid. Furthermore, the impact of decay during the years of World War II, the inability to build or improve housing from lack of materials, and the movement into the cities because of war production caused additional housing problems. At the end of the war, the birth rate expanded, marriages increased sharply, and the demand for new houses rose at an unusually rapid pace. Veterans moved to the suburbs with their large families, creating a new type of housing pressure in areas where people had previously lived on farms. The inadequacy of many of these new dwellings built immediately after the war, the inability of the soil to handle the enormous amount of on-site sewage, and the lack of proper water supplies contributed to a new wave of serious housing and environmental problems.

Poverty and poor housing are mutually inclusive; one feeds on the other. The poorer the individual, the more crowded the living quarters. Poverty further reduces the chances of attaining a better education, occasional recreation, and occupational skills. These limitations create feelings of insecurity and helplessness and deprive the poor of alternatives. In many cases, the poor have great difficulty maintaining proper sanitary conditions in their dwellings. There is either a serious problem before they arrive or they are unable to physically, mentally, financially, or emotionally cope with the upkeep of their dwellings. The situation tends to grow worse instead of better. Housing and poverty are not simple matters. In addition to poverty, the education background, lifestyle, health practices, outlook on life, age at marriage, understanding of marital responsibilities, and the single-parent syndrome must be understood. The housing of the poor leads to situations that intensify the potentially serious hazards that exist within their neighborhoods.

Urbanization and suburbanization have become the predominant population shifts in the United States. Lower income groups left the rural areas of Appalachia and the South to crowd into already overcrowded cities. This, coupled with increasing population growth, caused a severe housing problem, since the supply of new homes in these areas was minimal and the older properties continued to deteriorate at a rapid rate. Many of the people in the urban areas moved to the suburbs after attaining a higher socioeconomic status in the hope of finding a better way of life for themselves and their children. This has resulted in a huge suburban sprawl that has created innumerable environmental problems.

The inner city has, in many ways, become a jungle. Not only do people live with rodents and roaches and under the most severe conditions, but a new hazard has arisen. The arsonist is creating a hazard with a pricetag of 10–15 billion dollars annually. Residents are suffering injury and death because individuals either out of hate or for profit are setting fire to old structures. In Boston, 22 people were arrested including 6 attorneys, 11 real estate brokers, 4 public insurance adjustors, 1 police officer, and 1 retired fire chief for arson. These individuals were arraigned in Suffolk County Superior Court on charges varying from fraud to bribery and murder. These arrests were just an indication of the kinds of fire problems occurring in ghetto areas. Unfortunately, in some areas the only way owners could get any money out of their property was to have the property destroyed and to cash in on the insurance. New York City increased its force of investigators from 77 to 152 to try to cope with the arson problem. The destruction of property was increasing the serious housing problem that already existed.

Inflation is also placing a strain on housing. The American dream of securing a good education and a good home is very difficult to realize. Homes that 10 years ago sold for $40,000 are now selling for $75,000. In some areas homes that were selling for $40,000 5 years ago are now selling for $65,000. In other areas homes that sold for $38,000 14 years ago are now selling for $120,000. Due to this rise in cost, many individuals are unable to move to better living quarters and are forced to remain in overcrowded run-down dwellings. In existing homes, owners are having difficulty maintaining their property. Many homeowners have had to obtain second mortgages to simply repair or maintain their property. Unfortunately, the current mortgages carry high interest rates. Many individuals become burdened with large debts and in many cases have to move to less expensive housing to survive. In the coming months and years, the situation is not expected to improve but rather to deteriorate. The federal government continues to put billions of dollars into housing, but still has not found a way to utilize this money effectively.

Current land-use patterns show a tendency toward further urbanization, which is consuming a significant amount of valuable land for agriculture and wildlife. Some communities have begun to renovate the older well-built structures found abandoned or under-utilized in the inner cities; however, most communities are still ignoring this vital resource. Improper land use, improper use of existing structures, and improper growth are serious environmental problems in our society today.

Indoor Air Pollution

During the last 25 years there has been a sharp increase and concern by the public and various levels of government about the quality of the air we breathe inside homes, schools, and where we work. In the 1970s energy conservation measures were widely supported by the federal government through income tax breaks and also by state and local government and the public in general. The cost of energy had become extremely high. Energy conservation meant the development and use of new insulation materials and "tighter buildings." Formaldehyde became part of the insulation material in many buildings. In the mid-1960s there was less understanding of potential cancer risks due to asbestos. Asbestos has been found in a variety of buildings, especially in schools. Radon gas was observed in the 1970s. Particularly high levels of radon were found in the Reading Prong geological formation that ran through New Jersey, Pennsylvania, and New York. Energy conservation also brought on the emerging problem of "sick building syndrome" or "tight building syndrome."

Indoor air pollutants were found in homes, offices, schools, hotels, restaurants, various buildings, buses, trains, and planes. In each of these environments, people were exposed to a wide variety of pollutants that came from smoking, building materials, rugs, furniture, appliances, pesticides, cleaning and deodorizing agents, etc. People were exposed to increased levels of carbon monoxide, nitrogen dioxide, metals, inorganic compounds, radioactive pollutants, and volatile organic compounds (VOCs). The best known inorganic compounds are mercury, chlorine, and sulfur. The particulates include fiber from asbestos and tobacco smoke. The volatile organic compounds include formaldehyde, benzene, and carbon tetrachloride.

SCIENTIFIC, TECHNOLOGICAL, AND GENERAL INFORMATION

Neighborhoods and Their Effects on Housing

The neighborhood is that area comprising all the public facilities and conditions required by the average family for their comfort and existence. Residents of a neighborhood share services, recreational facilities, and generally an elementary school and shopping area. Some neighborhoods were artificially established in older cities, by a particular ethnic, religious, or racial group. Neighborhood community facilities include educational, social, cultural, recreational, and shopping centers; utilities and services include water, light, fuel, sewage, waste disposal, fire and police protection, and road maintenance.

The size of the neighborhood is difficult to assess, as it varies with the types of building structures, the density of the population, the identification of the population as a specific group, and the various services available. To establish limits for the area or population seems impractical. Small developments on the outskirts of a particular neighborhood should be included within the neighborhood if they utilize its services. Logical neighborhood boundaries also include such things as rivers, topographical barriers, interstate highways, parkways, railroads, industrial areas, commercial districts, and parks.

Site Selection and the Neighborhood Environment

When developing a new neighborhood, it is essential to consider not only the aforementioned services, but also the land-use trends for the area; the presumed availability of transportation, public utilities, and schools; and the legal controls placed upon the area by the local governing bodies. A new neighborhood should contain between 2000 and 8000 persons, with the most desirable size at about 5000 persons; it should be within a maximum walking distance of 0.5 mi from the /local elementary school; it should not cover substantially more than 500 acres.

Site selection for the neighborhood is extremely important. The existing neighborhood may have to change, be razed, or be redeveloped to meet proper neighborhood and housing standards. The new neighborhood should avoid the problems of existing neighborhoods by locating on proper sites. In site selection, competent professional engineers should determine the geology of the land, the type of soil, the type of weather conditions prevalent in the area, whether disturbing conditions such as superhighways, heavy industry, or hog farms are nearby, and if the site is in a flood plain. Soil and subsoil conditions must be suitable for excavation, site preparation, location of utilities, weight bearing capacity, and so forth. Test borings should be made to determine the types of soils and subsoil conditions. Soil with a high clay content will shrink when extremely dry, causing cracks in foundations and within houses. The type of groundwater and drainage conditions should also be evaluated. Drainage conditions after construction should be forecast. Areas in which basement flooding may occur, swamps or marshes are present, or the groundwater table is so high that it will periodically flood the site should be avoided. It is also essential to plan a site that will not be flooded by surface waters from streams, lakes, tidal waters, or higher land. The land must not be too steep for proper grading and usage. Buildings should not be situated at an elevation above normal water pressure. The slope of the ground must also be evaluated. Adequate automobile and pedestrian traffic must be accessible to the community site. Sufficient land should be reserved for private yards, gardens, playgrounds, and neighborhood parks. The area should not be located in the vicinity of hazardous bluffs, precipices, open pits, or strip mines.

Water supply and sewage disposal are important keys to proper site selection. New sites in the community should not be approved unless public water and public or semipublic sewage are available. Today, due to poor planning or technical ignorance, millions of homes have overflowing on-site sewage disposal systems.

It is also necessary to determine the method of removal and disposal of solid waste. Since reasonably priced electricity is essential, the availability of electricity for the planned community must be determined before the site selection is approved. In addition, if gas is available, it should be included in these determinations. Telephone service must also be planned for the new site. The accessibility of firefighting crews and equipment, and the kind and amount of police protection, has to be considered before the site can be definitely approved.

Major accident hazards have to be eliminated from the site and adjacent areas. Accident hazards include aircraft, cars, trains, and so forth. The site should not be located near the use of fire and explosion hazards, such as oil or gasoline storage, or firearms. Dumps, rubbish piles, and sanitary landfills should not be adjacent to the site. The site should also be a safe distance from unprotected bodies of water, such as strip mines and quarries. It is essential that railroad crossings are well marked and automatic signals are used. Excessive noise, odors, smoke, and dust should be avoided. A site should be so selected that pollution from sewage plants, sewage outfalls, or farms will not contaminate any bodies of water running through the community. Adequate public transportation, and pedestrian and bicycle ways should be provided for proper recreation and access to community facilities out of the area of the new neighborhood. Other types of services, such as junior and senior high schools; large shopping centers; areas of potential employment; urban areas; outdoor recreation areas; public and private health services, including physicians, laboratories, and a hospital, must be accessible to residents. During the actual development of the neighborhood, proper grading techniques should be employed to ensure adequate surface drainage so that water does not stand and cause mosquito problems or flow in such a way to cause erosion.

Property Transfer Assessments

A property transfer assessment is a determination of those areas of potential legal and economic liability that a buyer may assume by purchasing a business or industry that has actual or potential environmental problems. There is a nationwide trend in the federal and state legislatures to hold individuals responsible for cleaning up environmental contamination on pieces of property that they own. This becomes a serious problem for the purchaser of any piece of property, since the environmental health hazards may be hidden or buried. These hazards may consist of spilt chemicals that are polluting the groundwater, asbestos insulation, or buried waste. These are three major phases in determining the problem. They are as follows:

1. Examine available records to determine the prior land ownership and use. Contact appropriate agencies and review any hazardous materials handling practices that may have occurred at the site.
2. If necessary, conduct a more detailed investigation to determine how far local contaminants have already spread through the environment.
3. Develop and evaluate alternative site cleanup plans and costs. Only after the audit has been completed and the determination made concerning hazardous materials, should the buyer consider purchasing the property.

These properties are very much a part of the community and could be located near housing developments. If there are hazards present, they may constitute an unacceptable risk for the community. At the time of sale of the property, the

community has an excellent means of bringing about change in the buildings if
needed.

The Internal Housing Environment

The internal housing environment, which consists of heat, light, ventilation,
plumbing, and so forth, affects the physical, emotional, and mental states of the
occupants. An individual needs to be protected against the elements of heat, cold,
disease, insects, and harmful chemicals. People need to know that when they
arrive home they can leave the pressures of society behind them and can relax in
safety and comfort in order to face the challenges of society again.

Heat

The ideal thermal environment for a given individual varies with age, sex,
conditions of health, and so forth. Most housing codes and air-conditioning
engineers recommend a minimum indoor temperature of 68°F for winter and 75°F
for summer, however, for some this temperature is too cold or too warm. The
recommended temperature must be evaluated in the light of the types of individuals, the amount of air flow, and the amount of energy available. With increasing
energy problems and rising energy costs, the entire concept of air-conditioning
may well change in the future.

The most important aspect of the thermal environment, regardless of the
selected temperature, is that the individual should not suffer from undue heat loss
and that the heat within the building is not so excessive as to be oppressive. The
temperatures within rooms will vary tremendously from floor to ceiling depending on the window location, the type of heating devices, and the way in which the
heating system is planned. Weatherproofing and insulating of buildings reduces
heat loss and retains the cool air provided by air-conditioning during hot weather.

Light

The sources of lighting in a dwelling are both natural and artificial. Few
dwellings have been built to maximize natural lighting from the sun's rays, which
serve not only to illuminate, but can control temperature as well. Light is essential
for cleanliness and avoidance of accidents, and it contributes to healthy mental
attitudes.

Ventilation

Proper ventilation is necessary to maintain a thermal environment that allows
adequate heat loss from the body, removes unnecessary chemicals, and permits
proper aesthetic sensibility within the home environment. Heat loss is controlled

by air temperature, relative humidity, air movement, and the mean radiant temperature of the surrounding surfaces. Cool moving air is valuable for promoting restful sleep. Odors, which are not removed in a poorly ventilated area, affect the well-being of the individual.

Electrical Facilities

The electrical code, the housing code, and, if these are not available at a local level, the National Electrical Code should be referred to during installation of electrical wiring in the home. Electricity is the conversion of mechanical energy into electrical energy, often by means of a generator. Electrical voltage increases by several hundred thousand to more than a million volts when it passes through a transformer. The high voltage is necessary to increase the efficiency of the transmission of power over long distance. This high voltage is then reduced to the normal 115- or 230-volt household current by a transformer located near the home Electricity is transmitted to the home by a series of wires called a service drop. In order for an electrical current to flow, it must travel from a higher to lower potential voltage. The current flows between hot wires, which are colored black or red, at a higher potential to the ground or neutral wire, which is colored white or green, at a lower potential. The voltage measures the forces at which the electricity is delivered. Electrical current is measured in amperes, which is the quantity of flow of electricity. Wattage equals the number of volts times amperes.

The Earth is an effective conductor. Near the house the usual ground connection is a water pipe of the city water system. Care must be taken that this ground is properly installed so that an individual will not be shocked by touching both sides of the pipe if the water meter is removed.

Plumbing

Plumbing is defined as the practice, materials, and fixtures used in installing, maintaining, and altering all pipes, fixtures, appliances, and appurtenances that connect with sanitary or storm drainage, a venting system, and the public or individual water supply system. (An in-depth discussion of plumbing can be found in Volume II.)

Structural Soundness

The structural soundness of a dwelling is related to the prevention of disease and injury, and the promotion of good health. The site selected should follow the requirements set forth under neighborhood site selection. The site must not be approved if a public sewage system is unavailable or the land is unable to absorb the fluid from on-site sewage systems.

The building must be placed on the site in such a way as to provide sufficient outdoor space, adequate air circulation, quiet, and safety. The location of the building may reflect certain conditions that cannot be corrected, such as the trajectory of the sun, the direction and velocity of winds, the general climate, and the nature of the water runoff from other properties or the potential for water problems created by heavy rains or snows. The ground slope is important, since this may well control the dryness and the safety of the structure. The site selection and the type of housing construction are dictated by local zoning, housing ordinances, and the amount of money available.

Daylight, direct sunlight, and the heat of the sun play an important role in the placement of the structure on the site. The building should be located on the site to maximize year-round exposure to sunlight in each of the rooms. This means at least 1 hr of sunlight on a clear day in each room. Daylight should be available to all habitable rooms during the daytime hours to reduce the additional energy requirements and also to provide a more natural setting.

In site selection and housing construction, it is essential to consider the free circulation of air around and through the structure. This helps to improve the indoor thermal environment and to reduce the cost of air-conditioning. During the winter, proper insulation will keep the cold air out of the house. In summer thermal comfort is a combination of air temperature, air movement, and relative humidity. The air movement within the building is a combination of the amount of air that penetrates into the building and the wind velocity. It is generally most comfortable if the effective temperature within the structure in the evening is not more than 72–75°F.

Site selection must also be based on the avoidance of excessive noise due to exterior problems such as factories, railroads, and highways. By orienting bedrooms away form recreational areas within the house and noisy outdoor areas, individuals are able to sleep uninterrupted. The noise level should not exceed 50 dB within the dwelling unit and 30 dB within study or sleeping areas.

In residential areas the site must be so oriented that the structure is placed an adequate distance from the road and other buildings.

Physical Structure

The basic physical structure of the house consists of the foundation, the framing, the roof, the exterior walls and trim, and roof covering. Since this chapter is not intended as a guidebook on building construction, many of the details that would ordinarily be covered are excluded.

Foundation refers to the construction below grade, such as the footings, cellar, and basement walls. It also refers to the composition of the Earth on which the building rests and other supports, such as pilings and piers. The foundation bed may be composed of solid rock, sand, gravel, or clay. Sand and clay are the least desirable, for they may shift, in the case of sand, or swell and shrink, in the case of clay, leading to sliding and settling of the building. The footings should distribute the weight of the building over a large enough area of ground to ensure that the foundation walls will stand properly. Footings usually are composed of concrete.

Foundation wall cracks occur for a variety of reasons, including the previously mentioned unfavorable subsoil, as well as improper construction or earthquake tremors. The foundation walls support the weight of the structure and transfer weight to the footings. These walls may be composed of stone, brick, or concrete. They should be moisture-proof, which involves the use of plastic sheeting joined with tar or asphalt and footing drains around the exterior of the walls to take moisture away from the property. The basement or cellar floor should have at least 6 in. of gravel beneath it and be composed of concrete. This protects the basement from rodents and flooding. The gravel distributes the groundwater moving under the concrete floor and reduces the potential of water penetrating the floor. Again a plastic shield should be laid before the concrete is poured. All basement doors and windows should be water-tight and rodent-proof.

HOUSING PROBLEMS

Health and the Housing Environment

There are approximately 12 million crowded substandard dwellings in the United States. This comprises about 15% of all available housing. Four million are in such poor condition that they cannot be refurbished without major repairs.

Twenty-five million home accidents occur in the United States each year because of faulty appliances, electrical connections, poor lighting, broken furniture and equipment, and stairs, floors, and walls in need of repair. There are an estimated 60,000 cases of rat bite reported each year. Since many incidents of rat bites are unreported, the problem is far greater than the actual statistics indicate. Six thousand cases of rat-transmitted diseases occur each year. Accidental poisoning due to inadequate storage facilities in the home and improper storage of hazardous materials and chemicals causes 3000 deaths and one million injuries. There are over 400,000 children with unacceptably high levels of lead in their blood. Lead comes from paint chips, plaster chips, and windowsills and other woodwork in old houses painted with lead-based paint. Improperly constructed, installed, or maintained home-heating devices cause at least 1000 deaths and 5000 injuries from carbon monoxide poisoning annually. In 1986, there were 20,500 home accident related deaths, and 3,100,000 disabling injuries, at an estimated cost of 16 billion dollars.

In many areas, houses are abandoned because it is cheaper to leave them than to repair them or pay the property taxes. As a result, houses become dilapidated, vandalized, and eventually become places where young children, teenagers, and adults can get into serious trouble. Not only are such buildings aesthetically unpleasant, but they also become dumping places for garbage, trash, and other junk; breeding places for roaches, mice, and rats; and a meeting place for alcoholics, drug addicts, and criminals.

Disease Problems

Infectious diseases, chronic diseases, and environmentally created diseases are more prevalent in poor housing than in better housing. While congestion within the house can lead to increased upper respiratory diseases because of the close contact with contaminated individuals, and the spread of diseases occurs often in kitchens where food is prepared along with other activities taking place, such as baby diapers being washed, it is difficult to attribute any given disease to the housing problem. A number of additional variables should be considered. These include the basic personal hygiene practices and habit patterns of individuals, their susceptibility to certain diseases, and a vast variety of additional stresses. We know that within a defective housing structure there are various stresses, such as noise, improper lighting, inadequate space, improper ventilation, the presence of myriad insects and rodents, and a variety of solid waste. Surveys indicate that individuals living in substandard housing with the aforementioned problems have higher infant mortality rates, a greater level of disease, poorer health, more nutritional and dental problems, and a variety of other health defects. Even though a specific disease may not be traced to a specific type of housing problem, with the exception of an outbreak of typhoid fever being traced to a typhoid organism found in a defective plumbing system, it is still recognized that disease rates are higher among substandard housing dwellers.

Environmental stress not only reduces our ability to fight off infectious disease, but also causes rather specific environmentally related diseases. For the chronically ill, environmental stress creates further problems, contributing to deterioration of their conditions. Noise is a specific environmental problem. Today humans live in shelters that have increasingly high levels of noise. The noise may be due to a lack of soundproofing in poorly or cheaply constructed buildings, and neighborhoods in or near commercially zoned areas and traffic routes. Noise causes nuisances, irritability, and loss of sleep. Eventually the noise may lead to actual, temporary, or permanent hearing loss. Some of the greatest sufferers from noise pollution are those who live within the vicinity of noisy industries or airports. The level of sound well exceeds that recommended for a home situation. Thirty-five decibels of background noise is acceptable at night, with occasional sounds reaching 45 dB. Forty-five dB is acceptable during the day, with occasional sounds reaching 55 dB. Annoyance occurs in residential areas when the average continually exceeds 50 dB of noise in the background. Some data suggest that 60 dB may cause autonomic changes. Surveys indicate that individuals may get headaches or become nervous when the sound levels exceed 50 dB on a 24-hr average.

Ventilation is an important part of stress or the stress reduction factor within shelters. A minimum of 5 ft^3/min/person of uncontaminated air should be provided within the residential environment. The use of mechanical ventilation today may be good or bad, and if fresh air is added rather than inside air recirculated, then home ventilation may be good. In recirculated air, as many as 100 identifiable contaminants have been found. These include pesticides, cleaners, bacterial contaminants, and viral contaminants.

The thermal environment helps reduce or increase stress, depending on the combinations of temperature and relative humidity present and the velocity of the air. A proper temperature within the home should range from 68–75°F with relative humidity ranging from 20 to 60%. Temperatures or relative humidities above these levels tend to cause discomfort, and in extreme cases may even lead to death if an individual is already subjected to serious diseases.

Illumination levels are essential in providing a good pattern of daily activity. When lights are too low or glaring, they cause fatigue and may lead to a variety of accidents or additional stress upon an individual.

There are numerous cases where the external environment has contributed sharply to the problems within the home. In Texas, where a substantial Mexican-American population lives, there are many areas where silt and dirt filter through the cracks in the buildings into the kitchens. This contributes to improper handling of food and may lead to disease. In other areas, migrant workers have been periodically sprayed with pesticides used for the control of pests in the fields. These conditions are not unique. They may be found in every major or small city in the country.

Indoor Air Pollution Problems

The serious concern over pollutants in the indoor air is due to the fact that indoor pollutants are not easily dispersed or diluted. The indoor air pollutant concentration is often many times higher than the outdoor concentration of these pollutants. At times indoor pollutants have exceeded the upper levels by 200–500%. Further, people spend as much as 90% of their time indoors, which subjects them to greater levels of risk. The degree of risk depends on how well the buildings are ventilated and the type, mixture, and amount of pollutants present in the structure. Improperly designed and operated ventilation systems add to these problems. Individuals complain of eye, nose, and throat irritations, fatigue, lethargy, headaches, nausea, irritability, and forgetfulness. Long-range health effects can include impairment of the nervous system as well as cancer. The harmful indoor air pollutants, in addition to the inorganic and organic compounds previously mentioned, include viruses, bacteria, and fungi.

Once a substance enters the indoor air environment, it is difficult to remove. These substances are in constant motion, because of the air currents and people-type activities occurring within the building. The movement allows the substance or microbe to come in contact with other substances and either interact or become attached to other substances. Because there has been a reduction in ventilation rates from approximately 10 ft^3/min to 5 ft^3/min in these buildings, the amount of air turnover or potential removal of these substances has been decreased sharply.

The various pollutants or microorganisms may produce either acute effects if the contamination concentrations are high enough, or chronic effects if there is a prolonged exposure to the contaminants at low concentrations over an extended period of time. *Sick building syndrome* is a general term applied to a group of illnesses where an acute exposure has occurred. This syndrome was rare before 1977, but by 1985 it was responsible for 13% of all of the complaints received by NIOSH. There are five different forms of this illness. The first form is an allergic response called humidifier fever. The symptoms are breathlessness, flu-like symptoms, and a dry cough. The cause of the illness is the growth of microorganisms in the water that is used to cool the air conditioning units, or microorganisms in the water of humidifiers. The second form includes common allergies. The symptoms are runny nose, sneezing, and irritated eyes. The causes are fungi, mites, pollens, and dust that have entered the indoor environment and are present in the air. The third type is infectious diseases. These diseases are caused by microorganisms that are found in cooling towers, ice machines, shower heads, and water condensers. The symptoms include coughing, fever, and pneumonia. Legionnaires' disease is an example of this. The fourth type is due to allergic reactions caused by synthetic material. The symptoms are rashes, irritation, itching, and eye problems. The causes are probably the release of chemicals within the buildings. The fifth type, the largest one of all, is due to unknown causes. The symptoms are mucus, irritation, sneezing, headache, sore eyes, and sore throat. The symptoms appear when a person enters the building and disappears shortly after he or she leaves the building. Two possible explanations for this fifth type include the release in very low concentrations of VOCs or a psychological reaction by the employee because of a dislike of his or her work.

Little information is available on the chronic effects of indoor air pollution. The problems of lung cancer caused by radon gas and the potential of cancer due to asbestos or health problems related to lead have been based more on what is occurring in industrial settings than that which is occurring in the pure indoor air environment of homes and businesses.

Age is a factor in determining how a person will be affected by indoor air pollutants. The groups most affected are the very young from birth to 10 years of age, and the elderly 65 and up. The reasons for this include the amount of time they spend indoors, the condition of their immune system, their physical condition, medications that they may be taking, and their current health problems. In small children there is a relationship between the weight of the contaminant, especially if it is a chemical or inorganic material, and the weight of the child. In the elderly, many of the individuals already have some form of upper respiratory distress, and therefore, are more affected by indoor air pollutants. Some individuals are highly sensitive and are allergic to the particular contaminant. When an antigen enters the body, there is an immune response, therefore, the greater the dose an individual receives, the greater the response, which in itself creates a harmful effect within the body.

The major route of entry for the indoor air pollutants is the respiratory system. The nasal passages are the major portals through which the air enters the body. Two filtering mechanisms are supposed to remove the particulate matter. The first is the nose, which includes its hairs. The second are mucous membranes that line the interior of the nasal passages. If these mechanisms do not work effectively, the particles may cause further harm. The oral route is the path through which the contaminants may go down into the gastrointestinal system and could eventually cause disease (such as cancer), depending upon the type of chemical ingested. The eye is also a portal of entry for indoor air pollutants. Various chemicals can be dissolved in tears and therefore move into the bloodstream.

Residential Wood Combustion

The problems associated with wood combustion include incomplete combustion, the type of wood, the type of combustion chamber, inadequate ventilation, and the burning of lumber that has been treated with chemicals. Carbon monoxide is probably the major pollutant emitted during the burning of wood. Nitrogen oxides are produced. They may be irritating to the eye and throat, and may cause coughing. Polycyclic Aromatic hydrocarbons (PAHs) may be produced and may be carcinogenic. Further, a large portion of the particulates from wood combustion are in the respirable particle range, which is less than 3 µm. These particles can be deposited in the deep alveolar region of the lungs and therefore help contribute to pulmonary disease.

Environmental Tobacco Smoke

Passive smoking is an involuntary second-hand inhalation of tobacco smoke by nonsmokers. The nonsmokers may receive the exposure either through the workplace or through the home. Two groups of individuals are within the category of nonsmokers. The first one is the exsmoker who has quit direct smoking but continues to be a chronic passive smoker, since he or she is still inhaling smoke from others. The second group are the true nonsmokers who are inhaling smoke from others. The two types of smoke entering the indoor environment are the "mainstream" smoke, which is inhaled directly through the butt of the mouthpiece during the pumping process and then exhaled into the environment. Second is the "sidestream" smoke that is produced while the tobacco is smoldering. The four places in which nonsmokers are subjected to passive smoking problems are in private homes, public building, public transportation, and office buildings. Passive smoking can result in at least seven different types of potential health problems. They are

1. Lung cancer — the risk of developing lung cancer is enhanced through passive as well as active smoking.
2. Chronic air flow limitation — passive smokers experience chronic air flow limitation when compared to a control group of individuals who have not been exposed to passive smoke.
3. Small-airway dysfunction — passive smoking increases the degree of small airway dysfunction.

4. Asthma symptoms — maternal smoking is a contributor to asthma development in the newborn child.
5. Acute cardiovascular response — nicotine is the primary constituent of tobacco smoke that affects the cardiovascular system. Acute cardiovascular alterations can result from the inhalation of smoke.
6. Pneumonia and bronchitis in infants — maternal smoking can increase the risk of infants developing pneumonia and bronchitis.
7. Carboxyhemoglobin affects on health — carbon monoxide is a gaseous constituent of tobacco smoke and it can combine with the hemoglobin in the blood to form carboxyhemoglobin.

Over 500 compounds are present in tobacco smoke. Proven animal carcinogens such as PAHs and N-nitrosamines are released from tobacco smoke. Carbon monoxide has been linked to arteriosclerosis and cardiovascular disease in people. Nitrogen oxides enhance the development of emphysema.

The potential dangers from the sidestream cigarette smoking to the nonsmoker are compounded by the amount of time of exposure to tobacco smoke, sex, age, ambient pollution exposure, size of family, occupation, drug consumption, alcohol consumption, and general health and immunological response of the individual.

Carbon Monoxide

In addition to cigarette smoke, which has been just discussed, carbon monoxide is also produced in the residential setting by improper combustion of fuel. On the average, the carbon monoxide levels appear to be higher in mobile homes than in regular housing. Improper ventilation in the tightly sealed mobile home and improperly functioning appliances are the two main reasons for higher carbon monoxide levels. One of the most common sources of emission is from the faulty furnace flues. The furnace is typically located very close to the living quarters, and therefore, any escape of carbon monoxide may enter the breathing space of the individuals very quickly. In addition, gas stoves and space heaters used in mobile homes may be hazardous. In the conventional house, carbon monoxide levels are increased by improper combustion and poor ventilation from gas stoves, gas water heaters, and gas furnaces. The carbon monoxide is produced by the incomplete burning of the gas, and once produced may be distributed rapidly through the forced hot-air system. If there is an attached garage and heating or cooling vents are present in the garage, carbon monoxide produced from the automobile may seep through the vents into the remainder of the house. One other exposure in the house would be the improper burning of fuel in a fireplace.

Radon

Radon is a naturally occurring gas that results from the radioactive decay of radium. The primary sources of radon gas in the home are from the soil and rocks upon which the house is built. Radon gas may also be present in well water. Occasionally, radon gas may come from building materials. The radon enters the buildings through cracks in the foundation. When inhaled, radon can adhere to particles and then lodge deep in the lungs. This increases the potential risk of cancer.

The EPA estimates that radon may be responsible for 5000–20,000 lung cancer deaths per year. Radon gas is a particularly bad risk for smokers who already have a health risk of cancer 10 times greater than nonsmokers. Radon-222 is a radio-active gas, with a half-life of 3.8 days. It is colorless, odorless, and chemically inert, but is fairly soluble in water. Radon gas is nine times heavier than air, and therefore, remains close to the ground, diffusing into the stagnant air of wells or crawl spaces under houses. Radon goes through a series of transformations, emitting alpha, beta, and gamma radiation. The radioactive elements that result from the decay of radon are called radon progeny or radon daughters. The radon daughters formed during the decay process include polonium-218, lead-214, bismuth-214, polonium-214, lead 210, bismuth-210, polonium -210, and finally the stable isotope lead-206. The amount of radon gas flowing into a home is primarily dependent on the radium content of the ground. Certain building mate-rials, including concrete, brick, cement, granite, gypsum wallboard, and concrete blocks, have been found to contain radon. Radon daughters, unlike radon gas, are chemically active metals. These elements will stick to anything that they come in contact with, such as particulates in the air, on furniture, and in the room. When radon daughters are attached to particulates and come in contact with the air passageways in the lungs, they may adhere to the surfaces. The biological effects of radon can be attributed mainly to its alpha-emitting products.

Radon concentrations in indoor air are influenced by four main groups of factors. They include the properties of the building material in the ground, building construction, meteorological phenomena, and human activities. Con-struction materials, such as concrete, brick, and stone, can emit radon. Since radon originates from the bedrock, the levels of radon at the Earth's surface is affected by the thickness of the unconsolidated materials, the type of unconsolidated materials, and the level of uranium concentration in the bedrock. The thicker the unconsolidated material and the further away the bedrock is from the surface of the ground, the lower the radon level. Course grain sizes of gravel and sand are highly permeable and therefore allow more radon to escape. Mud and wet clays have low permeability and, therefore, allow little or no radon to escape. If there is poor water percolation, then there is a low radon availability, unless the ground is the site of a high uranium yield area. The meteorological phenomena relates to the amount of air that flows through the area and therefore is available for ventilating the buildings. The human activities include the rate of ventilation in the building, as well as the rate of smoking and burning of wood within the building.

Radon enters the building through dirt floors, cracks in concrete floors and walls, floor drains, sumps, joints, and tiny cracks or pores in hollow-block walls. Radon may also enter the house in water from private wells and may then be released into the home when the water is used in showers, in the kitchen, or in bathroom sinks.

Formaldehyde

Formaldehyde is a probable human carcinogen. It is responsible for a variety of acute health problems, such as eye and nose irritation and respiratory diseases. People who have lung diseases or an impaired immune system, children, and the elderly may be affected by formaldehyde. Formaldehyde is used in furniture, foam insulation, and pressed wood products, such as some plywood, particle board, and fiberboard.

Formaldehyde is a colorless gas with a characteristic pungent odor. Eight billion pounds of formaldehyde are produced in the United States each year. Formaldehyde is also found in cosmetics, deodorants, solvents, disinfectants, and fumigants. Urea-formaldehyde (UF) is a resin that is produced from a mixture of urea, formaldehyde, and water. It is most commonly used as an adhesive in plywood, particle board, and chipboard. The emissions from the UF boards and other materials are due to unreacted formaldehyde that remains in the product after its manufacture. It also occurs from the breakdown of the resin by its reaction with moisture and heat. Urea-formaldehyde foam has been used extensively as a thermal insulation in the walls of existing residential buildings. It is injected into the wall cavities through small holes that are sealed afterwards. The insulation is accomplished by mixing the UF resin with a foaming agent and compressed air. The material then looks like shaving cream and can be pumped through a hose and forced into the wall cavity, where it will cure and harden. Particle board is made by saturating small wood shavings with UF resin and pressing the mixture at a high temperature into its final form. Particle board can emit formaldehyde on a continuous basis for several months to several years. In buildings in which these products are used, either for partition walls or furniture, the formaldehyde can reach concentrations that will cause eye and upper respiratory irritation. Where the air-exchange rates are low, the concentration can reach 1 ppm or more. In contrast, exposure to formaldehyde at a concentration as low as 0.05 ppm in sensitive people, under certain conditions of temperature and humidity, may cause burning of the eyes and irritation of the upper respiratory system. Concentrations higher than a few parts per million, may produce a cough, constriction in the chest, and wheezing. Menstrual disorders are one of the most frequent problems caused by exposure to formaldehyde.

The environmental factors mentioned earlier relating to temperature and humidity are important in determining the overall affect of formaldehyde on people. If the average indoor temperature is raised by 10°F, the levels of formaldehyde will double. If the average indoor temperature is decreased by 10°F, there will be an approximate 50% decrease. An increase of relative humidity from 30–70% will cause a 40% increase in the formaldehyde present. During very cold and windy days, if the house is not well insulated, the level of formaldehyde will drop because of the infiltration of air and the mixing and dilution that will occur. Generally, the greater the temperature difference, the more ventilated air is available for formaldehyde dilution. Formaldehyde levels indoors increases during the winter months and decreases during the summer months because of the amount of air that is available.

Other Volatile Organic Compounds

Other VOCs commonly found in the indoor environment include benzene from tobacco smoke and perchlorethylene emitted from dry-cleaned clothing, paints, and stored chemicals. VOCs can also be emitted from drinking water. Twenty percent of water-supply systems have detectable amounts of VOCs, although only 1% exceed the 1986 Safe Drinking Water Act standards for VOCs.

Pesticides may be used indoors or outdoors for termite proofing. Preservatives may be used for the treatment of wood. Even when pesticides and preservatives are used properly, they may release the VOCs. Wood preservatives, which may cause cancer, include creosote, coal tar, inorganic arsenicals, and penta-chlorophenol. Household cleaners, solvents, and polishes of all types may release VOCs.

Asbestos

Asbestos fibers have been shown to cause lung cancer and other respiratory diseases. Asbestos has been used in the past in a variety of building materials, including many types of insulation and fireproofing materials, wallboards, ceiling tiles, and floor tiles. When buildings containing asbestos materials are demolished, tiny asbestos fibers are freed, and these then can be inhaled by the individuals. Even during the normal aging process of materials, the deterioration may be released as asbestos fibers, which can be inhaled and accumulated in the lungs.

Buildings constructed between 1945 and 1978 may contain asbestos material. These materials were applied by spraying or trowling onto ceiling or walls. Asbestos has been used as a thermal insulator to protect walls, floors, and people from intense heat produced by stoves and furnaces. Some door gaskets and furnaces, ovens, and wood and coal stoves may contain asbestos. Pipe insulation manufactured from 1920 to 1972 contained asbestos to prevent heat loss from the pipes and provided protection from being burnt by hot pipes.

The reason why asbestos was used was because of its durability, flexibility, strength, and resistance to wear. It was also used because of its great insulating properties.

Asbestos fibers may remain airborne for periods of time depending on the fiber diameter, and to a lesser extent, the fiber length. An asbestos fiber will settle in stale air in a 9-ft high room in approximately 4 hr if it is 5 μm long and 1 μm in diameter. The same length of fiber with a 0.1 μm diameter will remain airborne up to 20 hr. If the air becomes turbulent, the fiber may stay airborne for long periods of time.

Asbestos has been found in office buildings where the practice used to be to fireproof these high-rise buildings with materials that contain 10–30% asbestos. It is still a major source of indoor-generated asbestos in these buildings, because the asbestos fibers are recirculated by the air-conditioning system. In schools the problem of asbestos has become acute. The EPA estimates that approximately 15 million students and 1.4 million teachers and other school employees use buildings that contain asbestos. School corporations are now in the process of trying to correct this problem.

Asbestos may cause four major types of health effects. Lung cancer, which may take as long as 20 years to develop, is known to cause 20% of all deaths among asbestos workers. The total amount of lung cancer present in these workers depends on their length of exposure, age, and how long they worked with asbestos. Smoking increases the risk of dying from lung cancer by 92 times as much as nonsmoking. The level of lung cancer among other individuals exposed at far lower concentrations needs to be explored. Asbestosis is a disease that involves a permanent scarring of the lungs caused by the inhalation of asbestos fibers. Asbestosis lung scarring may continue after the exposure to the asbestos ends. The symptoms include coughing, shortness of breath, and a broadening of the fingertips. Eventually heart failure may occur because of the lung obstruction.

Cancer can also develop in the digestive tract because of the consumption of asbestos in food, in beverages, and in the saliva that has been contaminated with cigarette smoke. Mesothelioma is a diffuse cancerous tumor that spreads very rapidly over the surface of the lung (pleura) and abdominal organs (peritoneum). This also may be due to exposure to asbestos.

Behavioral Problems

The density of individuals or the crowding factor within a dwelling can produce nervous irritation that is detrimental to mental and physical health. Crowding makes people more irritable, and causes interruptions and personal clashes that may end in deep-seated repressed bitterness or hatred. Crowding deprives individuals of personal space, which is a basic need of all animals. Privacy, another essential component for a healthy existence, is virtually impossible to obtain in crowded quarters when there is no place where an individual can go off and be alone. Sometimes it is necessary for people to "cool down" to avoid turning a misunderstanding into a disastrous argument. Even loving couples need time to be alone to assimilate what is occurring and to be able to redirect their thoughts and their desires.

Overcrowding increases environmental stresses by increasing noise levels, decreasing ventilation, and increasing odors, poor housekeeping, and solid waste materials. Behavioral problems resulting from inadequate control in crowded dwellings may increase infectious diseases through lack of personal hygiene. This obviously amplifies the major oral-fecal, airborne, skin, and respiratory routes of transmission of disease.

Alcoholism is often a problem in overcrowded housing, not as a direct result of population density, but rather because the alcoholic, unemployed, or chronically ill person is frustrated by his or her inability to provide adequate facilities necessary for a healthy, happy family life. Where population density is high, children lack the privacy for proper study, which creates additional problems in attempting to raise their educational level. Generally, the cycle that is formed by poor nutrition leading to disease, and disease leading to poor nutrition, may also be applied to education and lack of ability to find adequate places to study or concentrate. Poor education may lead to poor housing and poor housing, in turn, may lead to poor education.

Types of Housing

There are numerous types of dwellings. Those under discussion here include the single-family home, multiple-family dwellings, trailers, transient housing, dormitories, and emergency housing.

Single-Family Dwellings

The single-family dwelling is defined as one unattached dwelling unit inhabited by an adult person plus one or more related persons. Single-family homes vary from small cabins to mansions. The significance of the single-family home is that it contains only family members or close friends, thereby reducing the potential for conflict and neglect of property. Single-family homes are subject to housing inspections, depending on the housing code, in like manner to other dwellings. They vary in upkeep from poor to excellent, and in environmental problems from great to none. Some single-family homes lack water, adequate sewage, proper waste disposal, adequate lighting, and plumbing. Others have excellent physical facilities, including sufficient space to accommodate family members or close friends.

Multiple-Family Dwellings

A multiple-family dwelling is any shelter containing two or more dwelling units, rooming units, or both. The multiple family dwelling varies from the duplex, which is a single structure with two complete and separate units, to apartment houses, which contain many separate dwelling units. Apartment houses vary from condominiums, where the occupants own their separate units, to reconverted houses, which are single-family dwellings converted to multiple-family dwellings. The latter group is of greatest concern to housing authorities. A subdivided single-family home once inhabited by four to seven people is converted to a dwelling inhabited by six to eight different families with as many as 50–60 family members.

Further difficulty arises from the enormous number of people crammed into a structure originally built for smaller numbers. These structures tend to deteriorate rapidly. The windows break, the wallpaper and paint peels, the stairs decay, and solid waste may be found everywhere — in the hallways, in the basements, and in the backyards. Unfortunately, the tenants of such a structure are generally poor and lack the capabilities to move to a better housing situation. The types of housing problems discussed earlier in this chapter are all found in these over-crowded converted structures.

Apartment houses, unless well protected by guard systems, may be very dangerous for residents. Burglars, muggers, and attackers frequent poorly super-vised apartment houses. Many high-rise buildings, built for the urban poor in the inner city on redeveloped land, have become places for crime, dirt, and destruc-tion. The concept of multistory inner-city housing is a disastrous failure in reality. In St. Louis, a development that cost many millions of dollars had to be destroyed because the occupancy rate had fallen so sharply. It is impossible for the inner-city poor to live under these conditions in a satisfactory manner.

Travel Trailers and Mobile Homes

The major environmental health problems are the same as in other forms of housing. In addition, mobile homes are being driven to out-of-the-way places and are being used as permanent dwellings. Specific problems relate to access to adequate water supplies and on-site sewage disposal. Travel trailers are being used as permanent residences in some areas. These units are neither large enough nor properly equipped to fill the needs and requirements of a permanent dwelling.

Transient Housing

Motels, hotels, and lodging houses are transient dwellings where occupants reside for a day, a week, and possibly longer. The problems emanating from this type of housing are similar to those found in hospitals, nursing homes, convales-cent homes, and other short-term medical facilities. The major difference is the individuals within transient housing units are supposedly healthy individuals and are under no form of supervision. Many individuals fall ill when traveling, or may be carriers of a variety of diseases and spread the microorganisms throughout the transient housing environment.

The motel or hotel has the same basic kinds of problems as the medical facilities involving housekeeping, laundry, physical plant, food, solid waste dis-posal, and so forth, but in a lesser degree.

Dormitories

College dormitories are basically similar to hotels, with the principal difference being that individuals occupy dormitories for several months rather than several days. The one problem that frequently occurs in dormitories is the spread of contagious or infectious diseases. In the event of illness, a dormitory resident should be removed to the student health center, placed in isolation, and properly treated.

Rehabilitation vs Redevelopment

In the past, redevelopment was the technique most widely used for the elimination of slum areas and for the reuse of land. Run-down and neglected structures were repossessed and razed by the local government using the right of "eminent domain." The land was then utilized for a variety of purposes, including the building of roads, parks, university buildings, and high-rise and low-rise housing for the poor. Several problems have developed as a result of redevelopment. The cost was extremely high. The families living in the area prior to redevelopment had to be relocated to neighboring areas, which in turn realized a devaluation in property and eventually were transformed into the same kinds of slums.

The high-rise multiple-family dwellings, commonly referred to as *projects*, quickly deteriorated, becoming sources of crime and dirt. Unfortunately, the occupants of these high rises, many of whom were relocated due to redevelopment, were not trained in the proper use of these new buildings and facilities. Some of the better properties, which were vital resources and which could, in fact, have been rehabilitated, were destroyed.

ACCIDENTS AND INJURIES

Accidents are the fourth leading cause of death among the total population and the leading cause of death for the age group of 1–34. Accidents cost more than 118 billion dollars a year in cash, lost time, medical treatment, and insurance. The home is the most frequent place of injury. One third of all nonfatal accidental injuries and one fourth of all fatal injuries occur in the home and its surrounding environment. Over 15 million people injured annually in home accidents need medical attention. The two most susceptible or highest risk groups are the very old and the very young. Accidental deaths and injuries occur in all socioeconomic groups and in all areas. The home environment has been altered tremendously over the last few years because of many new kinds of labor-saving and recreational devices. People have not kept up with the safety precautions necessary to use these products effectively. There are 3,100,000 disabling home injuries annually and 80,000 of these are permanently disabled. There are 20,500 deaths each year. The leading cause of physician contacts per year are due to injuries.

An accident may be defined as an event or incident resulting from the interaction of people with things and the environment. A home accident occurs within the home or its immediate environment. Injury is some hurt or harm done to the body as a result of the accident. A hazard is any source of danger.

Accidents are due to glass, gas, tools, pesticides, and so forth. Accidents do not occur spontaneously; people are involved in one way or another. People must understand the potential hazards of the home environment and the precautions necessary to avoid accidents.

The two most dangerous areas in the entire house are the kitchen and bathroom. In the kitchen, fires may be caused because of thoughtless situations or inoperative equipment. People leave dish towels, paper, and other combustibles on or near stoves. Frying foods are left without proper supervision, and if the temperature of the oils becomes too high, they may ignite to cause a grease fire. People wear clothing with long, loose sleeves, resulting in food being overturned or garments catching fire. Pot handles are turned outward and in reach of toddlers, who may easily upset hot liquids and suffer severe burns. Unfortunately, the kitchen becomes a storehouse for most aerosol-type cans. Although many of the labels clearly indicate that the cans should be stored away from direct sunlight and temperatures over 120°F, individuals do not read the labels, creating serious potential hazards. Since the aerosol can is under pressure, it will explode like a hand grenade at high temperatures and fragments, including flammable materials, may shoot out in all directions.

Houses using LP or any other gas must be checked carefully to ensure that there is no leakage. Gas leaks in stoves and ovens may be due to faulty or improperly cleaned equipment. Carbon monoxide poisoning is another danger that will occur when there is inadequate combustion of the gas in the stove. If the flame is irregular, slow, and yellow, the burner is not working properly and carbon monoxide may be produced. Other incidents of carbon monoxide poisoning occur when heaters are improperly serviced or chimneys or flues are clogged. All equipment used for heating other than that serviced by the central heating system should be checked carefully, since the equipment may not be vented properly to the outside. As houses become better insulated, the incidence of carbon monoxide poisoning will increase.

Poisoning is a particularly serious hazard within the kitchen, bathroom, and garage. In these areas, children often have access to furniture polish, waxes, all types of household cleaners, bleach, medications, gasoline, and a variety of other materials stored for household use. Within the kitchen, falls occur from spilled grease, water, or other liquids. Individuals also create hazards by standing on rickety boxes or chairs to try to reach objects on cabinet shelves. The bathroom is a notorious area for accidents. Very serious falls occur because of wet and slippery floors, groggy individuals entering the bathroom at night, and narrow spaces in many bathrooms. Poisons are mistakenly swallowed by half-awake individuals when there is little or no illumination.

In new homes there is special concern about fireplaces. The fireplace is a source of danger, especially to children. Individuals making fires must be certain that sparks do not ignite a room and that the children are kept at a safe distance.

Handrails, stairs, and carpeting within halls and on stairwells should be checked for good condition to ensure that they are not torn, worn, or made slick. Adequate lighting is essential for these areas. Particular attention should be paid to the presence of childrens' toys on staircases.

Garage and other storage areas are extremely hazardous. Unfortunately, flammable liquids, including gasoline, paint, paint thinners, turpentine, and other such things, are kept in glass containers, improper storage areas, and within reach of children. Other garage and storage area hazards include tool boxes and tools, pesticides, pesticide sprayers, garden equipment, lawnmowers, glue, poison, and other material that an individual saves, stores, or utilizes.

Epidemiology of Injuries

Epidemiology, which is the fundamental science for studying the occurrence, causes, and prevention of disease, can also be used on a theoretical basis to study the problems related to injuries. In order to do this it would be necessary to determine the time, place, person injured, age, sex, nature of the injuries, factors involved in injury causation, other environmental factors, and other human factors.

Injuries like diseases do not occur at random. The elderly and persons age 15–24 have the highest fatality rates. The risk of fatal injury is $2^1/_2$ times greater for males as it is for females. Males are also at greater risk for nonfatal injuries. Fracture rates are highest among older women who have osteoporosis or bone decalcification. Death rates, with the exception of homicides and suicides, are highest in rural areas, possibly because or differences in socioeconomic status, types of occupational and other exposures, and the lower availability of rapid emergency care. Socioeconomic factors influence the incidence of homicide, assaultive injury, pedestrian fatalities, and fatal housefires. These occur at the highest rate among the very poor. Among the wealthy, the greatest number of injuries are caused by falls and home swimming pools. Other factors causing injuries include high-risk jobs, poor housing, older cars, space heaters, and housing related problems. These are especially true among the poor. The use and abuse of alcoholic beverages have a huge influence on all types of injuries, especially among young teenagers. About half of the fatalities related to injured drivers or pedestrians are alcohol related. Alcohol is frequently detected in the blood of individuals who have been involved in a variety of different types of injuries. Usually the more severe the injury, the higher the level of alcohol. Other types of behavioral factors are very difficult to determine, since they may be very transient.

The environment may contribute to the level of injuries that occur. Poor weather, as well as improperly maintained equipment, may contribute to injuries. The speed of emergency medical care is essential in controlling the level of the injury that has occurred. This emergency medical care system helps reduce the levels of fatalities among individuals.

Electrical Hazards

There are several common electrical violations found during housing inspections. (1) The power supply is not grounded properly and does not have adequate capacity for major and minor appliances and lighting. (2) The panel box covers or doors are not sealed to prevent exposure to live wires. (3) The switch outlets and junction boxes are not covered to protect against electrical shock. (4) Frayed or bare wires are present as a result of drying out and cracking of the insulation around the wires, or constant friction and rough handling of the wires. (5) Electrical cords run under rugs or other floor coverings creating potential fire hazards. (6) Bathroom lights are not permanently installed in ceiling or wall fixtures and controlled by wall switches. (7) Inadequate light is used in stairwells and hallways and within habitable rooms. (8) Octopus outlets or several different appliances or lights are hooked into one outlet. (9) Excessive or faulty fuses are used. (10) Hanging cords or wires are used. (11) Long extension cords are used. (12) Temporary lighting is used in areas where the lighting comes from other types of electrical fixtures.

Heating Equipment

In the use of any of the fossil fuels it is important that carbon monoxide is not produced by faulty combustion. Coal also produces many excess volatile hydrocarbons and sulfur dioxide. Gas, which is colorless, may be detected by odors inherent or added to the gas. Gases form explosive mixtures when there is a leak, causing additional hazards. Oil is as hazardous as the other two fuels, and in addition where oil leakage occurs, it may be a fire hazard. Inadequate amounts of air added to any of the fuels creates hazardous byproducts.

Hot-water heaters may be especially hazardous, as gas can escape if the flame in the gas hot-water heater extinguishes.

Housekeeping

Areas where people live should be kept uncluttered and clean to promote good general emotional health and to prevent infestations of insects and rodents. A badly cluttered and dirty residence not only implies potential or actual infestation of roaches, mice, and rats, but also gives some indication as to the lifestyle and emotional state of the individuals living within such premises. Solid waste and clutter, although in themselves environmental problems, are also symptomatic of other types of problems within our society, within the family, and with the individual. Since this book is not intended as a housekeeping manual, it is only necessary to state that keeping property free of solid waste, improperly stored foods, and clean, using adequate detergents and friction, will satisfy general housekeeping requirements.

OTHER ENVIRONMENTAL PROBLEMS

Noise

Noise is unwanted sound. Within the home environment, sleep or comfort is disturbed by loud television sets, grinding truck gears, fan motors, children, radios, dishwashers, garbage disposals, vacuum cleaners, washing machines, knocking pipes, and so forth. There are a myriad of sources of noise within the home. Noise is transmitted directly through the walls or indirectly through walls, ceilings, and floors. Each person has a unique feeling about a comfortable sound level and the definition of noise. For this reason, special consideration should be made of neighbors and family members when creating loud sounds at varying times during the day and night.

Pests

Roaches, mice, flies, lice, ticks, rats, and mosquitoes are in the interiors and exteriors of housing units. The infestation may be brought to the unit by grocery bags, clothing, and equipment, or may be caused by migration of insects and rodents from their normal habitats.

Sewage

Sewage is overflowing in many housing developments in various areas of the country. Currently there are over 10 million homes with septic tank systems. Considerable research is needed to determine the proper techniques for sewage disposal for these properties. An in-depth discussion on individual sewage disposal can be found in Volume II.

Drainage

Pools of stagnant water accumulate in poorly drained properties. These are excellent areas for mosquito breeding and may become hazardous. Electrical hazards must also be considered when standing water is encountered. Excessive dampness or water causes wood to rot and structures to deteriorate.

Solid Waste

The storage and disposal of solid waste in proper manner is the responsibility of the occupant of the property. In many areas the local governmental unit provides adequate solid waste removal and disposal. In some areas the occupant must make private arrangements. In any case, where solid waste is not handled properly, insects and rodents, odor, and air pollution problems due to the burning of solid waste, may occur. (Further discussion concerning this subject will be found in the appropriate chapter on solid and hazardous waste in Volume II.)

Environmental Lead Hazards

Lead poisoning is one of the most common preventable public health problems related to children today. Lead paint, the major high-dose source for children, is present on an estimated 30–40 million houses in the United States. It will continue to pose a hazard to children for many years to come. In addition to lead due to paint, lead may be found in house dust, garden or backyard soil, flaking paint of automobiles, and industrial emissions. Children are constantly being exposed to lead hazards. Lead may also be found in water because of old plumbing systems where soldered lead joints were used. Lead can leach into the water from the joints or from lead pipes. Acidic or corrosive water will cause this leaching. Lead may be present in food such as canned fruits, vegetables, and juices. The lead may be taken up by crops, from the soil, or may be deposited on the crops from lead air contamination. Lead may also be available in pottery, fish sinkers, or lead weights, jewelry making, antique ceramic doll making, metal sculpture soldering, special powders, and Chinese Herbal Medicines.

The External Housing Environment

The exterior premises in many ways are as important as the interior premises. Improper solid-waste storage, poor pest and weed control, and accident hazards create housing problems. Exterior premises with piles of old lumber or other materials, dilapidated shacks or garages, abandoned automobiles, or other kinds of junk or waste are both hazardous and unsanitary. The exterior solid waste problem is often an indication of the interior housing environment.

Fires may be caused by improper storage or disposal of flammable liquids and by obstruction of electric power lines by trees, birds nests, and so forth. Other safety hazards on the exterior grounds include stepoffs, holes in the ground, rocks, broken glass, power equipment, and storage of LP gas.

Role of Government

The many policy inconsistencies that exist as a result of changes in federal programs and tax laws have adversely affected housing. For example, federal deductions are permitted for building depreciation. It might be wiser to allow a tax deduction for building maintenance to encourage investment in deteriorating buildings. The federal highway administration provides funds for the construction of radial highways that often cut through valuable urban neighborhoods. These plans might well be altered to avoid interference with neighborhoods, thereby reducing the potential for deterioration. In the past, federal housing policies have emphasized new building rather than the revitalization of existing buildings and neighborhoods.

Housing Code Enforcement

There are many problems related to housing code enforcement. These include absentee landlords, properties belonging to estates, effective court procedures, the court and tenant, political support, fiscal support, proper personnel, code enforcement, and relocation of families. Unfortunately, a high percentage of the difficult housing code violations involve absentee landlords or estates. The landlord may not correct the violations, either because of the cost involved or because previous corrections were destroyed by the tenants or the property simply serves as a tax write-off and it is simpler to take a tax cut than to spend capital on expensive repairs.

Tenants are also taken to court for violations of the housing code, since tenants may create as many problems as landlords. Where the tenant refuses or will not take care of the housing environment, the court needs to exercise its powers in the same way that it would exercise them over property owners.

It is one thing to notify the property owner of violations and quite another to secure compliance. Although many individuals will correct violations because they are naturally law-abiding citizens, there are those who will attempt to avoid doing this. Unfortunately, politics may enter the situation. Political leaders generally are very careful about supporting vigorous code enforcement, since the code enforcement may in effect cause endless complaints from citizens and eventually result in the loss of votes.

Code enforcement at times cannot be carried out because the individuals in violation lack the money to make the necessary repairs. In the past, federal agencies, and in some cases state and local agencies, were able to provide funds for some corrections. These funds, however, are very indefinite and tend to vary from year to year based on the current political situation in Washington or in the various localities.

As a result of the uncertain political attitude toward code enforcement, inadequate funding is provided for the health department or housing department to carry out necessary code enforcement programs. This tends to deprive departments of trained competent individuals capable of performing necessary tasks and also deprives the department of having adequate numbers of individuals to make a program workable.

One of the difficulties of code enforcement is that when a property is found to be unfit for human habitation, a family must be relocated. Necessary assistance must be provided to help these families relocate in a shelter that is acceptable for proper living. In some areas, the properties involved are so numerous that it becomes a very difficult task to try to remove the families from substandard housing and to relocate them in other areas. Either the new housing is unavailable or the family is unable to live its normal life in a new situation because of transportation and shopping difficulties, and difficulties in reaching friends or relatives.

Impact on Other Problems

Health statistics demonstrate a relationship between morbidity or mortality rates and physical surroundings. People living in economically depressed neighborhoods with poor facilities are less healthy than those living in the higher socioeconomic areas. Obviously inadequate nutrition, medical care, and stressful social conditions contribute to these health problems. However, there still appears to be a relationship of physical facilities to infectious disease and infant mortality. As the facilities and levels of cleanliness improve, the rate of disease occurrence seems to drop. The role of housing and neighborhood conditions on juvenile delinquency, psychiatric illness, mental retardation, and learning disabilities is seriously questioned. There certainly are concerns about overcrowding and the effect that it may have on individuals. Substandard housing also contributes to air, water, and land pollution and to solid-waste problems.

Economics

The economic costs of housing are related to high taxes for services, congestion, streams loaded with silt, air pollution, the destruction of open space, and the provision for a vast variety of services. The social costs are certainly difficult to evaluate but must be included. Other costs include the provision of schools, necessary streets and roads, and utilities. Noise, energy consumption, and erosion of land are added factors. Personal aspects must be considered in any cost analysis, including travel time to and from work, traffic accidents, crime, and the cost involved in resolving emotional or psychiatric problems.

POTENTIAL FOR INTERVENTION

The potential for intervention is very difficult to assess in the area of housing. It is obvious that the technique of isolation can be utilized through the tearing down of unsafe structures. Substitution can be utilized by providing low-rise structures for high-rise ones in lower socioeconomic areas to eliminate problems that occur in the high-rise structures. Shielding is not an effective technique. Treatment includes renovation or improvement of the structures and the neighborhoods. Prevention is the most effective approach, since land planning coupled with adequate housing inspections and code enforcement, as well as good zoning, can help prevent problems from occurring. As has been pointed out repeatedly, the federal government has spent an enormous amount of money yet still some 12 million crowded, substandard housing structures are standing.

RESOURCES

Scientific and technical resources include the National Architectural Accrediting Boards Incorporated, the Association of Collegiate Schools of Agriculture, various graduate professional schools of architecture, the American Public Health Association, the National Environmental Health Association, construction and general contractors, and so forth.

Civic associations consist of a vast variety of homeowners associations, landowners associations, block councils, and community councils found in many cities or county areas.

The governmental resources include local and state health departments, local and state housing departments, the National Bureau of Standards, the Department of Defense, the Department of Energy, the General Services Administration, the Department of Housing and Urban Development, the Law Enforcement Assistance Administration, the National Aeronautics and Space Administration, and the National Science Foundation.

An excellent book that may be used for a study of housing codes and enforcement procedures is *Basic Housing Inspection*, published by the U.S. Department of Health, Education, and Welfare, Public Health Service, Publication #(CDC) 80-1979. A resource on safety is the Office of the Secretary, Consumer Product Safety Commission, Washington, DC 20207.

The American Public Health Association Centers for Disease Control recommended minimum housing standards, as updated in 1986, may be obtained from the American Public Health Association at 1015 15th St., N.W., Washington, DC 20005.

The Marion County Health Department has an excellent housing program. Their ordinance is entitled *Housing and Environmental Standards Ordinance.* Information may be obtained from the Marion County Health Department, 222 East Ohio Street, Indianapolis, IN 46204.

The U.S. Department of Housing and Urban Development is responsible for large numbers of programs related to community planning and development and housing. The material from them may be obtained by writing to the U.S. Department of Housing and Urban Development, Washington, DC 20410-4000.

Philadelphia has been at the forefront of injury prevention in homes. The Philadelphia Injury Control Program material may be obtained from the Philadelphia Department of Public Health, Division of Environmental Health, 500 South Broad Street, Philadelphia, PA.

Injury prevention is a serious concern that has been addressed by the Childhood Injury Prevention Resource Center, Harvard School of Public Health, Department of Maternal and Child Health, 677 Huntington Ave., Boston, MA 02115. "Injury in America," a major report of the Board of the National Research Council of the National Academy of Sciences, the National Academy of Engineering, and the Institute of Medicine, has been published by the National Academy Press, 101

Constitution Ave., N.W., Washington, DC 20418. The Centers for Disease Control, U.S. Department of Health and Human Services, Public Health Service, Atlanta, GA 30333 has programs and have published reports on injury control.

Indoor air pollution, radon gas reduction, lead contamination, and other associated programs and problems are under the control of the U.S. Environmental Protection Agency. Information may be obtained from them by contacting the U.S. EPA, 401 M Street, S.W. Washington, DC 20460.

Additional resources for injury control include: the Surgeon General's 1990 Injury Prevention Objectives for the Nation and Model Standards; A Guide for Community Preventive Health Services; Developing Childhood Injury Preventions; and Administrative Guide for State, Maternal and Child Health Programs; Strategies for Injury Prevention; Injury Prevention in New England; and the Surgeon General's Workshop on Violence and Public Health Report. (Contact the Centers for Disease Control, Atlanta, GA.)

STANDARDS, PRACTICES, AND TECHNIQUES

Rehabilitation

It is true that in some areas redevelopment has been necessary, but in other areas a better concept, known as rehabilitation, has been attempted. In rehabilitation, existing residential properties that are deteriorating are renovated with the aims of (1) ensuring improved housing that is livable, safe, and physically sound; (2) ensuring low-cost housing; (3) providing an acceptable minimum level for housing; (4) encouraging innovation and improved technology to reduce construction costs; and (5) utilizing instead of destroying resources that need upgrading. Obviously rehabilitative construction standards are not the same as new construction standards, since the work is being done on existing buildings. However, all safety precautions do have to be taken into consideration and followed carefully. All rehabilitated properties must have access to the property from the road by emergency vehicles, living units that exist independently of other living units each with its own entrance and exit to the building, and practical living units. Under rehabilitation, all highly dilapidated properties that cannot be rebuilt efficiently are torn down and the ground space is given to the community for use as playgrounds, parks, or parking lots, as decided by residents of the community. Adequate night lighting and open space are site criteria that must be provided; also, the land must be made free of flooding, sewage problems, and solid-waste problems. The materials and products used for replacements or additions have to be of good quality, conforming to generally accepted practice. A building must be structurally sound. The exterior walls must be able to support the weight of the roof and also must be moisture free.

It is important in rehabilitation that individuals have sources for adequate food, hot water, baths, and proper sewage and solid waste systems. One must recognize that rehabilitated property may not look as good as new property. However, low-income families have the opportunity to live in good shelters that provide all of the essentials for prevention of disease, elimination of behavioral problems and accidents, and promotion of health.

Standards for Housing Utilities and Construction

Heat

The APHA-Public Health Service (CDC) housing code states that the temperature should be at least 68°F at a distance of 18 in. above the floor level. If the person needs a higher temperature because of age or physical condition, 70°F is required. Others feel that this temperature should be maintained at different points above floor level, such as at 5 ft. If carbonaceous fuel is used, the heating devise or hot water heater must be vented to the outside and receive adequate air for proper combustion of fuel. In any case, the local housing code requirement will be the one followed by health department personnel.

Light

In most communities, every habitable room must have at least one window or skylight facing outdoors, and the window area must be at least 8% of the total floor area of the room. Daylight should be used wherever possible in such a way that the amount, distribution, and quality of the light aids in, rather than detracts from, visual tasks.

Artificial light varies tremendously from area to area, and from room to room. Artificial light must always be provided in public halls and stairways, and should be adequate for all use within the house. The house that is properly illuminated enables the family to function more efficiently. Switches should be in convenient places, and the type of lighting should serve the various needs of the living day. Obviously, such areas as kitchens and bathrooms should be well lit to prevent accidents. Such places as furnace rooms, laundry rooms, and other areas where appliances are kept should have adequate lighting to make the task less demanding and to prevent accidents. Proper lighting is also very essential in preventing fatigue.

All electrical units, wiring, and appurtenances with electrical connections must be installed by electricians following electrical codes to avoid the occurrence of fires.

Plumbing

In the house, the piping for water service should be as short as possible, and elbows and bends should be reduced to maintain water pressure. The water line to the house should be at least 4 ft below the soil to prevent freezing. Valves are usually located outside the building so that the building supply may be turned off when it is necessary to service the building. Hot water heaters should be thoroughly evaluated and properly installed and ventilated. Usually a .75 in. pipe is the minimum size required for water that must rise from the basement level to other parts of the house. The drainage system should have a main house drain a minimum of 6 in. in diameter. The house drain has to be sloped toward the sewer, usually at a level of .25 in. fall per 1-ft length. Determination of the amount and size of drainage lines depends on the types of rooms within the house. Traps are used on sinks and toilets to prevent sewer gases from entering the property. All plumbing systems must be adequately ventilated directly through the roof. The most essential part of the plumbing is the proper size of the lines, adequate installation, proper ventilation, and proper drainage to avoid cross-connections or other situations in which the potable water supply becomes contaminated or the sewage backs up through the toilets and various drains within the house.

Ventilation

The usual requirements for ventilation is that 45% of the minimum window area be easily opened or an approved mechanical means of ventilation be installed. Every bathroom and toilet room must comply with the various housing codes and rules on adequate ventilation. Ventilation must be to the outside. In many housing codes, two to six changes of air per hour are required.

Space Requirement

One hundred and fifty square feet of total, habitable, room area per person is required by the APHA-CDC code. Where second-person occupancy occurs, the requirement is reduced to 100 ft². The code also requires that no more than two people are permitted in a habitable room. With the exception of rooms that have sloping ceilings, the habitable room should be a minimum of 7 ft in height. Where a room is less than 5 ft in height, it should not be used for living purposes. Individuals should not live in rooms, except in emergencies, that are totally below the ground surface. There should be some exposure to the outside. The exception to this rule is where the construction is waterproof, there is adequate window area, and the other rules and regulations relating to light and ventilation are followed. Pipes, ducts, or other obstructions must not be less than 6 ft 8 in. from the floor. Where two people share a room, at least 70 ft² of floor space is required for the first person and 50 ft² of floor space for the second. Living quarters must have easy access to a bathroom, and an individual must not have to pass through another sleeping room to get to a bathroom.

Water

All housing codes require that adequate quantities of running water be provided for a dwelling. This generally means 1 gal/min of hot and cold running water per each fixture in the house. Generally, 120°F is accepted as the maximum hot-water temperature. However, the temperature may need to be lowered to prevent burns.

Kitchen Facilities

All kitchen facilities must contain sinks, cabinets or shelves, stoves, and refrigerators. The sink must be large enough for kitchen use and not a small hand-washing sink. As the kitchen sink is used in the preparation of food and the cleaning of dishes, utensils, and equipment, it must have hot and cold running water under pressure and must be so built as to be utilized in an effective manner. The stoves must be adequately built, installed, and maintained to avoid electrical, fire, or carbon monoxide accidents. A refrigerator is needed to keep food under 45°F at all times. Freezers are highly desirable.

Physical Structure

Houses are framed in many different ways. However, all of them contain foundation walls and provide support for the outside walls of the building. The flooring system is made up of girders, joists, subflooring, and finished flooring, which may be composed of concrete, steel, wood, composition material, and so forth. Studs, which are usually 2 × 4 in. or 2 × 6 in., are used to provide a wall thick enough to allow the passage of waste pipes. Usually studs and joists are spaced 16 in. on center. All openings, such as windows or doors, must be framed with studs to provide support for that portion of the wall. The interior wall finish varies from plaster to wallboard, paneling, wood, and so forth. Interior stairways must be greater than 44 in. in width with a maximum rise of 8.25 in. and a minimum tread of 9 in. The handrail and all other parts of the stair enclosure should be no lower than 80 in.

The four basic types of windows are double-hung sash windows, which move up or down; casement windows, which are hinged at the side; awning windows, which are in panes on a horizontal axis; and sliding windows, which slide past each other. There are many types of doors and door finishes.

The roof framing consist of the rafters, which support the roof and also create place for roofing material. The exterior walls and trim may be made of a variety of materials. Their function is to enclose and weatherproof the building. Exterior walls also serve as weight-bearing walls. They are composed of wood, brick, stone, and so forth.

The roof covering must keep the house dry and intact. Roof covering composition includes asphalt shingles, asphalt built-up roofs, tar, slate, tile, and wood. Flashing is the metal that joins two portions of the roof where it forms a valley. This keeps water and air from getting into the building. Gutters are used to take water off the roof and away from the house.

Insulation materials is extremely important to keep the house cooler in the summer and warmer in the winter. Good insulation prevents the loss of vital energy.

Fire Safety and Personal Security

All dwellings should have at least two means of egress leading to a safe and open space at ground level. Individuals should not have to exit through someone else's dwelling unit. All entrance doors into a dwelling or dwelling unit should be equipped with a deadbolt or locking device. The entrance doors in a multiple dwelling should be equipped with a device that allows the occupants of the unit to see a person at the door without fully opening the door. All exterior windows and other means of egress should be equipped with locking hardware. Every dwelling unit should have at least one functioning smoke detector located on or near the ceiling in an area immediately adjacent to a sleeping area. These smoke devices should be tamper-proof.

Travel Trailers and Mobile Homes

The travel trailer is a portable structure used as a temporary dwelling for travel, recreation, and vacation. The body width must not exceed 8 ft and the trailer must not exceed 4500 lb in weight of 9 ft in length. Travel trailers vary considerably in size, types of sleeping accommodations, and actual living facilities. The trailer may simply be built for sleeping or may have self-contained hot and cold running water, full kitchen, baths, gas tanks, air-conditioning, and lighting systems. The travel trailer has to be fitted for hooking into sanitary stations for removal of sewage. The trailer has toilets aboard and must also be fitted for taking on water. The facilities used for travel trailer parking areas have to be carefully selected for good drainage, gentle sloping, and no obstructions. The roads around the parking areas must be curved to reduce the speed of the drivers. A separation of 15 ft between trailers is an absolute minimum. In trailer areas there must also be provisions for service buildings where individuals can take showers, do laundry, and also use toilet facilities. Solid waste has to be stored in tight-fitting containers that are waterproof, fly-proof, and rodent-proof. These containers must be removed daily to prevent pest infestation. When a travel trailer area plan is ready to be approved, it must indicate the area in site dimensions; the number, location, and dimension of the trailer spaces; the location and width of roads and walks; the location of the service building, sanitary station, and other facilities; the location of water and sewer lines; and the location of storm drains and catch basins.

The number of mobile homes has increased enormously within the last 15 years. Mobile homes may be on wheels or modular in nature. The mobile home is being used in many cases as low-cost housing. Mobile homes may be placed on individual sites or in mobile home parks. The home placed on an individual site must follow all rules and regulations of the health department for proper water, sewage, site locations, and so forth. The site has to be zoned for this particular use. It should be well drained and free from topographical hindrances. It should not be located near swamps, marshes, heavy industries, in flood plains, or on steep slopes. It should also be placed, if possible, in such a way that in the event of tornados, it will be protected as much as possible. The owner of the home has to either hook into an existing sewage system and obtain city water plus other utilities, or he or she must dig a well and provide an adequate on-site sewage disposal system. (Well-drilling and on-site sewage disposal will be discussed in further detail in Volume II.)

Where a mobile home court is developed, there has to be a community development on the same order as any other type of community. All the roads have to be convenient and of adequate width for the parking and traffic loads. Streets should be approximately at right angles. Street intersections should be at least 150 ft apart, and the intersection of more than two streets should be avoided. The grades of the streets should not exceed 8% whenever possible, although for a short run as much as 12% is acceptable. The streets must be of all-weather construction and of hard dense easily drained material. Off-street parking is necessary. Walkways at least 3.5 ft wide must be provided within the area.

A mobile home should be on a minimum size lot of 2800 ft^2. A double-wide unit should be on a lot which is a minimum of 4500 ft^2. The mobile home should not occupy more than one third of a total lot area. Small lots contribute to overcrowding and create a poor aesthetic image of the area. If an area is developed for five or more mobile homes, a recreational area that is safe and free of traffic hazards should be made available for use by children and adults. A recreational area can be located next to recreation or service buildings. The minimum space should be .66 acre. Where swimming pools are constructed, they must comply with all of the regulations of the various state and local health codes.

All buildings in a mobile home park must conform to the housing codes of the local and state areas. They must be kept in a sanitary way, be screened, properly ventilated, and properly lighted. Separate bathroom facilities have to be provided for men and women. If a laundry facility is present, it has to be of adequate size, properly installed, and the wastewater must go into adequate wastewater disposal units.

Before a mobile home development can be approved, the plans have to be submitted to the health department, zoning department, and other responsible authorities. Detailed construction plans, and the site and the location of all roads, buildings, and so forth have to be evaluated. It is essential, wherever possible, that the public water supply and public sewage disposal be used. If these are not available, a package treatment plant should be built. Access to firefighting equipment, police protection, and schools have to be determined before the site is approved.

In the mobile home park, a mobile home stand has to be provided for adequate foundation and anchoring of the mobile home. The stand must be properly graded and of such material that it can support the weight load, regardless of weather conditions.

The anchors must be cast in place in concrete in order that they will not be ripped out in the event of storms. Particular attention should be paid to this in those areas where tornados frequently occur. Since tornados will knock over mobile homes or cause damage to them, all water and sewage hook-ups and electrical and gas hook-ups have to be installed in such a way as to avoid hazards and to provide adequate service. Telephone service is also needed. Further discussion concerning water and sewage will be found in other chapters in Volume II.

Storm-water drainage systems have to be installed in such a way as to avoid flooding of the mobile home site. The drainage system must remove the heavy rain waters from the area. Solid-waste containers, storage containers, and collection techniques have to meet the requirements of local and state codes. Further discussion concerning this subject can be found in the chapter on solid waste in Volume II.

Insect and rodent control must also be considered in a mobile home park. Of particular importance is mosquito control, since many mobile home parks are in outlying areas where weeds or other breeding places, such as depressions in the ground, may be present. In addition, the areas must be free of cans, jars, buckets, old tires, and all other junk that serve as breeding places for mosquitoes. Fly- and rodent-control procedures are to be followed. It is essential that all garbage be properly stored and then disposed of. Insect and rodent control are discussed in further detail in other chapters of this book.

Electrical wiring, equipment, and appurtenances should be installed and maintained in accordance with local and state codes. Where these codes do not exist, it may be necessary to consult the National Electrical Code. If possible, distribution lines should be installed in underground conduits, which are placed at least 18 in. below the ground surface and at least 1 ft away from water, sewer, gas, or phone lines. If overhead power lines are installed, it must be done in such a way as to prevent accidents and should be available for service in emergencies. Grounding is necessary for all exposed, non-current-carrying metal parts of mobile homes and equipment. Adequate lighting has to be provided on exterior walkways, streets, and buildings.

The handling or storage of fuels, such as natural gas, liquified petroleum gas, fuel oil, or other flammable liquids or gases, have to meet federal or state ordinances or the National Fire Protection Association Standards. It is best to have central storage and underground distribution of fuel. All fuel oil containers or other fuel containers on the lot where the mobile home stands are to be securely fastened to prevent accidents or fires.

It is essential that adequate fire protection be provided in mobile home parks. Applicable state and local codes and laws should be followed. If they do not exist, standards established by the National Fire Protection Association may apply. A water supply of adequate quantity must be available for fire hydrants within 500 ft of all mobile homes, service buildings, or other structures. A minimum of two 1.5-in. hoses must be available for hooking up to the water supply. At least 75 gal of water per minute per nozzle must be available at a pressure of 30 lb/in.[2]. Where hydrants are not available, the types of equipment that can be utilized are limited to those that meet the National Fire Protection Associations' standards for adequate fire extinguishers. Class A extinguishers have to be provided for mobile homes.

Since carbon monoxide is a constant danger within the mobile home, it is essential that heating and cooking equipment be properly vented. If there is any question concerning this, the standards of the local or state authorities should be followed or the health department or fire department should make an evaluation of the equipment and the adequacy of the venting system.

MODES OF SURVEILLANCE AND EVALUATION

Housing Inspections and Neighborhood Surveys

Housing inspections and neighborhood surveys are used to determine the quality of housing and neighborhoods within a community. The American Public Health Association has done considerable work in this area. They have developed tools of evaluation, including comprehensive forms, and have established point systems for determining into which category a neighborhood falls. (For more detailed information on housing, the reader is referred to the latest methods, techniques, and publications of the American Public Health Association.)

The now defunct Bureau of Community Environmental Management of the United States Public Health Service developed a system of forms used by federal and local governments to evaluate neighborhoods and housing structures. The forms were entitled, "Neighborhood: An Environmental Evaluation and Decision System Interior Interview," form number 85-R-0134. Two additional forms, entitled "Neighborhood: An Environmental Decision System Exterior Premises Analysis," form number 85-R-0044, and "Neighborhood: An Environmental Evaluation and Decision Block Analysis," form number 85-R0044, can be utilized for an in-depth evaluation of neighborhoods, blocks, and houses. Since housing inspection is a long and complex procedure, the forms developed for this purpose could be utilized, or at the local level, new forms should be developed based on the federal forms to meet the individual needs of the community.

Indoor Air Pollution Surveys

Residential Wood Burning

Air samplers can be used to determine the level of particulate matter present in the air. Grab samples can be used for obtaining a small representative sample of the gases that may be present in the air from wood burning. The polycyclic aromatic hydrocarbons (PAHs), such as chrysene and benzo(d)fluoranthene, are found in residential wood smoke. These PAHs are adsorbed predominantly on particulate matter, which is collected on a filter during the air sampling process. The PAHs can be extracted from the filter. They are then analyzed by the use of column gas chromatography (GC) and high-performance liquid chromatography (HPLC). Further biological monitoring techniques are useful for detecting whether an environmental or occupational exposure has occurred to the PAHs. PAHs are present in blood, mammary, adipose, liver, and brain tissue from rats. PAHs may also be found in the urine of exposed humans.

Passive Smoking

Passive smoking can be determined by observing the smoking habits of individuals within the buildings and whether or not there is adequate ventilation or shielding of the nonsmokers from the smokers.

Carbon Monoxide

Carbon monoxide is measured by infrared spectroscopy. The test is sensitive in ranges from 0.1 to 50 ppm. Water vapor and temperature must be controlled in order to ensure accurate readings. There are three types of personal monitors that are currently being used. Two types have color indicators, while the third has an alarm system that indicates when the threshold limit value has been reached. Grab sampling can also be used to determine what is occurring at any given time. The sample should be taken near the floor, since carbon monoxide will sink because of its density.

Radon Gas

Radon gas sampling can be accomplished by using charcoal canisters or alpha track detectors. A trained technician may also test for radon gas levels within the home. Charcoal canisters and alpha track detectors require exposure to the air in the homes for a specified period of time and then are sent to the laboratory for analysis. For charcoal canisters, the test period is 3–7 days. For alpha track detectors the test period is 2–4 weeks. Technicians can do a direct radon measurement at the time of analysis. These include real-time radon measurement, continuous radon readers, and integrating readers. The biggest problem with measuring radon is that the levels of radon fluctuate considerably during different time periods.

The appropriate technique for using radon detectors is to first do a screening measurement. This gives you an idea of the highest radon level in the home. It also advises you if it is necessary to continue with radon testing. The screening measurement should be made in the least livable area of the home, probably the basement if one exists. All the windows and doors should be closed for at least

12 hr prior to the start of the test and kept closed as much as possible throughout the testing period. This will help keep the radon level relatively constant throughout the testing period. It is recommended that the short-term radon measurements be done during the cool months of the year. Next, determine the need for further measurements. In most cases the screening measurement is not a reliable measure of the average radon level to which the family is exposed. The screening measurement only indicates the potential for a radon problem, since radon levels vary from season to season and from room to room, based on the level of ventilation that is present. If in the screening measurement the following results occur, then these results act as a guide to further measurements.

1. If the screening measurement result is greater than about 1.0 WL (working levels) or greater than about 200 pCi/l (picocuries per liter), it is necessary to perform follow-up measurements as soon as possible. Expose the detectors for no more than 1 week. Consider taking immediate action to reduce radon levels in the home.
2. If the screening measurement result is about 0.1 WL to about 1.0 WL or about 20 pCi/l to about 200 pCi/l perform follow-up measurements. Expose the detectors for no more than 3 months.
3. If the screening measurement result is about 0.02 WL to about 0.1 WL or about 4 pCi/l to about 20 pCi/l, perform follow-up measurements. Expose detectors for about 1 year or make measurements of no more than 1 week in each of the four seasons.
4. If the screening measurement result is less than about 0.02 WL or less than about 4 pCi/l, follow-up measurements are probably not required.

Formaldehyde

Instruments used to detect formaldehyde include Infrared Spectrometer photoionization detectors and electrochemical sensors. A source emission test that is conducted is the extraction test and the static chamber test. Extraction tests measure the total releasable formaldehyde content under typically harsh environmental conditions. The static chamber test determines the steady-state formaldehyde concentration.

Other Volatile Organic Compounds

Sampling techniques used for other volatile organic compounds include the evacuated bomb and the sampling bag, which are two means of grab sampling. In both cases, the air is taken back to the laboratory for analysis. For a better study, a quantitative time-weighted average test should be conducted. This type of test would include charcoal adsorption, silica gel, and bubblers.

Asbestos

Air sampling is used to determine the amount and size of asbestos fibers present in the air.

CONTROL OF HOUSING

Zoning and Land Use

Zoning is a means of ensuring that community land will be used in the best possible way for the health and general safety of the community. To prevent disorderly patterns of growth, a community must plan. It would be very unfortunate if heavy industry and superhighways were placed in the center of residential areas. When a house is built where a zoning ordinance is in effect, the builders must comply with the zoning regulations, and they in turn are assured that the other houses in the area comply with the same regulations. A single-family dwelling, for example, cannot be converted into multiple-family units unless zoning laws permits this type of structure. It is important that public health officials, housing inspectors, developers, and planners have a good understanding of the plans of a community and the applicable zoning laws.

Zoning regulations have been in effect for several hundred years. They were originally passed to keep gunpowder mills and storehouses away from heavily populated areas of town. Later zoning was used to form fire districts and to prohibit certain types of highly flammable structures. Zoning has been used and misused in a variety of places. Proper usage would mean that zoning was tied to an overall community plan. Improper usage is to change the zoning by using variances, so that special interest groups or political groups can utilize structures or land as they see fit. The objectives of zoning are to regulate the height, bulk, and area of structures; to avoid undue levels of noise, vibration, and air pollution; to lessen street congestion through off-street parking and off-street loading requirements; to facilitate adequate provision of water, sewage, schools, parks, and playgrounds; to use areas that are not subject to flooding; and to conserve property values. Zoning cannot correct existing overcrowding of substandard housing or change materials and methods of construction, since these are controlled by building codes. Zoning laws do not affect the cost of construction, since this is based on the economy. Zoning cannot be used to develop special regulations or covenants, nor can it design and layout subdivisions, since this is controlled under other rulings. Zoning is concerned with lot size, usage, depth, and width; the amount of open space; and the types of buildings and alterations to buildings placed on the lot.

Role of Government

There are four major problem areas in the role of government in land use and housing. These include the proper use of federal government program and federal policy, state government in action, the local taking of land by right of "eminent domain," and adequate municipal financing. There is a need for coordination and consistency between the federal, state, and local governments in land use programs and policies. State governments should have greater input in land use matters affecting large local areas. The regional metropolitan units of government should be concerned with land use and its effects on outlying, as well as immediate, areas. It is important that a sharp distinction be made between public interest and the rights of private ownership.

Since local governments' major source of revenue is generally the property tax, it is important that land-use decisions be made in accordance with good, sound scientific and socioeconomic political decisions. Not only must the metropolitan area be concerned with proper land use, but also with misuse. Many unutilized buildings, if properly redesigned, could once again be used for shopping centers, office buildings, and so forth. Recycling of structures such as warehouses, garages, factories, and schools is not frequently considered. They are therefore abandoned or torn down. Recycling of these structures could possibly be done at a lesser expense than redevelopment, thereby aiding the rebuilding of the nuclear city, which is so necessary to the metropolitan area.

Since the Tenth Amendment to the Constitution grants the states the right to plan land use over nonfederal land, the states should devise uniform zoning laws and delegate their authority to proper local governments. They should also provide adequate supervision and evaluation. If this were done, instead of having an enormous number of local decisions made on zoning that affects land use, decisions might be made on a more rational basis and the decisionmakers would be responsible to some higher authority. Zoning would cease to be a political game and serve its proper function of providing adequate land use for the greatest number of people. Considerable legislative and judicial actions concerning the right of private property vs the public interest have been taking place. It is essential that this question be viewed in detail and that firm long-range decisions be made to provide the greatest good for the greatest number of people and still protect private interests.

It is also wise to consider whether property-tax revenue is an adequate base from which to obtain taxes to run the government. Should other major sources be utilized? If property does continue to be the principal source of taxation, then decisions must be made to provide proper land use and adequate taxing for a given area. Assessment practices vary from community to community and industries often receive tax breaks to attract them to a given area to provide jobs. All of these things must be considered in the evaluation of the proper use of land and the role of government in this use. The federal government would do well to integrate all of the various programs from housing through road construction to environmental controls in such a way that the states would have a better understanding of how land-use patterns could be properly determined. The federal government should coordinate land-use patterns with the states by the establishment of adequate interstate highway construction, pollution-abatement facility construction, home-mortgage guarantees, proper federal income-tax deductions, and the planning of areas surrounding federal lands.

The state governments should not only carefully plan and develop land-use and zoning laws, but should also establish standards for the use of flood plains, coastal areas, wetlands, agricultural lands, highways, and airports. They should further set up a governing body to coordinate the innumerable local government agencies involved in land-use planning. It is essential that land use becomes not a single problem of one governmental unit but a coordinated approach by all agencies at all levels of government.

Housing Codes and Housing Laws

The state may regulate the use and enjoyment of individual property for the good of its citizens. This does not usually conflict with the constitutional protections given to the owner or occupant of a property. The power that the state has in establishing housing codes is based on the state's police power, which has never been precisely defined. However, the general authority comes from the Constitution, where it is stated that the state has the power to protect the order, safety, health, morals, and general welfare of the society in which its citizens live. The courts have stated that municipalities may impose restrictions on individual property owners if these restrictions protect the public health, safety, or welfare. In some court cases, it has been clearly held to be constitutional that housing codes may require windows and ventilation, screens, hot water, and other necessities of life. Local housing codes have been generally upheld by the courts.

In order for someone to conduct a housing survey, it has been generally agreed by the Supreme Court that an individual search warrant is not needed in a systematic area inspection unless the individual refuses entry to the environmentalist. It is not necessary for the health department to establish due cause, providing that there is a reasonable need to conduct periodic areawide inspections based on the passage of time, the nature of the buildings, or the condition of the entire area. However, under the Fourth Amendment to the Constitution, warrantless nonemergency inspections of residential and commercial premises cannot be carried out without the occupant's consent, unless a complaint has been lodged because of conditions supposedly existing within a given dwelling. In all cases where entry is refused to the environmental health practitioner, the county or city attorney should be consulted on the proper procedure used to gain entry to conduct the housing inspection.

It is difficult to determine what is a typical housing code. Although the American Public Health Association and the United States Public Health Service has issued, from time to time, a book entitled "APHA-CDC Recommended Housing Maintenance and Occupancy Ordinance" (the 1986 edition is called "Housing and Health, Recommended Minimum Housing Standards"). Local government can and does determine what it feels is a proper housing code. Therefore, codes differ from one area to another. As the housing codes are amended and changed by the local authorities, the tendency is toward decreasing stringent requirements to meet the needs of the governmental units. Where codes have been made stricter, it is because of federal requirements for obtaining urban renewal or other housing funds. It should be emphasized that housing requirements are minimum standards and that they affect the owners or landlords more than the tenants, although the tenants are minimally involved.

The entire establishment of the housing code and its subsequent use and enforcement cannot be based on any given political area, since it is necessary to look at the geographic rather than political boundaries. However, in reality, political boundaries still control the types of housing codes that are in effect. Although houses outside the city boundary are on a street adjacent to the city boundary, the housing code may be completely different. Most housing codes are municipally centered and therefore provide for the control of properties within the municipality and not outside it. Unfortunately, the contiguous areas become blighted because of lack of proper control. Some states have attempted to impose statewide housing codes, including California, Maryland, Arkansas, and Georgia. Generally, these housing codes grant counties the authority to establish codes for their own area. A small number of counties have already done this. Unfortunately, the political nature of most counties and their varying subdivisions are such that conflicts arise and housing code enforcement becomes very difficult. An exception to this is Marion County, Indiana, with Indianapolis as it center city. The Marion County Health Department exercises the authority to regulate housing throughout the entire county. As a result of this, Marion County exercises greater control over housing code violations than many other areas of the country. Honolulu, Hawaii; Denver, Colorado; and Dade County, Florida have city-county housing programs. Urban renewal, through the use of federal funds, has helped establish housing codes in a variety of jurisdictions. As a result of this, many urban areas have had an opportunity to raze large blighted areas and to reuse the land for other purposes; examples include Kansas City, Philadelphia, Baltimore, and St. Louis.

Effective court procedures may be needed to ensure compliance with housing codes. The procedures should be brief and court hearings should be held immediately following code violations. Conclusive evidence should be presented to enhance the procedure. Difficulties occur from long delays in hearing cases and because certain courts or judges are not particularly concerned with housing code violations.

A specific housing court would be able to exercise careful and quick judgement when needed. The court's objective would be to correct the violation rather than to punish the individual. A special court could also provide specific time blocks that could be utilized for better total enforcement. The court could use the punitive measures as a last resort and court appearances as a final attempt at an educational approach to correct the situation.

Personnel enforcing housing codes should be well-trained environmental health practitioners, who are assisted by environmental technicians. The practitioners should be responsible for the overall area in housing surveys. The technicians should be responsible for the house-to-house surveys. It is necessary for the technicians as well as the practitioners to understand the role of the housing code inspection in the overall problem of the environment. They need to understand how to work effectively with people to avoid the kinds of problems that lead to securing a warrant to enter a property. Good community and public relations on a one-to-one and one-to-community basis are absolutely essential to obtain good results. The function of the environmental health technician is to attempt to improve the housing environment, not to be a punitive police officer, or to gather evidence to cause problems for a family. He or she is there to evaluate the environmental health problems and to attempt to instruct the individual and family on corrections needed to improve their housing environment. Good, sound personal communications coupled with scientific educational approaches presented in a simple manner derive far better results than any series of court procedures. Once a court hearing occurs, partial defeat is suffered. Time that should have been used for helping people has been lost in courtrooms, and the individual forced to act against his or her will is not prone to maintain his or her property properly unless repeatedly penalized.

Community Programs in Housing

Housing programs work best when the community is involved and when agencies other than the health or housing department are also involved. The other agencies include planning, zoning, welfare, low-rent public housing, building inspection, and urban renewal.

Community groups have grown in numbers and importance in the last 50 years. They were formed in the cities for the purpose of neighborhood or community maintenance. Many of these groups are either ethnically or racially homogeneous. Individuals are often property owners and the groups are called *Improvement Associations*. With the change in neighborhoods, particularly in the inner city areas, community organizations have assumed different roles. Some of them are comprised of individuals with a civic consciousness who want to help others. These individuals may or may not live within the specific neighborhood and may include business associations, clergy, educators, or other professionals. They may also include individuals who want to improve a neighborhood. Other neighborhood organizations form for specific interests, either because they are not getting something they want in the way of services or because they are seeking some means of providing a better place to live or a better social atmosphere. Unfortunately, neighborhood associations are often difficult to maintain. They are funded by contributions and often where such organizations are most needed, residents are unable to contribute. There also appears to be a general situation of public apathy. If 5% of any given block or area actively participates in a neighborhood association, a considerable amount can be accomplished. The active group will carry out the necessary implementation and planning, and with good public relations techniques, other residents will become involved in special projects for the neighborhood. It is difficult to establish a metropolitan or regional organization that is of value to a given neighborhood, since these organizations are made up of individuals who are the presidents, chairpersons, or executive directors of large agencies. They have a solid understanding of their own agencies, but generally lack adequate understanding of given communities.

In attempting to maintain or build a community group, it is necessary to have trained public health educators working with the leaders or potential leaders in the planning process and throughout the entire process of the projects undertaken within the community. It is important that these individuals maintain a low profile. They should be facilitators rather than leaders, consultants rather than decision-makers. They should help and guide the organization when guidance is needed or requested. In any community organization, it is important before any type of program is started, such as a housing program or rodent control program, that meetings of the community group be called to identify problems. Although the concern of the official agencies may be rodents or housing, the immediate concerns of the community may be police relations, recreational facilities, street repair, or better lighting. If the citizens receive an opportunity to voice their concerns and realize results, then they are more apt to work within their group to achieve the kinds of goals of interest to official agencies. The health educator should be involved in fact finding for and with the citizen groups, dissemination of information to the public, discussion of information among small groups, decision making on the basis of recommendations, testing decisions, establishing objectives, and helping to establish the program and the necessary follow-up and evaluation.

The community must identify its natural leadership and utilize this leadership in accomplishing ends established by the community and the official agencies. The community can identify those individuals who are most influential or powerful. These include church and business leaders, and professional, financial, governmental, educational, or labor leaders. These individuals can be utilized by the community in top coordinating positions. Other types of individuals, who actually get the work done, are the individuals to whom the community members turn when specific needs or problems arise. If these individuals exist in a community, they will be readily identified by other community members. When several individuals identify one person, a natural community leader has been found. This individual is probably most effective in helping get a program across, because he or she lives, works, and plays within the community, and is respected by its members. This individual can help determine the best approach to be used in establishing and implementing a program. It is wise to recognize and utilize such a person whenever and wherever possible.

INDOOR AIR POLLUTION CONTROLS

Residential Wood Burning

The impurities or pollutants of residential woodburning can be reduced by modifying the woodburning stoves or fireplaces for greater efficiency; by providing adequate air for proper burning; by using dry wood; by using smaller pieces of wood; by avoiding burning of any rubbish, garbage, plastic, or colored news-papers; and by cleaning the furnace, chimney, and fireplace regularly.

Passive Smoking

Passive smoking can be prevented or controlled by adequate sources of natural or forced ventilation. Further control will occur if there is a physical separation of nonsmokers from smokers. Many public laws today require that smoking take place in only certain very limited designated areas.

Carbon Monoxide

Carbon monoxide poisoning can be prevented in three specific ways:

1. Through proper ventilation, which will provide adequate air for burning, if burning is taking place, or will dilute the carbon monoxide that is present in the air.
2. Using proper equipment and methods to exhaust flue gases from the environment.
3. Physically separating all heating units and combustion appliances from people. Oil and gas heating units may be placed outside the house or vented to the outside.

Radon Gas

Radon gas can be controlled through natural and forced ventilation. This is done by ventilating the lowest level of the house. Either fans are used to force outdoor air into the home or ventilation grills are put into areas to allow outside air to flow in by itself. A major disadvantage of this is the increased cost of heating and cooling. Another technique used to control radon gas is to seal the cracks and openings through which the gas passes from the soil into the house. This would mean to seal all openings around utility pipes; the tops of concrete block walls, chimneys, joints; and any other cracks or openings. It would be impossible to find every crack or dent. Another technique is to heat recovery ventilation. In this case, replace the radon-laden air with outdoor air. The indoor air is warmed or cooled by the radon-laden air exhaust. This procedure is similar to convection where heat energy flows from one gas to another gas. There is still a loss of indoor heating and cooling energy. Another technique is through drain-tile suction. A drain title is installed completely around the house and a fan is used to draw the radon from the drain tile. A subslab suction system may be used to reduce the radon from the slab. There are also techniques utilized in block-wall ventilation and the preven-tion of house depressurization. In house depressurization, it would be necessary to consider discontinuing the use of wood stoves and fireplaces, since they lower the air pressure in a house by consuming air and exhausting it to the outside. One other technique would to be have more pressure in the house by actually pressur-izing the house and using fans blowing upstairs air into the basement.

Formaldehyde

There are seven different approaches that may be used in trying to alleviate the formaldehyde problem. They are as follows:

1. Use of interim measures, such as alternative housing or temporary housing, until the formaldehyde problem can be resolved. This is especially important for families with infants or young children, who are very sensitive to the irritating symptoms caused by formaldehyde. If the formaldehyde levels reach above 0.01 ppm, this measure should be considered.
2. The control of the indoor temperature and relative humidity can help significantly reduce formaldehyde levels. The recommended temperature levels for maximum formaldehyde reduction include lowering of the household temperature to 68°F during the heating season in northern states and the use of an air conditioner to maintain a constant indoor temperature of 70–72°F in the southern and southwestern states. The air conditioner also helps to dehumidify the air.
3. Source removal is the most effective control measure for formaldehyde. This would mean the removal of particle board, subflooring, and urea-formaldehyde foam insulation. This is very difficult and very costly. Other sources of formaldehyde emission include paneling, furniture, and cabinets. The removal of these sources are less costly and less difficult and can help reduce the amount of formaldehyde in the air.
4. Source treatment is an alternative to source removal but is generally less effective. This method consists of applying a coating material to the sources that emit formaldehyde. The coating material acts as a barrier and prevents formaldehyde release and also prevents moisture from entering the material, which helps promote the release of formaldehyde. Source treatment can be applied to unfinished wood products, particle board subflooring, plywood paneling, particle board shelving, cabinets and countertop undersurfaces. The two primary coating materials used are polyurethane and a nitrocellulose-based varnish.
5. Ammonia fumigation, if done properly, can cause a reduction of 50–60% of the formaldehyde. With this approach the whole interior of the house is enclosed and exposed to ammonia gas over a 12 hr period at an indoor temperature of 80°F. It is very essential that trained individuals do this, since ammonia is very toxic. At present this is the only method that is practical to reduce formaldehyde in environments that contain high concentrations. All surfaces must be vacuumed very carefully, since the ammonia combines with the formaldehyde to form hexamethylene tetramine, which is a fine dust.
6. Pure air purification systems remove formaldehyde from the air by absorption or adsorption. The effectiveness of the system is based on the chemical, the system design, the capacity to move air, and the levels of formaldehyde present. For adsorption, activated charcoal is relatively inefficient in reducing formaldehyde levels.
7. Forced ventilation systems will push fresh air into the contaminated building and therefore reduce or dilute the level of formaldehyde present. The amount of reduction is based on the amount of air flow.

Other Volatile Organic Compounds

Other volatile organic compounds used in a house, such as solvents, paints, polishes, household cleaners, etc., should be only utilized in areas where there is plenty of ventilation. The amount of material used should be strictly limited to that which is recommended on the container. Volatile organic compounds are hazardous. They should be treated as such.

Asbestos

Two techniques necessary for control of asbestos include

1. suppression
2. removal

Suppression is the best way to avoid dust, since the asbestos dust is not created. Removal is a complicated task that must be carried out by individuals who are highly trained. The EPA and OSHA have specific guidelines for the removal of asbestos from buildings. In any case, the asbestos materials are prepared for removal. They are wetted with a water and surfactant mixture sprayed in a fine mist, allowing time between sprayings for complete penetration of the material. Once the material is totally wetted, it is then removed from the building and placed in thick plastic bags and sealed in such a way as to make them leaktight. They are then doubled bagged and placed in plastic-lined cardboard containers or plastic lined metal containers and removed to the disposal site.

COMMUNITY PROGRAM AND INJURY CONTROL

Injury prevention is based on three general approaches. They are

1. Persuade the individuals at risk or injury to alter their behavior. An example of this would be the installation of smoke detectors.
2. Require that there be an individual behavioral change by law or administrative rule. An example of this would be a law that smoke detectors must be installed in houses or apartments.
3. Provide automatic protection by changing the product or by some form of design correction. An example of this would be the installation of built-in sprinkler systems.

The Philadelphia Injury Prevention Program is called the *Safe Block Intervention Program*. It is directed at reducing the occurrence, severity, and consequences of injuries in an economically disadvantaged community by creating safe blocks. Its goals are to conduct home modification programs, inspections of all homes in the target area, and education of the residents about injury prevention methods. The safe block concept includes the use of a public health team who inspects residences for problems and then has the environmental "modifiers" to help implement the prevention program. After the initial visit, a second visit will is made about 6 months later to determine whether the modifications remain intact, whether recommendations have been followed, and what the results are in injury control at this point. Obviously, data has to be gathered initially on the level of injuries, types of injuries, etc., or prior to the intervention program.

An intervention program consists of modifications of the home in the following ways:

1. Place stickers for regional poison control centers and 911 on telephone.
2. Provide fire extinguishers for the kitchen.
3. Provide smoke detectors for homes without the detectors. It may also be necessary to inspect existing smoke detectors and replace batteries.
4. Provide bathtub nonslip strips.
5. Provide ipecac and instructions for safe use if a poisoning occurs.
6. Reduce the temperature of the hot-water heater.
7. Provide light bulbs of higher intensity if there is inadequate lighting on the stairs.
8. Rat-proof the homes.
9. Install cabinet locks if young children are present.
10. Store poisons, medicines, and cleaning supplies out of the reach of children.
11. Install staircase gates at the top of the stairs and the bottom if children are present.
12. Inspect staircase hazards.
13. Inspect wiring or other hazards within the home.
14. Check space heaters and whether or not they are operational.
15. Inspect storage areas for newspapers, gasoline, and gasoline storage.
16. Provide education for the individuals in the home on how to avoid injuries from a variety of environmental sources.

This program is considered to be operating at a good level.

Role of the Federal Government in Housing

Congress stated in the Housing Act of 1949 that the general welfare and security of the nation and the health and living standards of its people require housing production and related community development to correct the serious housing shortage, to eliminate substandard and other inadequate housing through the clearing of slums, and to provide decent homes and suitable living environments for all American families. To obtain this national housing objective, Congress provided for private enterprise to carry out the major portion of the program; government assistance to private enterprise to carry out the major portion of the program; government assistance to private enterprise to be given where needed; establishment of positive local programs; government assistance to eliminate substandard and other inadequate housing through clearance of slums and other related areas; government assistance for decent, safe, and sanitary farm dwellings. Federal government departments were given the responsibility of facilitating the steps leading to this national housing objective. They were to do so by encouraging and assisting in the production of housing of sound standards of design, construction, livability, and size; reducing the cost of housing without sacrificing standards; using new designs, materials, techniques, and methods of residential construction; developing well-planned integrated residential neighborhoods; and stabilizing the housing industry.

The 1965 Department of Housing and Urban Development Act stated that, as a matter of the general welfare and security of the nation and the health and living standards of the people, the national purpose was to have a sound development of the nation's communities and metropolitan areas. To carry out this goal and in recognition of the importance of housing and urban development in the national life, Congress established an executive department to handle the problems of housing and urban development.

The 1968 Housing and Urban Development Act reaffirmed that national goal set forth in 1949 of a decent home and suitable living environment for every American family. Congress recognized that the goal had not been fully realized and that there were many suffering low-income families. They declared that special provisions be made to assist families with incomes so low that they could not decently house themselves. In August of 1968, under Title Sixteen, Housing Goals and Annual Housing Report, Congress declared that the supply of housing was not moving ahead rapidly enough to meet the national housing goal of 1949. It stated that there was a need for construction or rehabilitation of 26 million housing units, six million of which were for low- and moderate-income families. It stipulated that the President had to set forth a plan to be carried out for a period of 10 years from June 30, 1968 to June 30, 1978 to eliminate all substandard housing by the latter date. The Department of Housing and Urban Development Act, Public Law 89-174, granted the Department of Housing and Urban Development all of the functions, powers, and duties of the Housing and Home Finance Agency, the Federal Housing Administration, and the Public Housing Administration. In addition, the National Mortgage Association and other departments relating to housing were placed under the control of the Department of Housing and Urban Development. This department was to have vast powers to improve housing in the United States.

The Housing and Community Development Act of 1974 stipulated that there should be an allocation of housing funds to provide for local housing assistance plans. The Department of Housing and Urban Development, which received an extremely large budget from Congress, was responsible for the following areas: (1) the Council for Urban Affairs; (2) environmental and consumer protection; (3) rent supplements; (4) disaster assistance; (5) establishment of advisory committees; (6) civil defense, as it related to vulnerability to attacks; (7) handicapped individuals and accessibility to buildings; (8) assignment of emergency preparedness functions; (9) the Federal Council for Science and Technology; (10) coordination of federal urban programs; (11) the National Institute of Building Sciences; (12) housing renovation and modernization under Title 1 of the revised July 31, 1975 basic laws and authorities on housing and community development; (13)

mortgage insurance, Title 2; (14) war housing insurance, Title 6; (15) insurance for investment in rental housing Title 7; (16) armed services housing, Title 8; (17) national defense housing, Title 9; (18) mortgage insurance for land development, Title 10; (19) mortgage insurance for group practice facilities, Title 11; (20) FHA and VA interest rates; (21) the Rehabilitation Act of 1973; (22) closing of military bases and mortgage defaults; (23) special assistance programs such as emergency homeowner relief, housing for the elderly, college housing rehabilitation loans, rent supplements, urban homesteading, training and technical assistance; (24) community development assistance programs such as block grants, urban renewal, public works planning, public facilities loans, public facilities grants, model cities, historic preservation, Lead-Based Paint Poisoning Prevention Act; and (25) rural and other nonhousing and urban development community development programs, such as the Rural Development Act, Headstart, regional action planning commissions, and Appalachian regional development.

In 1974 the Emergency Home Purchase Assistant Act was passed. The Emergency Housing Act of 1975 included Emergency Home Owner's Mortgage Relief Title I. In 1976, the Housing Authorization Act was passed by Congress.

In 1977, the Housing and Community Development Act was passed by Congress. It included special titles on Community Development; Housing Assistance and Related Programs; Federal Housing Administration; Mortgage Insurance and Related Programs; Lending Powers of Federal Savings and Loan Associations and Secondary Market Authorities; World Housing; National Urban Policy; Community Reinvestment; and miscellaneous provisions.

In 1978, 1979, 1980, 1981, 1983, 1984, and 1987, additional housing laws were passed. However, in the 1980s, in reality, many housing programs were reduced in scope.

Programs of Housing and Urban Development 1988–1989

The Housing and Urban Development Department (HUD) is responsible for a huge multitude of housing programs to help the citizens of the United States. There are some 94 specialized programs handled in the Housing and Urban Development Department. They range from housing for the poor and the homeless, to the elderly, to nursing homes and hospitals, to public housing developments, to Indian housing. There are policy, development, research, and mortgage guarantee programs. The programs are extensive and extremely complex.

As can be seen, the federal government has made an enormous commitment to better housing for the American public. The ultimate goal has not yet been achieved. Considerable additional effort and money will be needed. But beyond this, considerable research is required in the area of attitude change and human behavioral patterns toward the housing environment. The best housing may be destroyed readily unless there is a thorough understanding of the effect of environment on human behavior. It is strongly urged that environmental health practitioners, other public health officials, citizens bodies, and legislators work toward obtaining proper funds for behavioral analysis and training. It will then be necessary to institute the training that will maintain the new environment in a proper manner for the individuals who will be living within it.

Emergency Housing

Emergency housing consists of the use of trailers provided by the Department of Housing and Urban Development, gymnasiums, churches, or any other available buildings for use when normal housing has been destroyed, made unsafe, or has been inundated by floods. Emergency housing should be closely supervised by environmental health and other public health authorities to prevent the outbreak of contagious or infectious diseases. Adequate water, food, clothing, and blankets must be provided. Generally, the American Red Cross is most helpful in this type of situation.

ROLE OF THE FEDERAL GOVERNMENT IN INDOOR AIR POLLUTION

The EPA has been very concerned with the problems of indoor air pollution. They are currently operating under a variety of environmental laws, including the Toxic Substances Control Act; the Federal Insecticide, Fungicide, and Rodenticide Act; the Safe Drinking Water Act; the Resource Conservation and Recovery Act; the Asbestos in School, Hazard Abatement Act of 1986; and the Uranium Mill Tailings Radiation Control Act. The EPA was given specific directions in establishing an indoor air quality program in the 1986 Superfund Amendments and Reauthorization Act.

Since 1982, the EPA has conducted research programs on indoor air quality. The research has been directed at increasing the understanding of personal exposure, emissions, health effects, and mitigation techniques. The Radon Gas and Indoor Air Quality Act of 1986 was directed at having EPA conduct research, implement a public information and technical assistance program, and coordinate federal activities on indoor air quality.

The EPA has taken regulatory action on asbestos, VOCs in drinking water, and certain pesticides. They require schools to inspect for asbestos, to prepare management plans, and to take action when the schools find friable (easily crumbled) asbestos. The EPA has issued maximum contamination levels for the VOCs in water supplies that serve more than 25 people. They have also acted to control indoor exposure to pesticides, as well as requiring childproof packaging for certain pesticides. They prohibit indoor applications of the wood preservatives pentachlorophenol and creosote.

Environmental Lead Hazard Abatement

Laws

The 1971 Lead-Based Paint Poisoning Prevention Act coordinated the efforts of health and housing people in dealing with the problem of lead-based paints in houses. It provided for the environmental management and control, as well as prevention, of lead-based paint illness. It also provided for screening and medical management of children who had ingested lead. In 1988 the Lead Contamination Control Act required the following:

1. The identification of water coolers that are not leadfree.
2. The repair or removal of water coolers with lead-lined tanks.
3. A ban on the manufacture and sale of water coolers that are not leadfree.
4. The identification and resolution of lead problems in schools' drinking water.
5. The authorization of additional funds for lead screening programs for children.

The reason for the enactment of this law is because of the very dangerous nature of the ingestion of lead from water for children. The harmful health effects include serious damage of the brain and central nervous system, kidney, and liver. Obviously inhalation of lead dust is also a serious problem.

Abatement Procedures

Abatement procedures consist of emergency repair and permanent repairs. Emergency repairs consist of the following:

1. scraping of all peeling, chipping, or flaking lead-based lead surfaces
2. sanding with a rough grade of sandpaper all lead-based paint surfaces
3. applying inexpensive contact paper to lead-based paint surfaced with wallboard, plywood, etc.

Permanent repair methods include the removal of all lead in paint, scraping and sanding, the use of liquid paint removers, the use of heat to remove leaded paint, and covering all areas where lead may be on the walls. Lead in dust may be removed by the use of proper dustless cleaning techniques. Lead in water may be removed by altering the plumbing system and removing the lead.

A Childhood Lead-Prevention Program

The City of Philadelphia has developed a childhood lead prevention program that includes screening of a large number of children in high-risk populations; a referral system that ensures a comprehensive diagnostic evaluation of every child with a positive screening test; a program that ensures identification and elimination of the child's lead exposure source and a system that monitors the adequacy of the treatment of the child. Children from 6 months to 5 years of age are screened for undue lead absorption and or lead toxicity. These children come from high-risk areas. Medical management includes evaluating the medical aspects of the program; evaluation of the timeliness and appropriateness of referrals; evaluation of case records; consultation with physicians; and conduction of educational programs. All individuals in high-risk areas are followed up after the initial determination has been made concerning lead poisoning.

Environmental management is carried out through:

1. Investigations to determine if lead-based paint is present or other sources of lead are present.
2. Abatement through the removal of lead hazards, especially paint. Abatement is difficult because of the total number of houses that have to be taken care of and the cost of doing this. Cost estimates for removing lead-based paint for an average three bedroom house range from $8000 to $15,000 per house.

An education program is carried on by health professionals to teach parents the problems of lead, to make the general community aware of lead problems, and to provide an outreach program to assist individuals, hospitals, and physicians, as well as day-care facilities, in understanding the risks of exposure to lead. The program that has been identified is operating in a successful manner. Other communities have also developed childhood lead programs.

SUMMARY

The indoor environment is a complex human environment. It encompasses facets of all other environmental problems. The indoor environment may include anything from an eight-family, three-story dwelling to a huge home with a swimming pool. The environment has an impact on the shelter, and the shelter has an impact on the environment. In addition, to fully understand the indoor environment, one must appreciate the problems of poverty, joblessness, welfare, social status, racial and ethnic origins, and the feeling in America that an individual's home is his or her castle. A house in good condition provides a satisfactory environment for psychological, physical, and mental well-being. The house and neighborhood in poor order may furnish the impetus for crime, poor education, poor health, and fires, and creates a society that is helpless in its dealings with the rest of the world. Further, indoor air pollution has caused additional stress and potential of actual disease. Injuries are an overwhelming concern.

There are many new theories on the relationship between housing and health based on nonspecific responses, in which a given cause may create a variety of health patterns and a given syndrome of physical or emotional maladjustment may stem from a variety of causes. It is necessary to consider the causes and to carry out an analysis incorporating all of them. New housing development, including new technology, materials, construction, and a variety of consumer products, have added complexities to the housing environment. In some areas this has provided for the greater good of the individual. In other situations, this has created additional problems, since the new product, for example, when broken must be serviced, and the service cost in many cases exceed the dollars that an individual has to take care of the problem.

Once again, it must be emphasized strongly that despite everything that the federal, state, and local governments, and various community groups, may do to improve the indoor environment, the basic changes that occur will be brought about by the resident. It is recommended that the resident obtain a copy of "The Effect of the Man-Made Environment on Health and Behavior," by L. E. Hinkle, Jr. and W. C. Loring, Center for Disease Control, Public Health Service, U.S. Department of Health, Education, and Welfare, Atlanta, GA, 30333 (Department of HEW, publication number (CDC) 77-8318, 1976). This publication may be obtained from the Government Printing Office, Washington, DC 20402, stock number 017-03-00110-8. In the future, conferences must be held between the specialists that examine the people-made environment and members of this environment for a fruitful exchange of ideas that we may move forward to the goal proclaimed by Congress in the Housing Act of 1949 — the ultimate goal of a decent home and suitable living environment for every American family.

RESEARCH NEEDS

It is necessary to thoroughly evaluate the existing types of planning used in community development and redevelopment. From this evaluation, measures should be formulated to determine the best approaches to housing development and redevelopment. Research techniques are needed to determine the manner in which improved housing can reduce social problems. Behavioral studies of individuals living in poor housing should be conducted to determine the underlying sociobehavioral disturbances that occur in these environments and also to determine why certain individual will move into good housing and lower the physical structure to the level of their previous housing. Behavioral studies are needed to determine the best means to educate individuals to improve their housing environment and also to learn to live within an improved housing environment. Studies are needed to determine the critical variables in establishing new communities.

Indoor air has become a serious concern within the home as well as within institutions. The indoor air may contain a variety of pollutants created within the property or outdoor pollutants that have entered the property through cracks and openings, especially in older structures. Carbon monoxide, particles, sulfur dioxide, and oxidants enter buildings. A variety of odors created by external pollution sources or indoor pollution sources contribute to the air problem. The indoor pollutants consist of particulate matter from activities such as cooking, tobacco smoke, and a vast variety of cleaning materials, synthetic floor coatings, polishes, cosmetics, pesticides, paints, vapors, glues, and other contaminants that come from the workshop or other parts of the property.

It is necessary to conduct research and demonstration projects on how to evaluate the techniques that are used to reduce radon levels in new and existing homes. It is also necessary to develop new and more effective techniques for determining if radon gas is present within the home. Additional research should be conducted on effective and inexpensive techniques for the removal of existing indoor air pollutants.

Surveys are needed to determine the patterns of polluting equipment and materials used in homes. Research priorities should then be developed to determine which areas need to be studied. Maximum exposure levels need to be determined and this information should be correlated with the individual's occupation to determine if he or she is receiving additional stress from the home environment. Studies are needed of the best techniques to control indoor concentration of air contaminants by improving air-cleaning devices and developing inexpensive techniques for the removal of air pollutants at low air pressures.

Injury control in the indoor environment is essential. Further research is necessary on the causes of accidents and injuries in the indoor environment, their controls, and how best to limit the degree of the injury. Research emphasis must be placed on an economical approach to altering the environmental conditions leading to accidents and injuries, as well as the best techniques for product modification to enhance safety. New public health education tools need to be developed and evaluated to determine if they are in fact reducing the levels of injuries in the indoor environment.

Chapter 8

INSTITUTIONAL ENVIRONMENT

BACKGROUND AND STATUS

The successful design and practice of an environmental health and safety program in an institutional setting is a comprehensive undertaking. It involves the coordination and integration of a broad range of people, both professional and nonprofessional, and their diverse and complimentary resources and skills. The institution is in effect a small community and therefore contains the numerous environmental, community health, social, emotional, and psychological problems attributed to or found in any small community. In addition, there are a considerable number of people congregating in a limited amount of space. The typical institution continues to grow in size and complexity. The budget of the institution and means of financing have become problems, since costs rise as new techniques and equipment are developed and as the inflationary economy continues to spiral. Without any technical changes, the dollar in 1967 now must be replaced by about $3.65 in 1990 to purchase the same goods and services. The actual cost of health care is much higher than inflation.

People congregate in institutions. This congregation comprises a shifting population that may use the facilities for parts of a day, the entire day, or for an entire 24-hr period. As in any other situation where groups of people gather, the potential for the spread of disease is increased. Institutions contribute to air and water pollution, and solid-waste problems. Further, each institution has its own unique set of occupational hazards and other types of safety hazards. Some problems in institutions are limited to the basic environmental health problems, including water, air, solid waste, food, and shelter. Others have many shops, or even the equivalent of comprehensive industries, and therefore face all the occupational hazards found in these settings, including chemical, microbiological, and radiological hazards. A further complication is the fact that many institutions receive federal and state financing.

Institutions include primarily hospitals, nursing homes, convalescent homes, old age homes, schools, and prisons. (For a detailed presentation of the institutional setting, the reader is referred to "Environmental Health and Safety," by this author, Pergamon Press, 1974, now out of print.)

Nursing and Convalescent Homes

Nursing homes exemplify in many ways a failure of public policy, public concern, and private enterprise. Although a good proportion of the patients are aged and have a variety of infirmities, including senility, there are still many patients with active minds who should be receiving the type of care to which they are entitled. It is much easier to medicate individuals to conform their behavior to the standards set by the nursing home than to allow them to maintain their dignity as human beings. Some nursing homes today are as bad as the pesthouses of the 18th century. Overcrowding, unsanitary conditions, and potential firetraps exist. Each year a serious fire occurs in which unfortunate older people die or are injured. The basic causes behind this dilemma are the social, health, and economic trends of the past 85 years.

The United States became a rapidly growing industrial nation during the early part of the century. People moved to the cities to find jobs and often to smaller houses with fewer rooms. They were unable to keep with them the grandparents or other family members who in the past had a secure place to live in their old age. Industrialization created more jobs that were filled by women and younger people. Older persons were frequently left out of the job market. If they became disabled, their opportunity to get a job was almost totally lacking. This situation with the aged survives today and grows worse, since long-term inflation has sharply reduced or totally eliminated the savings of a group of the aged, thereby creating additional problems.

Unfortunately, in the past, nursing homes did not evolve in an orderly manner but rather erupted, almost in the way an earthquake causes disruption within a community. In order to provide homes for the aged or sickly, many ill-equipped reconverted private houses were pressed into operation. Unfortunately, this contributed to the severity of the nursing-home problem. Nursing homes were set up in run-down residential areas, in poor and inadequate facilities, and in difficult areas for family members to reach, unless they had access to private transportation. The nursing home movement received further impetus from the Social Security Act of 1935. This act provided for the elimination of public poorhouses and public assistance for residents of these institutions. The residents went to private institutions, since these were available.

Today there are an estimated 20,000 nursing homes in the United States of which some 77% are proprietary or operating for profit. About 7% are government sponsored and about 16% are run by various nonprofit organizations, including religious groups. In some of these private nursing homes, individuals do not necessarily receive care commensurate with the amount of money that they pay. There are many fine proprietary nursing homes in the United States. However, there are others interested primarily in making profit and not in providing adequate care for the patients. Congressional hearings in New York City demonstrated that, despite federal regulations for skilled nursing-home facilities that were required by Medicaid and Medicare, innumerable abuses were found. Obviously, because the average age of the people in institutions is approximately 80 years old, they seldom challenge their treatment and care.

Nursing homes, in effect, are medical facilities that should have the same type of personnel and programs as hospitals, but to a lesser degree. Obviously such things as operating rooms and emergency rooms do not exist. However, nursing homes should provide good patient care, medical and nursing supervision, physical therapy, occupational therapy, dental services, eye services, necessary drugs, and effective humane care. Nursing homes should be able to handle emergencies to save the lives of patients. The facilities should meet the requirements of skilled nursing home facilities under Title 19 of the Social Security Act. These requirements are updated from time to time by the Department of Health and Human Services. For current requirements, the Department of Health and Human Services should be contacted.

Old Age Homes

Old age homes encompass every type of shelter from a simple house where a few elderly people live to extensive beautiful facilities where the individuals not only have shelter and food, but also adequate recreational facilities, medical care, and occupational and physical therapy. The old age home is an outgrowth of the same kinds of conditions that helped to create the nursing home.

Schools, Colleges, and Universities

Healthy school living requires considerable effort to provide good physical, emotional, and social well-being. The school or university administrator, along with the Board of Trustees and the various other administrative support and professional staffs, have as a goal the proper education of students at all levels. It is necessary to provide an environment that is wholesome and supportive of learning. This environment includes all of the problems, programs, and services found in other institutional environments, but with some exceptions. In grade schools, the young age of the students increases the likelihood of accidents. In high schools, the students not only face potential accidents from various play areas, but also from shops and laboratories. In colleges everything that applies to the high school continues to be a problem but is multiplied manyfold by the sheer numbers of students, by increasingly sophisticated and complex laboratory experiments and chemicals used, and by the fact that many of the students reside on the university grounds.

Prisons

Each correctional institution should provide a healthy place for a person to live. Since a healthy environment includes all of the aspects of a good institutional environment, and as custody means more than possession, the prisons of our country have a definite obligation to fulfill in taking care of prisoners. In the past there have been considerable problems in correctional institutions because of overcrowding, numerous existing health hazards, and a poor environment.

The conditions in all types of institutions continue to deteriorate. Specific types of concerns are discussed under the section on problems. To avoid repetition in this chapter, most of the material discussed will relate to the most complex type of institution, which is the hospital. In other institutions where problems, techniques, and so forth vary from the hospital setting, they too will be discussed.

SCIENTIFIC, TECHNOLOGICAL, AND GENERAL INFORMATION

Hospitals

The hospital today is more than a place where people come who are terminally ill. The use of intensive care units, cobalt therapy, artificial kidneys, organ transplants, inhalation therapy, and so forth has made the hospital significantly different from its predecessor. It is a continually and rapidly changing environment in which diagnostic and treatment facilities are improving, staffs are better trained, and costs are soaring. The functions of the hospital are to give care to the sick and injured, provide necessary health education, conduct realistic research, and promote community health.

Hospitals, generally, are either governmental, proprietary (run for profit), or voluntary nonprofit. The voluntary nonprofit hospital accounts for about 48% of the nation's hospital beds. These hospitals are usually brought into operation by community groups, churches, or other organizations. Governmental hospitals, which have about 40% of the hospital beds, are operated under federal, state, or local budgets. Proprietary hospitals, which contain about 12% of the hospital beds, may either be individually owned or owned by a partnership or corporation. Hospitals may also be listed as general, special, short term, or long term.

Organization

The Board of Trustees or Board of Directors is the highest authority within the hospital. This board determines the direction that the institution will take and establishes policies that are consistent with its goals. It usually represents a cross-section of businesspeople, educators, health leaders, and other distinguished individuals. They dedicate a significant part of the working time to the institution, usually without pay. Boards vary in size, although an adequate-size group probably would be a maximum of seven to nine individuals. The board forms committees and so forth. The board evaluates and approves the hospital budget and also gives guidance to the chief executive officer. It is responsible for providing a bridge between the administration, the medical staff, and the public.

The chief executive officer, who may have a variety of titles from director to administrator, is the day-to-day chief administrator of the hospital. His job involves managing, coordinating, and directing the activities performed by the employees. He presents the hospital's problems to the board, implements the board's decisions, and interprets the board's actions to the hospital community. It is essential that the chief executive officer be a fine administrator, keep in close touch with all parts of the institution, and have a good working relationship with the many chiefs and supervisors of the various parts of the institution.

The medical staff directs patient care. It is responsible for giving professional care to the sick and injured, and for making careful and proper diagnosis. It is also responsible for maintaining its own efficiency, governing itself, and choosing new members. It participates in educational programs and gives advice to the chief executive officer and the Board of Trustees. The medical staff is divided into a series of committees that oversee the various operations of the hospital, including medical records, the surgical department, pediatrics, radiology, and so forth.

For the hospital to operate properly, it is essential that the three major groupings — the Board of Trustees, medical staff, and chief executive officer — cooperate and work together efficiently. The other professionals, such as the nursing staff, environmental practitioner, and safety officer, generally work through the chief administrative officer and therefore are extensions of his administrative and supervisory control.

Electrical Facilities

Electrical energy is the major source of power on which almost every part of the hospital depends. Electrical supply must be adequate and always dependable. It is essential to have electrical supplies that switch on automatically when there is a loss of normal service. Electricity must be available on a 24-hr basis.

Heating, Air Conditioning, and Laminar Flow

A thermal environment should be maintained within the hospital to permit adequate heat loss but prevent excessive heat loss from the human body. At the same time, this environment must maintain an atmosphere that is reasonably free of chemical and bacterial impurities and odors, and one that is conducive to the best physiological and psychological comfort of the individual.

The function of the ventilation system, apart from providing adequate temperature and humidity control, is the control of the number of microorganisms in a given area. These microorganisms come from other areas into the target area, come from equipment and supplies, from outside air, from individuals within the immediate environment, from dirty bandages, other solid waste, blankets, or dirty linens. The function of mechanical ventilation in removing these organisms is carried out by literally blowing the organisms away from the areas that are to be kept free of contaminants. An example of this would be the wound or operating site on a patient. There have been several methods proposed to remove the air that may contain microorganisms from these critical areas. The first method directs a stream of air either horizontally or obliquely toward the operating table to create turbulent air flow directly over the patient. The second method introduces air at the ceiling in such a way that only mild turbulence is created by the supply air mixing with the room air, and then it is exhausted. This creates a constant dilution of the air. The third method is the same as the second, except that it produces minimal turbulence when mixing with the room air and is then removed by downward displacement of the air to ports located on the walls near the floor. The most effective technique for removing microorganisms appears to be the third system. The turbulent system, although it removes a considerable amount of contaminated air, also brings more contaminated air across the wound site.

Laminar flow of air provides the best total control of the air environment, which includes temperature, humidity, cleanliness of the air, and direction of flow. Laminar flow is simply the introduction of large volumes of clean air through a very large diffuser or perforated panel, which reduces the velocity of the incoming air, thereby preventing agitation and reintroduction of settled contaminants. It also decreases the contamination blown off of personnel and equipment. At the opposite side of the room, the air is removed through a perforated area of the same size as the inlet diffuser. This is a single-pass technique in which the air either comes from the wall or the ceiling, passes through ultra-high efficiency particulate air filters, moves uniformly across the room and through the outlet to return ducts, which then go through the final filter and back through the room again. The amount of air flowing depends on the size and shape of the room and the height of the ceiling.

Heat and proper air control of temperature is important. Heat balance in the human body is affected by the rate of metabolic heat produced in the body, the rate of storage or change in body heat, the rate of heat loss or heat gain by convection, the rate of heat loss or heat gain by radiation, and the rate of heat loss by evaporation of sweat. It is essential that the ventilation, rate of flow of air, and the actual heating and cooling of the areas be done in such a way that the individual will be comfortable. Comfort is difficult to define, since it is not only based on the previously mentioned items, but also on the sex, age, and physical condition of the patient, and the sex, age, and type of activity of the worker. Generally speaking, within the hospital a temperature of 75°F would be considered a comfortable temperature for most patients. The room temperature, however, may have to be adjusted to the individual.

Housekeeping

The function of the housekeeping service is to provide and maintain a clean, safe, and orderly environment that helps prevent disease and accidents, and promotes health. The individual, whether a hospital employee, visitor, or patient, is affected by aesthetic surroundings. It is important that all areas appear clean and in fact are clean to avoid hazards. Industry practices proper cleaning by using adequate procedures and techniques, proper task-oriented detergents, disinfectants, and proper equipment. It is important for institutions to follow this example.

Maintenance

The maintenance department is responsible for the maintenance and operation of the physical plant of the hospital. The size of the department and its functions vary with the size and complexity of the hospital. Maintenance usually includes the operation of the power plant, electrical service, plumbing service, heating, and maintenance; maintenance personnel include refrigeration mechanics, welders, carpenters, brick masons, roofers, gardeners, locksmiths, plasterers, painters, and air-conditioning specialists. The maintenance department is also responsible for all keys, especially the special keys that fit the narcotics cabinets, medicine cabinets, psychiatric areas, and prison areas. The proper operation of the maintenance department and the proper maintenance and repair of all equipment and parts of the physical structure is an essential part of a good environmental health program.

Plumbing

Hospitals probably have more piping systems than any other type of building. The common piping systems include cold potable water, chilled and recirculated drinking water, distilled water, fluid suction systems, vacuum cleaning systems, oxygen, fire sprinkling and stand pipes, lawn irrigation, air-conditioning, refrigeration systems, recirculated cooling water, drainage systems, soil and waste systems, vent systems, stormwater systems, and building sewers.

Radioactive Materials

The use of radioactive isotopes in hospitals and other institutions has greatly increased in the last 35 years. Radioactive materials are found in dental clinics, nuclear clinics, pharmacies, diagnostic and therapeutic radiology laboratories, and patient rooms where radionuclides are used.

Water Supply

The water supply for hospitals and other institutions must be of high quality, physically, chemically, biologically, and radiologically. The potable water must meet the drinking water standards of the U.S. Environmental Protection Agency-1986 Water Quality Standards. Water must be provided in adequate quantities under proper pressure. It may then be further purified within the institution, if it is used in various processes. The water may be distilled, it may be turned into steam, or it may be sterilized. It is essential that the plumbing utilized in the distribution of water within the system be set up in such a way as to avoid any cross-connections or submerged inlets.

Emergency Medical Services

The initial emergency medical service program began in 1966 when the Highway Safety Act provided federal grants for this purpose. Its primary objective was to establish self-supporting systems at the state and local levels. Each state provided a commission to receive the federal highway safety funds and to utilize them by establishing standards for ambulance services, their equipment and personnel, and for institutions engaged in the training of ambulance personnel. The current laws, which were initially funded in 1975 and are currently being funded, also provide funds for the certification and training of emergency medical technicians. Any person who serves in an ambulance and who administers emergency care to patients must be certified as an emergency medical technician by the commission. (See Chapter 8, Volume II, for more information.)

Fire Safety Programs

Fire prevention, protection, and suppression are an integral part of an institution's health and safety program. An effective fire prevention program includes early detection, prompt extinguishment, and evacuation of people. To accomplish this, the institution must define its needs in conjunction with skilled fire prevention specialists. It also must assign responsibilities for each part of the program to appropriate staff members; establish practical regulations; establish uniform procedures for fire prevention, fire protection, and fire suppression; give emergency instructions; provide appropriate training courses and educational programs; have adequate fire design control; make necessary inspections and corrections; and carry out fire drills.

Hand Washing and Hospital Environmental Control

Nosocomial infections (hospital-acquired infections) occur in approximately 6% of all patients admitted to acute-care hospitals in the United States. The cost of these infections run into the billions of dollars. The CDC put out guidelines concerning hand washing and hospital environmental control in 1982 and then updated them in 1985. The guidelines contain six sections. They include hand washing; cleaning, disinfecting, and sterilizing patient care equipment; microbiological sampling; infectious waste; housekeeping; and laundry.

Direct contact is considered to be the primary means of transmission of nosocoimial infections. Hand washing is considered to be the most important means of preventing these infections. The hand washing materials are extremely important, since they aid in the cleaning of the hands and the removal of not only soil, but also microorganisms. The mechanical friction that is used in hand washing is also of considerable importance. The use of antimicrobial agents are of particular significance when personnel are dealing with newborns, between caring for patients in intensive care units, before caring for patients whose immune system has been depressed, after caring for patients who have nosocomial infection, before labor and delivery, before operation, and before any type of procedure where the body will be invaded in any way.

The cleaning, disinfecting, and sterilizing of patient care equipment varies with the three categories of equipment being used. These categories include equipment for critical patients, semi-critical patients, and noncritical patients. Disinfectants used in the health-care setting include alcohol (to be used on patients only), chlorine and chlorine compounds, formaldehyde, glutaraldehyde, hydrogen peroxide, iodophors, phenolics, and quaternary ammonium compounds. These disinfectants are not interchangeable, and therefore the user must be aware of how the disinfectant works in different settings. Sterilization, which is the complete elimination or destruction of all forms of microbial life, is accomplished either by means of steam under pressure or dry sterilization. Ethylene oxide, which has been used as a cold sterilization gas, is now listed under CERCIA as a potential carcinogen. However, it still may be used if proper control techniques are put in place.

Under the current CDC guidelines, the only microbiological sampling that is recommended is the placing of spore strips in sterilized materials to determine if sterilization has occurred. In addition, the water used to prepare the dialysis fluids and the dialysate needs to be checked on a monthly basis. However, microbiological sampling could still be an effective technique if used when there is a massive outbreak of nosocomial infections in an institution and there is no known cause for the outbreak. It might be worthwhile to take air samples in the area where patients are becoming infected.

The area of infectious waste will be discussed later. Typical nosocomial pathogens include *Pseudomonas aeruginosa*, *Enterobacter* sp., *Klebsiella* sp., *Group B streptococci*, and *Staphylococcus* sp.

Housekeeping is an essential area for the reduction of nosocomial infections. The cleaning of hospital floors and other horizontal surfaces helps remove microbial contamination on these surfaces. Appropriate cleaning is another measure used to prevent the microorganisms that have landed on the surfaces from being reintroduced into the air environment.

Hospital laundering will reduce significantly numbers of microbial pathogens. The laundry then may be sterilized if used as drapes or other necessary items in specialized situations, such as the emergency room, operating rooms, delivery rooms, and other areas where sterile sheeting is required.

Human Immunodeficiency Virus and Hepatitis B Virus

The Health Omnibus Programs Extension Act of 1988 requires that guidelines be developed for the prevention of transmission of human immunodeficiency virus (AIDS) and hepatitis B virus for health-care and public-safety workers. These guidelines are essential to reduce the risk in the workplace of becoming infected with the etiological agent for acquired immune deficiency syndrome and for determining the circumstances under which exposure to the etiological agent may occur. The mode of transmission for hepatitis B virus (HBV) are similar to those of human immunodeficiency virus (HIV). The potential for HBV transmission in the occupational setting is greater than that for HIV. There is a larger body of experience relating to the control of the transmission of HBV in the workplace than the transmission of HIV. The general practices used to prevent the transmission of HBV will also minimize the risk of transmission of HIV. Other blood-borne transmissions of disease will be interrupted by adhering to the precautions used for protecting the workers against HBV and HIV.

The workers involved not only include hospital personnel, but also fire-service personnel, emergency medical technicians, paramedics, and law-enforcement and correctional-facility personnel.

HBV and HIV transmission in occupational settings occur only by percutaneous inoculation or contact with an open wound, nonintact (chapped, abraded, weeping, or dermatitic) skin, or mucous membranes, blood-contaminated body fluids, or the concentrated virus. The probability of infection by the health-care worker is related to the potential for the organism being present in the patient's blood, etc. The risk of infection changes rapidly with the type of patient who is being dealt with. The normal risk of HBV infection from the general population varies from 1 to 3 per 1000, whereas in the drug or homosexual population, the risk ranges from 60 to 300 per 1000 population.

In 1987 the CDC estimated the number of HBV infections in the United States to be 300,000 per year with approximately 25% of infected people developing acute hepatitis. Approximately 6–10% will become HBV carriers and will be at risk of developing chronic active hepatitis, cirrhosis, and coronary liver cancer, and spread the infections to others. The CDC has estimated that 12,000 health-care workers whose jobs entail exposure to blood will become infected with HBV each year and that 500–600 of them will be hospitalized with the infection and that 700–1200 of those infected will become HBV carriers. Of the infected workers, approximately 250 will die. The degree of risk not only to the health-care workers, but also to the emergency medical and public-safety workers varies with the amount of exposure to blood that these individuals come in contact with.

As of July 31, 1988, 1200 health-care workers have been tested for HIV antibody by the CDC because these people were exposed to the HIV infected patients and had received either a needle stick or splashes of blood to the skin or mucous membranes from these patients. Under the 1985 CDC strategy of "universal blood and body fluid precautions" all patients could be assumed to be infectious for HIV and other blood-borne pathogens.

PROBLEMS IN INSTITUTIONS

Hospitals

Since the institution is in effect a small community, all environmental health problems and necessary resulting services that would be found in any small community are found within the institution. These include air, food, housekeeping, insect and rodent control, laundry, lighting, noise, solid waste, water, and liquid waste. In addition, because of the uniqueness of the institution, other areas may develop specific environmental problems. These areas include the laboratories, the surgical suites, and emergency rooms. Unsafe and/or careless practice of environmental protection in the institution not only endangers individuals within the institution, but may also endanger the community at large.

Air

A significant difference between the normal population and the patient population is the increased susceptibility of the patient to infection and also the increased risks of spread of infection from the infected patient to others. This susceptibility to infection varies tremendously with the patient, the disease process, the chemical and radiological therapies, and the other techniques of treatment used. Pathogenic microorganisms, chemicals, and other contaminants are transmitted through the air. The transmission may be from humans or operating techniques to susceptible individuals. The organisms are found on clothing, bedding, and floors, and may readily become airborne. The rate at which the particles settle out is determined by their size. It has been observed that particles released from basements of buildings travel quickly trough the ventilation ducts to the upper stories of the buildings. This would indicate that microbiological or chemical contamination, which may occur anywhere, could easily be distributed to any or possibly all parts of the institution. The types of air contaminants include dust, particles that range in size from 0.1 to about 100 μ; fumes, solid particles formed by condensation of vapors from metals ranging in size from 0.001 to 2 μ; smoke, solid or liquid particulates produced by the incomplete combustion of fuel ranging in size from 0.1 to 30 μ; mists, small liquid droplets produced by atomizing, boiling, sneezing, and so forth, ranging in size from 0.5 to about 10 μ; vapors, gaseous phase of liquids such as gasoline; gases, produced by chemicals found in the institution, ranging in size from 0.0001 to 0.001 μ; odors, caused by a variety of processes or illnesses, ranging from 0.00001 μ to a much larger concentration; and microbial contaminants, ranging in size from 0.005 to 40 μ.

All of these contaminants are produced by some form of personnel activity or operation of a variety of pieces of equipment. Further, these contaminants may readily be caused by carriers of specific organisms or shedders or dispersers of organisms. Some examples of patient equipment or other hospital equipment as the source of air contamination include microbial contaminants found in humidifier water, microbes found on cooling coils and other parts of the air-conditioning and ventilation system (*Pseudomonas aeruginosa*), nebulizing equipment, and inhalation equipment. Other air contaminants include skin scales, textile fibers, and dust particles.

The use of lasers may cause a potential hazard within the institutional environment. The laser vaporizes or fragments harmful viable or nonviable substances and may cause an airborne health hazard or a direct hazard to tissues. The principle indirect hazards and sources of air contamination are solvents used in the cleaning of equipment and gases, and fluids associated with the use of the equipment. Some of the toxic substances used in the laser facilities include benzene, carbon disulfide, carbon tetrachloride, cyclohexane, nitrobenzene, pyridine, toluene, xylene, mercury, and chlorine. The direct sources of air contamination vary with the type of laser. The carbon dioxide laser has a high potential for generating air contamination. The beam can readily melt, vaporize, and burn a wide variety of materials, depending on the power density and the ability of the material to absorb power. The hazards are very similar to those produced in welding. Some of these hazards include the production of nitrogen oxide and ozone.

Some additional potential environmental health air hazards involve the degree of filtration of exhaust air from vacuum cleaners; aerosols produced from solid-waste compactors; aerosols produced from bedpan flusher units; aerosols from centrifuges, blenders, and other laboratory devices; aerosols from hydrotherapy tanks; air flow around doorways of isolation and reverse isolation rooms; airborne contamination from housekeeping materials, including polishes, solvents, detergents, disinfectants, and pesticides; and airborne concentration from rehabilitation and occupational therapy areas, including glues, solvents, and materials such as fiberglass, dust from grinders, shavers, and sanders.

Equipment Design and Construction

Equipment can readily be contaminated by patients, professional staff, or hospital employees. The equipment may then become a source of infection for other individuals. Additional problems occur because of improper design and/or construction, making the equipment difficult to clean, disinfect, and sterilize. Faulty design of equipment may also contribute to occupational hazards or cause hazards to the patients or visitors. Because of the great variety of equipment found in an institutional setting, it is impossible to discuss all of the potential problems.

Therefore, only a few pieces of equipment that may contribute to the spread of infection or cause other hazardous conditions will be discussed.

Hemodialysis machines have been associated with the spread of hepatitis to patients and the potential spread of hepatitis to staff and employees. Bronchoscopes, cystoscopes, and endoscopes are pieces of equipment that are not easily cleaned or decontaminated. Many types of electronic equipment also present a cleaning and decontamination problem.

Other problems related to electronic monitoring devices include the effects of the devices on patients and staff when they have been exposed for long periods of time. Plastic furniture and drapes are rapidly being introduced into the hospital environment. There is a poor understanding of what might occur if the drapes and furniture would start to smolder. Fumes from smoldering equipment mixed with gases utilized within the patients' rooms could present a toxic hazard to patients, staff, employees, and visitors. Hydrotherapy tanks are a constant source of potential infection problems. The tanks are difficult to clean because of the design and construction of the turbine pump and the agitator. The so-called bedpan sterilizer is actually a bedpan flusher unit that is inefficient, tends to become inoperative, is hard to clean, and becomes a serious potential source of infection. Bedpans, after they go through the flusher unit, may be more contaminated than before the cleaning operation.

Solid-waste compactors push enormous quantities of contaminated waste together. They are difficult to clean, often leak, and become a source of insect and rodent contamination, and air contamination. Fiber optic instruments are built in such a way that they are not readily sterilized. Inhalation therapy equipment and nebulizing, anesthesia, and infant resuscitation equipment have been known to cause pulmonary nectrotizing infections.

Food

The dietary department is similar to a large commercial food operation. In addition, it has the problems of preparing diets for sick and debilitated individuals, and for preventing massive outbreaks of food-borne disease. (An in-depth discussion on food appears in the chapter on food protection.) This section is concerned with some of the problems that can be found in food operations within hospitals and other institutions. They are as follows: (1) inadequate supply of light, contributing to worker fatigue, inadequate removal of soil and wastes, and increases in insects and rodents; (2) poor housekeeping resulting from inadequate management of the large food service area that must operate on a least a 12-hr minimum basis and in some cases up to 24-hr a day; (3) accumulation of food particles, dirt, dust, and other soil of food equipment, such as cookers, stoves, ovens, deep-fat fryers, slicers, can openers, conveyors, sinks, cabinets, shelves, storage areas, pots, and pans; (4) coating of surfaces with greasy residues; (5) dirty and corroded refrigerators with worn gaskets on the doors; (6) poor ventilation in dishwashing rooms; (7) corroded and clogged dishwashers and spray arms within the dishwashers; (8) inadequate steam pressure for washing, rinsing, and cleaning; (9) potential chemical food poisoning due to inadequate and improper storage of chemicals, insecticides, cleaning materials, and other poisonous materials; (10) inadequate and improperly placed hand-washing facilities; (11) food handlers who are working although they have or have recently had diarrhea, vomiting, severe upper respiratory illness, or skin infections; (12) carrying over of leftovers from one meal to another; (13) improper thawing of turkeys, meats, and other foods outside of refrigeration; (14) improper cleaning of equipment, such as grinder heads and tenderizer blades; (15) reusing disposable water carafes and glasses; and (16) improper care and cleanliness of food carts that have refrigerator and heating units on them.

Other problems revolve around the preparation, storage, and use of infant formulas and tube feeding. Great care should be given to adequate cleaning, evaluation, and supervision of all food areas to prevent an outbreak of food-borne disease. Unfortunately, too many times in the past, thousands of patients have become ill from avoidable outbreaks of food-borne disease.

Hazardous Chemicals

There are many thousands of raw materials used in industry today. The toxicity of these substances, the ways in which they are used, the chemicals to which they may be bound, the degree of susceptibility of individuals, and the methods of control are all intimately related to the possibility of toxic responses. The National Institute of Occupational Safety and Health lists over 8000 substances as toxic. It has been noted that over 630 different chemicals are used in hospitals. Of this number, about 300 are of unknown toxicity. About 30 are safe and about 300 have hazardous properties. These chemicals can either be toxic or carcinogenic, flammable or explosive. They include acids, alkalis, ammonia, assorted oxidizers, organic and inorganic peroxides, oxides, permanganates, nitrates, nitrites, toluene, ether, lithium, potassium, sodium, and more.

Areas of the hospital that have the greatest usage of chemicals and the greatest potential for hazardous situations include the departments of anatomical pathology, clinical pathology, renal dialysis, and various storage ares. The laboratories contain many solvents, fixants, and volatile hydrocarbons. In the patient rooms chemicals are used for cleaning, degreasing, and decontamination. These many affect the patients and the staff and personnel. The housekeeping department uses pesticides, polishes, solvents, and detergent disinfectants. The physical plant stores gasoline, paints, paint thinners, and a variety of materials used in the repair process.

The potential hazards of chemicals are enormous. It is necessary to determine the kinds of chemical mixtures that are in the air. Will the mixtures be explosive, flammable, toxic, or carcinogenic? There is, for example, a very serious potential problem related to anesthetic gases. Anesthesiologists have an unusually high incidence of headaches, fatigue, and irritability. Further, the incidence of spontaneous abortion and a high incidence of abnormal pregnancies have occurred among females in the operating room. It has also been postulated that an exposed male can transmit the defect to his unexposed wife. There has been an increase in the incidence of embryotoxicity, mutagenesis, carcinogenesis, and liver disease among females working in the operating room. There is a lack of knowledge as to whether these anesthetic gases will cause problems in patients receiving repeated doses of anesthesia who are receiving immunosuppression therapy or coronary care.

Housekeeping

Housekeeping procedures are related to the direct or indirect transfer of pathogenic organisms from person to person. Many special problems arise within the institution. The housekeeping activities must be adapted and scheduled in such a way as to avoid the activities of medical staff, nurses, and other auxiliary personnel. Some rooms are overcrowded because of the use of essential equipment. Other rooms contain food spillage, dripping urine, broken glass, silver nitrate, chewing gum, infected or contaminated bandages, linens, and other materials. The housekeeping attendant, his equipment, and any other material that he may utilize can become a vehicle for the transmission of disease.

Since the institution is a vast conglomeration of buildings, floors, corridors, windows, and assorted rooms used for a variety of functions. the housekeeping task becomes even more difficult. The equipment and techniques utilized in bathrooms certainly must be modified for use in kitchens. X-ray areas, storehouses, laundries, garages, outside grounds, intensive-care units, cardiac-care units, and surgical units are all independent and have unique problems.

Insect and Rodent Control

Insects and rodents cannot be tolerated in an institutional setting. Not only are conditions favorable for their entrance and multiplication, but also their normal propensity for transmitting disease is sharply increased by a substantial quantity of virulent organisms present in the environment. Insect and rodent control is an important part of the overall program of preventive medicine within the institution. These pests, which include German and American roaches, bedbugs, ants, flies, rats, mice, and fleas, may be found in food service areas, housekeeping areas, janitorial closets, nurses stations, clinical areas, patient areas, animal laboratories, conveying equipment and shafts, morgues and autopsy rooms, and the laundries. A complete description of the life cycles, problems, and controls of these insects and rodents appears in other chapters of this book.

Structural problems are of concern, since wherever the maintenance department is involved in correcting plumbing or other types of repairs, holes may be left open in walls, floors, and ceilings. These openings provide a natural pathway for the invasion of rats, mice, or insects.

Food areas are a primary harborage for insects and rodents. These areas include preparation, serving, storage, dishwashing, and equipment. Food residues, crumbs, and other food materials may accumulate under sinks, around walls, on the bottoms of shelves, under tables, under equipment, and behind and under all other types of standing equipment within the preparation and serving area. In storage areas, there are frequently accumulations of spilled flour, sugar, beans, and other foods. Storage areas provide an excellent source of food for insects and rodents, since these areas are generally warm and have sufficiently enclosed spaces for harborage. Insects and rodents breed in the residues from the evening meal, visiting-time snacking, encrusted step cans, uncovered wastebaskets, and baby formulas or foods dumped into sinks. Visitors frequently bring food into the institution and at times may even bring roaches in with their clothes or in food. The nurses stations, clinical areas, informal staff rooms, and other places where people congregate tend to be collection stations for food and become primary sources of insect and rodent problems. Animal laboratories are of special concern, since animal food is always available for the feeding of experimental animals. Roaches and rodents have a constant source of food, water, and harborage in these areas. Conveyors, tray carts, laundry carts, elevator shafts, plumbing, and electrical shafts are also sources of harborage and paths of travel for insects and rodents. Insects and rodents find powerful attractants in the laundry, since soiled and blood-encrusted linens are a food supply for insects and rodents. The solid-waste disposal areas are frequently nothing more than open dumps and become sources of insect and rodent problems

Laboratories

In addition to the fire hazards occurring in laboratories there are also biohazards. Biohazards are more prevalent due to the considerable amount of experimental work conducted today. Two key points to recognize are that it is necessary to identify and classify all biohazards, because any accident occurring in the presence of biohazard materials may result in infection, and many biological agents have unknown or incompletely understood etiologies. The environment may impinge upon laboratory animals or workers, on biohazardous agents or insects, on physical or chemical substances, and vice versa. Reported laboratory acquired infections only represent a fraction of those that have actually occurred.

Laundry

The laundry is a large service unit within the total hospital whose function is the collection and processing of soiled linens and the distribution of clean linens. Laundry and linens flowing to and from the laundry room can readily become a source of infection or a source of insect and rodent infestations. A well-run, reliable laundry service helps create an aesthetically clean environment. A poorly run service may become dangerous. There are various ways in which linens are processed. They may be cleaned within the institution or in commercial laundries. The institutional cleaning problems include the actual facility, which may be dusty and dirty. This facility may also have large quantities of lint on equipment, walls, and ceilings. The areas for sorting of soiled and blood-encrusted linens, if improperly kept, become a powerful attractant for insects and rodents. In many of these laundries, adequate screening is lacking. The physical flow of linens is very important. Where laundry chutes are used, the exhaust air driven back into the nursing units by the piston effect of a falling laundry bag can contain staphylococci and other organisms. Trash chutes create the same type of problem. Where the linens are taken by cart to the laundry, the carts tend to become dirty and highly contaminated. In special areas where the linens are coming from rooms where the individuals may have infections, the linens need to be kept separate from the rest of the laundry, transported separately, and then washed in the bags before they are sorted and rewashed. Studies conducted in various hospital areas indicated that the blankets, mattresses, pillows, bed, and linens are highly contaminated by varying patient discharges including urine, blood, feces, and saliva. Bed-making and linen-handling procedures, when carried out in a vigorous manner, tend to circulate these organisms, creating a potential airborne hazard.

Commercial laundries may cause a link of infections from hospital to community to hospital to community and to other types of institutions. The commercial laundries have to have separate delivery and receiving trucks for dirty and clean linens to prevent contamination. The hospital linens must be washed separately from other commercially washed linens; otherwise there is an opportunity of infections.

Lighting

Illumination is a very complex subject of exceeding importance to any environment. It is well known that poor lighting results in fatigue, an increase in the number of accidents and injuries, poor housekeeping, and increased risk of infection. Much of our available light is not necessarily lost because of poor design, but rather because of poor maintenance of fixtures and lack of replacement of light bulbs. The color and shade of paint on walls, ceilings, and floors affects illumination. Dark colors absorb and decrease light; light colors reflect and utilize light properly.

Noise

Noise is unwanted sound. Environmental noise may cause temporary or permanent hearing loss, physical and psychological disorders, interference with voice communications, and disruption in job performance, rest, relaxation, and sleep. The increased use of therapeutic equipment and poor maintenance of electrical and mechanical equipment, coupled with lighter weight construction materials, tend to increase noise problems. Normal human activities increase the noise level. Sources of noise in hospitals or other institutions include autoclaves, communications systems, compactor units, dishwashing machines, electrical equipment, heaters, ventilators, air conditioners, and transportation systems. In addition, noise is created by comatose patients who are suffering from specific pains or who may be upset by the strange environment. On the exterior of the hospital, noises come from automobile traffic, compactor trucks, equipment, and emergency vehicles.

Patient Accidents

Many patients have accidents within institutions. The most frequent of these accidents are falls occurring within the immediate area of the patient's bed.

Plumbing

Although various plumbing systems are installed properly during hospital construction, changes are made frequently as hospital space is either utilized for different purposes or as problems occur within the institution. The vast number of activities undertaken within the hospital and the formidable kinds of microbiological and chemical contamination present makes the hospital a potentially hazardous area. Cross-connections, submerged inlets, and back-siphonage problems occur. Plumbing fixtures become corroded, chipped, cracked, and blistered. As new hospital equipment is placed into operation, hospital plumbing surveys must be conducted to determine whether the equipment will put such demands on the plumbing system that it will create improper pressure differentials and excess loads of waste that cannot be handled by the existing plumbing system. The potentially hazardous fixtures and equipment include bathtubs, laundry tubs, sinks in all areas — including the scrub rooms and operating rooms, hydrotherapy tanks, soup kettles, drinking fountains, laboratory sinks, urinals, laundry machines, development tanks for X-ray and other films, and all tanks used for dilution. (An in-depth discussion of plumbing appears in the chapter on plumbing in Volume II.)

Radioactive Material

There are numerous concerns about the use, storage, and disposal of radioactive materials found in body wastes. In the event of fires, special care must be taken to avoid contaminating firefighters and the general public. Other problems include improper shielding of medical X-ray and fluoroscopic machines, improper size of the X-ray beam, and improper beam filtration and alignment. Most of the newer dental X-ray machines have excellent built-in protection mechanisms. However, in the older machines there is still considerable stray radiation. Other sources of radiation are lasers, ultraviolet tubes, microwaves, and ultrasound.

Solid and Hazardous Wastes

Hospital wastes are discarded materials and byproducts of normal hospital activities. The principal types of solid wastes are garbage, paper, trash, other dry combustibles, treatment-room wastes, surgical-room wastes, autopsy wastes, noncombustibles, such as cans and bottles, and a variety of biological wastes, including wound dressings, sputums, placentas, organs, and amputated limbs. In the last 25 years the number of disposable items have increased from a few to hundreds. They range from paper bedding to cardboard bedpans. This has caused a sharp increase in the volume and weight of solid wastes that must be disposed of on a daily basis. The average patient contributes an estimated 8.5 lb or 0.7 ft^3 of solid waste daily. Containers and equipment into which this waste goes are unclean and are not easily cleaned. The waste is usually dumped into plastic bags and then removed to either trash rooms or outside to large metal containers. The containers tend to be dirty, filled with organic material, difficult to clean, and frequently left improperly cleaned. Unfortunately, hospital personnel leave open the doors to the metal units, creating open dumps, instead of a desirable means of temporary storage of solid waste. Numerous additional problems are created by the presence of syringes, needles, examination gloves, catheters, enema basins, and petri dishes. This material is frequently contaminated or contains sharp items, causing safety and infection hazards. The solid waste of individuals with infectious or contagious diseases become considerably more dangerous to the other patients, the community, and the infected individual. Transportation and disposal of hospital solid waste and research animals to community incinerators and landfills deserves investigation. Approximately two thirds of the communities in the United States have unsatisfactory incineration systems or landfills.

Rubbish or trash chutes and trash rooms present additional hazards, since solid waste falling down a chute creates a piston effect, which forces air back into the patient areas. Further, as plastic bags hit the floor of the trash room they burst open and the organisms spew into the air, readily capable of recirculating through the institution.

Hazardous waste that exhibit ignitability, corrosivity, reactivity, and EP toxicity may be found in the institution. (See chapter on solid and hazardous waste in Volume II.)

Surgical Suites

The surgical suite in the modern hospital is potentially more hazardous today than in former years, because of the number of new surgical techniques and the increased risk of infection due to modern drug therapy. Today many hospitals perform various organ transplants, such as kidneys, hearts, and parts of the arteries. It is critical that infection is prevented, patients and personnel are safe-guarded from technological hazards, shock and trauma to the patients are mini-mized, physical comfort and emotional support is provided, and the staff delivers effective patient care in the surgical suite.

Some of the specific problems regarding the patients and the prevention of infection are elimination of cross-traffic or back-traffic of staff, patients, or soiled materials; separation of clean from dirty cases; preparation of operating rooms for handling general surgical procedures; establishment and observation of rules concerning aseptic techniques; immediate change of operating room garb by surgical staff; elimination of the possibility of contamination through air systems; provision of restricted flow zones; and proper cleaning and decontamination of operating rooms.

Hospital-Acquired Infections

There are three basic kinds of infections related to the hospital. The first is the infection of the patient present upon admission to the institution. This infection can spread from patient to patient. The second type of infection appears during hospitalization and most frequently is hospital acquired. This may also spread from patient to patient. The third type of infection appears after the patient is discharged from the hospital. The incubation period for the infection occurs during the hospital stay. The individual, once released, is in a position to spread the infection to members of his family, the community, and of course from the community back to the hospital. An example of the third type of infection would be some types of staph infections where the postpartum woman does not appear to have a breast abscess until she arrives home.

The hospital environment presents a special risk of infection to employees, patients, visitors, and the community. The emergency room, admission areas, outpatient clinics, surgical, medical, and obstetrical units are potential foci of infection. Patients, visitors, employees, supplies, equipment, and the physical structure of the institution may either aid in harboring pathogenic organisms or introducing them to the environment. The organisms are transmitted by improper patient care, mishandling of contaminated materials and equipment, poor house-keeping, and inadequate physical facilities.

Hospital-acquired infections differ depending on hospital size, type, and pa-tient usage; usage of antibiotics; employee supervision; time of year; and the quantity of virulent organisms that are introduced into the hospital environment. Some general types of infections found in institutions include gram-negative bacteria from urinary catheter-associated infections and infections following prostat-ectomies, burn infections, and infections of the newborn; *Staphylococcus aureus* infections from nurseries, surgical suites, surgical recovery areas, and other areas of the institution; infant diarrhea; infections caused by salmonella and shigella or other enteric organisms; *Klebsiella* infections due to a gram-negative bacteria found in thoracic surgery, intensive care units, and on floors and other surfaces;

Serratia marcescens infections due to organisms found in hexachloraphene dispensers; aerosols from ultrasonic cleaning devices; *Pseudomonas aeruginosa* found in the air and on surfaces; and *Bacillus anitratum* found in frozen plasma, water baths, respiratory equipment, and in hospital food-service, solid-waste, and laundry operations.

The extent of these infections and the susceptibility of the patients, whether they are burn patients, under special chemotherapy, or newborn, makes it necessary to understand the principles of epidemiology of the spread of these organisms, the means of determining potential hazards, and the institution of environmental, personnel, and patient controls.

Hospital-Wide Problems

1. Lack of training and inadequate supervision are often principal factors in the spread of infection.
2. Movement of personnel and supplies contribute to the dissemination of infectious organisms.
3. Failure to appropriately mark the chart of each infectious case can lead to contamination of unaware personnel.
4. Improper transportation and storage of sterile supplies can cause contamination.
5. Improper transportation and storage of infectious waste can cause contamination.
6. Failure to properly segregate clean and dirty linen, and failure to properly clean and disinfect bedding may increase the risk of infection.
7. Careless handling of any waste may introduce pathogenic organisms into the environment.
8. Since hospital food is often ground, creamed, and prepared in large quantities, the opportunities for contamination through improper handling and storage is increased.
9. Bedpans and bed urinals improperly removed, handled, stored, washed, and disinfected may cause a cross-infection.
10. Supplies, instruments, and equipment may directly or indirectly contaminate patients.
11. Failure in any detail of isolation technique creates a risk of infection from employees, visitors, and other patients.
12. Failure in any detail of proper catheterization technique or follow-up care can cause urinary tract infections.
13. Other procedures, such as inhalation therapy and intravenous cut-downs, may be contributors to infection.
14. The flow of traffic in special areas is of unique importance, since the mere movement of traffic can lead to outbreaks of disease through the distribution of microorganisms.

Routes of Transmission

Many factors influence the spread of infections within the institution. Therefore, it is frequently impossible to determine the actual causes surrounding the spread of a given disease. However, epidemiological studies should be conducted in all cases, since vital information can be gathered that is of value in efforts to reduce the numbers of organisms in the environment and in the patients. The routes of transmission of disease include droplets or droplet nuclei from person to person, enteric or cutaneous contact, formites, food, water, insects, and rodents. The disease organisms may come from draining wounds, lesions, and secretions. They may come from patients, staff, visitors, or indirectly through the environment. It is known, for instance, that strep and staph infections come from nasal discharges. Staph infections also come from draining wounds, boils, infected decubitii ulcers, sores, hangnails, and infected pimples. Enteric organisms, which cause salmonellosis or shigellosis, are spread by fecal material. Most disease organisms within the institution are probably spread through improper or poor medical nursing techniques. A smaller, but still a significant proportion, are spread through the environment.

Occupational Health and Safety

Hospitals have a real potential for occupational health and safety hazards, patient hazards, and hazards to visitors. The hospital contains not only all of the problems and services of a small community, but also the additional hazards brought to the institution by the need for special types of equipment, gases, chemicals, and operating procedures. Some specific problems include excessive heat from laundries, kitchens, boiler rooms, and furnace rooms; inadequate illumination in numerous areas of the institution; variety of chemicals, many of which are toxic, explosive, carcinogenic, or possible fire hazards; considerable noise from operational procedures in the physical environment; radioactive materials found in a variety of treatment and diagnostic areas; and research areas where animals contribute to disease and injuries. An in-depth discussion of particular safety hazards would take many pages. It is best to understand that the hazards exist and that they are found not only in the settings discussed, but also in any setting within the institution (see Chapter 10 in this volume).

Fires may be caused by problems related to housekeeping, smoking, electrical appliances and their installation, flammable liquids, heating units, explosive materials, painting units, compressed gases, and so forth. Fires are caused by poor housekeeping and storage of solid waste, including combustible debris, oil- or paint-soaked rags, and wood scraps. Smoking is hazardous, Another serious hazard within the institution is the use and storage of flammable and explosive liquids and gases.

Nursing Homes

Special Environmental Problems

Sanitary conditions within nursing homes may change very rapidly as a result of inadequate budgets, insufficient numbers of employees, poor administration, or inadequate building structures and facilities. Personnel are often lacking proper in-service training. There is a need to improve supervision of personnel, establish standard operating procedures, and develop a better understanding of cleaning techniques and environmental controls. In some instances, the quality of food is poor, not only because of inadequate preparation but, more importantly, because cheap food is purchased and may come from areas where there had been salvage operations. Salmonellosis, staph food poisoning, and other types of food-borne diseases appear regularly within the nursing home. Contributing to the environmental health problem is the nature of the patient and his or her mental and physical condition. More than half of the patients in proprietary nursing homes are disoriented at least part of the time. About 20% are in a state of confusion most of the time. The confused patient not only contributes to the accident problem, but also contributes to the spread of disease and infection. An estimated one third of the nursing home patients suffer from urine or feces incontinence or both. Obviously these wastes are excellent sources of contamination within the institutions. When incontinency combines with mental confusion and lack of adequate staff supervision, the chance of spreading environmentally related diseases increases sharply. Further, in many of the older homes, there is a problem with insect and rodent infestation. Even in the newer homes, infestations occur and can be spread because of lack of personnel and personnel training, and hoarding of food, providing a constant source of insects and rodents.

Special Safety Problems

The average age of the nursing home resident is 80 years. Residents generally have debilitating or crippling diseases, making them more prone to accidents. About 40% of the patients have cardiovascular conditions, and about 75% have some heart or circulatory difficulties. Sixteen percent have the residuals of paralytic strokes, and about 10% are receiving care for fractures, primarily hip fractures. All of these conditions may be coupled with arthritis, mental disorders, and diabetes. The physical conditions of the patients, plus their frequent intake of medications, make them prime subjects for accidents within the nursing home.

Other Special Problems

There is a definite shortage of nursing personnel. This causes unskilled staff to assume duties requiring training. The general quality of patient care is affected by this shortage. Unskilled labor is used within the nursing home to take care of many of the tasks of running the facility and for patient care. The individuals who fill these positions receive low pay, are overworked, improperly trained, and unfortunately are often improperly screened to determine their temperament and emotional stability.

A lack of communication exists between hospitals and nursing homes. Records are not transferred quickly enough or in enough detail to the nursing home for the nursing home personnel to be able to properly care for the patient. Physicians, because of large workloads, make infrequent visits. When emergencies do occur, it is difficult to reach the necessary physicians to provide adequate patient care.

Problems arise with the patients, since they are prone to chronic illness and multiple impairments and disabilities. Many feel rejected by friends and family. Patients may feel they have come to the nursing home to die and will lose the will to care for themselves and to recover and utilize their physical resources in the best way possible.

Enforcement is difficult because of small numbers of trained public health personnel, lengthy court procedures, frequent changes in federal regulations, budgetary problems, and general confusion regarding proper operation of the nursing homes. The high cost of nursing-home care is alarming. The number of state and federal dollars supplied is substantial.

Old Age Homes

The types of problems found in the old age home vary with the number of people and the condition of the facility. They may be simple problems of inadequately prepared food, insects, and rodents, up to the many complicated problems found within the nursing-home situation. It is often difficult to determine when an individual simply needs assistance with everyday life and when an individual needs nursing care. Since the population of the United States will increasingly fall into the 65 and over category, proper evaluation of old age homes will be essential to determine if they are performing their necessary function in our society.

Prisons

Environmental health practitioners have recognized the problems in prisons for many years. Prison riots, such as those in Attica, New York, have further emphasized the problems and needs. The United States District Court for Alabama in the case *James vs Wallace* held that the conditions of confinement within the Alabama penal system violated the judicial definition of "cruel and unusual punishment," and therefore improvements had to be made within the prisons. The improvements included reducing overcrowding, improving general living conditions, improving food service, education, recreation, and rehabilitation programs. The court not only defined the problems, but also stipulated that the prisons comply with the recommendations of the court or the court would order the prisons closed.

Several groups, including the APHA-NEHA Jails and Prisons task force, the Federal Bureau of Prisons, the federal government, and the state of Kentucky, have conducted studies and made recommendations to improve penal institutions. A model code proposed include specific details on food, water, air, liquid waste, solid waste, vectors, recreational environment, housing, accidents and safety, industrial hygiene, public facilities, and emergency plans. It was obvious from the discussions and studies that an environmental health specialist was needed in penal and correctional systems to assist in changing the environment of the institution.

Bedding

Frequently, bedding is unclean or in bad repair. Many institutions do not provide sheets or pillows. Linens and blankets, when available, are often dirty.

Food

Food problems are intimately related to the food-service facilities and to the prisoners, who are the food handlers. Food is often contaminated from improper purchasing and storage. Inadequate training in food preparation, storage, food hazards, and personal hygiene. The floors, walls, ceilings, and ventilation hoods are in poor repair in many institutions and contribute to the overall lack of cleanliness and the potential spread of food-borne disease.

Heat and Ventilation

Many buildings, because of their age and lack of repairs, provide inadequate heat and improper ventilation. Windows behind the bars have been smashed or are nonexistent. Proper heating and ventilation of buildings is necessary in providing a reasonable environment.

Insect and Rodent Control

Insects and rodents are found in about 25–35% of the institutions. Again this is due to the poor maintenance of buildings, to the age of the buildings, and to the nature of the institutions. In many cases, inadequate control procedures are used because of lack of proper understanding of insect and rodent control.

Laundry

Laundry service is not available in certain types of institutions unless friends or relatives bring clean clothing to the prisoners. Where laundries are present, they tend to be extremely hot and poorly ventilated.

Lighting

About one half of all the institutions surveyed in the state of Kentucky had inadequate illumination. About 45% of the light fixtures were in poor repair. Poor lighting was typically found in many institutions.

Plumbing

Plumbing problems abound within the penal institutions because of their age, overuse, and overcrowding. There is a serious lack of toilets and sinks. In many cases, the toilets, sinks, and shower facilities are in bad repair, dirty, or almost totally inoperative.

Safety

Many institutions lack adequate fire extinguishers or the proper techniques available to take care of prisoners in the event of fires or accidents.

Occupational Health

Occupational health practices need improvement to conform with OSHA regulations.

Schools

Environmental Health Problems

Many schools ranging from elementary through college are substandard because of the existence of environmental health problems.

Food

The basic food protection problems found in the hospital applies also to the school. The additional concern that exists for children in substandard urban areas is that the school lunch may be the only nutritious meal received within the course of a day. For this reason, it is even more important that food-service personnel provide a wholesome, safe meal in an attractive manner.

Within schools, mass feeding leads to leftovers, and leftovers and poor food-handling practices lead to serious outbreaks of food-borne disease. Bagged lunches become a particular hazard, since they are stored in school lockers until lunchtime. If food is perishable, the student may contract a food-borne disease. Often dust, dirt, roaches, and mice may get on the food bag or into the food. In universities, students have refrigerators in their rooms or take food back to the room for later snacking. This also leads to problems of potential food-borne disease and insect and rodent infestations.

Gymnasiums

Many accidents occur as a result of poor supervision, poor equipment, improperly trained individuals, poor lighting, horseplay, and so forth. Locker rooms present hazards, particularly through slipping, and a source for the spread of fungal and bacterial infections.

Laboratories

The problems of school laboratories are similar to those of hospitals, except that in junior and senior high schools fewer dangerous substances are used. The hazards are not only related to the types of substances used and the byproducts of these substances, but more importantly to the inexperience of the students.

Plumbing

The special problems encountered in schools include the students stuffing the toilets with toilet paper, paper towels, and other objects and wanton destruction of the facilities.

Roughly half of the schools lack adequate hand-washing facilities, soap, and single-service towels. About 10% lack adequate toilet paper. Since hands are probably the most common means of transmission of infectious diseases, it is important that these items be provided in the bathrooms.

Shops

Depending on the type of school, there are a variety of shops being used. These shops constitute a hazard for the students.

Ventilation, Heating, and Air Conditioning

Air heating, air cooling, humidity control, and air distribution is essential for the removal of body odors and for the comfort of the individual. Where there is inadequate air circulation, temperature of the room is raised by body heat, humidity is raised by occupants' breath and perspiration, organic matter is released into the room causing odors, oxygen content is reduced, and carbon dioxide is increased. All of this contributes to stuffiness and odors, and to an uncomfortable environment.

Water

In the school a specific problem occurs when water fountains are misused. All water fountains should be of an angular rather than bubble type. A student should not have to put his or her mouth directly on the water source since this becomes a vehicle of contamination for other students.

Accidents

Accidents are the leading cause of death in all student age groups. About 16,000 students of school age lose their lives each year. Motor vehicle-related deaths account for 58% of these deaths; drownings, 15%, and fires and burns, 7%. Of these accidental deaths, 43% are associated with social life. About half of these occur in school buildings. The most hazardous areas are the gymnasium, where one third of the school-related deaths occur; the halls and stairs, where one fifth of the school-related deaths occur; shops and laboratories, where one fifth of the deaths occur; and other kinds of classrooms, where one seventh of the deaths occur. About 40% of the deaths occur on school grounds. Football is the most hazardous of sports.

In a study of 1,900,000 students, it was determined that 24% of the injuries occurred within school buildings, 28% on school grounds, 5% going to and from school, 20% at home, and 23% in other types of situations.

The college is a complete, highly active community. It contains all of the community safety problems, all types of industrial safety problems found in laboratories and industrial shop areas, and various types of home and recreational safety problems. In a study conducted in 22 colleges and universities serving 207,000 students, roughly 15,000 injuries were reported to the student health services. Of these injuries, 75% were incurred by men and 25% by women. Fifty-two percent of the campus injuries were either in the athletic or recreational facilities, 20% were in the residence halls, 15% in the academic buildings, 11% on the campus grounds, and 2% in motor vehicles. It is obvious from these studies that schools, regardless of the level, can be dangerous places.

Bus Safety

Each year school buses are involved in accidents and in fatal injuries. This occurs because of the numbers of school buses traveling on the roads, poor roads and road conditions, bad weather, and in some cases inexperienced bus drivers. There are many school buses that are old, improperly maintained, and therefore, hazardous. Constant checks should be made of the equipment by utilizing systematic maintenance programs. All buses should be checked for levels of carbon monoxide at the back of the bus; carbon monoxide may seep in through the emergency door.

Hazardous Materials and Equipment

The storage and disposal of this material should be controlled by knowledgeable individuals. Laboratories present a special area of concern because of the lack of training of the student and the presence of potentially flammable liquids, explosive reactions, compressed gases, electrical units, and toxic substances.

POTENTIAL FOR INTERVENTION

The potential for intervention is based on isolation, substitution, shielding, treatment, and prevention. All these techniques can be utilized to a greater or lesser extent in the institutional environment. The potential for intervention in hospitals is very good. In nursing homes and old-age homes, the potential varies with the type of facility, the kind of staff, and the amount of administrative control. In schools the potential for intervention should be excellent. In prisons, it is poor.

Within all these areas isolation is a technique used to separate the contaminated or infected individual from the rest of the population. Isolation also refers to the separation of hazardous substances and gases, such as radiological material, hazardous chemicals, and explosive gases. Substitution is the use of a less hazardous chemical or technique for a more hazardous chemical or technique. Shielding is the utilization of special cabinets with special hood systems where highly dangerous or infectious materials are being used. Shielding also refers to personal protective devices, such as gowns, masks, caps, and safety glasses. Treatment is utilized in all institutional settings for exposure to disease processes and chemical, physical, and radiological hazards. Prevention is the principal technique that should be used within the institutional setting.

RESOURCES

Scientific and technical resources include the national, local, and state chapters of the American Medical Association; the American Public Health Association; the American College of Hospital Administrators; the American Hospital Association; the Catholic Hospital Association; the Joint Commission of Accreditation of Hospitals; medical schools within all of the states; and a variety of environmental health practitioners.

Civic resources consist of the various auxiliaries to the medical associations and to the hospitals, and a variety of foundations that provide funds for hospitals, such as the Ford Foundation and the Rockefeller Foundation.

Governmental resources include state health departments, the National Bureau of Standards, the Department of Defense, the National Institutes of Health, the Food and Drug Administration, the National Science Foundation, the Veterans Administration, the Department of Health and Human Services, and so forth.

Numerous codes and standards should be utilized in the actual construction of an institutional facility or in the modernization of existing facilities. These codes and standards may be obtained from the following sources:

1. The Superintendent of Documents, U.S. Government Printing Office, Washington DC 20402 (for all government documents)
2. American National Standards Institute, 1340 Broadway, New York, NY 10018

3. American Society for Testing and Materials, 1916 Race Street, Philadelphia, PA 19103
4. American Society of Heating, Refrigerating, and Air Conditioning Engineers, United Engineering Center, 345 East 47th Street, New York, NY 10017
5. Compressed Gas Association, 500 Fifth Avenue, New York, NY 10036
6. International Conference of Building Officials, 50 South Los Robles, Pasadena, CA 91101
7. National Association of Plumbing-Heating-Cooling Contractors, 1016 Twentieth Street N.W., Washington, DC 20036
8. National Council on Radiation Protection, 4210 Connecticut Avenue N.W., Washington, DC 20008
9. National Fire Protection Association, 60 Batterymarch Street, Boston, MA 02110
10. Underwriters Laboratories, Inc., 207 East Ohio Street, Chicago, IL 60611
11. National Bureau of Standards, U.S. Department of Commerce, Washington, DC 20234
12. American Hospital Association, 840 North Lakeshore Drive, Chicago, IL 60611

For an in-depth discussion of biohazard control, it is recommended that the reader obtain the book entitled National Institutes of Health Biohazards Safety Guide, U.S. Department of Health, Education and Welfare, Public Health Service, National Institutes of Health, Superintendent of Documents, U.S. Government Printing Office, Washington, DC 20402.

Very little has been written in the area of penal institution environmental health. It is recommended that the reader explore the publication "Corrections" by the National Advisory Commission on Criminal Justice, Standards, and Goals, Washington, DC, published January 23, 1973, as a base of study. The most knowledgeable source on prisons is the Federal Bureau of Prisons, 320 1st Street, N.W., Washington, DC 20534.

STANDARDS, PRACTICES, AND TECHNIQUES

Hospitals

Physical Plant and Site Selection

The site of the medical facility must be well drained and free of obstructions or other natural hazards (see site selection for communities in chapter on the indoor environment). Its location should be accessible for fire apparatus, ambulances, and other types of service vehicles. Public transportation systems should be available to the patients and staff. Competent medical and surgical consultation from private practitioners or from medical schools should also be readily available. Paved roads need to be provided from the main entrance to the buildings and the emergency section. Paved walkways are necessary for pedestrian traffic. Off-street parking should be provided and easily accessible to patients, staff, and visitors. Special curbing facilities and walkways are needed for wheelchair patients and the physically disabled. Parking spots located close to the entrance should be reserved for the handicapped. All public facilities, including bathroom facilities, should be designed for use by the handicapped. The medical facility should be accessible to major highways and conveniently located within reasonable distance of interstate highways to expedite the flow of patients or emergency vehicles to the institution. Heavily trafficked areas should be avoided. In planning the location, it is necessary to consider the path and proximity of existing air-polluting industries, future construction of industries, and the possibility of flooding and natural disasters due to fault lines, underground mines, and other potential hazards.

According to the National Environmental Policy Act — Public Law 91-190, the site and the project of the hospital must be approved before construction can begin. An impact statement should be prepared, and any possible adverse environmental problems should be shown and techniques of control explained.

Interior Structural Requirements

The following items are requirements for the various parts of the typical hospital. The general nursing unit should include:

1. Patient rooms should have a maximum capacity of four patients.
2. Minimum floor area of 100 ft^2 for a single bedroom plus 80 ft^2 for each additional bed.
3. Each room should have a window that opens and is above the floor or grade level. The openings should be limited or screened to prevent accidents.
4. A nurses' call system should be provided in each room and in the bathrooms.
5. Each patient room should have its own lavatory or be attached to a bathroom so that the patient does not have to go into the central corridor to utilize the toilets or sinks.
6. Room should be provided for wardrobes, lockers, or closets.
7. Provision should be made for visual privacy of the patients.
8. A series of service areas should be provided for the nursing unit to function properly, including a nurses' station; nurses' office; storage and supply room; convenient hand-washing facilities; charting facility; staff lounge and toilet facilities; staff lockers; conference rooms; patient examining rooms, with a minimum floor space of 120 ft^2; clean workrooms and holding rooms; dirty workrooms and holding rooms; drug distribution stations; clean-linen storage closets; nourishment stations; equipment storage; parking areas for stretchers and wheelchairs away from the normal traffic flow; patient bathing facilities; and emergency equipment storage facility.

The nursing facility also needs isolation rooms that are completely and totally separated from all other patients and patient activity rooms. Isolation rooms must have private toilet facilities, sinks, and a dressing area for change of clothing and decontamination of soiled isolation gowns, masks, and so forth. Each hospital needs rooms for emotionally or mentally disturbed patients.

The intensive care unit has critical space requirements, since it is used for seriously ill medical, surgical, or coronary patients. These patients are acutely aware of their surroundings and may be easily affected by them. As a result, it is necessary to eliminate all unnecessary noise, to provide individual privacy in keeping with the medical needs of the patient, and to provide adequate observation by the staff. Patients should have some outside view through a window. The window should not be higher that 5 ft above floor level.

The intensive care unit should include the following:

1. Coronary patients should be housed in single bedrooms.
2. Medical and surgical patients should be housed in either single or multi-bedrooms.
3. All beds must be under direct visual observation by the nursing staff.
4. The clearance between beds in a multi-bedroom must be a minimum of 7 ft.
5. In a single bedroom there must be a minimum clear floor area of 120 ft^2 and at least 10 ft as a minimum width or length.
6. A single bedroom has to be provided for each patient for medical or surgical isolation.
7. Viewing panels, which can be covered by curtains for privacy, must be installed in the doors and walls to provide adequate visual coverage by the nursing staff.
8. Intravenous solution supports must be provided for each patient so that they do not hang over the patient.
9. A hand-washing sink is required in each patient room.
10. A nurses' calling system is needed and should be required.
11. Each coronary patient must have direct access to a toilet facility.
12. The location of the nurses' station should permit direct visual observation of the patient.
13. Hand-washing facilities should be conveniently located at the nurses' station.
14. Charting facilities must be separated from the monitoring service.
15. The staff restrooms should contain toilets and adequate hand-washing facilities.
16. Storage areas must be close to nursing personnel
17. A clean workroom and system of storage and distribution of clean and sterile supplies is mandatory.
18. A soiled or dirty workroom for proper storage of soiled equipment and supplies is necessary.
19. Bedpan flushing and sanitizing units are required.
20. A convenient 24-hr distribution system and station for medications is needed within the nursing area.
21. A clean-linen storage closet is needed.
22. A nourishment station should be provided.
23. A storage room with easy access is needed for emergency equipment, such as inhalators and "crash carts."
24. Patient lockers must be furnished for patients' personal effects.
25. A separate waiting room, including toilet accommodations, seating accommodations, and telephones, is needed for visitors.

Newborn nurseries can be a serious hazard to newborn infants if the design is poor and special precautions are not taken to protect the infants. The following criteria are essential in a newborn nursery:

1. One nursery should not open directly into another.
2. Nursery should be conveniently located to the postpartum nursing unit and obstetrical facilities.
3. Nurseries should contain hand-washing sinks at a ratio of one to eight bassinets; emergency nurses' calling system; observation windows to permit viewing of the infants in public areas; charting facilities; full-term nurseries containing a maximum of eight bassinets, or as many as 16 if the bassinets are isolets; the minimum floor area for a bassinet should be 24 ft^2; special-care nursery for high-risk infants and infants in distress; a minimum of 40 ft^2 per bassinet in the special-care nursery; and workroom space.
4. Examination and treatment rooms or space for infants containing workcounters, storage, and hand-washing sinks.
5. Infant formula rooms should be provided if on-site formula preparation is carried out. However, it is recommended that, for safety and also for economic reasons, commercially prepared formulas should be utilized.
6. A housekeeper's closet for storage of all housekeeping supplies.

The pediatric and adolescent unit is separated from the adult units to provide a quieter setting for the adults and to prevent the spread of childhood disease to the adult population and conversely. The pediatric nursery follows the same requirements as the other nurseries. The pediatric rooms conform to the same requirements as the other rooms within the general nursing facilities, with the additional need for an area used for dining, educational, or play purposes. Also, special storage closets and cabinets for storage of toys (that can be sanitized), and educational, or recreational equipment should be installed.

The psychiatric unit should be constructed to provide a safe residence for patients and one in which patients are unable to hide, escape, injure themselves, or commit suicide. It must be flexible to care for the ambulatory inpatients and also meet the needs of various types of psychiatric therapy. The unit should convey a noninstitutional atmosphere to facilitate recovery. The following requirements for construction are as follows:

1. Maximum of four patients to a room.
2. Minimum room area of clear floor space for single rooms should be 100 ft^2; for multiple-patient rooms, 80 ft^2 per bed.
3. Each room should have a window with limited opening space and security screens.
4. Sinks must be provided in each patient's room.
5. Bathrooms must be provided in such a way that the patient does not have to go into the corridor to go to the bathroom. The bathroom door should unlock from the outside.
6. All other usual facilities found within a normal nursing unit, with the exception of all items that could be hazardous. These should be locked away and out of reach of the patients.
7. A special space for dining, recreation, and occupational therapy, with a minimum floor space of 40 ft^2 per patient.

The number and types of operating rooms are based on the anticipated surgical load for the hospital. The surgical-suite construction should be designed to prevent the possibility of contamination. The surgical suite must be located in a separate traffic area to exclude individuals unassociated with surgery. Surgical suites should meet the following requirements:

1. A minimum clear floor area of 360 ft^2, exclusive of all cabinets and shelves, per room, with the minimum dimension of 18 ft.
2. An emergency communication system.
3. Two X-ray film illuminators for each room.
4. Special storage space for splints and traction equipment in orthopedic surgery rooms.
5. A minimum clear floor area of 250 ft^2 for cystoscopic and other surgical endoscopic procedures.
6. Recovery rooms arranged for visual control of patients, drug distribution, handwashing, and charting.
7. Additional recovery space for outpatient surgical patients.
8. Special service areas containing control stations, for visual surveillance of all traffic entering and leaving the operating suite; supervisors office and station; sterilizing facilities with high-speed autoclaves, conveniently located near the operating rooms; drug distribution station; two scrub stations adjacent to and contiguous with the entrance to each operating room; dirty workrooms for all soiled equipment and materials; fluid-waste disposal facilities convenient to the operating room; clean workrooms and supply rooms for clean and sterile supplies; anesthesia storage facilities, where flammable gases can be stored without causing a hazard; anesthesia workrooms for cleaning, testing, and storage of anesthesia equipment; medical gas storage areas; special equipment storage areas; emergency equipment storage areas; staff locker rooms, with adequate space for cleaning, changing, and resting; outpatient surgery change areas; holding areas for patients prior to surgery that are under visual control of the nurses; stretcher storage areas out of the direct line of traffic; staff lounges; housekeeping closets for storage of housekeeping supplies.

Maximum capacity needs must be anticipated when constructing delivery and labor rooms to provide adequate facilities. Construction requirements for obstetrical facilities are as follow:

1. Delivery rooms should have a minimum clear floor area of 300 ft^2 with minimum dimension of 16 ft.
2. Rooms should have emergency communication systems.
3. Resuscitation facilities for newborn infants, including oxygen, suction, and compressed air.
4. Labor rooms with a minimum floor area of 80 ft^2 per bed.
5. A minimum of two labor beds per delivery room.
6. Hand-washing facilities within the labor room.
7. Adequate visual observation of the patients.
8. Recovery rooms with a minimum of two beds and all ancillary facilities.
9. The usual service areas, as stipulated in the surgical suites.

The emergency suite is an extremely critical part of the entire hospital. When an emergency occurs, this is the place where a life is saved or lost. The emergency suite is a combination outpatient clinic, operating room for minor injuries, admissions area, and surgical suite. It is essential not only that the emergency room staff is properly trained and efficient in dealing with patients, but also that the facilities are designed to facilitate emergency care. The emergency room facility should contain the following:

1. A well-marked entrance at grade level that is sheltered from the weather and easily accessible to ambulances and pedestrians.
2. A reception and control area conveniently located near the entrance.
3. Public waiting areas with bathroom facilities, telephones, and drinking fountains.
4. Stretcher and wheelchair storage areas.
5. Treatment rooms equipped with hand-washing facilities, containers, cabinets, and work counters.
6. Treatment rooms containing medical suction, storage of emergency equipment such as emergency trays, defibrillators, cardiac monitors, and resuscitators.
7. Staff work and charting areas.
8. Clean supply storage areas.
9. Sterile supply areas.
10. Daily workroom areas for supplies and equipment.
11. Patient restrooms.
12. Communications equipment and an area where adequate communications between physicians at the hospital and ambulances can be provided.
13. Adequate emergency surgical care and treatment in the event of a serious immediate surgical need when the individual cannot be taken to the operating room.

The communications system is of vital importance, since the physician can provide guidance to the emergency medical technicians (EMTs) or paramedics on the care of a given patient. It is also useful in alerting the hospital to the type of arriving patient and the specific needs, in order that the physician and nurses may be prepared to immediately start treating the patient.

The hospital also provides outpatient areas, clinics and other services, including diagnostic clinics, laboratories, physical therapy suites, occupational therapy suites, mortuary and autopsy rooms, pharmacies, administrative areas, public areas, medical records units, central supply, linen services or laundry, housekeeping, food areas, and solid-waste disposal areas.

Electrical Equipment

All electrical equipment and facilities must adhere to the standards of the Illuminating Engineering Society, The Institute of Electrical and Electronic Engineering, Inc., and the codes and standards of the National Fire Protection Association. The hospital should have at least two separate sources of electricity, so that in the event that one fails the second one can immediately go into effect.

The emergency system may be either operated as a separate service from the electric company or operate on portable generators or storage batteries. It is important that the emergency system be tested regularly to ensure working order should the regular system become inoperative. Since motors operate at a specific voltage, safety precautions should be provided in the event that voltage drops or increases sharply. The telephone and electric power conductors are not permitted in the same raceways, boxes, cabinets, and so forth. Any switchboards or panel boards should be protected to prevent shock. All hospital wiring must be placed in conduits to facilitate alterations and repairs. The conduits must be large enough to allow for expansion when needed. Receptacles should be installed in all required areas and at convenient heights. All receptacles should be grounded and explosion proof in locations where anesthetic gases are used. Patient rooms should have at least three double receptacles for a single bedroom with two outlets near the head of the bed. Multi-bedrooms should also have two double receptacles at the head of each bed. A variety of clocks are used in institutions; two types are wired and electronic. In the wired system individual clocks are controlled by a master clock; in the electronic system no control wiring connection is required. In locations where anesthetic or other hazardous gases are used, conductive flooring is required. The electrical resistance of these floors should be only moderately conductive to provide a conductive path for neutralizing static charges and to act as a resistor to limit current to the floor from the electrical system in the case of an electrical short. The possibility of electrostatic sparks causing ignition of the gases must be eliminated. As a further precaution, the electrical distribution system should be ungrounded in the operating and delivery rooms, with the exception of fixed nonadjustable lighting fixtures eight or more feet above the floor and permanently installed X-ray tubes. All operating room floors and delivery room floors have to be monitored periodically for the buildup of static electricity. Generally, improper cleaning or removal of detergents and detergent residues contribute to an increase in electrostatic electricity.

Lighting fixtures must be installed to meet the comfort and work requirements of the particular workspace. Since these requirements vary, it is important to obtain from the Illuminating Engineering Society the current lists of lighting levels recommended for hospitals or other institutions. The voltage supplied to the X-ray unit should be nearly constant so that images and pictures will be uniform and consistently reproducible. It is recommended that an independent feeder with capacity sufficient to prevent a voltage drop greater than 2% be used. Elevators should be manufactured, inspected, and maintained in strict excellent working condition. Skilled elevator mechanics should be utilized for repairs. In the event of an emergency, the alternate source of electrical service in the elevator should automatically be shifted to elevator usage. Only one elevator should be so arranged.

The communications system within the hospital may include loudspeakers, chimes, coded bells, radio communications, telephones, buzzer systems, light signals, closed circuit television, and independently carried receivers. Since the communications system is so essential to the operation of the hospital, it is necessary that emergency electricity be provided for its operation in the event that the normal system goes out of order.

In addition to the ones mentioned, modern hospitals have remote dictation services. Fire alarm systems are necessary in every hospital. A manually operated system should be available and utilized and the system tied in with the local fire department. Special electrical installations include the facilities of the electro-encephalographic areas and medical electronic equipment areas.

Heating and Air Conditioning

Heating and air-conditioning systems should be designed to provide the following temperatures and humidities: operating room, 65–76°F at 45–60% relative humidity; delivery room, same as operating room; recovery room, 75–80°F at 50–60% relative humidity; nursery (observation for full term), 75–80°F at 50% relative humidity; nursery (premature), 75–80°F at 50–60% relative humidity; intensive care, 70–80°F at 30–60% relative humidity; and all other occupied areas at 75°F minimum temperature.

The ventilation system must be mechanically operated, with a minimum separation of 25 ft from air intake to air exhaust. The outdoor air intake should be a minimum of 8 ft above ground level, preferably on the roof. If on the roof, it should be 3 ft above roof level. The ventilation system must be designed to provide the general pressure relationships required in various areas of the hospital. For example, the air pressure should be positive within the operating room and negative within the corridors; and negative within the isolation room and positive in the corridors. In the operating room, the fresh air supply should be delivered from the ceiling and the air exhaust at floor level. Corridors should not be used to supply or exhaust air, with the exception of bathrooms or janitors' closets, which open directly onto the corridors. The ventilation systems in the operating and delivery rooms should have final filters located downstream from the main coil systems and a 99.7% efficiency rating in removing micron-size particles. All other areas can have an efficiency rating of 80–90%. A manometer used in the central air systems will measure the flow of air through the system. All isolation rooms, laboratories, and other areas where potential environmental hazards exist must have ducts to the outside. Air should not be recirculated from these rooms to any other part of the hospital. In food preparation areas, the minimum exhaust rate is 100 $ft^3/min/ft^2$ of hood space area. The ventilation system in anesthesia and other rooms where hazardous gas is used must meet requirements of the National Fire Protection Association. Boiler rooms require sufficient outdoor air to maintain adequate combustion in the equipment.

Flammable and Explosive Liquids and Gases

In storing flammable and explosive materials, the following precautions should be used:

1. Gasoline, other flammable liquids, and gases should be handled and stored in specially designed containers and should be kept in well-ventilated areas. All flammable substances should be appropriately labeled.
2. Cleaning fluids should have high flash points.
3. Containers with flammable liquids should always be sealed when not in use, and caps should be replaced promptly after use.
4. All flammable liquid tanks should have vent pipes and accessible shut-off valves.
5. Flammable liquids and gases should not be kept in areas where individuals could trigger an explosion or fire.
6. All spills of flammable liquids should be cleaned immediately.
7. Approved fire extinguishers should be kept in areas where flammable liquids or gases are utilized.

Structural Features

The structural features of the hospital building must comply with established building, fire, and electrical codes. All footings must lie below the frostline, and the foundations must rest on solid earth or on pilings or piers that have been predetermined to be able to withstand the weight of the structure to avoid detrimental settling. The building materials must in all cases be fireproof. The minimum corridor widths must be 8 ft, 10 in., and the minimum door widths should be 3 ft, 10 in. All doors leading to patient rooms and bathrooms should be easily opened from the outside in the event of an emergency. Thresholds and expansion joint covers must be flush with the floor. Drinking fountains, telephone booths, vending machines, and other objects should not block corridors or exits. All parts of the building below the grade level must be waterproof and built to prevent the entrance of surface water. Storage rooms, patient bathrooms, and so forth must have a minimum width of 7 ft, 6 in. All patient rooms should have at least one window leading to the outside.

Boiler rooms, food-preparation areas, mechanical equipment rooms, and laundries need to be insulated and ventilated to prevent the floor surfaces from exceeding normal floor temperatures. It is important that the boiler rooms not be located underneath patient or patient service areas. In the event of a fire or explosion, this could cause serious loss of life or injury.

Adequate cooling systems are needed for all areas of the hospital. Indirect cooling is most desirable in surgery suites, delivery areas, nurseries, emergency, and intensive care areas. The exhausting of air from various critical areas should be done in such a way that it does not interfere with the air movement patterns of the rest of the institution.

Where solid-waste or linen chutes are used, they must comply with the applicable National Fire Protection Standards, which include approved-glass service openings; chute openings located in rooms with a minimum of 1-hr fire-resistant construction; the minimum diameter of the chute must be 2 ft; chutes must discharge directly into the refuse room, or in the case of linen into the linen room; the chutes must have sprinklers and provisions for washing the chutes and appropriate areas.

Housekeeping Services

One of the most effective techniques for organizing housekeeping services is to establish good work plans. The first step in planning work is to identify the type of surfaces to be cleaned and the amount of congestion, plus the potential hazards within the area. Once these have been identified and specified, time allotments are made for the amount of work to be produced by the average housekeeping person. The actual work scheduling is set up based on the competency of the individual, the kind of equipment available, and the type of problems encountered during cleaning. A daily cleaning schedule should be established for floor washing, dusting with treated cloths, and cleaning of bathrooms and other specialized rooms. This schedule should be followed on a daily basis. In the daily cleaning, a general-purpose, good detergent disinfectant should be utilized. Equipment used in a contaminated area should not be brought into other areas, as this is a direct means of spreading organisms from one area to another. Housekeeping personnel entering contaminated areas must take the same precautions in the use of caps, gowns, gloves, and masks as other personnel, to prevent the spread of infection.

Periodic cleaning or heavy duty cleaning should be performed in areas where patients can be moved or in areas where patients are not present at all times. This includes cleaning of windows, and walls; stripping of wax and sealers from floors; and cleaning of doors by the use of mechanical equipment. Electrical equipment should be utilized for this work, and the individuals should be specifically trained in the cleaning procedures and proper use of the equipment.

Special cleaning procedures should be carried out in isolation rooms or when terminal cleaning of the room is called for, after the patient with an infection or contagious disease has been transferred, discharged, or has died. Special cleaning techniques are also used in operating rooms, delivery rooms, nurseries, and in rooms where reverse isolation has taken place. In special cleaning, it is essential that all equipment and materials be fresh, clean, and free of contaminating organisms. The cleaning equipment must not be transferred from special cleaning areas to other areas of the hospital. All disposable materials and equipment should be marked, and placed in special plastic bags to be disposed of separately. Equipment should be washed, cleaned, and sanitized before reuse when possible. All personnel should perform a 2-min medical scrub before entering or after leaving these areas.

Lighting

The desired levels of lighting vary tremendously from the areas where social activities and gatherings may take place with a minimum of 10–15 foot candles, to the cystopscopic tables, where a minimum of 2500 foot candles of light is required. It is suggested that the environmentalist obtain lighting standards from the various Illumination Engineering Society reports that are current for a given time period. Since there are innumerable places within the institution with different lighting requirements, it is difficult to keep an updated list in any text.

Plumbing and Water

All sinks used for patient care or in-service areas must have a water spout a minimum of 5 in. above the rim of the fixture. In surgical areas and in patient isolation areas foot controls are necessary for the staff to hand cleanse before and after patient care or treatment. It is particularly important that water storage, such as distilling tanks, is tested frequently, since the water could become contaminated. Beside the normal water supply, institutions need a minimum pressure on the upper floors of 20 lb/in.^{2} during maximum demand periods. Additional water supply and water sources are needed for firefighting. These water supplies should be kept separate from the potable drinking water. (Further information concerning water, plumbing, sewage, cross-connections, and submerged inlets is found in Volume II.)

Emergency Medical Services

Typical ambulance specifications include front-disk power brakes, eight-cylinder engine, a speed of at least 70 mi/hr when fully loaded, three forward transmission gears, and heavy-duty double-action shock absorbers. Another requirement is ambulance rescue equipment, including a number 10 ABC dry chemical fire extinguisher, primer wrecking bar, ropes, bolt cutters, and so forth. A third requirement is emergency care equipment, such as portable suction apparatus, bag mask ventilation unit, and oral phyrayngeal airways for adults, children, and infants.

Training requirements include emergency medical technician courses successfully completed and a passing grade on a written and practical examination, 24 hr of in-service instruction each year, and U.S. Department of Transportation emergency medical technician refresher training courses.

Fire Safety

The fire program should come under the control of a fire marshall or fire director trained by the fire department. The fire marshall should be a member of the overall safety committee and should be responsible for evaluating all fire hazards within the institution. The fire marshall, in conjunction with appropriate administrators, should determine the institutional needs and provide the following: supervision of fire prevention, protection, and suppression programs for the staff; administrative supervision and support for all subareas within the institution; coordination of all activities of all subareas; final authority on all matters related to fire prevention, protection, and suppression; ensure that periodic inspections are made of all parts of the building and that all equipment, facilities, procedures, and regulations are in keeping with good fire prevention policy; and act as administrative head should fires occur.

The subunit fire staffs, comprised of individuals from various operating departments of the hospital, have responsibility for working directly in all program activities or emergencies under the fire director; conducting training of fire protection and suppression staffs; understanding the effects of extinguishing agents, such as cooling, smothering, anticatalytic, or inhibiting; scheduling and conducting fire-prevention inspections; reviewing and monitoring all activities that constitute fire hazards; holding regular conferences with all individuals involved when fire incidents occur; reviewing and approving plans and drawings for construction and alteration of buildings to ensure adequate fire protection; reviewing and approving storage plans for flammable materials; approving requests for open flame operations, such as welding; making technical investigations concerning fire incidents; preparing and publishing seasonal fire protection and prevention material; enforcing fire standards and rules in various work areas; helping develop, coordinate, and present plans for emergency evacuation of personnel, fire control, and salvage of property; conducting fire and evacuation drills; evaluating workers' knowledge of fire extinguishers, fire practice, and how to report fires; developing and posting operational instructions for extinguishers and fire hoses; posting appropriate posters and signs concerning fire problems; and working closely with the fire department.

In the event of a fire at an institution, a specific code should be used to alert the institutional personnel concerning the whereabouts and severity of the fire. At the same time, the fire department should automatically be notified of the existence of a fire and its location. Immediate action should be taken to remove individuals from the fire area and to extinguish the fire if possible. Generally, however, in the event of serious fires, fire department personnel should be the ones responsible for extinguishing the fire. Evacuation plans should be fully understood and personnel well trained in necessary evacuation procedures.

Fire equipment on the premises should be checked periodically to ensure that it is in proper operating condition. This should be done by institutional personnel and by the fire marshall's office.

Typical fire prevention rules in an institution include keeping stairwell doors closed at all times; keeping room and corridor doors closed; discarding all solid waste promptly; placing waste chemicals in special safety cans and removing them through specialized hazardous waste collection companies; providing adequate ashtrays; and keeping corridors free of obstructions.

Patient Safety

Patient safety rules include keeping the adjustable height beds in the low position, except during actual treatment; the use of safety or siderails on the bed; the use of safety vests that restrict but allow movement; proper indoctrination of the patients to the hospital environment; a good communications system, including signals within the bathrooms; proper positioning of bedside tables and other furniture; proper use and instruction in use of patients aids such as walkers, wheelchairs, and canes; installation of shower bars; stools within showers; adequate illumination during the day and night; special attention to patients most frequently prone to accidents, such as those on medication, the elderly, and those with diseases causing dizziness; proper medication at proper times to avoid medication errors.

Radiation Safety

X-ray equipment should be monitored by competent individuals for stray radiation and improper functioning. Factors in radiation safety include the X-ray beam size, X-ray machine filtration, shielding, scatter measurements, tube-housing leakage, and fluoroscopic exposure rates. Surveys are conducted to determine the level of leaking radiation and to suggest recommendations for necessary corrections. The handling of radioactive material is of grave concern. The materials must be handled in such a way that the operator and the patient are not contaminated. Several specific controls are involved, including restricted areas for use of radioactive material; elimination of pipetting by mouth of radioisotopes; use of impervious gloves; proper cleaning techniques; storage of radioisotopes in containers that limit exposure to 100 mrem/hr maximum at a distance of 1 ft from the container. Hot radioactive materials must be transported in shatterproof containers and shielded, depending on the type of radiation. Radioactive wastes are stored in metal cans and removed to restricted areas, where they are disposed of using special techniques. Animals containing radioactive materials are tagged, and all waste material, including their excreta, are collected and disposed of as all other radioactive material would be. The carcasses are wrapped in absorbent material and disposed of in a safe manner. Human sources of radioactive material include patients containing radionuclides that are used for treatment or diagnostic purposes. The internal beta-gamma emitters must be kept at the lowest practical level. Patients should not receive a total dose exceeding 10 rad in any 12-month period. External beta-gamma emitters must not exceed the levels of the internal beta-gamma emitters. Since internal alpha emitters are particularly hazardous, special consideration must be given by a committee on radionuclides before they are used on humans. Nursing care for patients receiving radioactive isotopes must be carried out with the greatest caution, since the individual or the nurse may have contaminated skin, be contaminated by inhalation or ingestion of the radioactive materials, or may have their body irradiated by these materials. These problems are largely eliminated by good housekeeping, proper hand-washing techniques, and clean work habits. In addition, equipment must be handled with great care.

Infectious Waste Management

An Infectious Waste Management program provides protection to human health and environment from infectious waste hazards. An infectious waste management system should include the following:

1. Designation of infectious waste.
2. Handling of infectious waste, which includes segregation, packaging, storage, transport and handling, treatment techniques, disposal of treated waste.
3. Contingency planning and staff training. The program needs to be designed for the particular situation. Such things as the number of patients, the nature of the infectious waste, the quantity of infectious waste generated, the availability of equipment for treatment on site and off site, the regulatory constraints, and the cost are important factors in planning. The method of treatment will vary with the waste type. The waste must be evaluated for its potential to cause disease. Such characteristics as the chemical content, density, water content, and microbiological content must be evaluated. Steam sterilization may be used for laboratory cultures, whereas incineration would be used for pathological waste. When possible, it is best to handle all infectious waste in the same way in order to cut down on cost. In any planning program, it is necessary to determine the local standards for incinerators, regulations for water quality, the regulations and standards relating to chemical pollutants, thermal discharges, biological oxygen demand, total suspended solids, regulation of sharps, and any other environmental regulations at the local or state level.
4. A responsible person at the facility should prepare a comprehensive report that outlines the policies and procedures for the management of all infectious and hazardous waste in the facility. Reports should contain detailed procedures for all phases of the waste management program at the facility, including research, clinical laboratories, autopsy rooms, and other types of infectious waste from a variety of areas of the institution.

It is necessary to determine which of the waste at the institution will be considered to be infectious waste, and specific policies and instructions concerning this waste need to be presented in written material. The infectious waste should then be separated from the general waste stream to make sure that these wastes will receive appropriate handling and treatment. The infectious waste should be segregated from the general waste stream at the point at which it is generated, either in the patient areas or the laboratories. Infectious waste should be discarded directly into containers or plastic bags and marked with the universal biological hazard symbol, which is three circles with a gap overlaying a smaller circle. The symbol is fluorescent orange or orange red in color.

The infectious waste should be packaged to protect the waste handlers and the public from possible disease and injury. The packaging must be preserved through handling, storage, transportation, and treatment. In selecting the packaging container, it is necessary to determine the type of waste, the handling and transport of the packaged waste before treatment, the treatment technique, special types of plastic bags, and the package identification. Liquid waste should be placed in capped or tightly stoppered bottles or flasks. Containment tanks can be used for large quantities of liquid waste. Solid or semisolid waste, such as pathological waste, animal carcasses, and laboratory waste, may be placed in plastic bags. Sharps should be placed directly into impervious, rigid, and puncture-resistant containers to eliminate injury. The clipping of needles is not recommended, unless the clipping device contains the needle parts. Otherwise, an airborne hazard may occur. Sharps should be placed in such containers as glass, metal, rigid plastic, wood, and heavy cardboard. The sharps' containers should be marked with the universal biohazard symbol. A single bag for containment of infectious waste is not adequate. Either a double plastic bag should be used or a single bag should be placed in a bucket, box, or carton that has a lid or a seal. Containers of sharps and liquids can be placed in other containers for transportation and storage. When other plastic bags of infectious waste are handled and transported, care must be taken to prevent tearing of the bags. Plastic bags containing infectious waste should not be placed in a trash chute or on a dumbwaiter. The proper practice for the handling of these plastic bags include loading by hand, transporting in such a way that minimal handling occurs, and putting the plastic bags in rigid or semirigid containers to prevent spills or breakage of the bags. Where recycled containers are used for transport in the treatment of bag waste, they should be disinfected between use. If the waste is to be incinerated, the containers must be combustible. If the waste is to be steam sterilized, the packaging materials must allow steam penetration and evacuation of air. Interference with the steam sterilization treatment may occur because of high-density plastics, which prevent effective treatment of the air and therefore will not allow the contents of the bag to reach the appropriate temperatures. Low-density plastics will enhance steam penetration and allow air evacuation from the waste.

There are several types of plastic bags available for waste disposal. The quality of the plastic bag and its suitability to contain infectious waste is based on the raw materials used during manufacture; and the product specifications suitability is based on the fitness and the durability of the plastic. It is most important that the plastic bas be tear resistant until the waste is treated. Tear resistance can be improved by not placing sharps, sharp items, or items with sharp corners in the bags or overloading the bag beyond its weight and volume capacity.

Although it is preferable to treat the infectious waste as soon as possible, it may not be practical to do so. Therefore, the waste will have to be stored in an appropriate manner. There are four important factors related to the storing of infectious waste. They are the packaging, storage temperature, duration of storage, and location and design of the storage area. The packaging must provide containment of the waste throughout the waste-management process and must deter rodents and insects from invading the packaging, and thereby potentially becoming vectors of disease. As the temperature increases, the rate of microbial growth and putrefaction increases. This will cause bad odors to be associated with the waste because of the retained organic matter present. Time is also a major consideration here. It is suggested that the waste be stored under refrigeration in special units at 34–45°F for a maximum of 3 days. The storage areas should be extremely secure, have limited access, be free of rodents and insects, and display the universal biohazard label.

The infectious waste when transported on or off site should be in appropriate packaging, which has already been mentioned, and in containers that are leakproof, rigid, and appropriately marked.

The infectious waste is treated to reduce the hazard associated with the presence of infectious agents. This will be discussed further under Controls.

Once the infectious waste has been effectively treated, it is no longer biologically hazardous and may be mixed with and disposed of as ordinary waste, providing it does not pose any other hazards that are controlled by federal or state regulations. Treated liquid waste may be poured down the drain to the sewer system. Treated solid waste and incinerator ash may be disposed of in a sanitary landfill. Needles and syringes must be rendered nonusable before disposal. Treated sharps can be ground up, incinerated, or compacted. Body parts should be incinerated under special incinerators or buried, depending upon religious law.

The institution should set up a contingency plan, which would provide for the following situations:

1. Spills of liquid infectious waste and the cleanup procedures utilized by personnel. Also, procedures should be available for the protection of the personnel and disposal of the spilled residue.
2. Rupture of plastic bags or other loss of contaminants and the necessary cleanup procedures, protection of personnel, and the repackaging of the waste.
3. Equipment failure and the alternate arrangements for waste storage and treatment.

Staff training is essential in order that all technical, scientific, housekeeping, and maintenance personnel will be protected. The training programs should be used for all new employees whenever infectious waste methods and practices change and when a new program is put into operation. Continuing education, including refresher courses, should be given periodically to protect the employees of the institution.

The Medical Scrub

The medical scrub, which should take a minimum of 2 min should be practiced by all personnel when hand washing procedures are indicated. The medical scrub consists of the following steps:

1. Remove all jewelry, rings, watches, etc.
2. Turn on the water (running water is essential).
3. Wet hands.

4. Soap hands and wash them. Use considerable friction. Begin with hands, lead to wrists, then to elbows (be sure to clean well between fingers and around thumb).
5. Rinse (rinse hands, arms to elbow, letting water run off the elbow.)
6. Clean nails under running water. Use orange stick and discard.
7. Repeat steps 4 to 6.
8. Dry hands. Wipe water from hands to elbow with one paper towel. Use second paper towel to finish drying.
9. Turn off faucet. Use second paper towel if knee-action control is absent. Use paper towels to turn off the faucets.

Essential Practices Related to AIDS Virus and Hepatitis Virus

The essential practices related to the prevention of the transmission of the HIV and HBV viruses include the following:

1. classification of the work activity
2. development of standard operating procedures
3. provision of training and education
4. development of procedures to ensure and monitor compliance
5. workplace redesign

Further, the worker needs to have available a program of medical management. This program should reduce the risk of infection by HBV and HIV. All workers involved in any exposure to blood or other body fluids should be vaccinated with hepatitis B vaccine.

Once an exposure has occurred to blood or other infectious body fluids and the person has not been adequately protected, a blood sample should be taken from the individual (with the individual's permission) to determine if the surface antigen of hepatitis B or the antibody to HIV is present. If the individual has not previously been given the hepatitis B vaccine, a single dose of hepatitis B immune globulin is recommended within 7 days of exposure. The individual should also receive the vaccine series. If the person is possibly exposed to AIDS, it is recommended that the Centers for Disease Control in Atlanta, Georgia be contacted and the most recent techniques approved by the Centers for Disease Control be followed.

Employees need to take precautions to prevent injuries from needles, scalpel blades, and other sharp instruments or devices that they use during various procedures and that they may handle during cleaning, disposal, or removal of these objects.

Schools

Physical Plant and Site Selection

Site selection is based on present and projected student population; distance from areas served; distance from sources of noise, air pollution, and other hazards; type of surface drainage; type and nature of industrial or highway construction; amount of acreage needed for athletic fields, parking, landscaping, and other recreational areas; access to good highways; suitability for a given neighborhood. The site is also determined by the type of soil, soil conditions, and the recommendations of school architects, engineers, public health officials, and community groups.

The physical structure or structures should be designed and built to facilitate learning and to reduce the potential for hazards. The building construction must take into account all necessary lighting and acoustics, heating and ventilation, water supply and waste disposal, laboratories and gymnasiums, workshops, and classrooms. Before the plans are approved, the school architect submits them to an advisory committee of school personnel, community groups, and finally to the state departments of education and health.

Housekeeping

A clean environment is conducive to a happy school day and therefore promotes good emotional health and the desire to learn. Many of the housekeeping tools and techniques described under the section on hospitals are appropriate for schools. (The book entitled "Environmental Health and Safety" by this author mentioned earlier contains detailed housekeeping procedures.) Proper organization of the housekeeping or custodial staff, adequate scheduling, proper training of personnel, and the use of proper cleaning agents and detergent disinfectants are essential to a good housekeeping program. The program should include, at the minimum, the following responsibilities for housekeeping personnel: maintenance of a clean, safe, and aesthetically satisfying environment; maintenance of grounds to prevent accidents and to promote aesthetic appeal; maintenance and orderliness of all apparatus and equipment; prevention of fires and promotion of fire safety; proper operation of heating ventilation, and all other types of equipment; adequate storage of all materials utilized in construction; separate and proper storage of all hazardous equipment and materials, including chemicals and other potential fire hazards; and maintenance of proper records concerning housekeeping.

A variety of floors are used within the school as within other kinds of institutions. These floors include concrete, terrazzo, ceramic tile, rubber or vinyl tile, asphalt tile, magnesite, cork, and a variety of woods. Concrete floors are quite common in school buildings. They are easy to clean, yet they sometimes pit or crack. Adequate maintenance is important. If the concrete is painted, it should be thoroughly cleaned with solvents to remove all oils and greases. Terrazzo floors are a mixture of cement and marble, or cement and granite chips. They are strong and durable, but acids, abrasives, and strong alkaline cleaners cannot be used because they will deface the terrazzo. It is important to put a sealer on this type of floor to prevent penetration. Ceramic tile floors, rubber or vinyl tile floors, and asphalt tile floors are not resistant to grease, oils, turpentine, or petroleum-based waxes. These floors look very attractive and must be given special care. Magnesite floors are somewhat like concrete and should be treated similarly. Cork floors may be cleaned by sanding lightly with steel wool and applying light water-emulsion wax. Wood floors consist of a variety of woods, ranging from very hard to soft. Although they are very pretty if maintained properly, many wood floors have difficulty withstanding the constant pressure of footsteps.

A special problem within schools are chalkboards. Black slate boards are difficult to clean and pose a problem for use in classroom instruction. It is preferable to use green or blue boards or other kinds of special materials. Erasers are best cleaned by using a special electric vacuum eraser cleaner. This should be carried out one or twice per week.

The following lighting standards are recommended: 10 footcandles of light for open corridors and storerooms; 15 footcandles for auditoriums, cafeterias, locker rooms, and stairways; 20 footcandles for reception rooms, gymnasiums, and swimming areas; 75 footcandles for classrooms, study halls, lecture rooms, libraries, shops, and laboratories; and 150 footcandles for drafting rooms, typing rooms, sewing rooms, and special learning rooms.

Maintenance is an essential part of lighting. A program should be established to include periodic cleaning of lamps and fixtures, replacement of burnt-out bulbs, and repainting of room surfaces where needed. Student seating should be arranged to eliminate shadows and glare. Window shades, drapes, or other types of window coverings should be provided to give balanced lighting and to eliminate glare problems.

Noise

Noise prevention starts during construction of the school. Site location, classroom design, and the use of special noise-control materials are important. Noise must be suppressed, since it is a source of irritation, distraction, and emotional strain, and fosters inefficient performance in school children, teachers, and employees. Noise levels should not exceed the following: classrooms, 35–50 dB; cafeterias, 50–55 dB; outside noise, 70 dB; other specialized rooms, 40 dB.

Space Utilization

Each elementary school should have a minimum of 12 acres of land for the first 200 students and 1 acre of land for each 100 additional students. In the kindergarten, 1050 ft^2 of space is needed for a maximum of 30 students. In grades one through three, 875 ft^2 are required for 30 students, and in grades four through six, 750 ft^2 per 30 students. In the high schools, which include junior and senior high schools, 12 acres of ground are required for the first 300 students, and 1 acre for each 100 additional students. Also, in the high schools, 85% of the classrooms must have a minimum of 750 ft^2 for each group of 30 students. Other classroom sizes vary with the types of activities. All of the above requirements are minimum standards that can and should be exceeded by the school system for special types of study or where more space is desirable.

Plumbing

There should be one sink for every 30 students in elementary schools, and one sink for every 40 students in secondary schools. In the elementary schools, a toilet is required for every 40 male students and one for every 30 female students. In the secondary schools, one toilet is required for every 75 male students and one for every 45 female students. There should be at least one shower head for every five students and one drinking fountain for every 100 students.

The recommended classroom temperatures during the winter should be at 68°F with slightly lower temperatures in gymnasiums. The temperature should be approximately 70°F in elementary schools. Fresh air should be added to the school at a rate of not less than 15 ft^3 of air per minute per person per room. Where the activity level is higher, the amount of room air should be increased to 20 or 30 ft^3 per minute per person. The air must be distributed within the room to reduce a chilling effect and to prevent discomfort. The velocity of air should not exceed 500 ft/min at the duct outlets. A recommended relative humidity would be 50%. The type of heating system varies with the school, its location, and the type of fuels available.

MODES OF SURVEILLANCE AND EVALUATION

Infections

There are several techniques used to determine the extent of possible hospital-acquired infections. The first technique is to make in-depth intensive microbiological studies of all critical areas and other areas of the hospital environment. These samples are taken through the air or by means of swabs and Rodac™ contact plates used on surfaces. As a general rule, the microbiological sampling technique is not effective and should only be utilized where a specific organism is being sought and the cause of the disease outbreak cannot be determined. Microbiological sampling is also used on a research basis to determine the levels of microorganisms in given types of operations or where special study techniques have been used and the sources of the outbreak of disease cannot be traced.

A second technique involves the development of a surveillance and reporting program in which a nurse epidemiologist regularly carries out the following functions:

1. reviews daily charts indicating that the patient has an infection or is suspected of having an infection

2. holds conferences on a daily basis with nursing supervisors and nursing units to determine potential infection problems
3. reviews all laboratory reports confirming infections and correlates the laboratory studies with the observations of the physicians and nurses
4. reviews special reports submitted by physicians or supervising nurses where suspected infections problems are occurring
5. lists and tries to determine the location and source of specific types of infections
6. evaluates necrospy reports to determine if infections not previously reported are found upon autopsy
7. maintains a close liaison with the employee health center and evaluates employee infections as they may relate to hospital acquired infections
8. repared monthly statistical reports indicating levels of infection as determined by charts and laboratory tests and pinpoints where the higher levels of infection occur

The nurse epidemiologist may work for the hospital environmental health specialist, whose responsibility would encompass the overall environmental health, infection control, and occupational health and safety programs of the institution.

The third type of technique is a programmed approach for a team comprised of the supervising nurse, the physician in charge of the unit, and the environmental health specialist, using specifically established study forms to determine the potential sources of infection, within the institution. This team approach, systematically carried out, helps to establish the potential sources of infection, thereby assisting the individuals within their own unit and also in the hospital at large in determining the kinds of potential infections occurring within the institution. The programmed approach and the study form are found in Diagram 8.1 and Figure 8.1.

Accidents

The hospital building should be carefully evaluated and supervised to minimize accidents and injuries. It is important to evaluate all accidents and illnesses to determine if any hazards were involved and to make the necessary recommendations to eliminate the hazardous situation or unsafe behavioral pattern. When investigating the incident, all possible facts should be collected. The equipment operation should be fully understood; processes, special conditions, the fatigue factor, and other environmental factors related to the individual should also be known. Always determine the direct causes of the accident, such as faulty operating equipment or the use of improper techniques, and the indirect causes, such as poor eyesight and ignorance. There may be additional contributory causes, such as inadequate standards, inadequate training, lack of policy, lack of enforcement, faulty design, faulty maintenance, or failure of supervisors to perform their functions. Reporting and recording of information notifies the institutional management of the incident, the results of the incident, the contributory causes, and in this way helps them conduct proper evaluations and take necessary actions to prevent future injuries. Inspections, special studies, and investigations before

Programmed Approach to Hospital Infection Control

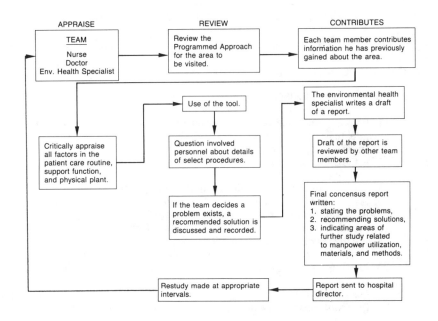

APPRAISE

TEAM
Nurse
Doctor
Env. Health Specialist

REVIEW

Review the Programmed Approach for the area to be visited.

CONTRIBUTES

Each team member contributes information he has previously gained about the area.

The environmental health specialist writes a draft of a report.

Use of the tool.

Critically appraise all factors in the patient care routine, support function, and physical plant.

Question involved personnel about details of select procedures.

Draft of the report is reviewed by other team members.

If the team decides a problem exists, a recommended solution is discussed and recorded.

Final concensus report written:
1. stating the problems,
2. recommending solutions,
3. indicating areas of further study related to manpower utilization, materials, and methods.

Restudy made at appropriate intervals.

Report sent to hospital director.

Diagram #1

accidents occur are of vital importance. A good inspection should determine specific unsafe or unhealthy acts and conditions; determine the need for specific safeguards for people, equipment, and materials; detect inefficient, inadequate, or totally lacking safety programs; determine the condition of work areas, structural components, and service facilities; determine the condition of equipment; determine job procedures; identify the various types of materials, chemicals, and other hazardous substances being used; identify the controls that should be employed to prevent accidents; determine if these controls are actually operable; determine all of the potential problems related to the environment, such as lighting, ventilation, heat, obstructions, slippery surfaces, and so forth; determine the kinds of existing hazards and the types of controls used for each type of operation.

Evaluation by the occupational health specialist includes routine surveys of chemical laboratories, storage, and handling areas; sampling and detection of airborne contaminants; evaluation of noise, lighting, temperature, ventilation, and so forth; regular testing of chemical fume hoods to ensure proper functioning;

_____ Hospital

PROGRAMMED APPROACH TO HOSPITAL INFECTION CONTROL

S = Satisfactory   I = Improvement Needed   U = Unsatisfactory

Building Name and No. _____ Floor _____ Date _____ Time_____

Names of Study Team _____

ADMINISTRATIVE PROCEDURES

_____ 1. Does any employee have a patient history of being a carrier of pyogenic infections?
_____ 2. Do all employees report pyogenic infections?
_____ 3. Is the employee who reports an infection immediately removed from patient care?
_____ 4. Is there a restriction of traffic within the given patient area?
_____ 5. Is there routine daily recording of infections on charts?
_____ 6. Is indiscriminate antibiotic use discouraged?
_____ 7. Are the interns given instruction in hospital infection control?
_____ 8. Are the residents given instruction in hospital infection control?
_____ 9. Are the nurses given instruction in hospital infection control?
_____10. Are nonprofessional employees given instruction in hospital infection control?
_____11. Are visitors restricted in time and number?
_____12. Do visitors observe special isolation procedures when indicated?

STERILIZATION PROCEDURES

_____ 1. Is each autoclave tested monthly for its ability to destroy bacterial spores?
_____ 2. Are autoclave indicators or dates used?
_____ 3. Are routine cultures made of utensils, trays, packs, instruments, catheters, gloves, sutures and parenteal solutions?
_____ 4. Are records kept of all isolations of significant bacteria such as "staph"?
_____ 5. Are all autoclaved items distinctly marked as sterile?
_____ 6. Are autoclaved items stored separately from nonsterile items?
_____ 7. Are autoclaved items protected from contamination while in storage?

LAUNDRY AND LINEN PROCEDURES

_____ 1. Are linens washed in such a manner that all pathogenic bacteria are destroyed?
_____ 2. Are bacteriological samples taken of the clean linen to insure that all pathogenic bacteria are destroyed?
_____ 3. Are the clean, ironed linens exposed to the dirty linens in such a way that there may be a transfer of pathogenic bacteria?

Figure 8.1.   Programmed approach study forms. *Source: Environmental Health and Safety* (Elmsford, NY: Pergamon Press, 1974), p. 140. With permission.

_____ 4. Are clean linens protected during transportation to ward areas?
_____ 5. Are clean linens transported on the same truck as dirty linens?
_____ 6. Are all linens, blankets, and other bed covers gently removed from beds to prevent air dispersion of bacteria?
_____ 7. Are all linens from infected patients placed in a specially marked laundry bag and closed at once?
_____ 8. Do all linens from infected patients receive special handling?
_____ 9. Are laundry bags from isolation placed in clean outer bags?
_____10. Are all blankets sterilized before reuse?
_____11. Are mattresses washed with a germicide between patients?
_____12. Are pillows washed with a germicide between patients?
_____13. Are beds washed with a germicide between patients?
_____14. Are nonporous covers used for mattresses and pillows?
_____15. Are the laundry trucks and carts cleaned daily?
_____16. When not in use, are the trucks stored in an area which is free of gross contamination?

TRASH AND SEPTIC WASTE REMOVAL PROCEDURES
_____ 1. Are infectious materials either autoclaved or placed in sealed plastic bags before being discarded?
_____ 2. Is food waste from infected patients placed in sealed plastic bags?
_____ 3. Are paper bags used in isolation rooms for discarded tissues, dressings, tongue blades, and other disposable items?
_____ 4. Are they immediately closed and placed in a sealed plastic bag?
_____ 5. Are bags containing infectious materials marked "Danger, Infectious Materials Within"?
_____ 6. Do employees who remove infectious materials discard their gowns at the completion of the waste disposal and then follow proper hand washing procedures?
_____ 7. Do employees who remove infectious materials wear gowns?

FOOD HANDLING PROCEDURES FOR INFECTED PATIENTS
_____ 1. Are all clean dishes, trays, and eating utensils protected against contamination during transportation or storage?
_____ 2. Are disposable single service paper units discarded after each use?
_____ 3. Are all remnants of meals scrapped into a plastic bag in the isolation room?
_____ 4. Are nondisposable utensils thoroughly and immediately disinfected after each use by infected patients?
_____ 5. Are all dishes, trays, and eating utensils washed and sanitized before reuse?
_____ 6. Is ice making equipment and ice storage equipment kept clean and sanitary?
_____ 7. Are there individual water containers and glasses for each patient?
_____ 8. Are these containers and glasses thoroughly washed and sanitized before reuse?
_____ 9. Are utensils for medications thoroughly washed and sanitized before reuse?

Figure 8.1. continued.

BEDPAN HANDLING

_____ 1. Are all bedpans, urinals, enema containers and other equipment for collection of excreta sterilized before reuse by other patients?

_____ 2. Are all such contaminated utensils handled in an aseptic manner after use by patients?

_____ 3. Are bedpans stored off the floor in a cabinet?

_____ 4. Is toilet paper stored off the floor in a sanitary manner?

EQUIPMENT HANDLING AND SANITIZATION PROCEDURES

_____ 1. Is there an individual thermometer for each patient?

_____ 2. Are thermometers stored in an adequate concentration of germicide to kill all bacteria?

_____ 3. When equipment such as portable X-ray units, flashlights, otoscopes, ophthalmoscopes, stethoscopes, ice bags, etc., are used for infectious cases are they cleansed thoroughly with a germicide solution before reuse?

_____ 4. Is there a Graduate Nurse assigned to the use of the dressing cart?

ISOLATION PROCEDURES

_____ 1. Are all patients with draining wounds isolated?

_____ 2. Is reverse isolation strictly enforced when indicated?

_____ 3. Are patients isolated in such a way and in such areas that proper procedure can be carried out?

_____ 4. Are isolated patients properly identified?

_____ 5. Is a suspected infected case reported the day it is noted?

_____ 6. Are clean gowns put on before entering isolation rooms?

_____ 7. Are the gowns discarded in containers in the isolation area?

_____ 8. Are hands washed with an acceptable technique before and after handling any contaminated material or patients?

_____ 9. Are all areas in which infectious cases have been treated thoroughly cleaned and sanitized before reuse?

CATHETERIZATION PROCEDURES

_____ 1. Are the genitalia thoroughly cleaned with an appropriate soap in preparation for the catheter?

_____ 2. Does the catheter come from a sterile package?

_____ 3. Does the syringe used for irrigation of the catheter come from a sterile package?

_____ 4. Are hands washed before donning sterile gloves?

_____ 5. Do the gloves used for sterilization come from a sterile package?

_____ 6. Does the urine drainage bottle overflow?

_____ 7. Is the urine bottle suspended above the floor?

_____ 8. Is the urine bottle autoclaved?

_____ 9. Is there a cotton plug in the air vent of the catheter?

_____10. Is an antiseptic present in the urine bottle?

DRESSING OF WOUNDS

_____ 1. Are dressings put on with gloved hands?

Figure 8.1. continued.

_____ 2. Are hands washed before and after dressing patients' wounds?
_____ 3. Are dressing trays and carts arranged to prevent contamination of sterile and clean equipment?
_____ 4. Are dirty drapes or dirty dressings kept on dressing cart?
_____ 5. Does cold sterilization take place on the dressing cart?
_____ 6. Are wounds dressed at specifically scheduled times?

SURGICAL FACILITIES
_____ 1. Are the operating suites separate from each other and other patient areas?
_____ 2. Is cross traffic through these units from other hospital areas eliminated?
_____ 3. Are the dressing rooms for physicians and nurses accessible directly from the rest of the hospital and do they open directly into the operating suites?
_____ 4. Is there adequate separation between utility areas used for clean and those used for contaminated equipment, supplies, or waste?
_____ 5. Are all surgical wastes placed in plastic bags for disposal?
_____ 6. Are sub-sterilizers (rapid instrument sterilizers) accessible to operating rooms?
_____ 7. Are facilities for scrubbing directly accessible to each operating room?
_____ 8. Are these facilities kept clean at all times?
_____ 9. Are locker rooms kept clean at all times?
_____10. Is there adequate space in clean areas for storage of sterilized items?
_____11. Are there facilities in the recovery room for isolation of infected cases?
_____12. Are there closed containers for all contaminated and soiled dressings and other contaminated material?
_____13. Are there adequate closed containers for disposal of gowns, masks, and booties in the locker room?

ANESTHESIA AND INHALATION THERAPY
_____ 1. Inhalation
         a. Is the equipment cleaned and sterilized after use by patients?
_____ 2. Intravenous
         a. Do the needles and syringes come in a sterile package?
         b. Does the plastic tubing come in a sterile package?
_____ 3. Spinal
         a. Is detergent used in cleaning needles?
         b. Is all equipment sterilized?
         c. Are solutions sterilized?
         d. Are contents of ampoules sterile?
_____ 4. Conduction Anesthesia (local)
         a. Are gowns, masks, gloves, worn when anesthesiologist does conduction anesthesia?
_____ 5. Is the anesthesia table stored outside of the O.R. within a traffic area?
_____ 6. Are endotraceal tubes cleaned and then sanitized in 70% alcohol plus Formalin? (1 part Formalin to 250 parts alcohol.)
_____ 7. Is the endotraceal tube then placed in a boiling sterilizer to boil off the Formalin?

Figure 8.1. continued.

_____ 8. Is humidification maintained during inhalation therapy?
_____ 9. Is the equipment cleaned properly?
_____10. Are the nasal catheters sterilized?
_____11. Is the humidifying solution sterile?

SURGICAL PROCEDURES

_____ 1. Do all personnel change from street clothes before entering surgical
          suites?
_____ 2. Are booties put on at the O.R.?
_____ 3. Are surgical room clothing and shoes worn outside of the operating room
          suite?
_____ 4. Are surgical gowns put on and discarded properly?
_____ 5. Are adequate surgical gowns provided for the surgeons?
_____ 6. Are surgical masks discarded after each procedure?
_____ 7. Are surgical masks discarded when wet?
_____ 8. Are hands washed before donning surgical suit, mask, and cap?
_____ 9. Are patients brought to the O.R. on beds?
_____10. Are they returned on their beds?
_____11. Does the operating table leave the O.R. suite?
_____12. Does the operating team scrub its hands thoroughly in an acceptable
          manner with soap or liquid hexachlorophene detergent prior to an opera-
          tion?
_____13. Is the patient "prepped" in an acceptable manner immediately prior to
          the operation and before the surgeon gowns?
_____14. Are the razors and razor blades used for preoperative shaving cleaned
          and sterilized after each use?
_____15. Are disposable razors or razor blades used for preoperative shaving?
_____16. Is the nurse responsible for sterility of surgical procedures? (In unusual
          cases the surgeon makes this determination.)
_____17. Does the surgeon inform the nurse about a contaminated case?
_____18. Are precaution gowns provided for the circulating nurse and
          anesthesiologist?
_____19. Are contaminated patients scheduled for a separate operating room?
_____20. Are the contaminated cases scheduled last?
_____21. Are special precautions taken for T.B. cases in the O.R.?
_____22. Is the questionably contaminated case treated as a contaminated case?
_____23. On contaminated cases, are the gowns removed before the gloves?
_____24. Do personnel wash hands after a contaminated case and before going to
          the dressing room?
_____25. Is dropped equipment resterilized before reuse?
_____26. Is talking kept to a minimum during an operation?
_____27. Is the number of people in the O.R. kept to a minimum?
_____28. Is the suture package opened in an aseptic manner?
_____29. Is the adhesive tape stored in a satisfactory manner?
_____30. When operating in two different sites (grafting procedure) are separate
          instruments used?
_____31. When operating in two different sites, are different personnel used?
_____32. When operating in two different sites, do the personnel rescrub and
          regown:

Figure 8.1. continued.

_____33. Are special precautions taken during an intestinal anastomosis? (Is a sub-set of instruments used?)

_____34. Are the proper types of drapes used?

_____35. Is the proper thickness of drapes used?

_____36. Are dry drapes used?

_____37. Do they get wet during the procedure?

_____38. Are instruments retrieved from below the patient level and used?

_____39. Is sheet wadding sterilized when used on open procedures?

_____40. Are gloves changed when needle punctures them?

_____41. Are personnel with significant respiratory illnesses or with overt lesions excluded from the operating and delivery suites?

_____42. Are Kelly pads being used on the O.R. table?

_____43. Is the O.R. table cleaned properly between operations?

_____44. Is the operating room equipment cleaned properly?

_____45. Is the recovery room clean?

_____46. Is the recovery room maintained with the same basic technique as the operating room?

_____47. Is the operating room thoroughly scrubbed and washed with a germicide solution and allowed to sit for at least 24 hours after an infected case?

_____48. Are the walls spot cleaned between all operations?

_____49. Are the floors and other horizontal surfaces cleaned between all operations?

_____50. Is a germicide solution used in the above mentioned procedures?

_____51. Is there routine bacteriological surveillance of aseptic techniques and sterilization in the O.R.?

_____52. Do surgeons wash their hands, put on masks, and don sterile gloves before performing intravenous cutdowns?

_____53. Are intravenous catheters inserted in an aseptic manner?

_____54. Are intravenous catheters removed promptly when not needed?

HOUSEKEEPING
See self-inspection sheet for Housekeeping Program in Housekeeping section.

COMMENTS:

Figure 8.1. continued.

evaluation and investigation of employee and student accidents; consultation services for special problems to make necessary recommendations for correction; routine surveys of shops, laboratories, and other specialized areas; special studies concerning special hazards.

The Food Service Self-Inspection Concept

Consistent unbiased evaluation of items requiring correction is essential to the improvement of a hospital food-service program. The best approach to be used by the food-service managers is the self-inspection program. The food-service manager, on a weekly basis, should systematically inspect his or her own facility; identify problem areas; determine whether improvements are being carried out each week; and record the facts, which can be used for program changes or additional budget requests for training and/or workforce. In order for the food-service managers to be able to carry out a good self-inspection of the facility, these individuals should by put through a training program, which would be conducted by the local health department, state health department, or an in-house environmental health practitioner. The recommended program would have 17 sessions of 2 hr each. Each session would be conducted as a seminar, with the trainer getting necessary preliminary information, citing case histories and solutions, and then bringing the entire group of managers into the discussion concerning actual problems at the institution and potential solutions for them. The recommended training program is as follows:

- Session 1 – introduction by hospital administrative staff as well as introduction of the program itself
- Session 2 – review of general hospital problems related to infections, injuries, and hospital practices
- Session 3 – review of general hospital problems related to infections, injuries, and hospital practices
- Session 4 – the self-inspection form and technique
- Session 5 – field experience in use of the form
- Session 6 – field experience in use of the form
- Session 7 – review of food-service department problems
- Session 8 – review of food-service department problems and start using forms on a weekly basis
- Session 9 – general housekeeping principles and practices and discussion of results of inspections
- Session 10 – general housekeeping principles and practices and discussion of results of inspections
- Session 11 – general food service equipment and utensil cleaning techniques, and discussion of results of inspections
- Session 12 – bacteriology and communicable disease control and discussion of results of inspections
- Session 13 – food preparation, storage, serving, and transportation and discussion of results of inspections

- Session 14 – insect and rodent control, and discussion of results of inspections
- Session 15 – supervisory techniques
- Session 16 – supervisory problems
- Session 17 – final exam

At the completion of the course, individuals should be given a certificate indicating completion or completion with honors (see self-inspection form in Figure 8.2).

Ventilation

To evaluate a ventilation system, it is necessary to evaluate the air pattern functions, air velocities, room air ventilation rates, and types of air supply and exhaust systems for moving the contaminated particles throughout the room. In this evaluation, the most important factor is the rapidity with which widespread contamination can be removed. The removal rate is determined by taking airborne samples of particles and microbial particles from various parts of the room during various operations.

Housekeeping

It is the function of the housekeeping department to determine the special housekeeping problems and, in consultation with the environmentalist, to deter-mine those matters most critical to the environment. In-depth studies should be made of existing housekeeping practices, work plans, work scheduling, activities of housekeeping personnel, employee training, and work and time evaluations. It is essential that there be understanding as to how organisms are spread, and the daily periodic and special cleaning techniques required to fight the spread of these organisms.

Supervisors, after gaining familiarity with the total cleaning program, will evaluate the work of the personnel and give them necessary additional in-service training when such training is called for. The function of the supervisor is one of inspection, evaluation, training, encouragement, and disciplinary action, in the event that such action is needed.

Various housekeeping evaluation techniques have been proposed. The most effective appears to be the use of housekeeping self-inspection forms, where supervisors evaluate their own areas and the information is then tabulated by the environmentalist. This is supplemented by the use of a white-glove technique on surfaces to determine if they are dirt free and the use of either contact or swab bacteriological samples or air samples. Since sampling is a very costly and inaccurate means of determining the level of housekeeping in an institution, it is recommended that its use be limited to the evaluation of an actual outbreak of disease. Under ordinary circumstances, the use of self-inspection forms and a wet, white towel smeared across surfaces would be excellent indicators of the degree of cleanliness of a given area.

FOOD SERVICE INSPECTION REPORT

(Submit Weekly to Sanitation Officer)
Environmental Health Practitioner

_____ (name) _____ HOSPITAL
ENVIRONMENTAL SANITATION

KEY:
2 – Satisfactory
1 – Improvement Needed
0 – Poor

BUILDING NAME AND NO.	FLOOR	DATE	TIME	TOTAL SCORE

POINTS	ITEM
	1. Perishable foods used within a 2-hour period of preparation or immediately refrigerated.
	2. Perishable foods are not to be reused at anytime.
	3. All refrigerators kept below 45°F.
	4. Ice making machines clean and operating properly.
	5. Food carts washed and sanitized at ward kitchens.
	6. Food protected from contamination: Preparation Storage Transportation
	7. Hands thoroughly washed and cleaned after use of toilets and before preparation of food.
	8. Clean outer garments worn at all times.
	9. Soap, paper towels, and waste receptacles available at all times at handwashing sink.
	10. Lavatory facilities clean and in good repair.
	11. Dressing rooms, areas, and lockers kept clean.
	12. All food handlers with diarrhea removed from duty until negative stool culture.
	13. Food utensils, dishes, glasses, silverware, and equipment scraped, pre-flushed, and soaked.
	14. Silverware, dishes, and glasses clean to sight and touch.
	15. Silverware, dishes, and glasses sanitized.
	16. Dishwasher wash and rinse temperatures checked daily *(insert temp. rinse)* *(insert temp. wash).*
	17. Food utensils, such as pots, pans, knives, and cutting boards clean to sight and touch.
	18. Grills and similar cooking devices cleaned daily.
	19. Food-contact surfaces of equipment clean to sight and touch.
	20. Tables and counters clean to touch.
	21. Non-food-contact surfaces of equipment clean to sight and touch.
	22. Washing and sanitizing water clean.
	23. Single service food articles properly stored, dispensed, and handled.
	24. Containers of food stored off floor on clean surfaces.

25.	No wet storage of packaged food.
26.	Sugar in closed dispensers or individual packages.
27.	Poisonous and toxic materials properly identified, colored, stored, and used; poisonous polishes not present.
28.	Clean wiping cloths used and properly restricted.
29.	All late trays served after 8 p.m. scraped, rinsed, and stacked in previously cleaned sink.
30.	Garbage and rubbish stored in metal containers with lids; containers adequate in number.
31.	All garbage removed from area at end of working day.
32.	Containers cleaned when empty.
33.	When not in continuous use, garbage and rubbish containers covered with tight-fitting lids.
34.	Garbage and rubbish storage areas adequate in size and clean.
35.	All outer openings screened.
36.	Absence of rodents.
37.	Absence of flies.
38.	Absence of roaches.
39.	Floors kept clean.
40.	Floors in good repair.
41.	Floors and wall junctures properly cleaned.
42.	Walls, ceilings, and attached equipment clean.
43.	Walls and ceilings properly constructed and in good repair; coverings properly attached.
44.	Walls of light color; washable to level of splash to shoulder height.
45.	Lighting adequate.
46.	Light fixtures clean.
47.	Areas clean; no litter.
48.	Ample quantities of cleaning supplies available.

REMARKS:

_____

Signature of Dietician or Nursing Supervisor

Figure 8.2. Food service inspection report. *Source: Environmental Health and Safety* (Elmsford, NY: Pergamon Press, 1974), p. 34–35. With permission.

All hospital problems and problems of other institutions are evaluated by trained environmental health practitioners, who use the techniques and study forms discussed in other areas of this book.

The housekeeping self-inspection program is similar to the food-service self-inspection program. The housekeeping training program for supervisors and managers would be similar to the one discussed under food service, but the specific topics would change in food to housekeeping. (See Housekeeping self-instruction form, Figure 8.3.)

INSTITUTIONAL CONTROLS

The institutional environmental health specialist is the staff person most involved in carrying out comprehensive studies and presenting reports with combined recommendations for various committees and individuals. The environmental health professional has specific training in epidemiology; a broad knowledge of biological, physical, and chemical hazards; extensive public health and environmental health training and experience; and an intimate knowledge of the institutional environment, its risks, and problems. The function of this specialist is to promote better patient care and better care of all individuals in institutions by reducing the risk of infection and the opportunity for accidents, and by limiting the extent of injuries and improving the physical environment.

Environmental Control Measures

Environmental control measures include architectural considerations; proper installation of equipment; control of airborne contamination; control of surface contamination from linens, and solid and liquid wastes; and control of other environmental factors.

A good part of environmental control of infections in hospitals is based on initial architectural considerations. Basic problems of hospital design and construction always include traffic patterns; systems used for handling materials, equipment, and liquid and solid wastes; ventilation systems; special airflow control; and the use of easily cleanable materials for surfaces. Traffic patterns are established in such a way as to keep the majority of individuals away from crucial areas, such as the surgical suites, obstetrical areas, isolation and reverse isolation areas, and nurseries. Further, the emergency room and its traffic patterns must prohibit movement of individuals to other critical areas of the institution, unless the patients are to be admitted to the institution. Even at that point, the emergency room, patients, visitors, and families should never enter the critical areas that have been identified.

Materials handling, which is of great concern within the institution, must be so organized to avoid the possibility of disease spread. Mechanical conveyors are designed for easy access for cleaning and inspecting to determine if insect and rodent problems may exist. Certainly where such materials as bedpans and their contents are moved, specific consideration must be given to avoid taking them into clean rooms and contaminating other areas and personnel. The ventilation system should be designed such that potentially pathogenic or even questionable organisms, dust, lint, respiratory droplets, and dirt will not be transmitted from dirty areas to clean areas. Laminar air flow is utilized in rooms where patients have or may have infections.

The materials used in construction should have good acoustical properties, durability, fire resistance, pleasant appearance, and be easily cleaned. Many other environmental controls are discussed in other sections of this book.

Control of infectious diseases for personnel and patients parallel each other. The reservoir of infection, that is, the individual who has the infection or the insect or rodent carrying the organism, the secondary reservoirs of infection, must be either treated, controlled, or removed from the presence of individuals who are susceptible to the infection. This is done using a surveillance program to determine the potential or actual carriers of disease; using medical aseptic precautions in the care of patients with communicable diseases, such as mumps or various staph infections; controlling secondary reservoirs of infection through good housekeeping, the removal of infected linens and solid waste, the destruction of rodents and various insects, the filtering and cleaning of ventilation systems, and the control of other fomites that may cause infections to spread throughout the institution.

Personnel are exposed to infectious disease, since they treat diseased patients or they are involved in either removing waste products or cleaning the environment. All personnel will be protected if adequate isolation techniques and procedures are utilized. Where masks are required, they must be clean, dry, and changed frequently. Masks are not an absolute barrier for droplets or aerosols. If gowns are donned and removed properly, and changed after leaving the patient's room, the opportunity for infection by organisms will be reduced. Hand washing is still probably one of the most effective techniques used for proper protection of personnel and patients. All staff members, including all housekeeping and ancillary personnel, must perform a medical scrub before and after they leave the room where a patient is isolated. Gloves should be used when indicated and should be disposed of properly to avoid contaminating personnel or patients. Equipment and supplies used in patient care should stay within the isolation room until time for disposal. At that time, all equipment that cannot be properly sterilized or adequately sanitized should be disposed of in special double-bagged sacks marked *isolation*. Linens coming from isolation areas should also be stored in double bags and clearly marked *isolation*. Employees can further protect themselves by reporting to the hospital infirmary as soon as they have a suspected infection of any type, including diarrhea, vomiting, skin disorders, and so forth. Personnel should receive vaccinations for specific diseases in those areas where they may be subjected to the diseases. They should also be tested frequently in highly critical areas such as tuberculosis areas.

HOUSEKEEPING CHECKLIST
PREPARE IN DUPLICATE

This form based on a design by Herman Koren, R.S., M.P.H., Chief of Environmental Health and Safety at Philadelphia General Hospital, Philadelphia Dept. of Public Health. Printed and distributed by Economics Laboratory, Inc., 250 Park Ave., New York, N.Y. 10017.

RATINGS

U – Unsatis. (0)
S – Satis. (1)
N – Not apply

TOTAL SCORE

$$\frac{S's}{S's+U's} \times 100 = \quad \%$$

BLDG. NAME AND NO.	FLOOR & NAME	DATE	TIME
SUPERVISOR MAKING INSPECTION (Sign)		CUSTODIAN OR HOUSEKEEPING AIDE (Sign)	

Fill in all parts of the form. If the item is "Unsatisfactory" show a "U" under the **RATING** and the code under the **AREA CODE**. If an item is "Satisfactory" show an "S" under the **RATING** and nothing under the **AREA CODE**. For example, if the floors are found dirty in the Nurses Station, Laboratory, and Supply Room, put a single "U" under **RATING** on line #2 (*Floors – dirty*). Then list N.R., Lab. #3, and S.R. Use "N" if the item is not applicable. Under **TOTAL SCORE**, count up all the S's, then divide by the S's plus the U's to give you the percentage grade.

AREA CODES:

W. – Ward – Give #	U.T. – Utility Rm.	S.A. – Storage Area
So. – Solarium	T.R. – Treatment Rm.	W.R. – Wait. Rm. & Lobby
P.R. – Patient Rm. – Give #	E.R. – Exam. Rm.	Of. – Offices – Give #
N.R. – Nurses Station – Give #	J.S.C. – Janitor Supply Closet	Co. – Corridor
B.R. – Bathrooms	L.C. – Linen Closet	Lab. – Laboratory – Give #
	S.S.C. – Sterile Supply Closet	L.R. – Locker Rooms

A.R. – Animal Rooms	L.D.R. – Labor & Del. – Give #	
K – Kitchen	O.R. – Operating Rm. – Give #	
S.R. – Supply Rm.		
I.C.R. – Instr. Cleaning Rm.		
A.U. – Autoclave Rm.		
M.R. – Miscellaneous Rm.		

FLOORS	RATING	AREA CODES	LAVATORIES	RATING	AREA CODES	RATING
1. Dust			32. Toilet Paper			
2. Dirty			33. Wash Bowls			
3. Litter			34. Soap			
4. Spillage			35. Paper Towels			
5. Stains			**HOPPER ROOM**			
6. Warning signs			36. Dirty			
WALLS			**JANITORIAL SUPPLY CLOSET**			
7. Dust			37. Adequate supplies			
8. Splattering			38. Clean			
9. Cobwebs			39. Equip. stored properly			
10. Dirty			**PAILS**			
WINDOW SILLS, VENTILATORS			40. Dirty			
11. Dust			**MOP WATER**			
12. Spots			41. Dirty			
RADIATORS			**MOP HEADS**			
13. Dirty			42. Dirty			

DOOR LEDGES & FRAMES, TOPS OF CABINETS
14. Dirty

WARD SINKS
15. Wash bowl
16. Wall
17. Soap
18. Towels

WASTE RECEPTACLES
19. Need emptying
20. Dirty

EQUIPMENT
21. Dust

FURNITURE
22. Dust
23. Spillage

LAVATORIES
24. Walls and windows
25. Partitions
26. Floors
27. Tubs
28. Showers
29. Urinals
30. Toilet Seats
31. Toilet Bowls

43. Poor Condition

REPAIRS
44. Repair reported below

INSECT AND RODENT CONTROL
45. Mice
46. Roaches
47. Ants
48. Flies

LIGHTS
49. Fixture dirty
50. Bulbs need replacement

WINDOWS AND SCREENS
51. Dirty

STAIRWAYS
52. Litter
53. Dust
54. Spillage
55. Needs repair
56. Needs bulbs

ELEVATORS (Give number)
57. Floor
58. Walls
59. Tracks
60. Doors

44. REPAIRS REPORTED	LOCATION
a. Wall needs repair	
b. Broken window	
c. Broken or missing window screens	
d. Ceiling needs repair	
e. Peeling paint	
f. Floor tile loose	
g. Floor tile broken	
h. Broken lighting fixtures or burnt out bulbs	
i. Stopped up floor drains	
j. Hopper broken or stopped up	

44. REPAIRS REPORTED	LOCATION
k. Wash basin faucets leaking	
l. Wash basin stopped up	

Figure 8.3. Housekeeping self-inspection form. *Source: Environmental Health and Safety* (Elmsford, NY: Pergamon Press, 1974), p. 52–53. With permission.

Control of Infectious Waste

Infectious waste may be controlled by steam sterilization (autoclaving), incineration, thermal inactivation, gas-vapor sterilization, chemical disinfection, and sterilization by irradiation. Incineration and steam sterilization are the most frequently used techniques for the control of infectious waste. Steam sterilization utilizes saturated steam within a pressure vessel (known as a steam sterilizer) at temperatures that will kill the infectious agents present. The two types of steam sterilizers are: the gravity displacement type, in which the displaced air flows out the drain through a steam-activated exhaust valve; and the prevacuum type, in which a vacuum is pulled to remove the air before steam is introduced into the chamber. In both cases, the air is replaced with pressurized steam. The temperature of the treatment chamber continues to increase as the pressurized steam goes into the chamber. The treatment of the infectious waste by steam sterilization is based on time and temperature. The entire waste load has to be exposed to the proper temperature for a defined period of time. The decontamination of the waste occurs primarily because of the steam penetration. Heat conduction provides a secondary source of heat. The presence of residual air within the sterilizing chamber prevents effective sterilization, since the air will act as an insulator for the material. Three factors that can cause incomplete displacement of the air include use of heat-resistant plastic bags; use of deep containers; and improper loading of the chamber. The type of waste, the packaging and containers, and the volume of the waste load is important to know when establishing how best to treat the waste during steam sterilization. Infectious waste with low density, such as plastic, is better for steam sterilization than high-density waste, such as large body parts and large quantities of animal bedding and fluids. Steam sterilization can be used for plastic bags of low-density, metal pans, bottles, and flasks. The infectious waste also include other kinds of hazards, such as toxic material, radioactive material, or hazardous chemicals. Steam sterilization should not be utilized to treat these wastes. The individuals who operate the steam sterilization process should be highly skilled in proper techniques to minimize exposure of people to the hazards due to the waste. Protective equipment, minimization of aerosol formation, and prevention of spillage of waste during loading of the autoclave are important factors to be considered. A recording thermometer should be used to make sure that the proper temperature is maintained through an appropriate period of time during the cycle. All steam sterilizers need to be routinely inspected and serviced. Steam sterilizers used for infectious waste must be in an area away from the central sterile supply and should never be used for sterilization of equipment or instruments.

Incineration is a process that converts combustible material into noncombustible residue or ash. The product gases are vented to the atmosphere through the incinerator stack. The residue may be disposed of in a sanitary landfill under proper conditions. The advantage of incineration is that it reduces the mass of the waste by as much as 95%, and therefore substantially reduces transport and disposal costs from the site of the incinerator. Incineration is especially good when used for pathological waste and contaminated sharps. If the incinerators are properly designed, maintained, and operated, the microorganisms present are killed in an effective manner. If the incinerator is operating poorly, then micro-organisms may be released into the environment. When incinerating waste, it is necessary to consider the variation in waste composition, the rate of feeding of the waste into the incinerator, and the combustion temperature. The amount of mois-ture contents and heating value of the waste affects its combustion. The rate at which the waste is fed into the incinerator affects the efficiency of the burn. The combustion temperature of 1600°F will allow for a complete destruction of the microorganisms in the infectious waste and will also allow for a considerable reduction of the material. The amount of air and fuel that is used has to be adjusted to maintain the combustion temperatures at the proper level, depending on the type of waste and volume. If the infectious waste has special hazards attached to it, such as the inclusion of antineoplastic drugs, then the waste should be incin-erated only in special incinerators that provide high enough temperatures and a long enough time to completely destroy these compounds. The plastic content of the waste is also important, since many incinerators can be damaged by tempera-ture surges caused by combustion of large quantities of plastic. Polyvinyl chloride and other chlorinated plastics are important, since one of the combustion products is hydrochloric acid, which is corrosive to the incinerator and may cause damage to the lining of the chamber and the stack.

Thermal inactivation includes treatment methods that utilize heat transfer to provide conditions that reduce the presence of infectious waste. The waste may be preheated by heat exchangers, or heat may be acquired by steam jacket. The amount of heat is predetermined, and the material is kept in a vessel for at least 24 hr. However, the temperature and holding time depends on the type of patho-gens present in the waste. After the treatment cycle is complete, the contents of the tank are discharged.

Dry-heat treatment may be applied to solid infectious waste. The waste is heated in an oven that is operated by electricity. Since dry heat is less efficient than steam heat, it is necessary to have higher temperatures to accomplish the same type of kill. The typical cycle for dry-heat sterilization is treatment at 320–338°F for 2–4 hr.

Gas-vapor sterilization is an option that could be used for treating certain types of infectious waste. The two most commonly used chemicals are ethylene oxide and formaldehyde. However, there is considerable evidence that both of these chemicals are probable human carcinogens, and therefore, considerable caution must be taken if the materials are used as gas-vapor sterilizing agents. In fact, it is best not to use ethylene oxide gas because of its toxicity and the fact that other options are available for treating the infectious waste. If formaldehyde gas is used to sterilize certain disposables, such as HEPA filters from biological safety cabinets, then only trained persons should do the work.

Chemical treatment is appropriate for liquid waste, but also may be used in treating solid infectious waste. To use chemicals effectively, it is necessary to consider the following: type of microorganism; degree of contamination; amount of protein materials present; type of disinfectant; concentration and quantity of disinfectant; contact time; temperature; pH; mixing requirements; and the biology of the microorganism.

A new technology involving the use of ionizing radiation sterilization includes nominal electricity requirements; no steam; no residual heat in the treated waste; and the performance of the system. The disadvantages include the high capitol cost; the requirement for highly trained operating and support personnel; a large space requirement; and the problem of the ultimate disposal of the decayed radiation source. The ionizing radiation must be properly monitored.

Biological indicators are used to monitor the treatment process when some form of heat is utilized. These indicators are used in steam sterilization, incineration, and thermal inactivation. Spores of a resistant strain may be used as an indicator. An instantaneous reading may be gained by using a chemically induced color change when a particular temperature has been reached. The chemical indicators are not good for the sterilization process, since the temperature is only part of the sterilization effort.

Control Measures for Transmission of HIV and HBV

Control measures for transmission of HIV and HBV for health care workers and public-safety workers include the general principles that have been discussed previously in the relationship to infectious body fluids and in the use of personal protective equipment. Precautions need to be taken with the individuals who are exposed to blood or body fluids through procedures related to CPR, IV insertion, trauma, and delivering babies.

Disposable gloves must be used by individuals in institutions, especially when coming in contact with patients where blood or body fluids are involved. This is particularly true in the emergency room. It is also true for fire and emergency medical service personnel. If public-safety workers are involved in any type of body searches, they ought to also wear disposable gloves. The disposable gloves should provide for dexterity, durability, and fit in the task being performed. Where large amounts of blood are likely to be encountered, the gloves must fit tightly at the wrist to prevent blood contamination of the hands around the cuff. Where multiple trauma victims are encountered, the gloves must be changed between patients. Contaminated gloves should be removed properly and placed in bags that are leakproof and can then be appropriately disposed of, as has been discussed earlier.

Masks, eye wear, and gowns should be utilized in institutions and in emergency vehicles as needed. The protective barriers will be necessary in order to protect the workers from diseases that are borne by blood and other body fluids. The masks and eyewear (safety glasses) should be worn where blood or body fluids will be splashed. Changes in gowns and/or clothing should be available if necessary.

Although HBV and HIV infection has not been shown to occur during mouth-to-mouth resuscitation, there still may be a risk of transmission of such diseases as *Herpes simplex* and *Neisseria meningitis* during mouth-to-mouth resuscitation. It is also a theoretical risk of HIV and HBV transmission during artificial ventilation of trauma victims. Disposable airway equipment for resuscitation bags should be used. Pocket mouth-to-mouth resuscitation masks designed to isolate emergency response personnel from contact with victims' blood and blood-contaminated saliva, as well as respiratory secretions and vomitus, should be utilized by emergency personnel.

Patient Control Measures

The techniques utilized in employee control measures plus proper environmental controls will be a major step toward proper patient infection control procedures. In addition, patients who are highly susceptible to infections should not be kept in the same areas with patients who have infections. If in doubt, remove the patient from the area. Patients should be taught good personal hygiene and body care to avoid recontamination by the discharged organisms. In hazardous situations, such as where blood-bank blood is used, special techniques must be employed to avoid contaminating the patient. Whenever special procedures are used, such as hemodialysis treatments, tracheotomies, intravenous cutdowns, or change of surgical dressings, careful aseptic techniques must be employed.

The Infection Control Committee

The infection control committee is a working committee of the institution. It is necessary that the committee consist of representatives of the medical staff, nursing staff, environmental health staff, infection control staff, and various members of the administrative staff. The function of the committee is to evaluate current and past levels of infection and to make necessary recommendations for changes in either medical-surgical techniques or other techniques used within the institution that may contribute to infections. They also act as the supervisors and recipients of all special studies related to infections that are conducted within the institution. Once they have evaluated the study, they forward it, along with their recommendations, to the executive director of the institution for implementation.

Occupational Health

The occupational health and safety program consists of program planning; establishing program objectives to reduce frequency and severity of occupational hazards, increase inspection coverage, charge supervisors with safety responsibilities, coordinate health and safety activities, plan and evaluate all safety activities, and plan and direct an adequate accident prevention program; safety and health training, education, and promotion; accident and illness investigation, reporting, and analysis; safety and health inspections; safety and health committees; safety and health engineering for hazard control; safety and health standards and regulations; fire prevention, protection, and suppression; motor vehicle safety; medical help and first aid; accident cost control; program design, application, and evaluation.

Health education is the process of transmitting knowledge and developing skills to produce safe practices. In order to accomplish this, the trainers must evaluate and the trainees must learn about hazardous conditions, safe and healthy practices, general rules of health and safety, special on-the-job problems, and job safety analysis.

The engineering approach is necessary in the initial layout and construction of the institution. Proper engineering will eliminate many of the hazards already mentioned and will provide not only a safe, but also an aesthetically comfortable environment.

All institutions should provide pre-employment and preplacement examinations. These examinations should not only determine the employee's medical history, but also determine if there are specific problems with specific types of environmental conditions. It is well to determine if the applicant can read and interpret written instructions and is able to react to posted instructions concerning safeguards. In addition, periodic medical examinations should be given to the employee to determine if he/she is currently suffering from any health problems and needs to be removed from his current environment and placed in a different type of job. An institutional occupational health program should include the diagnosis of occupational disease and injuries, the treatment of occupational disease and injuries, the treatment of nonoccupational illness and injuries, and the evaluation of the kinds of health problems, occurring over time, that may become serious health or safety problems. Special concern should be given to hearing, vision, the employment of pregnant women in special hazardous situations, the prevention and control of alcoholism, and the establishment and use of emergency control centers in the event of toxicity.

Biohazard Controls

Some of the measures employed to achieve maximum biohazard control include air-pressure differentials; filtration of supply air; filtration of exhaust air from laboratories, animal rooms, and so forth; air locks and pass-through autoclaves between clean and contaminated areas; change rooms; isolation and separation of utility systems between clean areas and biohazard areas.

Housekeeping

The selection of housekeeping personnel is important. Individuals should be selected for their desire to work, their intelligence, experience, age, physical makeup, character, and ability to read and understand housekeeping procedures.

makeup, character, and ability to read and understand housekeeping procedures.

Housekeeping training programs need specific formats. Individuals should be taught the various housekeeping techniques in the classrooms and in the institution. They should also be taught the basic fundamentals of the spread of infection and the seriousness of the use, storage, and handling of dangerous and hazardous materials and chemicals. There should also be instruction in isolation techniques from nursing personnel. By the completion of the course, new housekeeping employees should have a solid understanding of where they fit into the entire institutional process, what are their specific functions, how they should carry out the tasks, the materials and chemicals to be used, how they can prevent accidents, spread of infectious diseases, and promote better health and welfare for all. After completing a week of this intensive training, the individual should receive 6 to 9 months of probationary work under close supervision. He or she should take a refresher course of at least 1 day every 6 months.

Supervisors and administrators generally reach their positions because of the good work done at a technical level. It is important that the supervisors and administrators themselves receive training. They need to learn everything that their staff learns and, in addition, need a good understanding of how to best obtain an adequate quantity of quality work from the individuals working for them. It is one of the important functions of a supervisor and administrator to determine what the area looks like and how well the job is being done. This could be accomplished by using, on a weekly basis, the housekeeping self-inspection sheet found in Figure 8.3.

A good housekeeping program will integrate all housekeeping techniques with a self-inspection program, where the individual supervisor will have an opportunity to evaluate the actual work being carried out in a given area. The supervisors and administrators may also want to earn continuing education credits while learning about basic supervision and administration by taking corresponding courses written by the author. If so, contact Continuing Education Department, Extended Services, Alumni Center 220, Indiana State University, Terre Haute, IN 47809 or phone 1-800-234-1639, ext. 3077.

Financing and the Role of Government in Institutions

Nursing home care is paid for by a combination of funds from the individual, federal government, and private organizations. Nursing homes wishing to participate in federal government programs must adhere to the rules and regulations of the U.S. Department of Health and Human Services. Regulations exist in the areas of Medicaid and also Medicare. These encompass all types of nursing home activities, including environmental health activities. For example, standards have been established for the maintenance of sanitary conditions in the storage, preparation, and distribution of food. Unfortunately, out of a booklet of some 49 pages on specific requirements for nursing home facilities under the Medicaid program, only one quarter of one page was given to the environment, and this portion concentrated on food and wastes. All other areas of the environment seemed to be ignored.

SUMMARY

It is obvious that the previous discussions of topics within the institutional environment are, of necessity, limited in scope due to the nature of the material being presented. All these areas are evaluated by one or more public agencies, including the local and state health departments and the local and state welfare departments. In addition, various areas of the hospital environment are supervised and comply with the standards of the various professional societies that are specialists within the given area. Further, all hospitals must be accredited to operate. Accreditation comes from the Joint Commission on Accreditation of Hospitals, which evaluates physicians, facilities, techniques, and so forth. The Joint Commission, after duly studying a hospital facility, will make certain recommendations that must be carried out by the institution to maintain its accreditation. This accreditation is extremely important for the operation of the institution.

The institutional environment and its problems are so complex that it takes not only the environmental health specialist, but also a true team effort on the part of medical, nursing, administrative, and other personnel to properly operate the institution and to provide a better, safer and disease-free environment for all.

RESEARCH NEEDS

There is need for continued assessment of potential sources and transmission of institutionally acquired infections in individuals and staff. New techniques should be developed to rapidly assess the modes of the spread of infection. New accident-prevention techniques are needed. Studies to determine the impact of noise on relaxation and sleep are needed. Studies are also needed to determine the quantity and exposure to chemicals in institutions. These studies must include epidemiological, toxicological, monitoring, and laboratory analysis. The air levels of chemicals associated with housekeeping, laboratories, and other areas where solvents are used should be assessed.

Chapter 9

RECREATIONAL ENVIRONMENT

BACKGROUND AND STATUS

Recreation is a vital part of public health. Recreational activities provide necessary exercise for individuals to control weight problems, strengthen hearts, and provide an emotional release from the day-to-day cares faced in the home and at work. With increasing numbers of middle-class people, the development of the interstate highway system, and the overcrowding of many areas, individuals have sought to obtain their fun through recreational activities that take them into state parks, federal parks, various national forest reserves, and local park areas. Each of these areas, while providing the kind of necessary escape from tension and troubles, also provide an environment highly conducive to the spread of disease and injury. Individuals come from different areas, states, and even countries. They congregate in recreational areas and bring diseases from home. These diseases spread readily within the recreational areas because of existent environmental health problems. The public has increased sharply the number of visits it makes to various recreational areas. In 1950 approximately 110 million people visited state parks. In the 1980s over 500 million people visited state parks. The National Park Service in 1950 had approximately 25 million people visit their various park areas. In the 1980s over 125 million people visited national parks. The U.S. Forest Service had approximately 30 million people visit forest service areas in 1950. In the 1980s about 200 million people visited these areas. These figures do not include the millions of individuals who visit local and county park and recreational areas.

Water-oriented recreation has also increased at an unusually rapid rate. In 1960, 19 million people bought 23 million state fishing licenses. In 1970, 31 million licenses were purchased. In 1970, it was estimated that there were nine million boats of various types in use. Since the end of World War II, there has been an enormous change in recreational activities in all categories. In 1970, it was estimated that 150,000 seasonal homes were built at a cost of 1.2 billion dollars. Since then, there has been an even sharper increase in the number of individuals who wish to build homes on any type of water body. It has been estimated by the Bureau of Outdoor Recreation that by the year 2000, the summertime participation in swimming will increase 207% above 1965 levels, 78% in fishing over 1965 levels, 215% in boating over 1965 levels, and 363% in water skiing over 1965 levels. There is also an anticipated increase in camping of 238%, in picnicking of 127%; and in sightseeing of 156%. It is obvious from this smattering of data that water recreational areas have become of enormous value to our society from both the aesthetic and financial points of view.

The American public make an estimated 7.8 billion 1-day visits each year to a variety of recreation areas, including national parks, parks of other federal, state, county, and local agencies, and private-sector facilities. This is an estimated increase of 40% over 1970. The tremendous number of visitors to these areas has intensified sharply the various pollution, land, and wildlife problems, and problems at historical sites. The recreational environment is becoming an increasing source of concern to public health officials, ecologists, and private groups.

SCIENTIFIC, TECHNOLOGICAL, AND GENERAL INFORMATION

The scientific and technological background varies with the specific type of recreational environmental problem being discussed. See the appropriate chapters within this textbook and Volume II for each of the problem areas.

PROBLEMS IN RECREATIONAL AREAS

Because of sheer numbers of people, lack of adequate facilities, and subclinical carriers of a variety of diseases readily spread through the environment, it is essential that there be a better understanding of recreational environmental health. An example of the kinds of hazards encountered include improper drainage and soil permeability, inadequate sewage facilities, inadequate kitchen equipment and improper cleaning practices, improper refrigeration of food, contaminated water supplies, unsuitable solid waste disposal, vector problems, swimming pool hazards, other accident hazards, and inadequate training and physical fitness of individuals visiting recreational areas.

Recreational areas have peculiar problems that generally are not found in the average city or community. They include seasonal operation, problems of public behavior, vector and animal problems, noxious plants and weeds, remote locations, and protection of wildlife from humans and humans from wildlife.

Seasonal operation is one of the most difficult problems faced in a recreational area. As a result of this, equipment is outdated or inadequate. This equipment includes chlorinators, dishwashing machines, swimming pool filters, food facilities, sewage treatment, and water facilities. Further, it is difficult to get experienced personnel, since they are not anxious to work for a 3- to 4-month period during the year. As a result of this, many of the individuals are either college students with no specific training in the environmental areas or job hoppers who have difficulty keeping jobs and do not do an adequate job within the recreational area. The maintenance of equipment, even where good equipment exists, generally is poor, because the problems have to be discovered by the operators of the equipment and frequently are not, and because it takes time for maintenance personnel to come from some central area to make necessary corrections within the recreational area. Another difficulty caused by seasonal operation is that of inspectional procedures. Since the environmental health personnel have many other functions apart from the recreational areas, by the time the studies have been completed and corrections made, the season may be over. It is necessary then to start again the following year with new management and new personnel.

Public behavior is a difficulty because some visitors are highly irresponsible. Some people dump garbage outside of their cabins or trailers, leave solid waste scattered everywhere, and contribute to fly and mosquito problems. In addition, some visitors like to take souvenirs along to show where they spent their vacations. As a result of this, they cause damage that is expensive to repair. The money spent could have been used for improving environmental conditions.

In the recreational environment, the individual is exposed to animals, reptiles, and insects. Infection by rabid bats, ticks causing Rocky Mountain spotted fever, mosquitoes causing encephalitis, bees, and hornets, as well as poisonous snakes, become troublesome problems. In some areas, individuals have even been mauled by bears. The outdoor environment contains weeds causing hayfever and other allergic reactions, such as poison ivy and poison oak. Since in some areas electrical power is not available and roads are not accessible, it is difficult to obtain a proper water supply and to treat an injured person.

As has been mentioned, communicable disease is a very definite hazard, since the agent or reservoir of infection may be present or may have been brought into the environment. The vector or vehicle of disease is present in the form of insects, rodents, or inanimate objects. It is known that when large numbers of people gather together and associate closely, disease spreads rapidly. This, coupled with the lack of adequate environmental protection measures, contribute to the spread of communicable disease. When such a disease does occur, it should be immediately reported to the nearest health department and also to the park authorities.

Impact on Other Problems

The quality of the recreational experience is being eroded because of the vast number of individuals who are invading recreational areas. There are constant shortages of motels, campgrounds, stores, laundry facilities, gas stations, and space. The sewage systems and other utilities cannot service the increased numbers of people. Noise and physical damage has been caused by the use of light planes within the national parks. The alpine tundra close to access roads have been damaged by excessive use. Crime has increased sharply, and vandalism will eventually destroy our Indian cliff carvings and many of our natural wonders if it is not controlled. The national and state parks used to be free of pollution. This is no longer true. The automobile has brought air pollution to the park areas. Water pollution has increased sharply as a result of human interference with various watersheds, animal, and plant life. Congestion has become a part of the peak use periods of recreational sites. The increased numbers of people have created a transportation problem, since the existing systems cannot service them and there is a lack of adequate access to transportation. Problems have been created by haphazard private development of areas adjacent to recreation areas.

Economics

There is no real way currently of measuring the economic cost of misuse of the recreational environment. We do know that when conditions are poor, visitors will not return to the recreational areas, resulting in a tremendous loss of money to highly specific types of businesses. However, there is little understanding and information available concerning the impact of economics as a problem of the recreational area.

POTENTIAL FOR INTERVENTION

The potential for intervention in the variety of environmental problems varies with the given problem and with the number of individuals who literally invade a recreational area. Human impact on the recreational environment can only be reduced by either increasing facilities and constraints, which may have tremendous economic and/or political costs, or by reducing the number of visits that can be made to recreational areas. The full potential for intervention can only be understood when enough research is conducted to determine what can and cannot be done within this unique environment.

RESOURCES

The resources for each of the environmental health problem areas are identified by problem area in other parts of this book and in Volume II. See the appropriate problem area for the resources desired. In addition, many of the civic organizations, such as the Sierra Club, are concerned with recreational environmental problems. Governmental agencies include the Corps of Engineers, the National Park Service, the Department of the Interior, and the Tennessee Valley Authority.

STANDARDS, PRACTICES, AND TECHNIQUES

In planning for a recreational site, it is essential to gather and analyze all pertinent data concerning the land, the environmental systems present, the environmental controls needed, and the use of the land. This type of preplanning concerns not only the recreational agencies and the health agencies, but also the Department of Interior, the Department of Mines, and other agencies concerned with the health of individuals involved in the environment.

Although it is seldom possible to meet all desirable site criteria, a good many of these and, of course the most important ones, should be considered when developing a new campsite or altering a former recreational area. It is important that the area be well drained; gently sloping; free from topographical and other hindrances; accessible to sources of water supply, sewage disposal, and solid-waste disposal; away from heavy traffic and noises; conveniently located near major highways or helicopter landing strips to remove injured or seriously ill; away from swamps and marshes, where mosquitoes may breed and cause annoyance and disease; provided with adequate facilities for proper food storage and handling; and provided with adequate housing.

Water Quality

Water quality depends upon watershed management, water supply, water usage, sewage, and plumbing. Watershed management is concerned with the supervision, regulation, maintenance, and use of water in the drainage basin to produce the maximum amount of water of desirable quality and to avoid erosion, pollution, and floods. The condition of the soil, the use of the land, and the construction of such things as roads all affect watershed management. To protect water at its source and to ensure its proper usage, it is necessary to preserve undeveloped areas and to prevent abuse of the watershed. In other areas, the quality of the watershed can be restored by controlling these problems. Preservation of existing quality or improvement of existing quality in effect means control of adjacent areas involved in construction, logging, raising of livestock and wild game, mining, land development, and waste disposal.

The construction activities that affect the quality of water are roads, railroads, power transmission lines, mines, and dams. These activities increase the amount of silt or the quantity of chemicals, oil, gasoline, and solid waste. Soil erosion occurs when construction is not carefully preplanned and too many trees are cut down.

It is essential that logging operations be carried out in such a way that erosion will not occur. Products of logging activities, decayed vegetation, pesticides, oil, and gasoline must not enter the water sources. All areas where construction roads are built must be carefully controlled to avoid erosion and watershed contamination.

Erosion is also caused by overgrazing of recreational areas by livestock and game. Grazing should be supervised by the proper authorities and wildlife controlled by hunting-season programs.

Waste disposal of liquids and solids is a serious watershed problem. People who build cabins and do not install proper sewage systems cause sewage to seep into the watershed and watercourses. Garbage and solid waste are frequently carelessly discarded, becoming contaminants. Industrial operations must be controlled to avoid inadvertent contamination. If sanitary landfills are used, they should not leach into the watercourses.

Uncontrolled mining and processing of ores have destroyed many streams. The mines and the mine wastes not only spoil the land, but also drain into bodies of water. Abandoned mines must be protected to avoid seepage into nearby bodies of water. In addition, roads, strip-mining operations, and stockpiling of various chemicals used in processing contribute to the water problem. Mining may cause the streams to be clogged with silt, colored matter, minerals, acids, and chemicals. Abandoned strip mines should be leveled and covered with the original soil or new top soil, and then seeded with an approved vegetative cover.

Pesticides, as explained in a previous chapter, are toxic to humans and animals. Depending on the pesticide and its use, it may become a contaminant of the watershed. It is important that the persistence and concentration of the pesticide, as well as its lethal effects, be thoroughly understood before it is used. Usage of pesticides in recreational areas is necessary, but must be handled carefully to prevent contamination of the watershed areas. Where recreational areas exist in watersheds, it is important that careful environmental controls be followed to avoid human contamination.

Fires either caused by carelessness, lightning, cooking, or bonfires must be carefully controlled to avoid damage to the forested areas and the watersheds. The greater the quantity of trees, the better the chances of avoiding erosion and maintaining the watershed. Conservation activities are carried out by a variety of agencies, including various park services, fire services, corp of engineers, and bureaus of land management.

Water Supply

There must be an adequate supply of pressurized water for drinking and other household purposes. The water supply must meet the bacteriological, chemical, physical, and radiological requirements of the 1986 Safe Drinking Water Act. These standards stipulate that the water supply system must include all the collection, treatment, storage, and distribution equipment needed for a proper, safe water supply. For firefighting purposes, water may be utilized from streams or creeks. To meet the requirements of proper water supplies, water should be taken from wells, springs, or infiltration galleries that are not subject to contamination. The water should be examined regularly to determine the level of safety. When water comes from underground or surface areas and it has been determined that low degrees of contamination are present, the water must be chlorinated. Where water contains impurities that require coagulation, sedimentation, and filtration for removal, the water should receive complete treatment. In inaccessible areas, water must be hauled to the places of use in tank trucks. In this case, the water must come from a safe source and the following precautions taken to prevent contamination:

1. The tank interior should be devoid of all defects. It must be thoroughly cleaned and disinfected before use.
2. The tank lining must be nontoxic.
3. The tank must be used for no other purposes.
4. Chlorine should be added to the water in the tank to ensure that a residual of at least 0.4 ppm of free chlorine is available after the chlorine has been in contact with the water for at least 20 min.
5. Only one hose that is utilized strictly for filling and unloading the tank should be available. The hose should be stored in a covered container in the truck when not in use and then thoroughly flushed with clean water before each use.

Further discussion of water supply and water treatment is found in the chapter on water in Volume II.

In the event of emergencies, water suspected of being contaminated with bacteria may be used if any of the following procedures are followed:

1. Boil vigorously for 10 min.
2. Add five drops of 2% U.S.P. tincture of iodine to each quart of cold water and allow to stand for 30 min.
3. Add iodine tablets, one tablet for each quart of water and allow to stand for 30 min.
4. Add chlorine bleach at two drops to a quart of water and allow to stand for 30 min.
5. Add chlorine tablets at one tablet per quart of water.

It is essential that drinking fountains be constructed so that the individual's mouth does not come into direct contact with the water fountain itself.

Sewage

Safe disposal of human and domestic waste is essential in recreational areas. Improper sewage disposal contaminates the water, thereby ruining the water for its many recreational and drinking purposes. Sewage treatment systems must have adequate capacity and must meet all of the necessary requirements set forth in federal and state water pollution control laws. Water carriage systems must be kept in good repair to prevent leakage into the underground water supplies. In many areas, water carriage systems are available, but sewage treatment systems are not. Septic tanks and subsurface disposal systems are utilized. In these cases, it is necessary to keep the on-site sewage disposal system away from wells, streams, or other sources of water supply. It is important that they do not overflow into the creeks or other water areas draining the campsite. The septic tanks and systems must not be permitted to flow downhill into water sources. (Additional discussions on this subject will be found in the chapter entitled On-Site and Public Sewage Disposal in Volume II.) Nonwater carriage sewage disposal facilities are also necessary in recreational areas. These include such things as chemical and burn-out toilets and pit privies. Again it is essential that these units be so constructed that they do not contaminate the water supply nor become a source for insect breeding. It is important that privies be correctly built and be properly maintained. (Further information on this area will also be found in the chapter entitled On-Site and Public Sewage Disposal in Volume II.)

Plumbing

Plumbing includes all of the practices, materials, and fixtures used in installing, maintaining, extending, and altering any pipes, fixtures, appliances, and appurtenances. (Plumbing fixtures and materials, as well as an additional discussion on plumbing, will be found in the chapter on plumbing in Volume II.) It is known that back-siphonage and cross-connections cause contamination of potable water supplies. Plumbing systems should be installed carefully by registered plumbers who adhere to the minimum requirements of the National Plumbing Code. The systems should be checked carefully to ensure their proper functioning. Air gaps, non-pressure-type vacuum breakers, or back-flow preventors are utilized to prevent problems from occurring. Cross-connections and back-flow preventors are most prevalent in the following recreational situation: direct connection of the water supply to cooling or condensor systems in refrigerators or air-conditioning systems; fish ponds with submerged inlets; fire hydrants with drains connected to sewers; frostproof water closets; kitchen and laundry fixtures with common waste and supply lines; underground water sprinkling systems that do not have vacuum breakers; and ice-cube machine drainage lines connected to the sewer lines.

Comfort stations are mobile flush toilets used for the public. They are frequently used in recreational areas. It is necessary that comfort stations be constructed in such a way that any materials removed from them will not contaminate water supplies or ground areas.

Shelter

Shelter consists of the site locations, structures to be utilized, and various campgrounds, picnic areas, and facilities.

Site Selection

The site selected must be away from swamps and mosquitoes, and must be off the main thoroughfares to reduce potential accident hazards. The access roads should be constructed with turns and loops so that vehicles have to reduce their speed. If feasible, one-way traffic patterns should be utilized. The roadways should be of all-weather construction where possible. The sites should be removed from sources of noise, air pollution, odors, or other kinds of contaminants. Recreational areas should be located on gently rolling well-drained land that is not subject to flooding and natural hazards. If possible the camps should have an eastern exposure for the campers to receive the early morning sun and afternoon shade. Lightly wooded areas provide additional shade and help make the area more comfortable. If public utilities are available, they should be utilized.

Structures

The construction of buildings is important from a public-health viewpoint, since people spend many hours within them. All units should meet the minimum requirements of the proposed housing ordinance of the American Public Health Association. The following items are important for good and proper housing:

1. A habitable room should have at least one window or skylight facing outdoors.
2. Window or skylight should be readily opened.
3. Every bathroom or water closet should have adequate quantities of hot and cold running water.
4. Heating facilities should be available in those areas where temperatures are low at night. It is essential that the heating facility be well constructed, well engineered, and well ventilated.
5. All hallways and stairways should be properly lit.
6. Mosquito screening should be supplied in all housing units.
7. Foundations, floors, walls, ceilings, and roofs should be weathertight, watertight, and rodent-tight.
8. Plumbing fixtures and water and waste pipes should be properly installed and in good working order.
9. There should be at least 150 ft^2 of floor space for the first occupant and 100 ft^2 of floor space for every additional occupant.
10. Cellars should not be used as rooms.
11. The buildings should be provided with or have easy access to adequate water supplies and sewage disposal.
12. Fire protection devices should be provided.

Additional material on housing will be found in the chapter on the indoor environment.

Trailers

Trailer parking facilities must be so arranged that adequate water and sewage connections, and adequate shower and sanitary stations, are provided. The sites must be well drained and free of obstructions. Further information on trailers is provided in the chapter on the indoor environment.

Campgrounds and Picnic Areas

Campgrounds and picnic areas are part of the area of shelter. These need to be selected in such a way that they are well drained, free of heavy undergrowth, and on solid and gently sloping ground. The layout should be such that cars have to travel slowly. Individuals should be able to walk comfortably to all areas of recreation, water supply, and sewage disposal and solid-waste disposal. There should be adequate supervision within the park to maintain these areas in good working condition.

Food

The handling of food tends to be less stringent and the facilities less modern and well maintained than in other types of situation. In a recreational setting, food may either be prepared for one individual, a family, several families, or a very large camp. It is essential that all of the rules, regulations, codes and good practices associated with food service be followed in the camp situation. At times it is difficult to refrigerate in this setting, however, since proper refrigeration of food is so essential to avoid food-borne disease, this is an absolute must.

Food Service

To conduct an effective food service program in recreation areas, whether it be for a family or for a large group, it is necessary to start with good clean products, keep them well refrigerated, prepare them under proper conditions, and utilize them within adequate time spans. The whole area of food protection will be discussed in detail in the chapter on food. However, for the recreational area it is well to remember some of the most frequent food problems. Food must be prepared with clean utensils, pots and pans of nontoxic materials. It is essential to maintain proper cleanliness. Handwashing is a must for the people who prepare and serve the food. Food that is left open should be disposed of rather than saved for other meals. Food should be served either on single-service plates, which are then thrown away, or on dishes and plates that are thoroughly washed and then sanitized. Solid waste from food should be stored in tightly closed containers to prevent breeding of flies and rodent problems. Since vending machines are used extensively in recreational areas, particular care should be given to these units. Vending machines must comply with the requirements of the vending of food and beverages codes. If foods are refrigerated, the temperatures must be adequately low. The machines should be cleaned carefully on a daily basis.

Milk

Milk, milk products, and frozen desserts must come from certified sources. Milk must be stored below 45°F at all times. Once the milk has been poured into pitchers, it is essential that it be utilized completely. It should not leave the original container and then be kept in a refrigerator for reuse. All milk, milk products, and frozen desserts should be inspected by the various county or state health departments.

Solid Waste

Solid waste consists of materials and entrails from fish cleaning, stables, eating, and life-associated processes. Solid-waste handling is important in the prevention or reduction of flies, other insects, and rodents. People have a tendency, especially in recreation areas, to throw their solid waste on the ground or dispose of it in careless ways. It causes a hazard and a nuisance and may lead to fires, odors, and unsightliness.

Fish Handling

Fish cleaning should take place in screen-enclosed facilities with nonporous floors. The washings from the fish should go into trapped floor drains. The traps should be cleaned thoroughly and the floor drains flushed out. It is important that the fish cleaning is done on impervious, nonabsorbent sloping surfaces to ensure proper cleaning. Pressurized water is necessary for final cleanup. All the materials — entrails, fish heads, and so forth — should be put into garbage-grinding units or plastic bags that are fastened and then enclosed in metal cans with tight-fitting lids. The fish materials should be removed daily to an incinerator or a properly operated landfill. These materials should be buried as soon as they are deposited within the landfill. Fish-cleaning facilities must be scrubbed down very carefully to avoid attracting insects and rodents and becoming an odor nuisance.

Stables

Since horseback riding is popular in recreational areas, and since cattle may also be found in certain recreational areas, it is important that the stables are properly maintained. Manure must be collected on a daily basis and put into concrete manure storage bins. These bins must be tightly closed to prevent fly breeding and odors. A second technique is to have the stables located on well-drained, gently sloping sites where the floors are made of concrete. The floors in the horse stalls should be paved with wooden blocks and sealed with asphalt or another impervious material. The rooms should be hosed down and the drainage should either go into a special holding tank to be pumped out later and removed to sewage treatment systems or should be put into waste stabilization ponds away from the recreational areas. It is essential that all areas of the stables be kept extremely clean at all times and effective insect and rodent control be practiced. Stables should be screened and rodent-proofed.

Other Solid Wastes

The volume of garbage and other solid waste produced depends on the geographic location, the season, the number of people, the kinds of facilities, and so forth. It is essential that all solid waste be stored in watertight, rust-resistant, nonabsorbent, durable containers, which are covered with tight-fitting lids. The containers should be lined with plastic bags. The garbage, if not ground up in a garbage disposal unit, should be stored in the same way as other solid waste. It is important that the solid waste at campsites be removed on a daily basis to prevent rodents and other scavenger animals from knocking over the containers and scattering the solid waste. The material should be removed to either an incinerator or sanitary landfill. Further information concerning solid waste disposal can be found in the chapter on solid and hazardous wastes in Volume II.

Swimming and Boating

Swimming takes place in artificial lakes, natural lakes, and swimming pools. An in-depth discussion on all types of swimming pools can be found in a separate chapter on swimming pools in Volume II.

Swimming Pools

Swimming pools in recreational areas must be designed and constructed in such a way as to prevent accidents and the inflow of sewage or other types of contaminants. The water supply must be pure and feed into the pool through a system where an air gap exists. The sewer system must be separate from the swimming pool and in no way connected with it. Dressing rooms, toilets, and showers should be provided for the individuals using the pool. It is essential that all pools have lifeguards and that children or adults not be permitted to swim unless adequately trained lifeguards are present. It is also essential that the swimming pool water maintain a pH of 7.2–7.8 with a chlorine residual of 1 ppm. All pools should be clear at the deepest point and properly filtered. The types of records necessary, water quality, and so forth will be discussed in the chapter on swimming pools. Outdoor bathing places, such as lakes, streams, rivers, and tidal waters, must be carefully checked before they are used for swimming. It is quite possible that the coliform aerogens group of bacteria from human feces may be present in the water. It is also possible that many organisms are present from runoff that comes from animal feces and from agricultural fields. It is important in outside bathing areas that all potential sources of harmful pollution are identified and these sources blocked or the bathing areas closed. These sources include sewage from boats, outlets from dwellings and other establishments, public sewer systems, leakage from improper dumping of solid waste or landfills, and the runoff from land. In artificial areas, it is also essential to recognize that the bottom of the swimming area may be muddy, and therefore it may be difficult to view or it may have stepoffs. Proper safety is an integral part of swimming area programs.

Boating Areas

Boating has become one of the largest industries in the recreational field. There are more than 10 million pleasure boats used for recreation in the United States. Many of these boats have galleys and toilet facilities. It is necessary to ensure that debris and garbage from the galley and the human fecal material from the various toilet facilities does not end up in the lake areas. These materials should be taken back to the marinas and emptied into holding tanks, where they are removed to sewage disposal plants for proper disposal. Other problems created by boats include waste oil, fuel, and any other refuse material that may be purposely or accidentally discharged into the waters. In addition to the environmental problems involved in disposal, additional environmental problems occur in the obtainment of adequate quantities of safe water. It is important that the tanks holding pure water are not cross-connected to sewage systems within the boat. These tanks must be properly washed, rinsed, and chlorinated. Water should be obtained from safe drinking sources. Food must be stored aboard the boats in refrigerated lockers until used. In addition to the problems created by motorboats or boats driven by gasoline, there are many sailboats and rowboats on the waters. All boat users must have a complete understanding of boating safety requirements and know what to do in the event of an emergency. It is essential that all water areas be clearly marked for hazardous situations.

MODES OF SURVEILLANCE AND EVALUATION

See the appropriate chapters for surveillance and evaluation by environmental health area.

CONTROLS

Recreational officials and agencies must work closely with public health agencies in an attempt to reduce potential and existing hazards. Individuals traveling into remote areas should have proper training in camping and hiking to prevent additional health and safety hazards. (See appropriate chapters for given environmental health areas in Volumes I and II.)

Safety

Recreational safety encompasses all forms of safety problems including problems related to site selection, types of buildings and equipment, fire and electrical wiring, heating systems, structural hazards, campgrounds, playgrounds, food service, swimming pools, bathing areas, and refuse disposal. In addition to these hazards found in all phases of life, it is essential to realize that the individuals who go out into the recreational environment are out there for fun. Many of them are not equipped to operate efficiently, effectively, and safely within this environment. They take risks, they do things out of lack of understanding, and they lack proper equipment. Fire safety is a special concern, since many buildings are made of wood and it is difficult to get proper firefighting equipment into these areas. Individuals hike and climb without understanding proper techniques and get into trouble, resulting in lost lives or injuries. Many facilities in recreational areas have the kinds of structural hazards and poor electrical wiring that would never be tolerated within the home. The individual is on vacation and is unconcerned about it. However, it is just these types of hazards that lead to many serious accidents. Further, within the recreational environment, people hunt and many hunters get killed each year. People drown because they do not understand water safety.

SUMMARY

To summarize the problems of the recreational environment and to properly introduce the kinds of programs that are necessary to resolve these problems, it is important that the individual read an entire book on the basic principles of environmental health and apply each section as it fits into a segment of the recreational environment. The difference between the recreational environment and other environments is that the former is more dangerous.

Humans have made increasing use of the environment for recreational purposes. These recreational uses by large groups of transient individuals cause destruction to the environment and intensify the potential for the spread of disease and injury. Limitations in the number of individuals using a given area, along with improvement of facilities and good educational approaches, are needed to enhance the recreational environment for humans and to protect it for succeeding generations.

RESEARCH NEEDS

See appropriate chapters by area of concern for research needs.

Chapter 10

OCCUPATIONAL ENVIRONMENT

BACKGROUND AND STATUS

Occupational health and safety is primarily concerned with the detection, evaluation, prevention, and control of environmental health and safety hazards in places of employment. These hazards include chemical, physical, biological, and ergonomic stresses causing illness or injury to the employee. Health hazards have been associated with certain occupations since early history. Effective action has either been totally lacking or inadequate to protect the health of the worker.

The chemical stresses include liquids, dusts, fumes, mists, vapors, solvents, and gases. The physical hazards include ionizing radiation, nonionizing radiation, noise, vibration, and extreme temperatures and pressures. Biological hazards include insects, rodents, food and fur animals, bacteria, viruses, molds, yeasts, and fungi. Ergonomic hazards include unusual body positions, repetitive motion, fatigue, monotony, and boredom.

In the first century A.D., the hazard of working with sulfur and zinc was described and protective masks were used by workers. The Industrial Revolution in England and the resultant increased exposure to toxic materials and dangerous occupations were noted, but little was done to improve the industrial environment. Conditions continued to deteriorate as the push for industrial expansion increased sharply. In the United States, prior to 1911, an employee had to sue his or her employer to collect damages for illness or injury due to his or her job. In 1911, the first state compensation laws were passed. More recently, workers have received increased protection through the following federal acts: Walsh-Healey Public Contracts Acts of 1936, Maritime Safety Amendments, Public Law 85-742 of 1958, Construction Safety Amendment, Public Law 91-54 of 1969, and other acts and administrative orders.

The Williams-Steiger Occupational Safety and Health (OSHA) Act, the most comprehensive and far-reaching federal legislation, was passed in 1970. The basic law was amended in 1974, 1978, 1982, and 1984. In addition, innumerable administrative rules and regulations have been passed. In 1989, OSHA amended its existing Air Contaminants Standards. It made more protective 212 Permissible Exposure Limits (PEL); set new PELs for 164 substances not currently regulated by OSHA; and maintained other PELs unchanged. Changes included revision of the PEL; inclusion of Short Term Exposure Limits (STEL) to complement 8-hr time weighted average (TWA) limits; establishment of skin designation; and addition of ceiling limits as appropriate. This legislation covers about 75% of the civilian labor force. Further special provisions of the act pertain to the 2.7 million federal government civilian employees.

The OSHA Act is a general law providing a broad spectrum of powers for use by the Secretary of Labor to reduce exposure to hazardous conditions in the occupational environment. The secretary has the authority to promulgate, modify, and revoke safety and health standards, make inspections, issue orders and invoke penalties, require employers to keep adequate safety and health records, ask the courts to stop potentially dangerous situations, and approve or reject various state plans for OSHA programs under the act. The Secretary of Labor works with the Secretary of Health and other agencies in a variety of ways. The Department of Labor conducts short-term training sessions for their personnel and for personnel from various states.

Under the OSHA Act, the Department of Health and Human Services (HHS) has primary responsibility for conducting health and safety research, evaluating hazards, determining and listing the toxicity of various substances, and workforce development and training. The National Institute of Occupational Safety and Health (NIOSH) is the unit of HHS that is responsible for these programs. Research programs in toxicology, physical and chemical analysis, physiology, ergonomics, engineering, psychology, physical agents, and epidemiology as they relate to industry have been established by NIOSH. The data from these programs help develop or modify criteria used for recommending health or safety standards. Surveillance and technical service programs, including hazard evaluation, are conducted by NIOSH.

Industrial safety is an extremely complex area that requires an understanding of administration, law, public relations, records evaluation, instrumentation, special surveys and studies, technical processes, and training. The types and numbers of possible industrial accidents are enormous. It would be necessary to evaluate and discuss each and every industry to fully understand the industrial safety problem in our society today.

Since this is a specialty area, the environmentalist should work closely with the safety management or safety engineering expert. These specialists receive basic training in chemistry, physics, mathematics, technical drawing, management, environmental safety, industrial accident prevention, traffic and transportation safety, fire protection, safety, industrial hygiene, emergency safety and security, corporate safety, personnel management, manufacturing processes, and industrial organization.

Over 117 million men and women are gainfully employed today. All these individuals are exposed to occupational hazards to some degree. Although OSHA requires that occupational illness be reported, reporting has been poor. One hundred and twenty-five thousand new cases of occupational disease occur each year according to NIOSH. The figure may be understated.

In 1986, there were 10,700 accidental work deaths and 1,800,000 work-related injuries. Work days in excess of 75 million were lost because of disabling injuries, and an estimated time loss for future years for these workers is 100 million days more. The total cost of accidents in 1986 exceeded 34.8 billion dollars. Total work-related deaths have decreased from a high of 19,000 in 1937, to a low of 10,700 in 1986. Occupational injuries in 1986 occurred most frequently to the trunk (580,000) and least to the toes (40,000). Of the 1.8 million disabling injuries, 60,000 resulted in some permanent impairment.

SCIENTIFIC, TECHNOLOGICAL, AND GENERAL INFORMATION

Heat

There are four factors that influence the interchange of heat between people and the environment: (1) air temperature, (2) air velocity, (3) moisture content of the air, and (4) radiant temperature. The industrial heat problem is created when a combination of these factors produce a working environment that may be uncomfortable or hazardous to individuals because of an imbalance of metabolic heat production and heat loss.

Humans regulate internal body temperature within certain narrow limits. Blood flows from the site of heat production in the muscles and deep tissues, which is the deep region or core, to the body surface, which is the superficial regions where it is cooled. Normally, an individual can easily control this process, which is called internal heat load. Body temperature, which usually ranges from about 97 to 99.6°F, can be altered by emotions, hard exercise, cold weather, or sleep. External heat load, which is produced by the environment, affects humans in different ways, based on physical fitness, work capacity, age, health, living habits, and degree of acclimation to heat. The effective temperature the individual feels is a combination of the ambient temperature, his or her activity level, and the clothing worn.

The body produces sweat to help cool itself. The amount of sweat produced is a measurement of the amount of heat stress applied to the body. Heat stress is determined by the strain on the circulatory system, the heart rate, and the body temperature. When the individual is slowly acclimated to heat exposure, he or she can work in a hot environment for longer periods of time without distress.

Heat tolerance is also based on the relationship of the surface area of the body to the body weight. Obese and stockily built individuals suffer more from heat problems than others. Individuals with heart or blood-vessel diseases are affected by heat more so than other individuals of the same age. The water and salt balances must be maintained to prevent heat-related problems. The physically conditioned individual can tolerate heat more readily than one who is in poor physical condition.

It appears that individuals who drink an excessive amount of alcohol within hours or within a day or two prior to heat exposure have far less tolerance to heat-related diseases than others. Alcohol suppresses ADH, the antidiuretic hormone, leading to a loss of body water in urine and severe dehydration.

Humans exchange heat with the environment through conduction and convection, radiation, and evaporation. If the substance in contact with humans, such as air, water, clothing, or other objects, is higher in temperature than the skin, the body will gain heat; if the substance is lower in temperature than the skin, heat will be lost. The rate of transfer is based mostly on the difference between the two temperatures. Air and water accelerate the transfer. Conduction is the transfer of heat through solids. Convection is the transmission of heat by the motion of particles of matter. Thermal radiation is the transmission of energy by means of electromagnetic waves of long wavelength. Radiant energy, when absorbed, becomes thermoenergy and results in an increase in the temperature of the absorbing body. Evaporation is based on air speed and the difference in vapor pressure between perspiration on the skin and the air. In hot, moist environments, evaporation may be limited or stopped. Cold is the absence of heat.

Industrial heat exposures are either classified as hot-dry heat or as warm-moist heat. If the moisture content of the air is not excessive, the individual will have evaporative cooling due to the sweating that will go on. In the case of the hot-dry heat, the body may absorb more heat by radiation or convection than the ability of the body to cool itself through sweating. As a result of this, the body's temperature will start to increase. This additional heat load may cause a variety of health problems, leading eventually to heat stroke, which may be fatal or at least debilitating.

Illumination

Light in the industrial environment improves the visibility of objects and awareness of the work space. Lighting must be properly designed and coordinated with the space, the colors of equipment, and facilities. Proper cleaning and routine replacement of light bulbs is essential to properly utilize lighting in a particular work environment. Good lighting reduces mistakes, increases production, reduces accidents, and improves morale. The amount of light needed varies with the task, the contrast required for certain details, and the general lighting available within the work area. The smaller a task, the more difficult it is to see. Therefore, the illumination must provide visibility for small detail and yet not create a sharp difference between the light on the work and the surrounding environment. Standards of lighting for industry have been established by the Illuminating Engineering Society. A manual entitled "The American National Standard Practice for Industrial Lighting" is available through this society. It sets forth the specific task to be carried out and the number of footcandles needed for the task. In general, a minimum of 30 footcandles should be provided in all work areas, including receiving, opening, storing, and shipping areas. Footcandle levels may rise to as much as 2000 in certain examining areas. The quantity of illumination, which has been discussed, is very much affected by the quality of illumination. The quality of illumination is determined by the amount of glare present. Glare causes discomfort and interferes with the successful completion of some work. The amount of glare is determined by the design of the lighting fixtures, the types of surfaces on which the light falls, the color of the surfaces and the environment, and the amount of outside light coming into the industrial facility.

Vibration

Vibration affects the human body mechanically, biologically, physiologically, and psychologically. The entire body or part of the body is affected by vibration, depending on the type of industrial operation. The entire body is affected if the individual is riding in a truck without springs. A partial vibration occurs if the individual is using a piece of mechanized equipment such as a chain saw. Vibration slightly accelerates the rate of oxygen consumption, pulmonary ventilation, and cardiac output and inhibits tendon reflexes and regulation of posture.

Ergonomics

Ergonomics is the application of biological and engineering sciences to the working environment to obtain the best adjustment of humans to work. This includes the physiological and psychological demands of the job and the worker. The major concern is work stress, which is any action from an external source that affects the body. This creates a strain resulting in some physiological response to the stress. Stress and strain do not necessarily occur at the same points. Work stress and strain are not necessarily bad. It is only when they cause a serious response in the human being that they must be eliminated.

Noise

Noise is unwanted sound. Sound is vibration conducted through solids, liquids, or gases. Sound, which is a physical occurrence, is produced when minute pressure variations having certain characteristics reach the ear. These pressure variations may be produced by any object that vibrates in a conducting medium with the proper amplitude, frequency, and cycle rate. Noise usually includes a number of different sounds that vary in intensity, pressure, frequency, and duration. The louder the noise, the higher its intensity and the more damage it causes to humans. The longer the noise exposure, the greater the damage to human hearing. Frequency of sound is the rate at which complete cycles of high and low pressure are produced by the source of sound. The frequency is the rate at which wavelengths are compressed and released and is measured in cycles per second, which are called hertz (Hz). Amplitude is the quantity of sound produced at a given location away from the source of the sound. Loudness is perception of the amplitude of sound. A decibel (dB) is a unit for measuring the relative loudness of sounds equal approximately to the smallest degree of difference of loudness ordinarily detectable by the human ear, over a range of about 130 units, with one unit being the faintest audible sound. A 1-dB sound will move the eardrum a distance of one tenth of a diameter of a hydrogen atom. The *A Weighing Scale* was devised as a measurement in accordance with levels heard by the ears. The human ear is more sensitive to certain frequencies than others in the 20–20,000 Hz Scale. This is the *dBA Scale*.

Noise problems in industry include

1. intrusion of plant noise in nearby residential areas
2. intrusion of plant noise into adjoining office spaces
3. interference with speech communication and audible warning sounds by noise in the work area or hearing loss, and other detrimental health effects caused by long-term exposure to excessive plant noise

Electromagnetic Radiation

Ionizing Radiation

Nuclear radiation is a process that accompanies the transformation of atoms from an unstable to a stable state. An atom is the smallest existing unit of an element exhibiting the properties of that element. Atoms are comprised of three basic building blocks: the electron, which has an extremely small mass and is negative; the proton, which has a mass 2000 times as large as the electron and is positive; and the neutron, which has about the mass of a proton and electron combined. The nuclear forces are mediated by the exchange of pions and possibly other mesons, which are thought to be the nuclear glue. If a neutron is standing free, it will spontaneously split into a proton and an electron. The protons and neutrons form the central core or nucleus of the atom. The electrons orbit the nucleus. The orbiting electrons do not have exact positions and, therefore, are called a cloud. Different elements have different numbers of protons in their nuclei. Any atomic table in a basic chemistry textbook will supply the numbers of protons in each of the elements.

Atoms of the same element have the same number of protons (the same atomic number), but may have different numbers of neutrons (different atomic weights). These atoms are called isotopes. For instance, neon ordinarily has 10 neutrons and 10 protons in the nucleus. However, the isotopes of neon may have 11 neutrons and 10 protons, or 12 neutrons and 10 protons. These isotopes are all stable, and therefore, are nonradioactive. Strontium-90 is an unstable, radioactive isotope. The usual atomic weight of strontium is 87. All elements with more than 82 protons in their nuclei are unstable or radioactive. Radioactive isotopes are isotopes that have an unstable nucleus that emits energy and/or particles to become more stable. This often brings about a change from one element to another, which causes a change in the chemistry of the substance.

The three major types of radiation from atoms are alpha, beta, and gamma. An alpha particle contains two protons and two neutrons. It is ejected from the unstable atom with extreme velocity changing the atom into a new element. By emission of the alpha particle, uranium becomes thorium and radium becomes radon.

Fission, which is the process occurring in nuclear reactors, is the splitting apart of uranium into two smaller atoms. The two smaller atoms are called fission products. These atoms have a greater number of neutrons compared to protons than the stable atom with the same number of protons. Therefore, in order for the atom to become stable, the neutron spontaneously ejects an electron and becomes a proton. The ejected electron is called a beta particle and travels at extreme velocities. The atom is said to have decayed. The atom has now become a new element.

Alpha or beta particles may be accompanied by gamma radiation, which is similar to X-ray. Gamma radiation is the release of energy from a newly formed nucleus after the emission of the alpha or beta particle. Gamma radiation does not affect the atom in any way.

The rate of decay of unstable atoms in a radioactive mass is important, since it measures the rate of release of alpha and/or beta particles. This rate is known as the curie. A curie of radioactive material is that quantity of unstable atoms whose frequency of decay is 37 billion per second; a microcurie is one millionth of a curie and has a decay rate of 37,000 per second. A micro-microcurie, also called picocurie, has a decay rate of 2.22 per minute.

Half-life is the time required for half of the original unstable atom population to decay. The longer the half-life, the more dangerous the material to humans.

Each radionuclide decays with a specific half-life. During decay, it generally produces other radioactive substances. Radium-226 decays with a half-life of 1620 years. Iodine-131 decays with a half-life of 8 days. Sodium-24 decays with a half-life of 15 hr. The rate of decay is important to environmental health personnel.

The X-ray is part of the electromagnetic spectrum, which also includes visible light, radio waves, gamma rays, and infrared and ultraviolet radiation. X-rays, once emitted, will travel in straight lines until they interact with matter. X-rays are of short wavelength, high frequency, and high energy. X-rays travel through space at the speed of light. They possess no real mass charge. They are capable of exposing photographic film. X-rays are produced by bombarding a material with electrons which contain energy. Only about 1% of the electrons in the diagnostic region will produce X-radiation. Most of the electrons give up their energy as they hit solid material and increase the heat of the material. The difference between the gamma ray and the X-ray is in their origin. A gamma ray is emitted from the nucleus of the atom when the nucleus goes at a higher or lower energy level.

Radionuclides are naturally occurring or artificially produced. The most frequently used natural radionuclide is radium-226. It is used in medicine and industry. Artificially produced radionuclides are developed by the Nuclear Regulatory Commission for medical, biological, industrial, agricultural, and scientific research. They are used for medical diagnosis and therapy.

Nonionizing Radiation

Laws — including Public Law 90-602, the Radiation Control for Health and Safety Act of 1968, and the OSHA Act of 1970 — were in part passed because of the sharp increase in use of electronic products that give off ionizing and nonionizing radiation. Radiation, which is energy, is transmitted by moving particles into stationary targets or by wave motions. Waves have frequencies, which are the number of vibrations per second (f). Velocity of the wave is recorded as a small (v). The wavelength (λ) equals the velocity divided by the frequency at a given point in time. Certain waves consist of oscillating electric and magnetic fields. These are known as electromagnetic waves or radiation and include radio and microwaves, infrared, visible light, ultraviolet, and gamma rays. Figure 10.1 illustrates the levels of energy in the electromagnetic spectrum.

Infared Radiation

Infrared radiation extends from the visible red light region (0.75 μ) to the 3000 μ wavelength of the electromagnetic spectrum. Infrared radiation can occur from any surface that is at a higher temperature than the receiving surface. The transfer of energy or heat occurs whenever the radiant energy emitted from one body is absorbed by another body.

Lasers

A laser is light amplification and concentration in one direction. The light from a laser beam has the same wavelength and therefore reinforces itself. The most common laser is the ruby crystal, which is made from aluminum oxide in which some chromium atoms have replaced some aluminum atoms. Laser beams travel in straight lines and do not spread out.

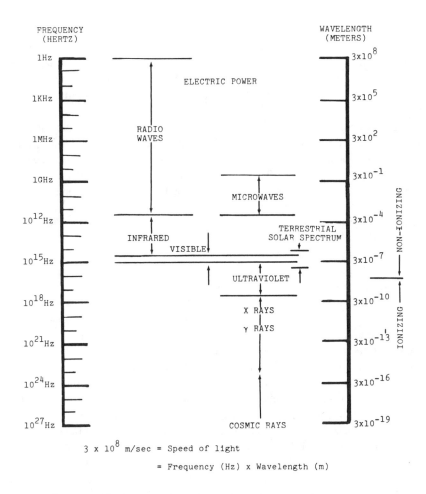

Figure 10.1.   Levels of energy in the electromagnetic spectrum.

Radio Frequency (RF) Energy

Radio frequency sealers have been used for more than 30 years in industry. These RF sealers are used to heat, melt, or cure materials such as plastic, rubber or glue. Specific uses include

1. the manufacture of many plastic products, such as toys, vinyl, loose-leafed binders, rain apparel, waterproof containers, furniture slip covers, and packaging materials
2. wood lamination processes, including glue setting
3. embossing and drying operations in the textile, paper, plastic, and leather industries
4. manufacture of various materials, including plasticizer, polyvinyl chloride, wood resins, polyurethane foam, concrete binder materials, rubber tires, and epoxy resins

Experimental data from animal studies suggest that there are potential problems of absorbing excessive amounts of RF energy. These problems may include changes in the eye, the central nervous system, conditioned reflex behavior, heart rate, chemical composition of the blood, and the immunological system. There have also been effects on reproduction and the development of offspring of women exposed during pregnancy.

Excessive amounts of RF energy absorbed by workers may produce adverse thermal effects resulting from heating of deep body tissue. These thermal effects may include potentially damaging alterations in cells caused by localized increases in tissue temperature. The body's heat sensors that are located in the skin are not activated when the RF energy is absorbed deep within the body tissues.

Industrial Toxicology

Introduction

Industrial toxicology is the study of human physiological effects of exposure to harmful materials in the industrial environment. Toxic substances may cause minor or major, reversible or irreversible changes in humans. It is important to know whether the toxic material is in the form of dust, fume, gas, mist, or liquid. Classification of toxic substances is based on the amount of hazard determined by clinical symptoms, the route of entry into the body, or the biological effects on the organs and systems. The clinical symptoms include acute reactions, subacute reactions, and chronic reactions. The route of entry includes inhalation, ingestion, absorption, and parenteral. There is also concern with the general and systemic effects of the hazardous material to the body and the way in which the body responds to the material. General effects include irritation; hypoxemia, which is a condition in which oxygen carried in the arterial blood is below the normal range; allergic effects; cancer-causing effects; teratogenetic effects; and possible changes resulting in mutation. Systemic effects include exciting or depressing the central nervous system or affecting the peripheral nerves or nerve cells; effects on the blood, the liver, the kidneys, other organs and tissues; and general effects.

When you evaluate the health hazard of a given substance, you determine the toxicity of the material, its physical properties, the probability of absorption of the material by the worker, the amount and time of exposure to the material, the control measures used, and the ability of the person to detoxify and dispose of the material. The worker's age, sex, physical condition, existing health problems, medications, and other stresses are additional factors in determining the individual toxicity of hazardous materials.

Fundamental Concepts in Toxicology

Toxicology is the study of how chemical substances, natural or those made by people, cause undesirable effects in living organisms. These chemicals that are found in the environment may pose a health hazard depending on the quality of the chemical and the toxicity of the chemical. All chemicals, including salt or water, may produce serious effects upon people if taken in large enough quantities. For example, salt can raise an individual's blood pressure and water taken in a large quality during certain heat problems may cause heat cramps to occur.

Three important terms in this area include

1. *toxicity assessment*, which is the process of characterizing the inherent toxicity of a chemical.
2. *exposure assessment*, which is the process of determining how much of a chemical is in the environment and may come in contact with people or other organisms.
3. *risk assessment*, which is the process of integrating available information in the inherent toxicity of the chemical with information on how much of the chemical may come in contact with the individual.

All living organisms are composed of cells. The cells in people and other higher organisms are specialized. The muscle cells perform motion, the red blood cells carry oxygen, and the retinal cells detect light. All cells also perform basic functions of generating the energy they need in maintaining and repairing themselves. Chemical reactions are necessary for all these specialized and basic functions. The chemicals reactions need to take place at a proper rate and must not be interfered with. All chemicals may result in adverse effects to the cell. Some chemicals react with the cellular molecules, changing their properties. Some chemicals render the cells ineffective or damage them. Other chemicals substitute for the normal body chemicals and produce products that may be toxic to the cells. Chemical toxicity is related to adsorption of the chemical into the organism through the skin, stomach, or lungs. It is related to the distribution of the chemical into the cells. It is related to the excretion of the chemical from the organism. How long does it take to excrete? Does it tend to accumulate over a period of time? It is related to the toxic effects of the chemical to the cells and whether or not the cells are injured or permanently destroyed. It is related to the sensitivity of the population of individuals related to their age, sex, or race. It is related to the mechanism of the chemical reactions occurring in the body. All of this essential toxicological information is needed to determine the toxicity assessment of a chemical.

The dose-response relationship of a chemical is the most fundamental concept in toxicology. The dose-response curve describes the relationship that exists between the degree of exposure to a certain dose and the magnitude of the effect, which is the response, by the exposed organism. No response is seen if the chemical is not present. As the amount of the chemical exposure increases, the response becomes more apparent and therefore increases. Many chemicals produce responses that show a threshold value. That is an exposure below which no response can be detected. The NOAEL is the No-Observed-Adverse-Effect Level. The lowest value where a significant adverse effect is first seen is the LOAEL, which is the Lowest-Observed-Adverse-Effect Level. NOAEL and the LOAEL values depend upon the effect (endpoint) being measured.

Some chemicals produce adverse effects that show a dose-response curve with no threshold. The reason for this is that the cells that are affected have little or no defense against the chemical and have little or no ability to repair or compensate for the damage that is done. It appears that there is no threshold for the effects of lead on the nervous system in infants and children. Chemicals that are carcinogenic are also considered to be part of the group that do not have thresholds. Therefore, any chemicals in these groups are considered to create a degree of risk at any exposure.

Dose-response curves also help show the toxic properties of the chemical and may be useful in comparing the toxicity of several chemicals. The midpoint of the dose-response curve or comparison of two chemicals is called the ED_{50}. This is the dose that produces 50% of the effect of the chemical. When a toxic effect is being measured, the term used is TD_{50}. When the lethal effect is being measured the term used is LD_{50}.

The slope of a dose-response curve is an important variable in assessing the toxicity of the chemical. The steeper the dose-response curve, the more cautious the individual must be in being exposed to a given chemical, because a small difference in the dose may produce a serious effect.

In order to assess the toxicity of a chemical, it is important to determine the route of exposure, the length of exposure, the species and individual characteristics of the exposed organism, and the nature or endpoint of the toxic effect being measured. In order for a chemical to show a toxic effect, it must first gain access to the cells and tissues of the organism. In humans, the major routes of exposure or entry into the body are through ingestion, inhalation, and dermal adsorption.

Ingestion brings the chemicals into contact with the tissues of the gastrointestinal track. The normal function is absorption of foods and fluids that are ingested. The gastrointestinal track is affected by absorbing toxic chemicals that are contained in the food or water. The degree of absorption usually depends on whether the chemical is easily soluble in water or easily soluble in organic solvents or fats. The lipophilic compounds, such as organic solvents, are usually well absorbed, since the chemicals can easily diffuse across the membranes of the cells that line the gastrointestinal track. The hydrophilic compounds, such as metal ions, cannot cross the cell lining in the same way as the lipophilic and have to be transported by systems in the cells. Many chemicals tend to bind to food and therefore are not absorbed as efficiently as when ingested in water. Further, some chemicals may be altered by digestive enzymes or intestinal bacteria, and yield different chemicals with altered toxicological properties, which may be carcinogenic

Inhalation brings chemicals into contact with the lungs. Most of the inhaled chemicals are gaseous or vapors of volatile liquids. Absorption in the lung is usually high, because the surface areas are large and the blood vessels are in close proximity to the exposed surface area. Gases cross the lung by means of simple diffusion. The rate of absorption depends on the solubility of the toxic agent in the blood. Chemicals may also be inhaled in solid or liquid form as dusts or aerosols. Liquid aerosols, if lipid soluble, will easily cross the cell membranes by passive diffusion. The absorption of the solid particulate matter depends on the size and chemical nature of the particles. The rate of absorption of the particles from the alveoli is determined by the solubility of the chemical in lung fluids. Certain small insoluble particles may remain in the alveoli indefinitely. Particles of 2–5 microns are deposited into the tracheobronchiolar regions of the lungs. They are typically cleared by coughing and sneezing, or are swallowed and deposited in the gastro-intestinal tract. Particles of 5 microns or larger are usually deposited in the nasopharyngeal region, where they are either expelled or swallowed.

Absorption of toxic chemicals through the epidermal layer of the skin is hindered by the epidermal cells. These cells are densely packed layers of horny keratinized material. Absorption of the chemicals through the skin will occur more readily if the skin is scratched or broken in any manner. The absorption of the chemicals by the skin is roughly proportional to their lipid solubility.

A more rapid exposure to the body occurs when the chemicals are accidentally injected or purposefully injected into the body.

The toxicity of many chemicals depends on the time of exposure and frequency of exposure. This time period is significant because some chemicals will accumulate in the body over long periods of time and will not be excreted. Toxic effects may depend on the duration of the exposure as it relates to the ability of these cells to repair themselves. Some adverse effects require an extended period of time to develop. A good example of this last item is the exposure of blood by young children and the resulting nervous system disorders, as well as the development of tumors that may become carcinogenic.

In toxicity assessment in the use of animals in comparison to human beings, it is important to understand that there is a difference in the toxic effect of a chemical in one species and in another species. The difference may relate to the absorption of the chemical, metabolism of the chemical, or the differences in anatomical function. The absorption of chemicals across the skin, lungs, or gastrointestinal tract is determined by the properties of the cells at the surfaces of these tissues. Metabolism of the chemical in the liver and kidney or other tissues is different in different types of organisms. The rate of metabolism of chemicals differs, as well as the kinds of metabolic byproducts.

Individual characteristics within a given group, such as human beings, may vary and cause different types of responses to chemicals. Some of these differences may be related to sex, race, or age. Nutritional status and dietary factors also may contribute to the differences in people and their reaction to different types of chemicals.

Toxicity assessment is important to determine the toxicological endpoints.

These endpoints may include liver toxicity (hepatotoxicity); chromosomal damage (genotoxicity); structural or functional abnormality (teratogenicity), usually to developing fetuses (fetotoxicity); and growth of malignant tumors (carcinogenicity). The criteria that are used for identifying the endpoint of a given chemical include sensitivity to the dose, the severity of the effect, and whether or not the effect is reversible or irreversible. In studying dose sensitivity, it is most appropriate to use the most sensitive endpoint. Toxicological endpoints include behavioral toxicity, where motor function coordination or sensory function learning and memory may be effected; carcinogenicity, where cancerous tumors are produced; hematological toxicity where there are changes in hemoglobin levels, erythrocytes, leukocytes, platelets, and plasma components; hepatoxicity, where there are malfunctions in the liver, lipid metabolism, protein metabolism, carbohydrate metabolism, and metabolism of foreign compounds; inhalation toxicity, where there is a gross or microscopic breakdown in the anatomy or function of the lungs; mutagenicity, where there are alterations in chromosomes, or DNA damage; neurotoxicity, where there is damage to reflexes, coordination, intelligence, nerve impulses, and other behavioral types or concerns; renal toxicity, where there is damage to the kidneys; reproductive toxicity, where fertility levels are lower and more stillborns occur; and teratogenicity, where there are functional or physical defects in the fetus.

Exposure assessments are based on several factors. They include what chemicals are present; if the chemicals present in air, soils, water, or a combination of these environmental areas; what concentration of each chemical is present; what living organisms are exposed to these chemicals; through what routes are these organisms exposed; and what degree of exposure is expected in the future. The data for the exposure assessments is collected by doing physical assessments of the site of the chemical exposure and the taking of appropriate samples for analysis. The major exposure-related factors that influence toxicity are the route of entrance, the duration of exposure, and the frequency of exposure. Samples may be taken of air, if the exposure is occupationally related in an industry, or of the air, land, and water, if the exposure is environmental or occupationally related at a hazardous waste site or other chemical release area. It is often necessary to consider the effect of being exposed to multiple chemicals at any given time, along with the other types of environmental conditions related to excessive heat, cold, pressure, inadequate lighting, dust etc.

Risk assessment depends on toxicological evaluation of the chemical, dose-response evaluation of the chemical, exposure assessment, and risk characterization. The toxicological evaluation and dose-response evaluation determine if the chemical has an adverse effect and in what quantity the chemical would be present to cause this effect. The exposure assessment determines the amount of exposure to a chemical by individuals. The risk characterization, which is the final step in risk assessment, estimates the amount of adverse health effects that will occur when the individual is exposed to certain chemicals for given periods of time. This must be divided into noncarcinogenic risks and carcinogenic risks.

Pharmacokinetic Factors in Chemical Exposure

Pharmacokinetics is the science of the quantitative analysis between an organism and a drug. It is also used as a means of determining the relationship between an organism and a chemical compound that may be harmful. It is a tool that is used to describe the physiological, biochemical, and physiochemical processes and mathematical relationships that determine xenobiotic and metabolic concentrations in various fluids and tissues of the body. A xenobiotic is a chemical compound, such as a drug, pesticide, or carcinogen, that is formed in a living organism. The purpose of pharmacokinetics is to study the time relationship of drug or chemical metabolite concentrations and amounts present in biological fluids, tissues, and excreta, and to construct models to interpret this data. The data that are collected are used to do the following:

1. estimate the rates of absorption, metabolism, and urinary excretion
2. estimate the available amount of absorption and rate of absorption of the chemical (this is known as bioavailability)
3. predict the blood levels of chemicals from multiple doses after measuring the amount and the blood level from a single dose, or relate the pharmacological response to the chemical by the individual by determining the cause of toxicity of the chemical

Biological Monitoring

Biological monitoring may include either measurement of the chemical or its metabolites in tissues or excreta, or pathological effects of the toxin on the person. Although monitoring devices can be set up close to the individual, they still do not provide the same kind of data as the actual exposure to the particular chemical that the individual is subjected to and the effect that it has on the body. The individual's blood, urine, breath, hair, neurological response, and physical condition can be determined to see whether or not exposure has occurred to certain chemicals and at what levels. The biological exposure index (BEI) can be established if conditions are standardized and biological measurements predict the environmental levels to which an individual is exposed. An example of this is the level of carboxyhemoglobin, from which the environmental level of carbon monoxide can be reasonably estimated. The level of carboxyhemoglobin is related to the air concentration of carbon monoxide that has been inspired. Other chemicals, such as toluene, xylene, styrene, and trichlorethylene, have had BEI values proposed for 8-hr exposures.

Types of Hazardous Materials

Dusts

Dusts are particulate air contaminants deriving from a wide variety of industrial processes. The chemical and toxic problem of dust is based on its concentration in the environment, the length of exposure by the worker, the worker's previous exposures, smoking habits, and physiological response. Dust causes bronchitis, pneumoconiosis, and may be a contributing factor in lung cancer. The hazardous dusts include silica, asbestos, talc, cotton dust, coal dust, fiberglass, fluorides, and metallic dust.

Dust is classified as inert, which mechanically blocks airways; toxic, which is usually metal compounds such as lead; allergenic, which causes asthma or dermatitis; and fibrogenic, which causes physical damage because of the character of the material.

Dust in the 8–10 μ range is trapped on the ciliated epithelium of the trachea, bronchi, or bronchioles. Dust particles less than 5 μ in size pass into the alveoli, where they have a detrimental effect on the cells or carry other hazardous materials, such as chemicals, bacteria, and viruses, into intimate contact with the alveolar membrane.

Fumes

Fumes are solid particles deriving from the condensing of the volatization of molten metals from the gaseous state. Usually, a chemical reaction such as oxidation occurs in the condensing of the gas. The solid particles are very fine, usually less than 1 micron in size.

Fumes cause a disease process known as metal fume fever. It is of short duration and produces an allergic response in the worker who inhales high concentrations of metal oxides. It can be caused by copper fumes in rolling mills, by iron fumes in welding, by cadmium fumes, and so forth.

Gases

Gases are formless fluids that occupy space and can be changed into a liquid or solid by increasing the pressure and decreasing the temperature. Gases are governed by the gas laws.

Vapors are the gaseous form of materials that are usually in the solid or liquid state at room temperatures and normal pressure. Gases and vapors are produced in industry and may be classified as follows: industrial, organic solvent vapors such as benzene, alcohols, carbon tetrachloride; upper respiratory irritants such as ammonia and sulfur dioxide; pulmonary irritants such as chlorine and nitrogen oxide; chemical asphyxiators such as carbon monoxide; simple asphyxiators such as methane and carbon dioxide; and other gases such as hydrogen sulfide.

Gases enter the body through the respiratory system, the digestive system, and the skin. The gases, which are primarily inhaled, are dissolved in the alveoli and diffused into the blood. The gases may in themselves or through metabolism form toxic substances, causing acute poisoning with rapid injury and possible death, or chronic poisoning, leading to destruction and impairment of vital organs.

Gases entering the digestive tract are absorbed in food and saliva. It is for this reason that eating and smoking should be prohibited in the working area.

Some gases enter unbroken skin and penetrate the blood. Gases that do not penetrate the skin can still cause dermatitis.

Mists and Aerosols

Mists are suspended liquid droplets in the air caused by the condensing of gases to the liquid state. Mists may be due to splashing, foaming, or atomizing.

Aerosols are liquid droplets or solid particles in the air that are small enough to remain dispersed for a period of time. Aerosols and mists cause irritation of the nose and throat, ulceration of the nasal passages, damage to the lungs, and damage to other internal tissues, depending on the chemical substance present. They also cause dermatitis.

Liquids

A liquid is a substance that is neither gaseous nor solid and flows freely. Liquids create numerous toxic effects, depending on chemical composition, concentration, time of exposure, and control techniques utilized, (See material in this chapter on solvents and hazardous materials.) Additional material on dusts, fumes, gases, mists, and aerosols, may be found in the chapter on air pollution in Volume II.

Classification of Toxic Substances by Action on the Body

Toxic substances act as convulsant poisons, central nervous system depressants, peripherally acting nerve poisons, muscle poisons, protoplasmic poisons, and poisons of the blood and blood-forming organs. Convulsant poisons act directly on the central nervous system. These poisons include strychnine and camphor. They produce convulsions, increased respiration rates, swelling of the heart, vomiting, and death. The common toxin nicotine may be fatal to a child in a 10-mg dose. DDT, aldrin, dieldrin, chlordane, and other chlorinated hydrocarbon insecticides are also causes of brain and spinal cord poisoning and death. Central nervous system depressants include ethyl alcohol, wood alcohol, ether, chloroform, trichloroethylene, and benzene. These chemicals depress the central nervous system, resulting in headaches, vertigo, coma, and death. The peripherally acting nerve poisons include lead and alcohol. (See the section on lead for further information.)

Muscle poisons include barium salts, benzene, digitalis, and potassium salts. These poisons cause a tingling of the neck and face, colic, vomiting, diarrhea, elevation in blood pressure, and death.

Protoplasmic poisoning is due to agents causing death to cells or cell damage resulting in an inflammatory change. It produces its effects by precipitating protein, altering cell membranes, and inhibiting enzymes. It includes an enormous number of irritating chemicals, such as ammonia, formaldehyde, dimethyl sulfate, arsenic, cyanide, mercury, phosphorus, chromium, zinc phosphide, turpentine, antimony, copper, detergents, gasoline, kerosene, bromine, chlorine, and ozone.

Poisoning of the blood and blood-forming organs, such as the bone marrow, may be caused by an enormous number of toxic materials. The materials include benzene, lead, radioactive compounds, and sulfonamides.

Routes of Entry

The routes of entry of toxic substances into the body include parenteral, or through injection, which occurs as an occupational hazard when an individual is injected with a contaminated needle or cut deeply with contaminated glass; oral, in which the individual swallows the material by eating contaminated food in the work area or by smoking a cigarette that is contaminated with the hazardous material; inhalation, in which the toxic materials are drawn down into the lungs and then onto the bloodstream; cutaneous absorption, in which the hazardous material reacts with the skin surface, causing irritation, is drawn into deeper cells causing toxicity, or enters the bloodstream; ocular, in which the toxic materials come in contact with the eyes and cause irritation or partial or complete loss of vision.

The parenteral route basically occurs with individuals working with research animals in a laboratory or in a medical-care institution. Glass contaminated with microorganisms may come from either location. The microorganisms introduced directly into the blood circulation enter at a maximum concentration. This can have an immediate effect on the worker. Chemicals introduced the same way have a much higher rate of toxicity than if they had been introduced into the body through some other means whereby the natural defenses of the body could reduce the concentration of the material.

By the oral route or by ingestion, the toxic substances penetrate the mouth, are ingested, are inhaled with soluble materials, or are ingested with contaminated fluids, food, and cigarettes. Many factors influence the absorption of toxic materials into the gastrointestinal tract, including physical and chemical relationships in the gastrointestinal tract, the time of retention of the food, and various physiological factors involved in adsorption and absorption. The type of material, its concentration, and the quantity of food absorbed with the toxic substance also affect the response of the body to hazardous materials.

Time of exposure with inhalation is very important, since most toxic substances are found in the air in the form of gases, particulates, mists, dusts, smoke, and fumes. The concentration of these materials must also be taken into account. The body clears the respiratory tract through the following techniques: ciliar reaction in the upper part of the respiratory tract involving agitation of the foreign material and discharge of the material outward by the mucus; ingestion of foreign material by giant phagocytes, or white blood cells, and discharge of the material upward by the mucus; direct penetration of the material through the alveoli into the lymphatic system.

The cutaneous or skin route is the most frequent means of entry of toxic substances in occupational disease. The skin, fatty tissues, and perspiration act as barriers to the hazardous material; however, skin absorbs some of the materials and, once damaged, provides an excellent path for invasion by chemicals and microorganisms. Fat or lipid-soluble materials, such as pesticides and organic solvents, easily enter the lipid membrane of the skin. Other materials penetrate through the skin and follicles.

The eye is particularly sensitive to toxic chemicals. Materials invade the eye quickly and can cause acute or chronic damage. The number of substances hazardous to the eye are almost limitless.

The distribution of the toxic substances determines whether local damage is basically the result. If the toxic materials enter the bloodstream, they affect any of the body organs or tissues. Gases and vapors dissolve in blood plasma and bind to the hemoglobin. Toxic substances are soluble in the body fluids, accumulating in the liver and other organs or in the bone or fatty tissue. In each of these cases, the concentration of the material causes additional hazards to the human. Toxic materials are usually partially metabolized or disintegrated and eliminated from the body through respiration, salivation, perspiration, urination, and defecation. Some of these substances cause damage to the organ of elimination during the elimination process (see Figure 10.2).

Effects and Physiological Responses

Materials having high chemical activity and high surface energy cause local effects, such as irritation, edema, inflammation, and possibly death of the tissue. The general effects of toxic substances that are absorbed and transferred by the blood to different organs include stimulation or inhibition of the organs, which may be reversible or irreversible; and intoxication, resulting in death of cells, tissues, organs, and the individual. The type of response by the body is determined by examination of the organs, either in their entirety or microscopically. The response is also measured by the growth rate of the person, the weight of the specific organ, such as the liver or kidney, and the response to a physiological function test. Tests may be run to determine the level of metabolites of toxic substance in the urine and blood. Additional tests may be run on animals to determine the reaction of the animal to the toxic agent, the animal's ability to

Routes of Absorption, Distribution, and Excretion of Toxicants in the Body

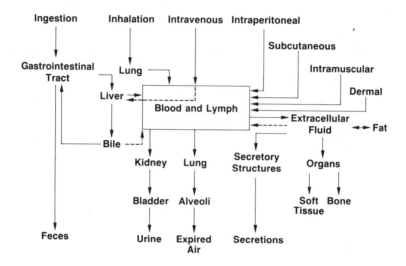

Figure 10.2.   Routes of absorption, distribution, and excretion of toxicants in the body. *Source: Environmental Toxicology and Risk Assessment: An Introduction, Student Manual* (Chicago, IL: United States Environmental Protection Agency, Region V), Visual 2.19.

reproduce, the type and number of malformations of the fetus, and the carcinogenic effects of the toxin.

Exposure Time and Concentration

The manner in which humans respond to a toxic material is influenced by the time of exposure and the concentration of the toxic substance in the environment. If concentration of material accumulates slowly, the body may be able to detoxify or excrete it before any damage is caused. The condition and age of the person, his or her state of health, previous exposure, other environmental factors, such as heat, and the individual's own response must be considered in an exposure time and concentration ratio.

The concentration of hazardous materials are measured in parts per million (ppm), millions of particles of a particulate per cubic foot of air (mppcf), milligrams of a substance per cubic meter of air (mgpm³), and micrograms of a substance per liter of solution.

To determine the amount of hazardous substance to which a worker can be exposed daily without any ill effects, the federal government developed the threshold limit value (TLV). The threshold limit values are used as guides in the control of health hazards. The biological threshold limit values are based on air analysis. This is a determination of the amount of the toxic material, such as lead or arsenic, or a metabolite found in the blood, urine, or exhaled air.

Detoxification by the Body

The human body detoxifies substances by oxidation, reduction, hydrolysis, and conjugation. Conjugation is the second phase of metabolism. When these processes occur, either a nontoxic or less toxic metabolite is formed. There is also the possibility that a more toxic metabolite is formed and that it will cause serious injury or death. The body detoxifies material to eliminate useless substances in the body.

OCCUPATIONAL HEALTH PROBLEMS

Occupational Disease and Injuries

An occupational disease is any abnormal condition or disorder caused by exposure to the work environment. It includes acute and chronic illnesses due to inhalation, absorption, ingestion, or contact. It excludes conditions due to injuries.

Occupational diseases include cadmium poisoning, mercury poisoning, chloride poisoning, asbestosis, cancer, emphysema, and dermatoses. These diseases vary in intensity and in time of onset from acute to chronic. An occupational injury is any injury resulting from a work accident or exposure to the work environment.

Disorders of Reproduction

A wide range of microbiological, physical, and chemical agents can adversely affect reproductive outcomes. At least 50 chemicals, including heavy metals, such as lead and cadmium; glycol ethers; organohalyde pesticides; organic solvents; and chemical intermediates, have been shown to produce impairment of reproductive functions in animals. The methyl mercury poisoning in 1970 of the residents of Minamata, Japan demonstrated the teratogenic potential for chemical exposures. Occupational exposures can produce a wide range of adverse affects on reproduction. Paternal exposure before conception may be shown in reduced fertility, unsuccessful fertilization, or implantation or an abnormal fetus. Maternal exposure after conception may result in death of the fetus or structural and functional abnormalities in newborns. It is estimated that there are 560,000 infant deaths, spontaneous abortions, and stillbirths each year in the United States. It is also estimated that 200,000 live infants are born with some birth defect each year. The causes of most of the adverse outcomes are unknown. No recognized anatomical or physiological abnormalities are found in 6–30% of the infertile couples. The causes for 65–70% of the birth defects are not known.

The studies of occupational reproductive hazards have been related to exposures to lead, ethylene oxide, and anesthetic gases. The absence of living spermatozoa in semen and the subnormal concentration of spermatozoa were reported among workers exposed to dibromochloropropane. Adverse effects have been found on the quality of semen in workers exposed to lead or ionizing radiation. NIOSH estimates that approximately 200,000 workers are potentially exposed to various glycol ethers, several of which show marked testicular toxicity in animals. An estimated nine-million workers are exposed to radiofrequency-microwave radiation, which has been shown to cause embryonic death and impaired fertility in animals. Approximately 50,000 personnel in hospital operating rooms are potentially exposed to waste anesthetic gases and 139,000 hospital and industrial workers may be exposed to ethylene oxide. Both of these chemicals have been linked to an increased risk of spontaneous abortions in humans.

Neurotoxic Disorders

Acrylamide has been shown to be neurotoxic in animals. In the 1950s it was also shown to be neurotoxic in humans. During the 1960s and early 1970s, dozens of cases of neurotoxic reactions occurred among Japanese and Italian workers exposed to solutions containing n-hexane during the manufacture of shoes. Serious human neurotoxicity has occurred among workers exposed to the chlorinated hydrocarbon, chlordecone, methyl-n-butyl ketone, and 2-t-butylazo-2-hydroxy-t-methylhexane. An estimate of workers exposed full time to one or more neurotoxic agents is 7.7 million individuals.

Dermatological Conditions

Of all occupational dermatological conditions, 20–25% result in lost time from work. Of all the requests that NIOSH receives, 10–15% are health hazard evaluations or because of skin complaints. The dermatological injuries are usually described as effects on skin that result from instantaneous trauma or brief exposure to toxic agents involving a single incident in the work environment. Skin injuries may constitute 23–35% of all injures, which means that an estimated 1–1.6 million dermatological injuries occur each year. Other kinds of dermatological conditions include contact dermatitis, infection, acne, and skin cancer.

Psychological Disorders

Psychological disorders may be caused by an unsatisfactory work environment. Factors contributing to this may include work overload, lack of control over the individual's work, nonsupportive supervisors or coworkers, limited job opportunities, conflict, rotation shift work, rapidly paced work, and environmental pollutants. Psychological disorders may result in anxiety; irritability; behavioral problems, such as substance abuse; sleep difficulties; psychiatric disorders, such as neurosis; somatic complaints, such as headaches; and gastrointestinal symptoms. Psychological disorders may also stress the immunological system and cause suppression of the system.

Heat

The following are heat-associated illnesses: heat stroke, caused by considerable exertion in a hot environment, lack of physical fitness, obesity, dehydration, and so forth; heat exhaustion, caused by extreme exertion and heat, lack of water, and loss of salt; heat cramps, caused by heavy perspiration during hot work, large intake of water, loss of salt; heat syncope, due to an unacclimated worker standing erect and immobile in the heat, causing blood to accumulate in the lower part of the body, which results in inadequate venous blood return to the heart; heat rash, due to unevaporated sweat from the skin; heat fatigue, caused by extreme heat for long periods of time; interference with the heat-regulating center in the brain, causing body temperatures to rise to 108–112°F, which may result in brain damage or death.

Certain safety problems are common in hot environments. Heat tends to promote accidents due to the slipperiness of sweaty palms, dizziness, or the fogging of safety glasses. Also molten metal, hot surfaces, and steam create serious contact burns. Heat lowers an individual's mental alertness and physical performance. It also promotes irritability, anger, and overt emotional reactions to situations, which may lead to accidents.

Heat-stroke symptoms include hot skin, usually dry, red, or spotted. The body temperature rises to 105°F or higher. The victim is mentally confused, delirious, possibly in convulsions, or unconscious. Unless the person receives quick and appropriate treatment, the individual will die. Individuals with these signs or symptoms of heat stroke need to be immediately taken to a hospital. Any emergency medical care prior to hospitalization should include removing the person to a cool area, soaking the clothing with water, and vigorously fanning the body to increase cooling.

Heat exhaustion includes several clinical disorders that have symptoms that may resemble the early symptoms of heat stroke. The person suffering from heat exhaustion still is able to sweat but experiences extreme weakness or fatigue, giddiness, nausea, or headache. In more serious cases the person may vomit or lose consciousness. The skin is clammy and moist, the complexion is pale or flushed, and the body temperature is normal or slightly elevated. In most cases, have the person rest in a cool place and drink plenty of liquids. If the case of heat exhaustion is mild, the individual will recover quickly. If the problem is severe take the individual to a hospital. Extreme caution must be taken concerning individuals who are on low-sodium diets and who work in hot environments.

Heat cramps are painful spasms of the muscles that occur in individuals who sweat profusely because of heat exposure. They may be drinking large quantities of water but not replacing the body's salt loss. The water tends to dilute the body's fluids while the body is still losing salt. The low salt level in the muscles causes cramps, which may occur in the arms, legs, or abdomen. Again, be most cautious concerning individuals with heart problems or those on low sodium diets.

Heat rash, which is also known as prickly heat, occurs in hot, humid environments, where sweat is not easily removed form the surface of the skin by evaporation and the skin stays moist. The sweat ducts become plugged and a skin rash then appears. Extensive heat rashes may become complicated by infection. An individual can prevent the problem from occurring by resting in a cool place each day and by bathing regularly and drying the skin.

Transient heat fatigue is the temporary state of discomfort and mental or psychological strain coming from prolonged heat exposure. If the person is unaccustomed to heat, he or she is particularly susceptible to the problem. The individual suffers a decline in test performance, coordination, alertness, and vigilance, which may lead to accidents. Transient heat fatigue can be lessened or eliminated by using proper techniques of heat acclimatization.

There are a myriad of hot jobs in industry. Some of these include work in foundries, laundries, kitchens, on trucks during the summer, boiler rooms, any furnace area, and on road or housing construction.

Cold

In order for the body to maintain a thermal homeostasis in a cold environment, it must limit heat loss and increase heat production by various physiological mechanisms. First there is a peripheral vasoconstriction, especially in the extremities, which results in a sharp drop in skin temperature. This helps reduce the body heat loss to the environment. However, this causes a chilling of the extremities, which results in the toes and fingers approaching freezing temperatures very rapidly. When the temperature of hands and fingers drop below 15°C, they become insensitive, malfunction, and the chance for accidents increases. The cooling stress is proportional to the total thermal gradient between the skin and the environment. The loss of heat through evaporation of perspiration is not significant at temperatures lower than 15–20°C. When vasoconstriction can no longer adequately maintain the body heat balance, shivering becomes an important mechanism for increasing the body temperature by causing the metabolic heat production to increase several times the resting rate. When proper clothing is used and the clothing does not become wet with water or sweating due to excessive work, then all parts that are protected by the clothing will help maintain the body temperature. Exposed parts such as the face or fingers, may become excessively chilled and frostbite may occur.

Frostbite occurs when there is an actual freezing of the tissues with the mechanical disruption of the cell structure. The freezing point of the skin is –1°C.

Increasing wind velocity will increase heat loss and cause frostbite to occur more rapidly. Once freezing occurs, it progresses rapidly. If the wind velocity reaches 20 mi/hr, the exposed flesh will freeze within about 1 min at −10°C. If the skin comes in contact with below-freezing-temperature objects, frostbite may occur, even though the environmental temperatures are warmer. The first symptom of frostbite is often a sharp, pricking sensation. Cold, however, produces numbness and anesthesia, which may allow serious freezing to develop without warning of acute discomfort. The injury from frostbite may range from a simple superficial injury with redness of the skin, to deep cyanosis and gangrene.

Trench foot may be caused by long, continuous exposure to cold without freezing, combined with a persistent dampness or immersion in water. Edema, tingling, itching, and severe pain may occur. This is followed by blistering, superficial skin necrosis, and ulceration.

General hypothermia is an extremely acute problem that results from prolonged cold exposure and heat loss. If a person becomes fatigued during physical activity, he or she will be more apt to have heat loss. As exhaustion approaches, the vasoconstrictor mechanism is overwhelmed and suddenly vasodilation occurs, with a rapid loss of heat and critical cooling. Alcohol consumption and sedative drugs increase the danger of hypothermia.

Light

Poor lighting is a major cause of fatigue, an underlying cause for a decrease in the resistance to disease and of accidents. Infrared radiation heats the skin, causing thermal burns. Extreme, excessive exposure of the eyes to radiation results in cataracts, characterized by opacity of the lens. Ultraviolet rays, from the sun or ultraviolet lamps, cause severe sunburn and damage to the lens of the eye. Ultraviolet radiation may occur around electrical arcs.

Sources of light include incandescent lamps, gas-filled and vacuum incandescent lamps, low-pressure mercury vapor lamps, high-pressure mercury vapor lamps, photochemical lamps, ozone-producing lamps, germicidal lamps, blacklight lamps, flash tubes, high-voltage neon lamps, and fluorescent lamps. The type of lamp used depends on the work environment, the period of exposure, the size of object, the contrast and brightness of the source, the brightness contrast between the source and the background, the location of the source and the field of vision, the total volume of light entering the eye, and the amount of exposure time to the brightness source.

Vibrations

Overexposure to vibration while using hand tools may cause neuritis, decalcification of the carpal and metacarpal bones, fragmentation, and muscle atrophy. An extreme condition resulting from the use of power tools is called Raynaud's syndrome or dead fingers. The circulation in the hand is impaired because of extended periods of using vibrating tools. The fingers become white, devoid of sensation, and mildly frosted. The condition usually disappears when the fingers are warmed for some time, but in some cases the condition is so disabling that the individual must seek new work. At present, there are no generally accepted limits for a safe vibration level. This is an area where additional research is needed.

Vibration is an oscillating motion of a system ranging from simple harmonic motion to extremely complex motion. The system may be gaseous, liquid, or solid; periodic or continuous. Vibration, which in effect is the motion of a particle, is characterized by the displacement from the equilibrium position, velocity, or acceleration. Some common sources of vibration include tractor operation, pneumatic tools, heavy equipment vehicles, vibrating hand tools, chain saws, stamping equipment, sewing machines, and looms.

Ergonomics

Work strain due to unusual postures or improper postural adjustment to the work results in back pain, headaches, nervousness, aching muscles, excessive perspiration, and depression. The most serious results of work stress are serious traumatic injuries, accumulative injuries, or death.

Stress is due to excessive heat, improper chairs, poor work space, or any other type of situation in which the individual performs his or her work in an unusual situation or in a continuous manner. Other environmental factors, such as illumination, climate, and noise, increase stress. An example of this would be writer's cramp, caused by continued use of the same muscles for long periods of time. Bed sores are another example caused by a patient lying in the same position in a bed for extended periods. All people experience restlessness after sitting in certain hard-back chairs for extended periods. Chairs must be designed to make sitting as comfortable as possible.

Noise

The ear is composed of the outer ear, the ear canal, the ear drum, and the inner ear. The outer ear collects soundwaves from the air and transmits them to the middle ear. The ear canal, which is part of the outer ear, contains hair and wax glands to trap debris and infectious materials. The middle ear contains the eardrum, which is a tough, tightly stretched membrane and three small bones. Soundwaves are transmitted by these bones to the fluid within the inner ear and in turn to the terminal ends of the nerve fibers in the cochlea. Each nerve fiber is tuned to a particular frequency.

Environmental noise may cause temporary or permanent hearing loss, physical and psychological disorders, interference with voice communications, disruption in job performance, and disruption of rest, relaxation, and sleep. Noise-induced hearing loss is due to damaged cell structures of the hearing organ. Physical and psychological disorders are due to a change in the response of the nervous system. These effects are annoying and may lead to accidents, but basically are not due to physical changes in the body. Noise is responsible for an increase in muscular activity; constriction in the peripheral blood vessels; acceleration of the heartbeat; changes in the secretion of saliva and gastric juices; an increased incidence of cardiovascular disease; an increase in ear, nose, and throat problems; and an increase in equilibrium disorders. A noise level of 100 dB will interfere with speech. One hundred and thirty decibels cause vibration of the viscera; 133 dB cause a loss of balance. The maximum 8-hr exposure is 90 dB. However, if this is maintained for long periods of time, loss of hearing can result. Any level above 130 dB is damaging. At 140 dB, the individual will start to experience pain. Annoyance due to noise is related to its loudness, frequency, and intermittency. As an example, if one were to sit in a quiet room and listen to water drip, it would eventually cause considerable annoyance and discomfort. (See Figure 10.2 for an illustration of the mechanical spectrum.)

Sources of noise include airplanes, highway vehicles, trains, cars, and industrial plants.

Ionizing Radiation

Radiation is an important human hazard, since absorption of energy in the body tissues causes physiological injury by means of ionization or excitation. Ionization is a process involving removal of an electron from an atom or molecule leaving the system with a net positive charge. Excitation is the addition of energy to an atomic or molecular system, changing the system from a stable to an excited state. Radiation causes cell damage by upsetting the cell chemistry or physics, and by disrupting the ability of the cell to repair itself. Radiation causes somatic damage, which means that it affects the cells and organs during the lifetime of the human. It also causes genetic damage, which affects future generations. The amount of damage depends on the amount of radiation received, the extent of body exposure, the part of the body exposed, the amount of exposure, and the type of exposure, whether external or internal.

External radiation occurs when radioactive materials pass through the body. Internal radiation occurs when radioactive material is inhaled, swallowed, or absorbed through breaks in the skin. A large dose of acute radiation exposure causes nausea, fatigue, blood and intestinal disorders, temporary loss of hair, and potentially serious damage to the central nervous system. This might happen in the event of a nuclear war or a massive nuclear accident. Long-term effects include increased incidence of lung cancer suffered by miners of uranium and pitchblende; bone cancer, which occurred in luminous watch painters years ago who had swallowed small quantities of radium; skin cancer and leukemia, which occurs in physicians and dentists from repeated exposure to radiation. Radiation also causes mutations in the genes of reproductive cells.

The biological energy in humans results from energy that is absorbed by the body tissues. The unit of measure used to describe this is called the roentgen (R), which is based on the amount of ionization made by a beam of gamma radiation or X-ray in air that produces 2.082×10^9 power ion pairs per cm^3 of standard air. Two other important units are the rad and rem. A rad, which is the radiation absorbed dose related to the chemical and biological changes in the exposed tissue, is a unit of energy absorbed by material such as body tissue. One rad equals 100 ergs/g. One roentgen of exposure will usually produce about 1 rad of absorbed energy in soft tissue. An erg is the force of one dyne acting through a distance of 1 cm or is equivalent to the energy required for an electron to ionize about 20 billion atoms of air. A rem is a unit of absorbed energy that takes into account the different relative biological effectiveness of different types of radiation. In 1984 the roentgen was replaced by the term *Exposure Unit* which was defined as the collection of one electrostatic unit (ESU) of charge in 1 cm^3 of air at standard temperature and pressure. The rad was replaced by the Gray (Gy) and the REM was replaced by the sievert.

The extent of the hazard to humans is based on the dosage, the body part, and the time of exposure. Large amounts of radiation to the whole body cause acute exposure. Large amounts of radiation to small parts of the body cause acute and chronic exposures. Small amounts of radiation to the whole body cause chronic exposure. Small amounts of radiation to limited portions of the body cause acute and chronic exposures.

Humans are exposed to radiation from the natural environment, including cosmic rays, the earth, and the atmosphere. Radiation from artificial sources include medical and dental sources, occupational sources, nuclear energy plants, radioactive fallout, radioactive dial watches, clocks, meters, radioactive static eliminators, television sets, isotope-tagged products, radioactive luminous markers, and certain household goods.

Presently, the largest artificial source of radiation exposure is medical and dental X-ray machines. Over 400,000 X-ray machines are used for the purpose. Over 400,000 medical and technical personnel are occupationally exposed to radiation from these machines. It is estimated that over 40 million roentgenographic films and over 28 million photofluorographic films are taken each year. In addition, an estimated 15 million fluoroscopic examinations are given annually.

X-rays are potentially dangerous when overused or misused. The average medical and dental X-ray emits 4.6 R of radiation, whereas fallout from atomic weapons testing averages 0.1 R. The possibility of acute radiation injury due to X-rays is minimal; however, long-term effects are not fully understood. These may include cancer, shortening of the lifespan, and genetic mutation. Individuals most affected by this potential hazard are doctors, nurses, and X-ray technicians, who most frequently use the equipment. The benefits of X-ray examination and therapy far outweigh the potential hazard, since X-rays are an essential part of medical treatment. Medical and nursing personnel should be properly shielded and protected from this potential hazard.

Radionuclides are used in medical diagnosis and treatment. They may be used in the following three ways:

1. Dilution techniques, where a radionuclide, such as iodine-131, is injected into the bloodstream and after a period of time a blood sample is taken to analyze the total dilution. This helps to determine blood volume, water volume, and extracellular body water. Red cell masses are also determined by the use of different radionuclides, such as sodium-24 or ferrous-59.
2. Cardiac output is measured using sodium-24. Peripheral vascular disorders, circulation time, and vascular constriction are also measured using sodium-24.
3. Biochemical concentration measures liver function, thyroid disorders, and the location and extent of malignancy using iodine-131.

Therapeutic radiation is another source of radioactive material from the use of X-ray machines, cobalt units, or radionuclides. Radionuclides may be selectively absorbed by certain tissues, as in the case of thyroid-absorbing iodine. Radionuclides are used in medical research. Such radioactive materials as sodium-24, carbon-14, ferrous-59, chromium-51, and phosphorus are used to study minerals, vitamins, metabolism, and cholesterol.

Industrial use of radiation produces additional sources of radioactive exposure. This includes the inspection of materials by means of radiographic inspection, fluoroscopic inspection, and continuous automatic inspection by X-rays to measure the thickness of material and the height at which something is filled. Personnel are checked for stolen articles, contraband, or weapons by whole-body fluoroscopes. The shoe-fitting fluoroscope has fortunately been eliminated in most states, since it was an important source of radiation, especially for small children. X-rays are also used to determine the diffraction pattern of crystalline materials.

Massive doses of radiation have been used in industry for sterilization of foods and drugs, killing insects in seeds, toughening polyethylene materials, and activation of chemical reactions in the petroleum process. One device used is called the Van de Graaf apparatus, which is an electron-beam generator. In the past, radium was used as a source for industrial radiography. It was also used in luminous paints and in the removal of static electricity generated by the movement of nonconducting materials at high speeds through machinery. The radium produced ionized air which removed the static charge. Today, other artificial radionuclides are used in industry, including cobalt-60, iridium-192, and cesium-137. Strontium-90 is used in some luminous paints and in the krypton switchlamp used by railroads in remote locations.

Atomic batteries are used to produce high-voltage, low-current sources in special instruments. Radionuclides are used to determine thickness, density, and depth of materials, and are used as tracers. Sources of radiation also include nuclear reactors, ventilation air, and cooling wastes. The latter come from nuclear power plants or nuclear ships and submarines.

Individuals may be exposed to radon gas in homes. (See chapter on the indoor environment.)

Thousands of mine workers are exposed to radiation in uranium mines and through milling. Fallout from atomic weapons research contributes additional radioactive materials to the soil, air, and water. Of serious concern is the production of strontium-90, which may be ingested by cows in their feed and then transmitted through milk to people.

Nonionizing Radiation

Ultraviolet radiation is due to sunlight, low- and high-pressure mercury discharge lamps, welding, and plasma torches. The maximum reddening of the skin due to ultraviolet radiation is at 0.260 μm. Ultraviolet light causes blistering, desquamation, thickening of the skin, and may also precipitate skin cancer.

Infrared radiation is used in industry in the drying and baking of paints, varnishes, enamels, and adhesives. It is also used to dehydrate textiles, paper, meat, vegetables, and for localized heating of objects. Infrared rays cause injury to the cornea, iris, retina, and the lens of the eye. It is also hazardous to the skin and the entire body.

Microwaves are created in the atmosphere from antennas associated with television, FM radios, and radar transmitters. Microwaves are also used in diathermic devices in medicine, in ovens for cooking, and in freeze-drying operations. The greatest increase in use has been with the microwave oven, in which only the food is heated and not the oven walls. The food is cooked throughout very quickly, rather than from the surface inward. Microwaves are also used in the transmission of power, and in basic chemical and in other types of research. Microwave energy is on the electromagnetic spectrum between radiowaves and infrared waves.

Microwaves cause an increase in heat in the whole body. Microwave absorption and heat generation is greatest in tissues having a high water content. Microwaves change polarity every half cycle, with the tissue molecules doing the same and thereby vibrating rapidly, which produces heat. They also cause damage to the lens of the eye and possible damage to the gonads. A particular problem in the microwave oven is the escape of microwaves from an improper seal on the oven door or a dirty oven. Proper housekeeping and maintenance will eliminate this problem.

Lasers have been used in the construction industry for measuring distances and welding, and by some surgeons to repair torn retinas. The hazard to the individual is due to the point source of great brightness. The laser may also injure the eye, skin, or possibly internal organs. The laser may oxidize chemicals in the ambient air.

Hazardous Chemicals

There are about 65,000 existing chemicals used in industry today. The toxicity of the substance, the way in which it is used, the chemicals with which it might bind, the degree of susceptibility of the person, and the method of control are all intimately related in the determination of toxic response.

In this section, we will discuss some of the hazardous chemicals and their potential effects on humans. The substances chosen are thought to be important in occupational diseases. Of necessity, many possibly hazardous chemicals have been omitted.

A hazardous chemical is any chemical whose presence or use is a physical hazard or a health hazard. A physical hazard is caused by a chemical for which there is scientifically valid evidence that it is a combustible liquid, a compressed gas, explosive, flammable, an organic peroxide, an oxidizer, a pyrophoric, unstable (reactive) or water reactive. A health hazard is caused by a chemical for which there is significant evidence, based on at least one study conducted in accordance with established scientific principles that acute or chronic health effects may occur in exposed employees, or acute effects or adverse effects on a human or animal that has severe symptoms developing rapidly and coming quickly to a crisis. Chronic effects are adverse effects on a human or animal body with symptoms that develop slowly over a long period of time or that reoccur frequently.

A combustible liquid is any liquid having a flash point at or above 100°F but below 200°F. A compressed gas may be one of three things: the gas or mixture of gases in a container within absolute pressure exceeding 40 psi at 70°F; a gas or mixture of gases in a container having an absolute pressure exceeding 104 psi at 130°F; or liquid having a vapor pressure exceeding 40 psi at 100°F. An explosive is a chemical that causes a sudden, almost instantaneous release of pressure, gas, and heat when subjected to a sudden shock, pressure, or high temperature.

A flammable chemical may fit into one of four categories: an aerosol flammable is an aerosol that when tested yields a flame projection exceeding 18 in. at the valve opening, or a flashback at any degree of valve opening; a gas flammable chemical is a gas that at ambient temperatures and pressures forms a flammable mixture with air at a concentration of 13% by volume or less; or a gas that at ambient temperatures and pressures forms a range of flammable mixtures with air wider than 12% by volume, regardless of the lower limit; a liquid flammable chemical is any liquid having a flashpoint below 100°F, except any mixture having components with flashpoints of 100°F or higher, the total of which make up 99% or more of the total volume of mixture; and a solid flammable chemical is any solid other than a blasting agent or explosive that is liable to cause fire, friction, absorption of moisture, spontaneous chemical change, or retained heat from manufacturing or processing, or that can be ignited readily and when ignited burns so vigorously and persistently as to create a serious hazard.

An organic peroxide is an organic compound that contains the bivalent -O-O structure and may be considered a structural derivative of hydrogen peroxide, as one or both of the hydrogen atoms has been replaced by an organic radical.

An oxidizer is a chemical other than a blasting agent or explosive that initiates or promotes combustion in other materials, causing fire either by itself or through the release of oxygen or other gases.

A pyrophoric is a chemical that ignites spontaneously in air at a temperature of 13°F or below.

An unstable chemical is one that tends toward decomposition or other unwanted chemical changes during normal handling or storage.

An unstable reactive chemical is one that in the pure state or as produced or transposed, will vigorously polymerize, decompose, condense, or become self-reactive under conditions of shocks, pressure, or temperature.

A water-reactive chemical is one that reacts with water to release a gas that is either flammable or presents a health hazard.

A carcinogen is a substance or agent capable of causing or producing cancer in mammals, including humans. A chemical is considered to be a carcinogen if it has been evaluated by the International Agency For Research on Cancer and found to be a carcinogen or potential carcinogen; is listed as a carcinogen or potential carcinogen in the annual report published by the National Toxicology Program (NTP); or is regulated by OSHA as a carcinogen. A chemical is considered to be toxic if it has a median lethal dose (LD_{50}) of more than 50 mg/kg, but not more than 500 mg/kg of body weight when administered orally to albino rats weighing between 200 and 300 g each. It may also be a chemical that has an LD_{50} of more than 200 mg/kg, but not more than 1000 mg/kg of body weight when administered by continuous contact for 24 hr or less if death occurs within 24 hr, if the substance is applied to the bare skin of albino rabbits weighing between 2 and 3 kg each. A chemical that has a LC_{50} (medium lethal concentrations) in air of more than 200 ppm but not more than 2000 ppm by volume of gas or vapors, of more than 2 mg/l but not more than 20 mg/l of mist, fume, or dust, when administered by continuous inhalation for 1 hr or less, if death occurs within 1 hr to albino rats weighing between 200 and 300 g each. A highly toxic chemical is one with an LD_{50} of 50 mg or less per kilogram of body weight when administered orally to albino rats weighing between 200 and 300 g each. It may also be a chemical with an LD_{50} of 200 mg or less per kilogram of body weight when administered by continuous contact for 24 hr or less if death occurs within 24 hr when applied to the bare skin of albino rabbits weighing between 2 and 3 kg each. It may also be a chemical that has an LC_{50} in the air of 200 ppm by volume or less of gas or vapor or 2 mg/l or less of mist, fume, or dust, when administered by continuous inhalation of 1 hr or less, if death occurs within 1 hr to albino rats weighing between 200 and 300 g each. A reproductive toxin is a substance that affects either male or female reproductive systems and may impair the ability to have offspring. An irritant is a chemical that is not corrosive but that causes a reversible inflammatory effect on living tissue by chemical action at the site of the contact. A corrosive is a chemical that causes visible destruction of, and irreversible alterations in, living tissue by chemical action at the site of contact. A sensitizer is a chemical that causes a substantial proportion of exposed people or animals to develop an allergic reaction in normal tissue after repeated exposure to the chemical.

Hepatoxins are chemicals that produce liver damage. The signs and symptoms are jaundice or liver enlargement. Examples of these chemicals include carbon tetrachloride or nitrosamines. Nephrotoxins are chemicals that produce kidney damage. The signs and symptoms include edema and proteinuria, which is an excess or serum proteins in the urine. Examples of these chemicals are halogenated hydrocarbons and uranium. Neurotoxins are chemicals that produce a primary toxic effect on the nervous system. The signs and symptoms include narcosis, which is a reversible depression of the central nervous system produced by drugs or chemicals marked by stupor or insensibility. Other signs and symptoms include a decrease in motor functions. Chemicals that may cause this are mercury and carbon disulfide. Agents that act on blood and the hematopoietic system decrease the hemoglobin function and deprive the body tissues of oxygen. The signs and symptoms include cyanosis and loss of consciousness. Chemicals that cause this include carbon monoxide and cyanides. Agents that damage the lung include chemicals that irritate or damage the pulmonary tissue. The signs and symptoms include cough, tightness in the chest, and shortness of breath. Chemicals causing this are silica and asbestos. Cutaneous hazards are caused by chemicals that affect the dermal layer of the body. Signs and systems include defatting of the skin, rashes, and irritation. Chemicals that may cause this are ketones and chlorinated compounds. Eye hazards are caused by chemicals that affect the eye or visual ability. The signs and symptoms include conjunctivitis and corneal damage. Chemicals causing this are organic solvents and acids. Target organ toxins are toxins that affect a specific organ of the body, such as carbon tetrachloride causing liver damage, as has been previously mentioned.

An example of a combustible liquid is a solvent or an oil. An example of a compressed gas is oxygen and chlorine. An example of an explosive is nitroglycerine. An example of a flammable liquid is gasoline. An example of a flammable solid is phosphorous. An example of an organic peroxide is benzoyl peroxide and peracetic acid. An example of an oxidizing agent is fluorine and ozone. An example of a pyrophoric is phosphorous. An example of an unstable (reactive) chemical is nitroglycerine An example of a water reactive chemical is sodium.

To summarize, hazardous chemicals are composed of a group that is ignitable such as solvents and oils, and corrosive, which may be acid waste or pickle liquors. They may be active, such as cyanide solvents, or toxic, such as pesticide waste, lead, mercury, and arsenic.

Acids

Inorganic acids are compounds of hydrogen and one or more other elements that break down in water or other solvents to produce hydrogen ions. Inorganic acids include chromic, hydrochloric, nitric, and sulfuric. Inorganic acids, when mixed with other chemicals or combustible materials, cause fires or explosions. They are corrosive in high concentrations and will destroy body tissue and cause chemical burns. They are especially hazardous to the upper respiratory tract, mucous membrane, and the teeth.

Organic acids include formic, acetic, propionic, chloracetic, and trifluoracetic. Organic acids may produce severe damage to tissues, including the skin and mucosal surface. They also injure the eyes and may injure or interfere with the enzyme systems of the body.

Alkalies

Alkalies are caustic substances that dissolve in water to form a solution with a pH higher than 7. These include ammonia, ammonium hydroxide, calcium hydroxide, potassium hydroxide, and sodium hydroxide. The alkalies in solid form or concentrated liquid solutions are more hazardous and destructive to tissues than most acids. They cause irritations of the eyes and respiratory tract and lesions in the nose. Potassium and sodium hydroxide, even in dilute solutions, soften the skin and dissolve skin fats. Alkalies cause coughing and painful throat and nose irritations.

Ammonia

Ammonia is an important part of many compounds containing nitrogen. It is used in making fertilizers, nitric acid, acetylene, urea, explosives, synthetic fibers, and synthetic resins. It is also used in producing paper products, photographic film, dyes, inks, glues, medicine, blueprints, some cleaning solutions, and in ice-making machines and commercial refrigeration.

Although ammonia has no additive effects on the human being, it can be extremely irritating to the eyes, throat, and breathing passages. Large concentrations of ammonia can produce convulsive coughing by preventing breathing and can therefore cause suffocation in a short time. It also dissolves in the skin's moisture, causing a corrosive effect on the skin. Contact with anhydrous and aqueous ammonia causes first- and second-degree burns of the skin and eyes. Blindness can be the ultimate result of this contact. Ammonia gas in the presence of oil or other combustible materials causes fires. Ammonia can combine with silver oxide or mercury to form a fulminate, which is an unstable explosive compound.

Aromatic Hydrocarbons

Aromatic hydrocarbons, which include benzene, ethyl benzene, toluene, and xylene, are commercial solvents and intermediates in the chemical and pharmaceutical industries. These chemicals affect the blood, bone marrow, central nervous system, eyes, respiratory system, skin, liver, and kidney. All of the aromatic hydrocarbons cause a central nervous system depression, as well as decreased alertness, headaches, sleepiness, and loss of consciousness. They cause a defatting dermatitis. Benzene suppresses the bone marrow function, thereby causing blood changes. Chronic exposure to benzene can cause leukemia. Aromatic compounds are frequently contaminated with benzene during distillation. Benzene-related health effects should be considered when exposure to any of these agents is suspected.

Inorganic Arsenic

Arsenic is found in small amounts in soils, water, and even food. It is usually present in the ores of metals as an impurity. Arsenic trioxide is found in the smelting process. It is also produced as a byproduct of the smelting of copper ores. Arsenic is used in pesticides, in the production of pigments, in the manufacture of glass, in textile printing, tanning, taxidermy, and in the control of sludge formation in lubricating oils. It is also used as an alloy in lead-based materials and to toughen copper and improve its resistance to corrosion.

Acute poisoning due to arsenic in industry is rare. Where it does occur, the individual has swallowed the arsenic and a reaction usually occurs 30 min to several hours after ingestion. However, chronic poisoning is a definite occupational hazard.

Asbestos

Asbestos is a chain silicate rock occurring in nature in fibrous form. The asbestos rock is found as chrysotile, which is hydrated magnesium silicate; amphidole, which is iron magnesium silicate, sodium iron silicate, or magnesium silicate. Chrysotile, which is about 90% of all asbestos processed, is resistant to heat but not to acids. It may be easily spun into asbestos cloth. Amosite is acid and heat resistant, and is used in bulk form for heat installation and molded into pipe insulation. Crocidolite is highly resistant to acid but has poor heat-insulation properties. It is used in acid-resistant cement pipe and in electric battery cases.

Asbestos is used in the construction industry for roofing, plastics, insulation, floor tiling, and cement production. It is also used in the automotive industry as friction material in brake and clutch linings. The textile industry uses asbestos for fire-resistant clothing, safety equipment, and curtains. It may also be used to strengthen materials that are exposed to heat.

Asbestos causes a variety of diseases. Since the fibers are so easily transported by workers in their clothing and picked up from contact with insulation materials, anyone, including the individual who produces asbestos fiber, the maintenance worker, the demolition worker, the family, and other individuals, can be exposed to asbestos and subjected to a variety of diseases.

Anesthetic Gases

It has been determined that an increased amount of miscarriages have occurred in operating-room nursing personnel. There is also concern for the effect of the gases on male personnel and the potential genetic problems that may occur when their wives conceive.

Benzene

Benzene, a constituent of petroleum, is a clear, colorless, noncorrosive, highly flammable liquid, with a strong, rather pleasant odor. Large amounts of benzene are used as a fuel, chemical reagent, and solvent in coke, gas, chemical, printing, lithography, paint, rubber, dry cleaning, adhesives, and petroleum industries and in chemical laboratories. Benzene is changed to nitrobenzene and the aromatic nitrocompounds for use in finished products or in aniline. The halogenated pesticides, such as DDT, use benzene in their formulation. It is also used in detergents, explosives, pharmaceutical, and dyes. It is a good solvent for rubber, plastics, paints, inks, oils, and fats.

Benzyne creates serious hazards because it is very toxic and highly flammable. It causes an acute narcotic effect, and it injuries the blood-forming tissues, even if exposure is in small amounts over long periods of time. Severe cases of benzene poisoning result in death. Although an individual may be removed from exposure to benzene, recovery takes a long time and there are frequent relapses.

Beryllium

Beryllium is a lightweight metal with high tensile strength that forms excellent alloys with steel, nickel, magnesium, zinc, aluminum, and so forth. Beryllium is extracted from several different forms. The primary form is beryl (3 BeO. $Al_2O_3.6SiO_2$), which is the most abundant of the minerals containing high amounts of beryllium oxide. Beryllium is extracted from the ore in either the sulfate or fluoride process.

Beryllium is used to reduce the speed of fission in nuclear reactors. It is mixed with uranium and used as a neutron source. Beryllium is also used in the automotive, computer, electronics, welding, communications, aerospace, optical instruments, propulsion, and plastics industries.

Exposure to beryllium compounds occurs not only within the production and manufacturing industries, but also in housekeeping, maintenance, salvage, and solid-waste areas. Where welding, burning, cutting, or other operations are performed on scrap metals containing beryllium, a serious health hazard exists. Unfortunately, once materials containing beryllium are added to the general scrap in solid waste, the beryllium loses its identity and may cause exposure problems to unknowing handlers.

Beryllium causes such diseases as contact dermatitis, beryllium ulcers, and injury to the conjunctiva. Chronic exposure causes bronchitis, pneumonia, right heart enlargement, and enlargement of the spleen and liver. Prolonged exposure to beryllium can cause death. An additional discussion of beryllium appears in the chapter on air pollution control in Volume II.

Carbamates

The carbamates include aldicarb, baygon, and zectran. They are used for pest control. They affect the central nervous system, liver, and kidney. Their potential health effects are the same as organophosphates, which are discussed later.

Cadmium

Cadmium is a naturally occurring element in the Earth's crust. It is most frequently encountered as cadmium oxide, cadmium chloride, or cadmium sulfide. It is used in metal plating, pigments, batteries, and plastics. Cadmium is widely distributed in the air, water, soil, and food. Smoking is an important source of cadmium. The largest sources of cadmium in the environment are the burning of fossil fuels, such as coal or oil, and the incineration of municipal waste materials. The use of phosphate fertilizers or sewer sludge increases cadmium levels in soil and in turn, in crops. Cadmium can enter the body by absorption from the stomach or intestines after ingestion of food or water. It can affect the kidney and cause kidney damage if the individual is exposed to excessive amounts of cadmium in the air or through the diet. It can cause such lung damage as fibrosis or emphysema. Animals that have been exposed to cadmium for long periods of time may also get lung cancer. Human studies suggest that long-term inhalation of cadmium can result in an increased risk of lung cancer.

Carbon Monoxide

Carbon monoxide is produced by the incomplete combustion of fuel. It is also produced in industry by the partial oxidation of hydrocarbon gases in natural gas or by the gasification of coal or coke. It is used in the manufacture of metal carbonyls. It is also a serious hazard in the chemical, iron, steel, pottery, automobile, and mining industries. Garage and tunnel workers are constantly subjected to excess levels of carbon monoxide. Carbon monoxide combines with hemoglobin, since it has an affinity for hemoglobin 200–300 times higher than oxygen. This causes oxygen starvation, resulting in acute poisoning. The highly sensitive brain cells will survive only about 8 min if deprived of oxygen.

Chlorofluorocarbon-113

Chlorofluorocarbon-113 (CFC-113), also known as freon-113, may cause death due to cardiac arrhythmia and asphyxiation. Fluorocarbons are halogenated hydrocarbons that contain chlorine. They are called chlorofluorocarbons (CFCs). These chemicals are colorless, noncombustible liquids that are used as refrigerants, propellants, degreasers, fire extinguishants, deicers, and agents used for cleaning electronic equipment and preparing frozen tissue for histopathology. Because of the high vapor pressure at room temperature, they can cause high ambient concentrations of vapor during use of a liquid. This produces a significant hazard in a confined space. The chemical's toxic effects include respiratory depression, bronchoconstriction, and impairment of psychomotor functions, such as manual dexterity, vigilance, and the ability to concentrate and breathe.

Chromium and Chromic Acid

Chromium is considered to be an essential nutrient that helps to maintain normal glucose, cholesterol, and fat metabolism. These are three major forms of chromium, which differ in their effects. The hexavalent chromium is irritating, and a short-term high-level of exposure can result in ulcers of the skin, irritation and perforation of the nasal mucosa, and irritation of the gastrointestinal tract. Hexavalent chromium may also cause adverse affects in the kidney and liver. Trivalent chromium does not result in these effects and is thought to be an essential nutrient. Long-term inhalation exposure of workers to low levels of chromium compounds has been associated with lung cancer. It is not clear which form of chromium is responsible for this effect in workers.

Chromic Acid

Compounds of chromium, such as chromic acid, chromate, and dichromate, cause harmful effects. Chromium is produced by reducing chromic oxide with aluminum powder. Chromium is used with iron and steel to form alloys. It is also use for chromium electroplating of equipment, such as automobile parts and electrical equipment. Chromic acid is a powerful oxidizing agent. It is used for hard chromium plating and decorative plating. It is also used in the oxidation of organic materials. Chromium compounds cause skin and mucous membrane irritation and corrosion, dermatitis, and ulceration, and are shown to be a contributing cause in the increase of lung cancer.

Chloroform

Chloroform is a colorless or water-white liquid. Most of the chloroform produced in the United States is used to make fluorocarbon-22. Much smaller amounts are used as a pesticide and solvent and in the manufacture of pesticides and dyes. Fluorocarbon-22 is used primarily as a cooling fluid in air-conditioners. The primary sources of chloroform discharged into the environment are pulp and paper mills, pharmaceutical manufacturing plants, chemical manufacturing plants, chlorination of wastewater from sewage treatment plants, and chlorination of drinking water. Chloroform can enter the body by inhalation, ingestion, or through the skin. Chloroform affects the central nervous system, liver, and kidneys. Short-term exposure to high concentration of chloroform in the air causes tiredness, dizziness, and headaches. Longer-term exposure to high levels of chloroform in air, food, or water can affect liver and kidney function, resulting in jaundice and burning urination.

Chlorine

Chlorine is a pungent, greenish-yellow gas that is more than twice as heavy as air. It is produced mainly by passing electricity through a solution of table salt. The gas is then compressed to a liquid.

Chlorine is used in the bleaching of materials and fabrics. It is also used to purify drinking water, disinfect swimming pools, and treat sewage. Large quantities of chlorine are used in the production of inorganic and organic chlorine compounds, including aluminum chloride and iron chloride. It is used in producing solvents, such as carbon tetrachloride, trichloroethane, and chloroform. Chlorine is also used in producing refrigerants, such as chloromethane, pesticides such as DDT, and polymers such as polyvinyl chloride and synthetic rubbers.

Chlorine is an irritant to the mucus membrane of the eyes, nose, throat, and lungs. It reacts with body moisture to form dangerous acids. Pure chlorine gas and liquid are not flammable. However, when mixed with hydrogen, hydrocarbons, alcohols, and ethers, explosive mixtures result. Sudden contamination by chlorine causes offensive odors. However, when a worker breathes chlorine at low concentrations over a period of time, he or she can no longer detect the odor of chlorine. It, therefore, becomes a potentially hazardous situation.

Coke Oven Emissions

The primary function of the coke plant is the production of metallurgical coke for use in a blast furnace. Chemical by-products are also recovered during the transformation. The by-products include coal tars, which may cause skin, lung, bladder, and digestive system cancers. Workers who are smokers have a much higher rate of cancer. This bituminous coal used causes black lung and other diseases in miners. There also seems to be an increased rate of bronchitis.

Cotton Dust

Cotton dust comes from the finely pulverized part of the cotton plant. The hazard is greatest in the handling of raw cotton during the carding stage, but hazards also exist during other stages in the cotton manufacturing process. Workers who inhale this dust acquire a respiratory disease called byssinosis.

Cyanide

Cyanides are naturally occurring substances found in foods and plants, and produced by certain bacteria, fungi, and algae. Cyanide salts are used in electroplating and metal treatment. The cyanide compounds that are of greatest concern to humans are hydrogen cyanide, sodium cyanide, and potassium cyanide. The sources of cyanides in the air are vehicle exhaust, emissions from chemical processing industries, iron and steel mills, metallurgical industries, metal plating and finishing industries, and petroleum refineries. Atmospheric emissions may also come from burning of some nitrogen-containing plastics, silk, and wool. Municipal waste incinerators will also release cyanide to the air. The most deadly release is a direct one to the atmosphere from any plant producing or utilizing cyanides in their operation such as the one that happened in Bhopal, India. Cyanides may be found in water and the discharges from organic chemical industries, iron, and steel works, and wastewater treatment plants. Small amounts may enter the water from stormwater runoff where cyanide salts are used. Cyanide may migrate from landfills into groundwater. Cyanide may be found in the soil where cyanide waste are disposed of in landfills. In the occupational setting, cyanide may be found in a multitude of industries, ranging from pesticide application to the manufacture of dyes and pharmaceuticals to photography and electroplating. Cyanide enters the body through the air or through water or food. Adverse health effects from cyanide include serious problems with the central nervous system, respiratory system, and cardiovascular system. Short-term exposure at high levels of cyanide can cause coma and/or death. Brief exposure at lower levels results in rapid, deep breathing, shortness of breath, convulsions, and loss of consciousness. If the cyanide does not remain in the body, the short-term effects may be reversed. Skin contact with the dust from cyanide salts can cause skin irritation and ulcerations. If a person has been exposed to low levels of cyanide over an extended period of time, there will be adverse effects to the nervous system and the typhoid gland. Breathing problems, visual problems, loss of muscle coordination, as well as retarded physical and mental growth in children, may occur.

Dioxin

Dioxin is an inaccurate colloquial name for 2,3,7,8-TCDD. It is a colorless solid with no distinguishable odor. It does not occur naturally nor is it produced intentionally by any industry, except as a reference standard. It is inadvertently produced in very small amounts as an impurity during the manufacture of certain herbicides and germicides, and during the incineration of municipal and industrial waste. The main environmental sources of 2,3,7,8-TCDD are the production and use of herbicides containing 2,4,5-trichlorophenoxy acids (2,4,5-T), as well as the production and use of 2,4,5-trichlorophenol. It is found during the production and use of hexachlorophene as a germicide. Small amounts are formed during the burning of wood in the presence of chlorine. It is also formed when there are accidental transformer-capacitor fires involving chlorinated benzenes and biphenyls. Leaded gasoline exhaust may also produce 2,3,7,8-TCDD. One other source of dioxin is the improper disposal of certain chlorinated chemical wastes. Dioxin may enter the body through contact with the skin and contaminated soils or materials. It may also enter through the ingestion of food that is contaminated. This food includes fish, cow's milk, and other foodstuffs. Breathing of contaminated ambient air may contribute a small amount of total body intake, however, the inhalation of particulates such as fly ash may constitute a major source of exposure. In humans 2,3,7,8-TCDD causes chloracne, which is a severe skin lesion that usually occurs on the head and upper body. It is disfiguring and may last for years after the initial exposure. There is suggestive evidence that 2,3,7,8-TCDD causes liver damage in humans as indicated by an increase in levels of certain enzymes in the blood. There is also suggestive evidence that the chemical causes loss of appetite, weight loss, and digestive disorders in humans. In rodents, the chemical, if administered during pregnancy, results in malformations in the offspring. 2,3,7,8-TCDD has been demonstrated to be a carcinogen in animals.

Ethylene Oxide

Ethylene oxide is used in sterilizers in institutions and also it is produced and used in chemical plants as an intermediate. At room temperature and atmosphere pressure, ethylene oxide is a colorless gas, whereas at higher pressures it may be a volatile liquid. It has a characteristic ether-like odor with a widely variable threshold of detection in humans. It is highly reactive and potentially explosive in the presence of alkali metal hydroxides and highly active catalytic surfaces or when heated. However, it is relatively stable in aqueous solutions or when diluted with carbon dioxide or halocarbons when used as a fumigant or sterilant. In order to reduce the explosion hazards, it often comes in mixtures such as 10% ethylene oxide and 90% carbon dioxide or 12% ethylene oxide and 88% halocarbon. Ethylene oxide is considered to be mutagenic and carcinogenic in animals, and is also capable of causing adverse reproductive effects. It also may cause chromosomal damage and has the potential for causing cancer and adverse reproductive effects in humans.

Ethylene Dibromide

Ethylene dibromide (EDB) has been produced on a commercial scale since the mid-1920s. It is used primarily as a scavenger in leaded fuels in combustion with ethylene dichloride. The scavenger are used to form volatile lead compounds during combustion that are more completely removed from the combustion chamber. EDB, now banned by the EPA, is also used as a grain fumigant, and is a solvent for resins, gums, and waxes. EDB exposure may result in a potential carcinogenic risk. It is highly toxic and increases the risk of adverse reproductive and other effects. Animal experiments indicate that these effects may include sterility, inheritable changes in offspring, teratogenesis, and adverse effects on the liver, kidneys, heart, and other internal organs. Where skin contact occurs with EDB, chemical burns, as well as systemic effects from the percutaneous absorption, may happen.

Epoxy Resins

Epoxy compounds are cyclic ethers containing oxygen attached to two carbon atoms. Since these compounds are chemically active, they are used as intermediates in the production of other chemicals. They form monoepoxides such as ethylene oxide and polyepoxides.

Epoxy resins are strong, resistant, lightweight, and join firmly to other materials. They are used in manufacturing, aircraft, automobiles, electrical appliances, and painting. Workers are exposed to hazards in occupations that entail mixing resins and hardeners, where the material may come into direct contact with the skin or the vapors may be inhaled; molding and casting, where the material comes into direct contact with the skin; tooling operations such as sanding, grinding or drilling, where the hardened epoxy resins or laminates form tiny particles that may become imbedded in the skin or may get into the eyes. Dermatitis is the major hazard to humans. However, the epoxy resins also cause eye irritation, cornea burns, and respiratory, nose, and throat irritations.

Formaldehyde

Formaldehyde is a highly reactive gas that is slightly heavier than air and combines readily with many materials. It dissolves well in water, alcohol, or ether. Formaldehyde is used as a germicide, fungicide, and preservative. It is used in the tanning and preservation of hides and furs, embalming, improving the fastness of dyes, waterproofing and strengthening fabrics, processing and preserving rubber latex, hardening paper products, developing photographic film, and refining gold and silver. The greatest industrial use of formaldehyde is the production of resins for plastic and synthetic fabric manufacturing. It is also used in the textile industry for permanent-crease and permanent-press clothing. It is used in a foam as an insulating material. Formaldehyde causes irritation of the eyes, nose, mouth, and throat, difficulty in breathing, and severe coughing.

Halogenated Aliphatic Hydrocarbons

Halogenated aliphatic hydrocarbons include carbon tetrachloride, chloroform, ethyl bromide, ethyl chloride, ethylene dibromide, ethylene dichloride, methyl chloride, methyl chloroform, methylene chloride, tetrachloroethane, tetrachloroethylene (perchloroethylene) trichloroethylene, and vinyl chloride. These chemicals are commerical solvents and intermediates in organic synthesis. They affect the central nervous system, kidney, liver, and skin. All of these chemicals cause a central nervous system depression, decreased alertness, headaches, sleepiness, and loss of consciousness. They also cause changes in the kidney that decreases the urine flow. Swelling occurs, especially around the eyes. These chemicals may cause anemia. The halogenated aliphatic hydrocarbons cause liver changes that lead to fatigue, malaise, dark urine, liver enlargement, and jaundice. Vinyl chloride is a known carcinogen, and several others in this group are potential carcinogens.

Heavy Metals

The heavy metals consist of arsenic, beryllium, cadmium, lead, and mercury. The heavy metals are used in a variety of industrial and commercial areas. The organs affected by the heavy metals include the blood, cardiopulmonary system, gastrointestinal system, kidney, liver, lungs, central nervous system, and skin. All of the heavy metals are toxic to the kidneys. The potential health effects of these compounds are listed under the particular heavy metal in this section.

Herbicides

The herbicides include chlorophenoxy compounds, such as 2,4,-D, 2,4,5-T, and TCDD. 2,4,-D is 2,4-dichlorophenoxyacetic acid. 2,4,5-T is 2,4,5-trichlorphenoxyacetic acid. TCDD is tetrachlorodibenzo-p-dioxin, better known as dioxin. Dioxin is a trace contaminant in 2,4,-D, and 2,4,5-T. Dioxin poses the most serious health risk of these chemicals. These chemicals are used for the control of weeds and unwanted plants. They affect the kidney, liver, central nervous system, and skin. They cause chloracne, and weakness or numbness of the arms and legs, and they result in long-term nerve damage. Dioxin causes chloracne and aggravates preexisting liver and kidney disease.

Industrial Solvents

Organic solvents are derived from hydrocarbons and are used to dissolve other organic materials. The solvents used in industry are divided into eight groups: hydrocarbons (aliphatic and aromatic), halogenated hydrocarbons, alcohols, ethers, glycol derivatives, esters, ketones, and miscellaneous.

Solvents are used in industry for extracting oils and fats, degreasing, dry cleaning, and printing inks. They are also present in a variety of products, including paints, varnishes, lacquers, paint removers, adhesive, plastics, textiles, polishes, waxes, thinners, chemical reagents, and drying agents. Many industrial solvents are highly flammable and are potential fire and explosion hazards. All organic solvents have some effect on the central nervous system and the skin, cause drowsiness, and cause damage to the blood, lungs, liver, kidney, gastrointestinal system, eyes, and respiratory system. Excessive exposure may result in death.

Hydrocarbons are powerful narcotic agents. Benzene is an example of this group that is extremely toxic to the blood-forming organ and the bone marrow. Halogenated hydrocarbons, such as tetrachloroethane or carbon tetrachloride, cause serious damage to the liver and kidneys. Alcohols such as methyl or *n*-butyl have low flashpoints and therefore are highly flammable. Their vapors are mildly narcotic. Ethers are highly flammable and have strong narcotic properties. Glycol derivatives, such as ethylene glycol, are highly flammable and have a toxic effect on the nervous system and the blood. Esters are flammable and cause irritation to the eyes, nose, and upper respiratory tract. Ketones, such as acetone, are quite flammable.

The miscellaneous solvents include the nitroparaffins, which are narcotic, irritate the mucus membranes, and cause irritation or damage to the liver and kidneys. Carbon disulfide is one of the most dangerous solvents used in industry. It is highly flammable and highly toxic, acting mainly on the central and peripheral nervous systems; it can cause insanity and death. The narcotic effects of these solvents are extremely important, since the worker becomes unable to use proper judgement and, as a result, may be involved in serious accidents.

Inorganic Lead

Lead is used in many industries. While the amount of lead in paint has decreased substantially, lead usage has increased in such products as storage batteries, brass, automobile radiators, glazes, solder, printing type metal, ceramics, plastics, chemical tank linings, and radiation shieldings. Lead is also mixed with other metals, such as antimony, arsenic, tin, and bismuth. It is added to alloys, such as brass, bronze, and steel. About 40% of lead is used as a metal, 25% in alloys, and 35% in chemical compounds.

Lead poisoning is insidious. Most frequently it is chronic. Lead accumulates in the body over a long period of time. In the home, a young child can eat lead-based paint or plaster. In industry, the worker may ingest the lead or breathe lead dust and fumes over an extended period of time. Ingestion of inorganic lead leads to mental retardation. (See the chapters on the indoor environment in this volume and air pollution control in Volume II for additional information.)

Mercury

Mercury is a silvery-white, heavy, mobile metal that is a liquid at ordinary temperatures. It combines readily with sulfur and halogens, and forms amalgams with all metals, except iron, nickel, cadmium, aluminum, cobalt, and platinum.

The more dangerous inorganic compounds formed include mercuric cyanide, mercuric oxycyanide, mercuric arsenate, mercuric phosphate, and mercuric thiocyanate. Metallic mercury and its inorganic compounds are used when working with gold and silver ores, the production of amalgams, and the manufacture of incandescent electric bulbs, radio valves, X-ray tubes, and batteries. It is also used in the agricultural, catalyst, paint, paper, pulp, and pharmaceutical industries.

The most common cause of industrial mercury poisoning is inhalation of the vapors. When spilled, it evaporates rapidly. The more frequent mercury poisoning in industry is the chronic one in which mercury accumulates in the body over a period of time.

Organochloride Insecticides

The organochloride insecticides are chlorinated ethanes, which include DDT, and the cyclodines, which include aldrin, chlordane, dieldrin, and endrin. They also include the chlorocyclohexanes such as lindane. They are used for pest control. They affect the kidney, liver, and central nervous system. The potential health effects include acute symptoms of apprehension, irritability, dizziness, disturbed equilibrium, tremor, and convulsions. The cyclodines may cause convulsions without any other initial symptoms. The chlorocyclohexanes can cause anemia. The cyclodienes and chlorocyclohexanes cause liver toxicity and can cause permanent kidney damage.

Organophosphates

The organophosphates include diazinon, dichlorovos, dimethoate, trichlorfon, malathion, methyl parathion, and parathion. These chemicals are used for pest control. They affect the central nervous system, liver, and kidney. All of these compounds cause a chain of internal reactions, leading to a neuromuscular blockage. Depending on the amount of poison that has been absorbed by the individual, the acute symptoms range from headaches, fatigue, dizziness, increased salivation, crying, profuse sweating, nausea, vomiting, cramps, and diarrhea, to tightness in the chest, muscle twitching, and slowing of the heartbeat. Severe cases can bring about a rapid onset of unconsciousness and seizures, leading to death. A delayed effect may be weakness and numbness in the feet and hands, with long-term permanent nerve damage possible.

Ozone

Ozone is a colorless gas with a pungent odor that is produced from oxygen by an electrical discharge. Ozone may be produced around high-voltage electrical equipment, electrical switches, photochemical reactions, arc welding, high-voltage spectrographic equipment, copying machines, ultraviolet lamps, and electronic air filters. Ozone is a powerful oxidizing agent and, therefore, can be a serious fire and explosion hazard. Ozone is a highly toxic, irritating gas causing inflammation and congestion of the respiratory tract. It may produce pulmonary edema, hemorrhage, and death. (See chapter on air pollution control in Volume II for additional information.)

Silica

Workers encounter silica and, therefore, the hazards of silicosis in underground mining, in quartz-bearing rock, cutting of granite, manufacturing of pottery, porcelain, abrasives, sand-blasting operations, stripping, mining, and steel and furnace making. The amount of danger is based on the concentration of the dust, the percentage of free silica in the dust, and the duration of exposure.

Silicosis, which is the disease caused by the inhalation of free silica, is the most important of the cancerous fibrogenic pneumoconioses. Silicosis frequently causes death. It also may be complicated by tuberculosis, inadequate air intake, and acute lung infections.

Vinyl Chloride

Vinyl chloride is a flammable, compressed gas used as a raw material for the production of polyvinyl chloride. Polyvinyl chloride is used as the basis for a wide variety of plastic products. Over 5.4 billion lb of vinyl chloride and 4.4 billion lb of polyvinyl chloride are used annually. Vinyl chloride is a safety hazard, since it is a flammable gas. It also has acute and chronic effects on humans. The chronic effects that are most apt to occur are very serious. Chronic toxicity from vinyl chloride causes damage to the liver, skin, vascular system, nervous system, and kidneys. In 1974, vinyl chloride was determined to be cancer causing. Several individuals exposed to vinyl chloride have died from cancer of the liver, which is a very rare disease in the United States.

Biological Hazards

Biological hazards vary with the type of occupation. The medical profession is exposed to staph and strep infections, viral infections, and many other types of organisms that their patients carry. Individuals working with solid-waste removal and disposal are also exposed to the same types of biological hazards. Research scientists are constantly working with viruses and are exposed to the potential hazard of diseases caused by these viruses.

The zoonoses, which are diseases transmitted from animals to humans, are another biological hazard to which the worker is subjected. These zoonoses include anthrax, brucellosis, tetanus, encephalitis, leptospirosis, Q fever, rabies, salmonellosis, trichinosis, bovine tuberculosis, tularemia, ringworm, and other common fungus infections. Workers also acquire roundworm, hookworm, and many diseases transmitted by fleas, ticks, and flies. Many workers, in addition, are subjected to various dermatoses. Migratory workers are subjected to infectious hepatitis and other diseases. AIDS, acquired immune deficiency syndrome, has an insidious onset with an unknown incubation period, and unknown communicability and susceptibility. People at greatest risk, besides those involved in sexual contact, purposeful needle contact, or necessary blood transfusions, are medical and other health-related personnel, as well as safety and security personnel.

Anthrax

Anthrax is an acute systemic disease involving the skin, gastrointestinal tract, or lungs. In most cases the organism enters the body through a cut or abrasion. The organisms are found in the carpet and leather-goods industries. They are present in hair, wool, hides, and skins. The incubation period is within 7 days, usually 2–5 days.

Salmonellosis

Large reservoirs of salmonella are present in chickens, human and animal foods, uncooked meat, vegetable products, and animal feeds. (See the chapter on food-borne disease for further information.)

Tularemia

Tularemia occurs when an individual is scratched by a piece of bone from a wild rabbit. Other animals, such as muskrats, woodchucks, squirrels, and skunks, harbor and transmit this organism to humans. Tularemia is also spread from animals to humans by wood ticks, dog ticks, and the blood-sucking deer fly. The incubation period is 2–10 days, usually 3 days, and is related to the virulence of the infecting strain and the number of organisms.

Q Fever

Q fever is a rickettsial disease of rats, cattle, horses, sheep, goats, and dogs. Cattle barns and holding areas for goats and sheep become heavily contaminated with these organisms. The organism is thought to be inhaled by the victim. Raw milk from infected cows has caused Q fever. The incubation period is 2–3 weeks, depending on the amount of organisms.

Aspergillosis

Aspergillosis is a fungus transmitted to humans through contaminated dust. It is a fungus disease of birds, ducks, chickens, and cattle. The incubation period is probably a few days to weeks.

Brucellosis

Brucellosis comes from infected cattle and swine. The disease is transmitted by direct contact with the diseased animals and by drinking unpasteurized milk or milk products. The incubation period is highly variable, usually 5–30 days, occasionally several months.

Dermatitis-Producing Agents

Bacterial, fungal, or parasitic agents in the environment may cause various types of dermatitis. Fungus infections are found among kitchen workers; bakers; food handlers; fur, hide, and wool handlers and sorters; barbers; and beauticians. Workers who handle grain or straw may also acquire parasites that cause grain itch and ground itch.

Viral Hepatitis

The onset is insidious with anorexia, vague, abdominal discomfort, nausea, and vomiting, often progressing to jaundice. The occurrence is worldwide among drug abusers, homosexual men, and patients and employees of various medical care institutions. The reservoir is people and the transmission is through blood, semen, saliva, and vaginal fluids. The incubation period is usually 15–180 days with an average of 60–90 days.

Laboratory Biohazards

Laboratory biohazards result from accidents in the use of biological materials or laboratory work done on a biological agent of unknown epidemiology or etiology. The worldwide literature reports an estimated 6000 cases of accidental infection have occurred in the laboratory. They include bacterial, fungal, parasitic, rickettsial, and viral infections.

Among the causes of accidents that result in laboratory-acquired infection are oral aspiration through pipettes, accidental syringe inoculation, animal bites, spray from syringes, centrifuge accidents, cuts or scratches from contaminated glassware, cuts from instruments used during animal autopsy, and spilling or splattering of pathogenic cultures on floors and table tops. A large number of these infections have not been traced, and therefore whenever hazardous infectious material is used, extreme caution must be taken.

Impact on Other Problems

Occupational disease and injury reduce the overall good health of the worker. When pollutants are removed from the occupational setting, they are released into the air and water, or buried in the soil.

Waste Disposal

Waste from industries contaminate the air, water, or land. Waste may be municipal, which is discussed in the chapter on sewage found in Volume II; agricultural, which is discussed in the chapter on solid and hazardous waste disposal found in Volume II; or industrial, which is discussed now and in the chapter on solid and hazardous waste disposal found in Volume II.

Industry generates considerable solid waste in various operations, such as quarrying, logging, farming, petroleum, mining, and extracting. Mining wastes, which are called tailings, are usually piled close to the mining source. The material is unsightly, pollutes the water through leaching of chemicals, and may also cause loss to human life. At times these piles of mining waste will slide, creating a serious accident hazard. In the quarrying industry, when stone, sand or gravel are removed, additional solid waste ranging from 0.5 to 5% remains, creating problems similar to mining and also a serious dust hazard.

The logging industry leaves behind about 30–40% of the weight of the tree. The material is scattered over a considerable area, where it becomes unsightly and a fire hazard.

Farming, depending on the type of operation, may generate an estimated 43–60 lb of solid waste per day per person. This includes a variety of odor-producing materials that cause insect and rodent problems. Manure, dead animals, and the organic material remaining from processing become an enormous problem. Odors, insects, rodents, litter, and air pollution are a result of the solid waste from the farming industry.

All of the above industries concentrate the waste material in specific locations, making disposal of this material a costly and difficult task.

The basic industries differ from the industries in which the material is extracted from the ground or raised in farming by utilizing the raw materials to produce finished materials. The metal industry produces slag, ashes, and trimmings from the product; residues from the refined products; and wastes from the materials used in aiding in processing, handling, and shipping wastes. The chemical industry produces an enormous amount of various solid wastes. The paper industry produces tree bark, wood fiber, paper pulp, inert filler, trimmings of paper, or paper board. The plastics industry produces trimmings from the product itself and waste due to improper operations. The glass industry produces slag from the purifying of glass sand, containers, residues of products, glass fragments, trimmings, crating material, and breakage. The textile industry produces waste from packaging and burlap, fibers, dirts, and other residues of the spinning, weaving, trimming, and dyeing operations. The wood product industry produces waste from tree bark, sawdust, shavings, broken wood, and trimmings. The power industry produces fly ash, bottom ash, and boiler slag. The packaging industry produces waste from excess materials from packaging, cuttings, and broken glass. The automotive industry produces trimmings and waste from the production of parts and from painting and upholstering. The electronics industry produces plastics, glass, wire, sheet metal scrap, various residues, packaging materials, and other wastes.

As can be seen, the amount of solid waste produced in a variety of industries varies by industry, quantity, possibility of recycling, method of concentration, and method of preprocessing before it enters a waste disposal plant. (A much more detailed discussion of waste disposal can be found in the chapter on solid and hazardous waste disposal in Volume II.) The major concern of solid and hazardous waste as an industrial control problem is to remove it in such a way that it will not be hazardous to the workers or to the community at large.

Solid waste that comes from a biological operation must be handled differently than that which comes from industrial and chemical operations. All biohazardous materials must be collected in impermeable containers, closed before removal from the work area, and incinerated. This is the only acceptable technique to prevent the spreading of the hazard to the workers or the community at large. All biohazardous liquid wastes must be removed and sterilized before the waste is permitted to go into the normal sewage system. (Waste causing a hazard to water will be discussed in the chapter on water pollution control in Volume II.)

Occupational Stress Problems

Occupational stress leads to disease and injury. Stress is the nonspecific response of the body to any demand made upon it. Stress may result in a fight-or-flight response to potentially dangerous situations. The body reacts through elevated heart rate and blood pressure, and a redistribution of blood flow to the brain and major muscle groups and away from the distal body parts. Subjective factors may play a much larger role in the experience of stress than objective factors. Job conditions that may result in stress include undue task demand, work conditions, or work situations; exposures to chemical and physical hazards; pressure to accomplish tasks; and suppressed anger at inappropriate supervision and management decisions. Job stress may be a combination of conditions at work and at home that interact with the worker and results in acute disruption of psychological or behavioral homeostasis. If these reactions or disruptions are prolonged, it may lead to a variety of illnesses that are stress related. These illnesses include hypertension, coronary heart disease, alcoholism, drug use, and mental illness. Lower resistance to infectious diseases may also occur. The individual may be more apt to have accidents that may result in injuries.

The stressors (those things that cause stress) may be sociocultural, such as racism, sexism, and economic; organizational, related to hiring policies, plant closings, and automation; the work setting, such as the time and speed necessary to complete tasks, supervision and management problems, and ergonomics; interpersonal, such as marital problems, death, and illness; psychological, such as improper coping skills, poor self-image, poor communication, addictive behavior, and neurosis; biological, such as disease, poor sleep, poor appetite, chemical dependency, and biochemical imbalance; and physical or environmental, such as air problems, climate, noise, poor lighting, poor equipment design, exposure to radiation, and exposure to toxic substances.

The distress cycle includes job stress, job dissatisfaction, organizational distress, stressful life changes, life and health risks, accident risks, and stress due to new technology.

Surveys suggest that 75–90% of all visits to primary care physicians are due to stress-related problems. These problems include backache, headache, insomnia, anxiety, depression, chest pain, hypertension, gastrointestinal problems, and dermatological problems. Stress may also lower your resistance to a variety of diseases and contribute to a slow healing process. Uncertainty, doubt, lack of recognition, pressure related to time, and insecurity all are causes of health problems, and all may be related to an inordinate amount of stress created on the job. This stress is not only unacceptable, but also should be corrected in order to create a happier person and thereby a healthier and injury-free individual.

Smoking and the Industrial Environment

Employees exposed in the workplace to toxic chemicals can receive additional exposure because of the presence of toxic chemicals in tobacco products. Cigarette smoking causes increased exposure to carbon monoxide. Workers are frequently exposed to carbon monoxide as part of their job. Therefore, the additional levels of carbon monoxide related to smoking can cause cardiovascular changes that are dangerous, especially to people who have coronary heart disease. Other chemicals found in tobacco that workers might be exposed to at their jobs include acetone, acrolein, aldehydes, arsenic, cadmium, hydrogen cyanide, hydrogen sulfide, ketone, lead, methyl nitrate, nicotine, nitrogen dioxide, phenol, and polycyclic aromatic compounds.

The heat generated by burning tobacco can transform the chemicals found in a workplace into more harmful substances. An example of this is an investigation of outbreaks of polymer fume fever, which is a disease caused by the inhalation of degradation product fumes of heated teflon. The disease is characterized by such effects as chest discomfort, fever, increased number of white blood cells, headache, chills, muscular aches, and weakness. Because these symptoms are similar to other diseases, such as influenza, polymer fume fever may not be diagnosed. Repeated attacks of this disease may lead to permanent lung damage. Tobacco products may become contaminated by chemicals used in the workplace and therefore increase the amount of toxic chemicals entering the workers' bodies. Smoking may contribute to an effect comparable to that which can result from exposure to toxic agents found in the workplace and therefore cause an additive biological effect. Combined worker exposure to chlorine and cigarette smoke can cause a more damaging biological effect than exposure to chlorine alone. Smoking may act synergistically with toxic agents found in the workplace and cause a more profound effect than the simple exposure that the individual is subjected to. Workers who were heavy smokers and installed asbestos had a higher level of cancer than those who were nonsmokers. Radon daughters also act synergistically with tobacco smoke. Smoking may contribute to accidents in the workplace. The accidents apparently are due to a lack of attention, preoccupation of the hand for smoking, irritation of the eyes, and coughing. Smoking can also contribute to fire and explosions where flammable or explosive chemicals are being used or stored. Smokers have poor lung construction and a higher incidence of urinary abnormalities when they are exposed to cadmium. A chronic cough and expectoration occurs when individuals are exposed to both smoking and various ethers.

Economic Factors

The cost of occupationally related disease and accidents not only must include the loss of dollars due to death, disease, or injury, but also the insurance, effect on society, and the cost of loss of financial stability for the families involved. An additional cost involves the economic impact of OSHA standards and the corrections that must be made in equipment, facilities, and workers' procedures. OSHA must consider the technical feasibility and economic cost of each of its major proposed standards. The cost of implementation can be determined with some accuracy, but the resulting benefits cannot always be estimated. Overall, the loss of manpower and production, as well as the securing of new equipment and facilities, has cost billions of dollars. Additionally, the cost of operating government is extremely high. It has also been estimated by the Commission on Federal Paperwork that the government cranks out enough documents each year to fill 51 major-league baseball stadiums.

POTENTIAL FOR INTERVENTION

The potential for intervention varies with the substance or physical factor affecting the individual. Intervention can be brought about through isolation, substitution, shielding, treatment, and prevention. An example of isolation is keeping all radioactive material or all very hazardous materials in special units that are handled mechanically rather than directly. In substitution, one hazardous chemical is replaced by a second, less hazardous chemical. In shielding, the worker wears protective garments. In treatment, the worker is given adequate training, uses equipment with proper guards, and utilizes adequate controls. Many tools of intervention are known for various physical hazards. Difficulty occurs in the use of chemicals, since there is poor understanding of the long-range problems related to the use of many chemicals. Cancer and diseases of the liver, kidney, nervous system, heart, and skin may occur. The potential for intervention can be best determined by conducting further research on these chemicals.

RESOURCES

Scientific and technical resources include the American Medical Association, the American Public Health Association, the National Environmental Health Association, the American Academy of Physical Medicine and Rehabilitation, the American College of Preventive Medicine, the American Occupational Medical Association, the University of Cincinnati, the University of Michigan, and Wayne State University.

Citizens groups include a variety of labor unions, along with their hospitals and research programs. Available governmental resources include the U.S. Department of Labor, the National Institute of Occupational Safety and Health, the National Institute of Environmental Health Sciences, the National Center for Health Statistics, the Social Security Disability Program, and Workman's Compensation Programs in various states.

The following federal agencies are useful in responding to occupational health and safety problems: The Department of Labor, Coordinator of Consumer Affairs, Department of Labor, Room 1032, Washington, DC 20210; the Mine Safety and Health Administration Department of Labor, Ballstow Towers #3, Arlington, VA 22203; the Office of Information and Consumer Affairs, Occupational Health and Safety Administration, Department of Labor, Washington, DC 20210; and the National Institute of Occupational Safety and Health, 4676 Columbia Parkway, Cincinnati, OH 45226. Research centers for Occupational Health include Harvard University, Kresge Center for Environmental Health, 665 Huntington Avenue, Boston, MA 02115; Institute of Occupational and Environmental Health, 1130 Sherbrooke St., West, Montreal, H3A2M8 Canada; New York University Center for Safety, Washington Square, New York, NY 10003; and University of Michigan Institute of Environmental and Industrial Health, School of Public Health, Ann Arbor, MI 48109. Additional professional associations interested in occupational health are American College of Toxicology, 9650 Rockville Pike, Bethesda, MD 20814; American Industrial Health Council, 1330 Connecticut Avenue, N.W., Suite 300, Washington, DC 20036; Industrial Chemical Research Association, 1811 Monroe, Dearborn, MI 48124; the National Association of Noise Control Officials, 53 Cubberly Road, Trenton, NJ 08618; National Council on Radiation Protection and Measurements, 7910 Woodmont Avenue, Suite 1016, Bethesda, MD 20814; and Society for Occupational and Environmental Health, P.O. Box 42360, Washington, DC 20015-0360.

STANDARDS, PRACTICES, AND TECHNIQUES

It is extremely difficult to discuss standards, practices, and techniques, since at present there are over 400 interim standards established by the OSHA Act under the NIOSH criteria documents. In addition, there are some 16,000 other chemicals that are not covered by standards at this point. Practices and techniques are, at present, a tremendous problem, not only to the industry, but also to the Departments of Labor and Health. As a result of OSHA's attempt to protect the workers, it has developed a code book that is 7 ft thick. OSHA regulations are so numerous and contradictory, that there have been many cases in which the Department of Agriculture has issued one ruling and OSHA has issued the opposite, or the Environmental Protection Agency has issued one ruling and OSHA has issued the opposite. Until such time as some sense can be made of the overall problem of the occupational environment by type of occupation, it will be impossible to reasonably discuss standards, practices, and techniques. For special information concerning any given area, see the resource section for the appropriate government department to contact for assistance.

MODES OF SURVEILLANCE AND EVALUATION

National Occupational Exposure Survey

NIOSH, Division of Surveillance, Hazard Evaluations and Field Studies Surveillance Branch, conducted the National Occupational Exposure Survey (NOES) from 1981 to 1983. Four thousand, four hundred and ninety establishments were surveyed in 98 geographic locations throughout the United States. These facilities were representative of virtually all of the nonagricultural, nonmining, and nongovernmental businesses covered under the OSHA Act of 1970. Its predecessor was conducted in 1972–1974. The survey was designed to provide the date necessary to describe potential exposure agents and to profile health and safety programs in American workplaces. The survey provided data on potential occupational exposures to chemical, physical, and biological agents and allowed for an analysis of the workplace. A manual used for the training of the individuals carrying out the study and for the actual study technique, entitled "National Occupational Exposure Survey Field Guidelines," was utilized for the purposes of carrying forth this necessary program. Observed potential exposure, as well as inferred potential exposure were recorded.

Potential exposure to physical agents included air pressure variation; temperature variations causing heat and cold stress; and potential exposures resulting from lasers and masers (microwave amplification by stimulated emission of radiation; a maser is a device that utilizes the natural oscillations of atoms and molecules for amplifying or generating electromagnetic waves in the microwave region of the spectrum); X-ray radiation; infrared radiation; ultraviolet radiation; microwave radiation; radiofrequency radiation; continuous noise; impact noise; ultrasonic noise; vibration-whole body; and vibration-segmental. Potential exposures to biological agents, such as viral, rickettsial, bacterial, fungal, and parasitic organisms, were recorded. Potential exposure to components of biological systems, such as blood, urine, sputum, or feces, were noted.

The functioning or nonfunctioning of intended controls were determined and recorded. These controls included local exhaust ventilation (LV); natural ventilation (NV); local gravity ventilation (LG); dilution ventilation (DV); respiratory protective devices; welding, brazing, soldering, and thermal cutting.

Chronic trauma hazards were identified. These would include chronic trauma injuries due to repetitive physical or mental activities occurring on a continuous basis. This type of cumulative injury resulted from the wear and tear of the repetitive acts. Symptoms of this problem include chronic sore forearm and elbow (tendonitis), low-back problems, shoulder soreness, neckaches, and headaches. Other problems measured included body positions or movements, transport motions, and hand manipulations.

Heat Surveys and Instruments

A heat survey includes the measurement of ambient temperatures, air motion, humidity, and radiant heat sources. Air temperatures are measured by mercury or alcohol in glass thermometers, thermoelectric thermometers, themocouples of copper and constantin, which is a copper and nickel alloy, and thermistor thermometers. An anemometer is used to measure air motion in distance per unit of time. It is important to understand air motion, since it affects the rate of evaporation and heat transfer by convection. Anemometers include the Wilson thermoanemometer, the Alnor thermoanemometer, the Anemotherm, and the kata thermometer.

The psychrometer is used to measure the amount of water vapor in the air, since the water vapor or humidity controls the rate of evaporation of water from the skin, lungs, respiratory passages, and eyes. Relative humidity is the amount of moisture in the air compared with the maximum amount that the air could contain at the same temperature. Relative humidity may be determined by use of the sling psychrometer, the motor-driven psychrometer, and the air hygrometer.

Radiant heat or the mean radiant temperature of the solid surroundings is measured by means of a Vernon globe (black globe).

To understand the thermal stress, the effective temperature of the environment must be determined. Effective temperature is a combination of globe temperature, wet bulb temperature, air speed, and metabolic rate. The wet bulb globe temperature index (WBGT) is used by NIOSH in its standard for heat in the environment. The wind-chill factor is a combination of the given temperature and the wind velocity in miles per hour. However, the wind velocity measurements are not necessary in determining the WBGT if the following formula is used. WBGT = 0.7 NWB = 0.3 GT for indoors. For out of doors WBGT = 0.7 NWB = 0.2 GT = 0.1 DB. NWB = natural wet bulb temperature, GT = globe temperature, and DB = dry bulb temperature. The globe thermometer is most widely used for determining the radiant heat load.

Light Surveys and Instruments

Light measurement is called photometry. The instruments used for measuring light are usually footcandle meters with light-sensitive barrier-layer cells. Light meters most accurately measure the light falling directly on the meter cell, rather than the light coming at an angle. Since the lighting outputs of lamps vary, the lamp must be lit for a period of time before an accurate reading is taken. Several terms must be defined now. The candela is the international reference standard for the unit of candlepower. A lumen is the unit of light flow, or the rate at which light falls on a one square foot surface one foot from a point source whose intensity is one candela. Footcandle is a unit of illumination. A footcandle is one lumen, uniformly distributed over a one square foot area. A foot lambert is a unit of brightness and is defined as one lumen uniformly emitted or reflected by an area of one square foot.

Since the visual environment is affected by the illumination level and the brightness ratio, these must be considered together in a lighting survey. The equipment must be corrected for temperature, stability, and so forth.

Vibration Surveys and Instruments

The most common method of measuring vibration is to use a vibration pick-up to transform the mechanical motion into an electrical signal. An amplifier enlarges the signal, and an analyzer measures the vibration in specific frequency ranges and in calibrated vibrational units. Some of the instruments used include the accelerometer, which is a vibration pick-up; the preamplifier, which enhances the signal; the analyzer, which is a vibration-measuring instrument; and the vibration recorder, which charts the vibration.

Noise Surveys and Instruments

The decibel (dB) is the unit that is used for expressing sound pressure levels. To be accurate, the decibel is the sound pressure level equal to 20 micropascal. This is a universally recognized pressure base and therefore is not actually quoted. In practice, however, a sound level meter is calibrated to read decibels relative to 20 micropascal. (One pascal is defined as one nm^2. The current reference point is 2×10^{-6} pascal or 20 micropascal.) An individual's response to sound is dependent on the frequency of the sound. People hear best at frequencies around 500–5000 Hz. Frequency distribution is an important part of any kind of sound level study.

Typical sound-level meters have three different frequency-weighing networks. They are identified as the a-, b-, and c-scale networks. The high-frequency noise passed by the a-weighing network correlates well with annoyance effects and hearing damage effects of the noise on people. Therefore, most of the studies that are done are read on the a scale. When the sound-level meter is switched to the

"A" position, the meter gives a single number reading that adjusts the incoming noise at the microphone in accordance with the a filter, and this number now is the dBA value that will be read. OSHA states that occupational noise exposures should not exceed 90 dBA for an 8-hr work period. The level of equipment noise is not the same as that which is received at the ear, so in doing the study you must be concerned with that which is received by the ear. Sound levels are additive, therefore, appropriate studies would include all sound levels the individual is exposed to during the course of the day. However, since the decibel scale is a log scale, simple arithmetic will not determine the overall dBA that the individual is exposed to. It will be necessary to use special formulas or charts for combining the decibel levels. These charts are typically found in handbooks of noise measurement.

An occupational noise and hearing survey measures and determines the noise exposure level of workers. It also describes the hearing status of the workers exposed to the industrial environment. The data collected include noise measurements, personal background information, medical and otological data, and audiometric data. Noise-level measurements are made at different points in the plant. Tape recordings are used to record noise characteristics that are later analyzed in the laboratory. Numerous types of sound-level meters, impact noise analyzers, and oscilloscopes are used to determine decibel levels, noise spectrums, fluctuating noises, noise direction, and the duration of noise bursts. The equipment is calibrated at least once each day. The microphones are placed alongside the workers to determine total daily noise exposure. This includes the general background noise and the specific noise of a given operation. An attempt is made to determine if the noise is impulsive, at a steady rate, at a low or high frequency, and whether it is continuous, fluctuating, or intermittent. Sequential samples are taken at 15-sec intervals during a 10-min period to get a good view of noise levels.

Ionizing Radiation Surveys and Instruments

Humans cannot detect radiation by means of their senses, therefore, instruments must be used. The principal means of detection are through gas ionization, photographic emulsions, scintillation media, semiconductors, chemical decomposition media, and radiophotoluminescent media. All of the methods of detection are based on radiation causing ionization. (See Figure 10.3 for an example of ion production.) Radiation produces charged bodies from neutral atoms and molecules. The various instruments differ only in the medium in which ionization takes place and in the method with which this ionization is detected and indicated.

Survey instruments are designed to measure the rate of ionization in milliroentgens per hour or counts per minute. Personnel monitoring instruments are designed to measure the total cumulative radiation exposure in units related to the absorbed dose. Laboratory instruments are designed primarily to measure the activity of a sample in units related to the curie.

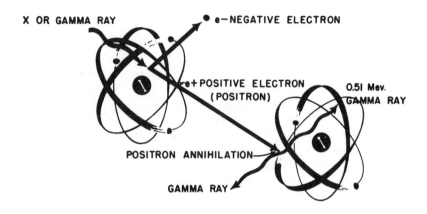

PAIR PRODUCTION
(HIGH ENERGY PHOTON, > 1.02 Mev.)

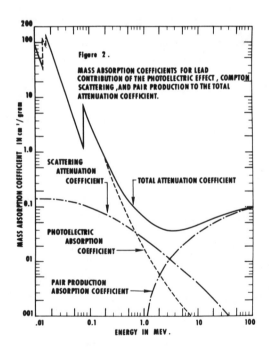

Figure 10.3.   An example of ion production.

Gas ionization instruments are based on the principle of collecting ions formed by the action of ionizing radiation in a gas. There are five regions of instruments response in gas ionization:

1. The region of recombination where the ions tend to recombine with each other rather than migrate to the electrodes. This region operates at a very low voltage.
2. The ionization chamber region, where the electron or negative portion of the ion pair is accelerated toward the anode in the chamber. The positive ion is accelerated more slowly toward the cathode. At some point as the voltage increases, a saturation flow of ions becomes equal to the number of ions produced by the radiation entering the chamber.
3. The proportional region, where when the voltage is increased above region two, the collection of ions increases. The additional ions are produced by the electrons at high velocity causing secondary ionization in the gas.
4. The Geiger-Mueller region, where the amount is relatively independent of applied voltage. This region is extremely sensitive to any radiation.
5. The region of continuous discharge, where the further increase in voltage causes a continuous state of discharge of ions.

The first range used in an instrument provides low sensitivity and only measures primary ionization. The proportional region instruments can distinguish between alpha and beta radiation. The Geiger-Mueller instruments are extremely sensitive but act for a very short time at a low range.

The darkening of a photographic plate by radiation occurs when the radiation interacts with the silver halide in an emulsion and causes an ionization. Radiation causes the darkening of the film.

A scintillation detector operates on the principle of energy being transferred from radiation to a substance, which in turn produces visible or near-visible light that may be picked up on a photosensitive vacuum tube and developed into an electrical pulse.

The chemical decomposition indicators are based on the fact that ionization produced by radiation in a chemical system is an indication of the amount of radiation received. Radiophotoluminescence is the fluorescing of certain materials after exposure to ionizing radiation. Instruments may either measure cumulative doses, dose rate, or disintegration rate.

Survey instruments include ionization chambers, such as the Cutie Pie, condenser, and r-meters. They are used in X-ray survey work, beta-particle and gamma-ray survey work, and if properly modified, in neutron monitoring. Geiger-Mueller instruments are also used in survey work for low-level beta-particle and gamma-ray evaluation. Scintillation survey instruments are used to detect alpha, beta, gamma, X-rays, or neutrons. However, there is not a single scintillation material that detects all of the above.

Personnel instruments include the film dosimeter, which gives a rough estimate of X and gamma radiation, the chemical dosimeter, the pocket dosimeter, and the pocket chamber.

When film badges, dosimeters, and pocket chambers are combined, they give a reasonable reading of human exposure to radiation. It should be remembered that various tissues and organs of the body are not affected equally by equal irradiation. The most sensitive organs are the blood-forming organs (red bone marrow), the lens of the eye, and the gonads. Therefore, proper interpretation of the amount of radiation exposure to the individual is important. Laboratory instruments for alpha and beta counting are composed of two basic components: detection media, such as gases, solids, and liquids, and electronic circuitry.

Radionuclides can be determined by measuring the half-life and the maximum beta energy, or through chemical separation techniques. Radionuclides, which are gamma emitters, can be determined through chemical separation or by the use of a gamma scintillation spectrometer. This spectrometer works on the basis of X or gamma rays interacting with certain types of crystals to cause pulses of light.

Ultraviolet radiation is measured by photoelectric cells, photoconductive cells, photovoltaic cells, and photochemical detectors. Selective filters are placed in front of the measuring device to isolate the ultraviolet spectrum portion of interest.

A microdek is used to measure the amount of microwaves escaping from a microwave oven.

Measuring Biological Effects in Humans

The biological effects of chemicals in humans can be monitored in a variety of ways. Breath monitoring may be used for an acute exposure to volatile substances, such as ethanol and hydrocarbon solvents. Blood and urine may also be utilized for these purposes. Hair and nails are useful for the analysis of metals, such as lead, cadmium, and arsenic. Although sampling tissues such as fat, bone marrow, liver, and kidney would be very useful, the potential risk would be great. These tissues could be sampled at an autopsy.

Blood lead levels are used as a monitoring device for lead exposure. This test is easily done in conjunction with a measurement of the pathophysiological effect, usually the erythrocyte protoporphyrin level. The combination of the two tests can give consistent results concerning monitoring of the individuals. Biological monitoring by the measurement of specific agents or metabolites is only useful when nonexposures are available.

The biological effects following the toxic exposure may either be structural or functional in nature. A blood cholinesterase inhibition may be found in toxicity due to organophosphate exposure. Changes in skin texture, pigmentation, vascularity (changes in the channels in the body carrying fluids), hair, and nails, as well as the existence of lesions, can be detected with physical examinations. Neurological examinations can be used to detect disorders of intellect, memory, coordination, reflex changes, and motor and sensory problems. The EEG (electroencephalogram) records activity in the brain and the response to visual, auditory, or tactile sensations. This is particularly useful in determining any limiting of the function of the brain.

The respiratory system can be examined and analyzed by the use of pulmonary function tests. These tests can help differentiate between the normal lung volume and the restricted lung volume. The restrictions may be due to a chest-wall dysfunction or a disease such as pneumoconiosis. A reduction in the flow rate would be consistent with asthma or chronic bronchitis. By measuring the air remaining in the lung after a maximum expiration of air, it is possible to make some determination about emphysema being present.

Chest X-rays are useful in measuring chronic pulmonary disease and cancer once the condition has occurred. Saliva may contain abnormal cells, which may be a means of determining if "asbestos bodies" are present in the saliva. Although this is not a diagnostic procedure that will confirm asbestosis, it still indicates exposure to the asbestos. Elevated GGT levels (gamma-glutamyl transferase) indicate liver problems that may be due to chemicals or alcohol. Kidney function may be determined by abnormal elevations of normal blood constituents, such as urea and creatinine. However, the kidney function will have to be severely restricted before these abnormalities can be measured. It is possible early in the toxic process to determine the presence of small molecular weight proteins such as microglobulins, which may be found in the urine. These proteins are normally reabsorbed by the kidney tubule. However, this function may fail early in the toxic process. A standard blood count that evaluates the number and distribution of blood cell types and morphology helps access the effects of chemicals on the blood and blood-forming tissues. This test is relatively insensitive for most toxic effects. However, a bone marrow examination may be more informative, but it is very traumatic and expensive.

Instrumentation for Particulates, Gases, Vapors, and Aerosols

Hazards are determined by the use of a variety of instruments, which may capture or trap the hazardous materials or which may be used for a direct reading of the hazardous materials. It is important to determine the amount of material, or dose, that the worker receives during a specific time period and also from the general environment during the course of his or her working day. Instruments should be placed in the worker's breathing zone, on his or her skin or wherever the hazardous material may enter his or her body. It is important to determine the peak doses of materials and to get a general average of exposure. Sampling devices include samplers for particulates, gases, vapors, and aerosols.

Air Sampling for Particulates

The particulates, which include solids and liquids, are suspended in the air and inhaled by the worker. They include dust, mists, and fumes. Samples are taken at different times and in different places to get a good overall picture.

Collection devices include particulate separators, which trap particles larger than 5 μ; filters made of various materials, such as cellulose, glass, and asbestos, that trap particles down to 0.5 μ; membrane filters, which trap particles between 0.01 and 10 μ; impingement on flat plates, which traps particles down to 1 μ; and electrostatic precipitators, which trap particles smaller than 1 μ.

There are many sampling devices on the market. The available one, or the one best handled by the individual collecting the sample material, should be used. Impingers are frequently used for dust collection. (Additional information on air sampling is available in the chapter on air pollution control in Volume II.)

Air Sampling for Gases and Vapors

The two basic methods for collecting gaseous samples are grab samples, in which an instantaneous sample is taken in an evacuated tube, and collection of the gas sample from a known volume of air through absorbing or adsorbing medium. Grab sampling is used to identify contaminants such as mine gases, sewer gases, carbon dioxide, and carbon monoxide. Adsorbing and absorbing mediums are used for ammonia, benzene, hydrogen sulfide, nitrogen dioxide, sulfur dioxide, toluene, and diisocyanate.

Whereas removal of the particles in particulate sampling is easy, in gas and vapor sampling it is not. Therefore, it is important to find a medium that will be efficient, retain the collected gas or vapor in a stable form, and release the sample readily. Continuous sampling of the work environment is needed when the contaminants are not released uniformly. It is important to determine the amount of air available, the time involved, and the position of the worker in relationship to the operation.

Direct Reading Instruments for Aerosols, Gases, and Vapors

Direct-reading instruments are used to find the sources of emission of hazardous substances. They are also used to determine if existing air standards are being exceeded and to check the performance of control equipment. They may be used as a safety system to alert the workers if a process control fails. The on-site evaluation of hazardous materials is extremely important to immediately correct or close an operation that is injurious to the workers. These instruments should be calibrated carefully, frequently, and accurately.

Direct-reading physical instruments are based on the following principles:

1. Aerosol photometry, or light scattering — an electrical impulse is generated by a photocell and detects the light scattered by a particle.

2. Chemiluminescence — a distinct colored glow is produced by oxidation, and the glow can be measured for the oxidizing agent. This is particularly good for ozone or the nitrogen oxides.

3. Photometry — the relative radiant power of a beam of radiant energy is measured as it passes through a gas-air mixture, or a suspension of solid or liquid particles. The energy source may be in the visible, ultraviolet, infrared, or electromagnetic spectrum. This is good for measuring mercury vapor, ozone, benzene vapor, and so forth.

4. Electrical conductivity — a gas-air mixture is drawn through a liquid and those gases that form electrolytes produce a change in the electroconductivity of the solution. This is useful in continuous monitoring of sulfur dioxide in the ambient air.

5. Thermoconductivity — the specific heat of conduction of a gas or vapor in a carrier gas, such as air, is a measure of its concentration.

6. Coulometry — there is a precise measurement of the quantity of electricity that passes through a solution during an electrochemical reaction. This method is capable of a high degree of precision. It is useful for monitoring ozone, nitrogen dioxide, and sulfur dioxide.

7. Flame ionization — the gas is fed into a stainless steel burner mixed with hydrogen and then burned with air or oxygen. A loop of platinum is set about 6 mm above the tip of the burner. The current carried across the electrode gap is proportional to the number of ions generated during the burning of the gas. The technique is useful in measuring hydrocarbons.

8. Gas chromatography — the components of a complex mixture are separated physically because of the varying affinities of the components for different packing materials. This technique is useful in providing extremely fine separations at very low levels of concentration.

9. Spectrophotometry — similar to photometry, except that prisms made of glass are used in the visible range, quartz is used in the ultraviolet range, and sodium chloride or potassium chloride is used in the infrared range. Visible light spectrophotometry is useful in analyzing lead concentrations in blood or urine. Manganese is also determined in the urine. Samples are analyzed for iron, manganese, arsenic, chromium, aldehyde, and sulfate concentrations. Ultraviolet spectrophotometry is used in determining benzene, toluene, xylene, phenol, and cresol concentrations. Infrared spectrophotometry is used in determining complex solvents, airborne mineral oil, carbon monoxide, sulfur dioxide, and ammonia concentrations. Fluorescent spectrophotometry is used in determining airborne particulate matter, such as beryllium and polynuclear aromatic hydrocarbon concentrations. Atomic absorption spectrophotometry is used in determining mercury and lead concentrations in water, sewage, and biological specimens.

10. Polarography — a sample solution is analyzed through electrolysis. This is useful in determining metallic ions.

11. Radioactivity — radioactivity (alpha, beta, and gamma) is measured in gases, liquids, or solids.

12. Direct reading colorimetric devices — the gas reacts directly with a reagent to produce a color. This technique provides a rough indication of a large amount of toxic chemicals.

Sampling

The environmentalist must decide where to take the sample, over what period of time, and the number of samples that should be taken. These decisions should be based on sound statistical understanding of the environmental problem and what is to be determined.

Personal samplers, which are located on the individual, are best able to give a direct exposure measurement. Other types include hand-held samplers, which are placed as close to the breathing zone of the worker as possible, and general air samplers, which are placed near the source of exposure, near the breathing zone of the. worker, and at other points within the environment.

Samples may be taken for a full 8 hr and averaged, or they may be taken for select periods during the times when the worker is exposed to a peak amount of the contaminant. Grab samples are used to determine, at any given time, the immediate level of exposure.

Air-Flow Measuring Devices

Air-flow measuring devices include the Pitot tube, the deflecting vane anemometer (velometer), the rotating vane anemometer, the thermoanemometer, the heated thermocouple anemometer, the hot-wire anemometer, the Kata thermometer, and the ionization anemometer.

Air-quantity measurements are also made using an orifice meter, a critical-flow orifice meter, a flowmeter, a Venturi meter, or a rotameter. Rough measurements of airflow systems are made through the use of smoke tubes, titanium tetrachloride, or smoke candles. The Pitot tube is the standard instrument for measuring the velocity of air. The velocity is determined by the following formula:

$$V = (2 \times {}^{p}/_{d})\ {}^{1}/_{2}$$

where V is the velocity of the air stream, p is the velocity pressure measured by the Pitot tube, and d is the density of the air.

To get an accurate reading, the duct opening should be divided into 16 equal areas, a velocity measurement made at the center of each of these areas, and the readings averaged. The Pitot tube is basically used for measuring the air velocity in ducts or at high velocity, from the supply-air or exhaust-air openings. The velometer is used as a direct-reading instrument, ranging from 50 to 24,000 ft/min. The instrument is operated by the pressure of air against a spring-loaded swinging vane. The rotating vane anemometer has a propeller connected through ears to a dial that counts the rotations. The dial reading divided by time equals feet per minute.

Thermoanemometers are based on the principle that the rate of heat loss of an object at elevated temperatures is a function of the air movement over the object. The orifice meter is a restriction in a pipe between two pressure systems. A flow meter is any meter that has a restricted opening and can calibrate the flow of gas or liquid. A Venturi meter is an instrument that has a constricted opening that starts to open outward again. This differs from the orifice plate where there is a sudden constriction and sudden flow. The usual Venturi meter has a 25° contraction at its smallest point and a 7° reexpansion further along the flow in the pipe.

Detailed discussions concerning the principles behind the operation of various meters used in industrial ventilation testing can be found in the references on industrial hygiene. Appropriate formulas for conversion of meter readings are also found in these books.

A rough estimate of the flow of air may be made by using smoke tubes or smoke candles to determine if the smoke is being drawn into the exhaust system. Titanium tetrachloride is used for this purpose as well. If the material is placed on the periphery of the hood, the resulting smoke will flow out of the hood if there is a problem along the periphery or if the air flow is not controlled properly.

Surveys of the Airflow System

A survey of airflow systems must contain a sketch of the system, including the hoods, elbows, branches, air cleaners, fans, stacks, supply ducts, plenums, and diffusers. Measurements must be made of the airflow, velocity, and static pressure. These measurements include the static pressures recorded at a point upstream and downstream from the air cleaner and the fan. Airflow must be measured in ft³/min at the elbow, branches and mains, and upstream and downstream from the fans. The velocities must be measured throughout the system. The fans must be evaluated for speed, efficiency, and total pressure.

It is necessary to determine if the fans and duct system are operating properly. To do this, smoke tubes or candles are used and samples are taken at the breathing area of the operator. Once the system is in operation, the whole area should be retested. If there seems to be a change from the original results during subsequent airflows, the system should be checked for obstructions; filters should be checked for excessive buildup; fans should be checked for excessive wear or belt slippage; and the entire operation should be checked to determine if new processes have been added that have created additional burdens on the systems.

The entire discussion on ventilation systems has been quite limited from an engineering standpoint, since the environmentalist is primarily concerned with taking samples and making a gross determination of possible environmental problems. Studies and general commentaries should be given to the ventilation or industrial engineer if problems are noted, and further studies made to set up good exhaust systems. (Additional information on ventilation is found in the chapters on air pollution in Volume II and food protection in this book.)

Recirculation Systems

The necessity for reducing energy costs in industry has brought about a change in recirculation systems, which could lead to an increase in contaminants penetrating the air cleaners and then recirculated back into the workplace through the return air. Secondly, there may be a reduction or cessation of airflow through the system, which would result in lower hood efficiency and contaminant loss. This area of recirculation systems, air cleaning and monitoring equipment, as well as other studies and reports on recirculation of exhaust air, use of respirators and industrial respiratory protection has been covered in a series of documents prepared by the National Institute for Occupational Safety and Health. It is suggested that the reader contact NIOSH to keep up to date in these vital areas of protection of the industrial environment.

Monitoring Devices

Monitoring devices must be set up and operated on a routine basis to determine the effectiveness of controls being used to protect workers. Special studies should be conducted periodically to evaluate not only the process and the worker, but also the entire environment of the processing or industrial plant. It is essential that the individuals who are monitoring have a solid understanding of the use of the equipment, can properly calibrate it, conduct adequate surveys, and interpret the data and write concise, accurate reports.

Reporting

The reporting of accidents and incidents on appropriate forms to the safety supervisor, industrial hygienist, or environmentalist is extremely important in order that the experts may determine the cause of occupational accidents and disease. These specialists can utilize the information to try to prevent another incident from occurring.

Reporting of accidents is also necessary to satisfy various federal and state laws and to obtain worker's compensation, where necessary, for the victim of an accident or disease.

Surveys

Before entering the industrial plant, the environmentalist must have a solid understanding of the type of industry under study. He or she must have a familiarity with all of the processes within the industrial plant, the chemicals or substances used, the toxicity of raw materials, byproducts, products, and any spillage that occurs during the process. He or she must have some understanding of the potential air contaminants and all of the physical stresses, including heat, illumination, noise, vibrations, ionizing radiation, nonionizing radiation, and ergonomics.

The survey itself consists of an evaluation of the general sanitation or housekeeping level within the plant; a determination of the variety of actual raw materials, products, and byproducts used within a plant; and any additional mixtures that may occur when the chemicals come into contact with each other under industrial situations, the numerous sources of potential air contaminants, the physical agents actually in use, the visible safety hazards, and any alteration in processes within the industrial plant.

After conducting the initial study, the entire series of processes within the industrial operation should be put on a flow chart, and beneath each item the following information should be listed: potential chemical hazards, potential air contaminants, exposure time and concentrations, the average age and health condition of the workers, the number of workers, the potentially hazardous physical agents, the control techniques used, the personal protective techniques used, the type of supervision of the process, the results of air-sampling devices, the results of sampling for physical agents, and the housekeeping level at each step in the operation. (See Appendix A at the end of this chapter for an example of a preliminary industrial hygiene survey form.)

After the survey has been completed, the industrial hygienist or the environmentalist should carefully review the medical records of the workers to determine if any of them have had acute or chronic diseases or have had any accidents due to any portion of the process within the industrial environment. After this information has been reviewed, environmental sampling is utilized.

Sampling can be carried out on a research basis when the various governmental agencies are trying to determine standards of maximum exposure to any given hazard or whether a given hazard over long periods of time may have effects on workers.

To determine if a given part of an operation within the industrial plant or the entire operation is a hazard to workers, samples should be taken where the suspected problem occurs. The industrial hygienist, environmentalist, engineer, safety specialist, and physician should make a determination concerning where to sample, how to sample, what instruments to use, how long to collect the sample, how many samples are needed, and over what time period the samples should be collected. Usually sampling is done at the worker's breathing zone, at the operation itself, and in the general room air. If sampling will be used to help identify physical hazards, the sample should be taken in the general environment, at the operation itself, and at the worker.

After the surveys have been completed, a report should be written for the industrial plant. The report should include a synopsis of the report; an introduction that discusses the type of plant, its operation, the number of employees, and the potential hazards; a description of the problem in detail; a description of the facility; the existing environmental conditions; the safety hazards found during the survey; a description of the sampling devices used; a summary of the data; a discussion of the findings; and recommendations for needed corrections.

The National Institute of Occupational Safety and Health (NIOSH) has developed and implemented a national occupational hazard survey. NIOSH determines such things as types of chemical exposures, duration of time exposed, concentration of materials, potential exposure, actual exposure, inferred exposure, organic liquid hazards, inorganic liquid hazards, unknown liquid hazards, aerosol cans, carbon arc lamp exposures, thermal decomposition, welding hazards, nitrogen and carbon dioxide asphyxiant hazards, barrier creams, various hand cleaners including solvents, and products of combustion.

The National Institutes of Health has published a survey manual, which may be obtained through the U.S. Dept. of Health and Human Services, Public Health Service, Center for Disease Control, National Institute for Occupational Safety and Health, Rockville, MD.

OCCUPATIONAL HEALTH CONTROLS

Heat

There are numerous controls recommended for workers in hot environments, including acclimatization, proper work and rest periods, distribution of the work load, assigning younger and better equipped personnel to hot work, scheduling hot jobs for the coolest part of the day, regular breaks, frequent physical examinations, adequate quantities of drinking water, replacement of salt, personnel education concerning work problems related to heat and the effect of alcohol and drugs, protective clothing, shielding from the sun, adequate ventilation, and shielding from sources of radiant heat. Acclimatization is so extremely important that a specific work regime must be set up to avoid difficulty for the worker. The individual should be given a period of 6 days to become acclimated to the environment. The first day he or she should perform 50% of the total work load and work during 50% of the total time exposure. There should be a 10% increase each day thereafter. A worker who goes on nine or more consecutive calendar days vacation should undergo a 4-day period of reacclimatization.

Light

A pleasant visual environment is the result of adequate quantities and proper quality of illumination. When a new industrial plant or industrial process is being planned, the engineers must give considerable attention to the lighting that will be needed and used. Since the most common faults of lighting systems are low levels of illumination, glare, shadows, and poor brightness ratios, these factors must be carefully considered during the planning phase. The size, contrast, brightness, and time involved in visual activities must also be considered. Any glare present within the field of vision may cause visual discomfort. This glare may either be direct or indirect. Shadows should be eliminated, since they interfere with effective vision and are very annoying. Care must be taken to provide proper brightness ratios between the visual task and the immediate surrounding environment. For example, a student should never sit in a darkened room with only a single source of light on his or her textbook. Whenever he or she looks up his or her eyes must adjust from the brightness of the task to the darkened environment. This causes fatigue and reduced the student's ability to concentrate and absorb the material he or she is studying. The environment itself must be clean and pleasant. It is always better, if possible, to use pastel color on walls and ceilings.

Another important control is the use of proper housekeeping and maintenance. The well-designed visual system is disrupted by bulbs burning out in electrical fixtures and by dirt accumulating in the fixture. The most efficient and least expensive way to maintain a good system is to routinely remove all fixtures in a given area, dispose of all of the bulbs whether or not they are still operating and wash the fixtures at the same time. The best way to determine such a schedule is to obtain a rating of the average light available in hours of the bulbs or tubes from the manufacturer.

Where it is impossible to eliminate a source of glare, the worker should be shielded from it and, if necessary, should wear protective goggles.

Vibration

Vibration is reduced by isolating the disturbance from the radiating surface, reducing the response of the radiating surface, reducing the mechanical disturbance causing the vibration, using materials that will deaden the vibration, reducing friction, and absorbing the energy created by the vibration. Proper design of equipment, adequate balancing of rotating machinery, and proper maintenance of machinery also act as effective controls.

Ergonomics

Tools and work situations must be selected to reduce as much work stress as possible. If the individual has difficulty reaching operating equipment, or suffers from poor posture, discomfort, or emotional strain, then physical illness or accidents may result. Proper design of the employee's work space is also essential. Care must be taken to eliminate fatigue. An evaluation must be made of the individual to determine if he or she can effectively handle the physical and mental loads of the assignment. The worker should be trained for the task and then checked by a knowledgeable supervisor who has the ability to communicate with him or her.

Noise

Noise is controlled by use of proper engineering, proper administrative techniques, and personal protective equipment. When replacing equipment, total noise levels for the area should be considered and pieces of equipment with lower noise levels should be utilized wherever possible. The proper maintenance of equipment is essential, since worn or imbalanced parts, improper adjustments, inadequate lubrication, and improperly shaped tools contribute to the noise problem. Belt drives on machines should be substituted for gears wherever possible. Vibrations may be dampened, or decreased, by increasing the mass or stiffness of the equipment or material and by using rubber or plastic linings. The supports of machines should be strengthened. Flexible mounts, hoses, or pipes should be used, if possible. Mufflers at either the intake or exhaust on internal combustion engines and compressors should be checked regularly and replaced when needed. The noise source and the operator should be isolated when possible.

Proper administrative controls reduce excessive noise exposure by adequate arrangement of work schedules, removing the worker to a low noise level when he or she has worked a high level during the day, dividing the work at high noise levels into several days or among several different people, running a noisy machine a small portion of the day, and running high-level noise machines when a minimum number of employees are present.

Personal protective equipment includes the use of plugs and earmuffs that reduce the sound level from 25–40 dB. Further, a hearing conservation program should be implemented. The individuals subjected regularly or frequently to 90 dBA or above should be tested. A complete medical examination and previous work history should be completed for each worker and maintained on a regular basis. Individuals showing signs of physical damage or having spent long periods of time at high noise levels in the past should be assigned to a much quieter environment.

Ionizing Radiation

The type of controls used for ionizing radiation are based on the intensity and energy of the radiation source and whether it is an external or an internal hazard. X-rays and gamma rays are the most common type of external radiation hazard, for they can penetrate the entire body. Beta rays may or may not be an external hazard, depending upon their energy. If the beta rays penetrate to the basal layer of the epidermis, they are hazardous. Neutrons, because of their high penetrating powers, are also external radiation hazards. Prevention of these hazards is accomplished through distance, proper shielding, and exposure time. The inverse square law applies to the reduction of radiation intensity due to point sources of X-ray, gamma ray, and neutron radiation. Good shielding includes proper enclosure of the radiation source and the use of diaphragms and cones to limit the beam to avoid stray radiation. Radiation will scatter, so it is necessary to limit this beam and have proper absorptive shielding. The exposure time should be kept to an absolute minimum. The total exposure time should not exceed 0.1 rad per week. Other controls include the use of protective aprons and gloves, wearing of film badges, carrying of dosimeters, and evaluation of machines to ensure that optimum voltage is used in the operation of the machine and the fastest film is utilized for best exposure.

Internal exposure is the result of the deposit of radioactive material within the body through inhalation, ingestion, and absorption. The hazard, which is created by the radionuclide in the body, depends on the amount of radionuclide in the organ, the energy of the radiation, the relative biological effectiveness of the radiation, the uniformity of distribution, the size and importance of the organ, and the effective half-life. Alpha emitters can cause great damage in a vital organ. Beta emitters, although they travel further, have about one-twentieth the effect of the alpha emitter. Since the radioactive material may enter the body through inhalation, it is important to wear proper masks in potentially dangerous areas and to contain the radioactive material in hoods with special exhaust systems. Food and water must be thoroughly monitored to ensure that individuals do not ingest radioactive material. Even a small amount entering the bloodstream is highly toxic. The skin should be protected by the use of protective clothing. Absolute cleanliness must be practiced at all times and double containers should be used for all waste or cleaning materials.

If a spill occurs, all doors and windows should be immediately closed, and all fans, air-conditioners, and other vents, including ventilation ducts, should be shut off. The individual should leave the room as quickly as possible, discarding his or her shoes and clothes at the door. The doors should be locked and all cracks sealed with tape. Supervised decontamination should take place and the area should be monitored before anyone is permitted to enter it.

The federal government has several special programs to determine if radiation is a community environmental problem. Air sampling is carried out in 74 stations throughout the 50 states to determine the levels of fresh fission products that may be inhaled or ingested with food. Daily 24-hr samples are collected by means of a high-volume air sampler using a carbon-loaded cellulose dust filter. The field measurements of radioactivity are evaluated against a known source.

Fresh milk is gathered weekly throughout the United States to determine the amount of radiation present. Strontium-90 is the most important radioactive material evaluated.

A food survey is conducted in a variety of institutions to determine the possibility of contamination by radionuclides. Human bone samples of people under 25 years of age are evaluated for the presence of strontium-90. Samples are taken in slaughterhouses to determine whether radioiodine is present in the thyroids of cattle.

A series of special radiological safety projects are carried out by the United

States Public Health Service. These studies are concerned with the construction and operation of nuclear-powered ships, the investigation of river and port environments close to atomic reactor operations, and monitoring in the vicinity of testing areas.

Transportation

The transportation of radionuclides is controlled by the Interstate Commerce Commission, the Coast Guard, the Civil Aeronautics Administration, and the Nuclear Regulatory Commission. All radioactive materials are classified into the following three groups: (1) gamma-ray emitters — including isotopes that emit gamma rays or have gamma emitting daughters; (2) neutron sources — including all mixtures, such as radium; and (3) negligible external radiation sources.

The necessary protection for these radioactive materials include proper seal, proper dimensions, proper bracing, shielding efficiency, proper shielding requirements, proper labeling, proper loading and storage, and proper record keeping.

Disposal of Waste

Although the amount of waste from various users of radionuclides is fairly small compared to the amount produced by the nuclear energy industry, certain problems still exist. The small users are not experienced in proper waste disposal techniques and generally do not have special facilities for radioactive waste treatment. Further, a number of individual users in a community may be discharging their liquid waste into the same sewer system, thereby causing an additive effect that may be potentially hazardous. Dilution and dispersion are the principal techniques now used for waste disposal.

Airborne waste may be discharged into the atmosphere from hoods. If the quantity of waste is small, there will be adequate dilution to prevent a hazard. Before this material is released, the gases should be scrubbed to remove as much of the radioactive material as possible. The concentrated radioactive waste removed from the air would then constitute a liquid-waste or solid-waste problem.

Liquid waste of low activity may be diluted and then disposed of in the sewage system. If the liquid waste has a high activity level, it should be stored. Only water-soluble radioactive material can be disposed of in the sewage system. Storage is a practical technique if the radionuclide is short lived. The material could then be stored until the activity has been greatly reduced. Chemical methods to concentrate radioactive materials in the liquid waste include coagulation, ion exchange, and absorption on selected media. This concentrated material would then be stored for long periods of time in protected shelter until adequate radiation decay occurs.

Solid wastes include laboratory glassware, other equipment used in processing radioactive materials, workers' clothing, building materials from demolished buildings in which radioactive materials were handled, and animal carcasses. Where possible, the material should be decontaminated. Combustible materials may be burned, providing care is taken to prevent the escape of volatile and particulate radioactive matter.

Where the material cannot be decontaminated or burned, it must be stored for long periods of time or permanently. Storage must take place in a concrete block or vault. If buried on land, care must be taken to consider the conditions of groundwater, soil acidity or alkalinity, potential earthquakes, and so forth. In all cases, the burial grounds must be closely supervised.

Nonionizing Radiation

Individuals should be protected against ultraviolet rays by minimizing the time of exposure, keeping a maximum distance from the source, and using proper shielding. Glasses should be used when needed.

Since ventilation does not control radiant heat, other techniques must be used. These techniques include installation of shields that reflect the radiant heat or absorb it without reradiating it to the individual and the use of reflective clothing, such as aluminized coats, pants, coveralls, and helmets. An individual should only be permitted to work within a hot area for short periods of time, wearing a breathing apparatus that prevents inhalation of hot vapors. Another means of protection is maintenance of proper distance between the source of heat and the worker.

The controls for microwave hazards include limiting the time and amount of exposure, and providing proper shielding and distance from the source.

The controls for lasers include proper engineering to prevent exposure to the laser beam, proper procedures to avoid errors, and proper personnel equipment coupled with periodic eye examinations.

Biohazards

Special controls are needed when working with biological hazards. These controls include air-pressure differentials, filtered supply air, filtered exhaust air, special airlock and pass-through autoclaves, ultraviolet barriers at throughways and in special laboratory areas, special treatment of contaminated sewage and waste, and isolation and separation of all of the water, steam, natural gas, and other utilities. Additional controls are discussed in the section on controls of hazardous and toxic materials.

Control of the Occupational Environment

The environmentalist is concerned primarily with the control of the occupational environment to reduce the potential for disease, injury, and stress, and to promote better health of the worker. The basic controls used involve the industrial process, the means of transmitting a hazard (air, water, solids), the general environment, and the worker. The controls include substitution; isolation; ventilation; education; personal protective devices; housekeeping; waste disposal; exposure time and concentration; identification of hazardous areas; repair, maintenance, and construction of equipment; monitoring devices; and proper reporting.

Substitution

Sometimes it is possible to substitute less hazardous materials, equipment, or even processes for more hazardous ones. Solvents such as dichloromethane and methylchloroform have been substituted for carbon tetrachloride. Toluene or xylene have been substituted for benzene. It is safer to dip an object into paint than to spray paint. This is an example of a process change.

Safety cans are much safer for storing flammable solvents than glass bottles. Another example of change in equipment would be the use of safety glass as opposed to regular glass in the fume hood and in glasses used by workers.

The substitution of a wet method of operation for a dry method, where dust becomes a serious hazard, is another technique in good environmental control. An example of this would be the wetting of floors before sweeping up dust or the use of treated mops rather than regular dry mops in the hospital. The use of water drills by dentist has not only reduced the heat generated by the drills and the potential pain, but has also reduced the number of particles that could fly into the dentist's eyes or injure his or her hands.

Isolation

Isolation is the physical separation of the worker from the hazard by use of a barrier, by increasing the distance from, or decreasing the time of exposure to, the hazardous material. Stored materials may be isolated physically from other materials and the worker or they may be stored in small units in specially ventilated rooms or under protective coverings. Radioactive materials may be stored in concrete vaults or in lead pigs. When the material to be mixed or processed is extremely hazardous and must be mixed by hand, either remote guidance controls should be used with mechanical hands within a specially protected container or the individual might use a glove box with his or her arms and hands in protective gloves and the rest of his or her body shielded from the area of material preparation.

Extremely hazardous equipment, such as that which is under high pressure, may be separated from the worker by reinforced concrete, steel, or armor plate. The work is performed by remote control and the worker observes the process through television cameras, mirrors, or periscopes. Remote-process control is used when the process itself is extremely hazardous. An example of this is automatic sampling analysis in petroleum processing plants, censors used on the production line, and processed equipment controlled through the use of a computer.

Isolation also includes the use of protective clothing by the worker and the proper use of local exhaust ventilation. Enclosures are another method of isolation. Enclosures are used in such operations as sandblasting, heat-treating, mixing, grinding, and screening. Enclosures are also a useful tool in noise control.

Ventilation

Ventilation is used to provide a comfortable thermal environment and to keep dangerous materials from the breathing area of the worker. It is also part of the air-conditioning system when heating, cooling, and humidifying processes are controlled to provide a good thermal environment.

Ventilation is brought about when air is removed from a given area and fresh air replaces it. The exhausted air carries with it heat and possible odors that have been generated by people, processes, or equipment. The quantity and type of distribution of air by means of ventilation systems depends on the amount of heat produced, the kinds of contaminants getting into the air, the quantity of contaminants, and the ultimate air quality desired. Proper ventilation dilutes the concentration of a hazardous material and prevents fires and explosions. The various types include natural ventilation, forced general ventilation or dilution ventilation, general exhaust ventilation, local exhaust ventilation, and comfort ventilation. In natural ventilation, air flows into the room through cracks, crevices, and open windows. Since it is difficult to control natural ventilation because of variations in temperature, pressure, wind velocity, and general weather conditions, this process is not very practical for an industrial environment.

Forced general ventilation or dilution ventilation is accomplished by blowing fresh air into a workroom. This may be used to control vapors given off by organic solvents at room temperature or to reduce the heat load within the working area.

Exhaust ventilation is accomplished by mechanically withdrawing large quantities of air from the work area. The air is then replaced by fresh air being sucked or blown into the work area. However, the movement of large quantities of air can cause drafts and may be uncomfortable for the worker. When this technique is used, it is preferable to have several exhaust fans of low velocity.

Diluting the contaminant in the air, either by forcing in large quantities of air or exhausting large quantities of air in the work area, does not reduce or eliminate the source of hazardous material introduced into the workroom. This is an important consideration when the hazardous material can cause a serious health problem for the worker.

Local exhaust ventilation utilizes less make-up air, and therefore reduces costs due to heating or cooling. It helps eliminate nuisance complaints, odors, and accumulation of particulates and vapors, and it helps reduce housekeeping problems. General dilution ventilation is acceptable if the toxicity of the contaminant is low, the amount of material is produced at a low rate, the kind of material at low concentrations does not have serious effects on the individuals, and the material can be diluted adequately with a large volume of air.

Local Exhaust Ventilation

Local exhaust ventilation, which must be located in the point of production of the hazard, consists of hoods or enclosures, duct work, exhaust fan, and an air-cleaning device to prevent pollution of the outside air. If the system is designed properly, the material is completely captured. This eliminates the hazard to the worker and may also be a financial asset to the firm if the material is costly. Material is trapped in the air-cleaning device, reprocessed, and reused.

Air flow through a duct is a composite of static and velocity pressures. Static pressure is the force that produced the initial air velocity and overcomes the resistance in the duct system that is caused by the friction of the air against the duct walls. It also overcomes the turbulence in the air caused by the change in direction or velocity of the air. Static pressure is the same as potential energy, because it exists when there is no air motion.

The pressure due to velocity of the air flowing in the duct is similar to kinetic energy. This pressure only exists when the air is in motion and it acts in the direction of the flow of the air.

Total pressure equals static pressure plus velocity pressure. Therefore, if the total pressure stays the same and the static pressure increases, the velocity pressure must decrease and vice versa.

Because the friction and turbulence within the duct has a direct bearing on the speed with which the air will flow through the duct, and since the friction might be increased because of a dirty duct or filter, then the total pressure must be increased to maintain the same rate of flow of air through the duct work.

$$Q = AV$$

Where Q is the quantity of air flow in (cfm) ft^3/min, V is the velocity of air flow in feet per minute; A is the area of duct work in square feet; and V = SP + VP. This formula applies when the cross-sectional area can be determined.

It must be recognized that there will be friction losses due to resistance of the surfaces, dynamic losses due to turbulence when the airflow changes direction, acceleration and hood entrance losses when the air is accelerated from its normal flow to enter a duct, pressure drop through the duct work due to the friction, and dynamic losses within the duct work when the duct work changes direction. When the environmentalist makes ventilation studies, he or she will be using a series of formulas and tables found in any reference book on industrial hygiene.

Education

The educational process is used in the industrial environment for training workers in job performance, retraining workers in their job assignments, or teaching new procedures, and hopefully, changing attitude toward the utilization of good personal practices. Although the machinery may be the best engineered, contains all available safeguards, and the process and materials may not be causing a problem within the environment, lack of good attitudes on the part of the employees may lead to hazardous situations.

Employees may improperly utilize equipment, remove guards, wear unsafe clothing, or allow their minds to wander during the work process. Good educational procedures should help workers develop a significant understanding of accident and health hazards. They should help workers to be flexible in their work habits. They should learn to work at a normal rate of speed, not intentionally violate existing rules and regulations, and carefully assess the potential problems of their unique environments.

Educational programs are directed by top management, the engineers constructing and designing the equipment processes, the supervisors controlling each of the work processes, and the workers themselves. The educational program should consist of formal lectures, informal seminars and talk sessions concerning problems, and think tanks. Outside experts, such as officials from the various federal, state, and local programs in occupational health, may be of considerable value to the organization. The National Institute for Occupational Safety and Health provides specialized training in occupational environment areas. The U.S. Department of Labor, through the Occupational Safety and Health Administration, provides training in inspectional techniques for their own employees and state and local employees.

Personal Protective Devices

Personal protective devices consist of respirators, protective clothing, shoes, helmets, eye and face shields and goggles, earmuffs, plugs, and showers.

Respiratory equipment is used to protect against inadequate oxygen, gases, toxic contaminants, airborne particulate matter, or any combination of these hazards. The type of respiratory equipment used is based on the kind of hazardous situation in which the individual will work. It must be determined if there is adequate oxygen in the atmosphere, the types of contaminants in the air, the effects of the contaminant on the body, how quickly the effect will occur, the concentration of the material, and the length of time the worker will be present within that environment.

There are three basic types of respiratory equipment. They are air-purifying respirators containing special filters or chemical cartridges, air-supplied equipment that have hoses and masks and are hooked up to a central system, and self-contained equipment that provides oxygen from a cylinder carried by the individual.

The worker should have an understanding of the equipment; should know how to use it properly; when to use it; how to adjust the face piece and supply valves; when and how to replace filters, cartridges, canisters, and cylinders; and know how to react in the event of an emergency.

Protective clothing includes the use of gloves, aprons, boots, coveralls, and other articles made of impervious materials. Their function is to prevent chemicals from coming in contact with the skin. Special types of reflective aluminum clothing are used in areas where there is considerable radiant heat. Helmets or hard hats are needed in any type of construction or in such industries as the steel industry.

Eyes and Face

The eyes and face are protected against corrosive solids, liquids, and vapors by using safety goggles, face shields, and so forth. It is important to protect the eyes against numerous hazards, such as glare, ultraviolet light, and other kinds of physical or chemical hazards. Such industries as grinding are extremely dangerous, since the particles may easily penetrate the eye or cornea, or cut the eyelid. Chemical burns in the eyes due to splash may cause temporary or permanent loss of vision.

Ears

Ear protection is provided by the use of earplugs or earmuffs. It is essential that the plugs and muffs are kept very clean to prevent ear infection, and the process should be evaluated, since the noise level may exceed the protection afforded by the muffs or plugs.

Showers

Special protective clothing should be changed when the individual leaves a very hazardous area. The individual should shower before donning fresh protective clothing or his or her own clothing. This is especially important where the worker may be exposed to virulent hazardous aerosols of a potentially infectious agent.

Cold showers are used in the event that a hazardous material is spilled on the individual. Showers should also be used where the individual may accumulate any hazardous material on any part of his or her skin or hair.

Housekeeping

The satisfactory environment is a complex interrelationship between the physical facilities, color, light, ventilation, aesthetics, dirt, chemicals, and microbiological hazards. The dark, shabby, dirty, littered environment is depressing and leads to accidents and poor occupational health practices.

The procedures and techniques of cleaning are extremely essential within the industrial environment. Attention must be paid to dustless removal, by the use of treated cloths, mops and vacuum cleaners with filters, of inert or hazardous dusts that may fall on the floors, ledges, equipment, and partitions. Wet or damp sweeping is an important technique in the removal of this dust, provided it does not cause a slipping hazard or the chemicals present do not react in the presence of water.

It is essential to use proper detergents, disinfectants where needed, and equipment capable of performing a good housekeeping job. The personnel must have proper training in the use of the equipment. They must also have an understanding of the types of materials being removed from surfaces. It is extremely important that housekeeping personnel be well supervised in order that a minimum of contaminants may be spread and a maximum of contaminants removed during the housekeeping process.

Where spills of toxic materials have occurred, previously established techniques for removal of the toxic material must be used. The workers must be trained to utilize these techniques and to know which techniques to use to protect themselves during the clean-up procedure. The housekeeping must be performed on a daily or shift schedule and also on special schedules depending on what must be done.

The types of equipment and floors to be cleaned vary with the industrial operation, making it difficult to list specific cleaning procedures. However, in all cases, materials used for cleaning must be evaluated so that they do not in some way combine with the hazardous or inert material to cause new hazards. Each industrial plant should be evaluated for its particular housekeeping problem by environmental or industrial hygiene consultants. Specialized programs should be established for each plant. A housekeeping evaluation form should be used by the supervisors and a self-inspection program implemented. The housekeeping studies should be made in each area of the plant at least once a week.

Where spills of hazardous materials occur, they should be recorded on a special incident reporting form to the environmental health specialist or safety supervisor. The information recorded should include the type of material spilled; when, where, and how it was spilled; and by whom. The quantity of material should be roughly determined and the potential hazardous effects from the spillage listed. The cleaning technique used to remove the spill should be included and also where the hazardous material was disposed of. The time of day and the day of the week should also be recorded. This information should be kept on file by the safety officer and reviewed periodically to determine the possible causes of spillage and what could be done to prevent further problems.

An effective industrial health program depends not only on the worker and medical personnel, and proper equipment and techniques, but also on the techniques used in housekeeping.

Exposure Time and Concentration

Each of the hazardous materials discussed in this chapter and all other hazardous materials are listed in the outline of hazardous materials put out by the National Institute of Occupational Safety and Health. Assigned to each are specific exposure times and concentrations for employees working an 8-hr shift over long periods of time. In addition, for certain hazardous chemicals, maximum peak times of exposure to certain higher levels of concentrations of the material are established. In order to understand the time and concentration relationship, it is necessary for the environmentalist to fully investigate the process under study and to determine in advance the kinds of hazardous materials or conditions and the established safety standards for the worker. Further information concerning this topic can be obtained from the National Institute of Occupational Safety and Health or the National Safety Council.

Identifying Hazardous Areas

All hazardous areas should be clearly identified by special signs, markings, and colors. They should read either *Biohazard Restricted Area*, *Radiation Restricted Area*, or *Hazardous Chemicals Restricted Area*. In all cases, outside of the restricted area there should be posted instructions on what should or should not be done and who should or should not enter the area. Instructions should also be posted for what to do in the event of an emergency.

Only authorized individuals, properly trained and properly protected, should be permitted to enter hazardous areas. The buddy system should always be used when an individual works within a hazardous area in the event that one of the individuals is harmed.

Repair, Maintenance, and Construction of Equipment

It is extremely important to ensure that all equipment is in perfect working condition. The industrial plant should maintain a knowledgeable staff of individuals who can repair and maintain all equipment. If the equipment is highly specialized, a contract should be issued to an equipment firm to provide 24-hr service, 7 days a week if the equipment may become hazardous when not operating properly. The engineering and construction of the equipment is only as good as the efforts to maintain it in excellent working condition and the repair work done when the equipment breaks down.

The Occupational Health Program

The occupational health program has as its major objectives protection of workers against health and safety hazards in the work environment; proper placement of workers according to their physical, mental, and emotional abilities; maintenance of a pleasant, healthy work environment; establishment of preplacement health examinations; establishment of regular, periodic health examinations; diagnosis and treatment of occupational injuries or diseases; consultation with the worker's personal physician, with the worker's consent, of other related health problems, such as heart disease; health education and counseling for employees; safety education for employees; establishment of research to identify hazardous situations or find means of preventing hazardous situations; and establishment of necessary surveys and studies of the industrial environment for the protection of the worker, his or her family, and the community.

These services are provided by physicians, nurses, environmentalists, engineers, industrial hygienists, chemists, and safety management personnel.

A good occupational health program not only benefits the worker but is also a value to the industry. It reduces the cost of worker's compensation insurance, reduces the cost of hospital insurance, reduces absenteeism and labor turnover, helps satisfy the legal requirements set forth in the OSHA laws, and other federal laws and state laws, and creates a good working relationship between management and workers.

Occupational Stress Management Programs

An occupational stress management program should help the employees cope with the occupational stresses, as well as the personal and societal stresses. Many companies have not only developed stress management training programs, but further have developed an "employee assistance program." The function of these programs is to have individuals establish a mechanism to reduce stress, plan alternatives to problems, and assist individuals in planning techniques to deal with stress when the problems can not be relieved.

This management program should encourage each person to become more responsible for his or her life and actions, and should focus on the dynamics of the individual, as well as the dynamics of the family and the organization where he or she works. The first step in planning the program is to determine from the employees the kinds of problems that exist and the level of intensity at which they exist. Under the section in this chapter on problems, which have been previously discussed, many of the stress-related situations have been mentioned. It will be appropriate to develop a tested instrument to measure these stress-type situations and then to evaluate with this instrument the results after behavior modification. The results of this study would assess and measure each of the employees' stress levels, determine the cause of the stress, and determine whether or not the stress is an individual or group problem.

The actual program could consist of lectures, TV tapes, small group dynamics, and referrals to physicians when specific problems of a personal nature need to be determined and treated. All stress management programs should include techniques of relaxation therapy, good nutrition, preventive measures, and control of habits that may lead to stress. Above all else, a good supervision and management program to train the supervisors and managers would be extremely useful, since good management can help reduce stress levels among employees. Employees expect their supervisors and managers to be thorough, fair, tactful, show initiative, have good emotional control, and be enthusiastic. The supervisor and manager need to use time, material, and money in an appropriate and effective manner. Employees want to be recognized for what they do rather than what they don't do or what they do poorly.

There are many approaches to management. They may be environmental controls, they may be supervision and management improvement, they may be removal of many other types of stresses related to home and family, or they may be related to the improvement of the individual and resolving of the individual's personal problems.

HAZARD COMMUNICATION PROGRAM

The Hazard Communication Program is required by the OSHA standards and by standards established for Federal Employee Occupational Safety and Health Programs. The program consists of three elements. They are hazard warning labels, material safety data sheets (MSDSs) and employee training programs. The standard which was issued in 1983 by OSHA, requires chemical manufacturers and importers to make a comprehensive hazard determination for the chemical products they sell. The manufacturers, importers, and distributors must provide information concerning the health and physical hazards of these products. This is accomplished by means of warning labels. The manufacturing sector that uses the chemicals to turn out additional products must protect their employees by meeting the following standards:

1. Preparing a written Hazard Communication Program, which includes an inventory of all hazardous chemicals used in their facilities.
2. Obtaining an MSDS for each hazardous chemical they use.
3. Displaying appropriate facility placards and warnings.
4. Preparing a hazard communication training plan.
5. Providing training to employees who are potentially exposed to hazardous chemicals in their facilities.

The elements of the Hazard Communication Program should include preparation of a written hazard communication plan; identification and evaluation of chemical hazards in the work place; preparation of a hazardous substance inventory; development of a file of MSDSs; provision for access to MSDSs for employees; ensurance that incoming products have proper labels; development of a system of labeling within the facility; development of the training program; identifying and training employees who are potentially exposed to the hazardous chemicals; and evaluation of the programs and redirection where necessary. The written plan must describe how complaints will be handled, labeling, how MSDSs will be obtained and made available, and how information and training will be provided. The inventory of all the toxic chemicals in the workplace will have to be cross-referenced to the MSDSs. Provision will be made for employees to deal with accidental spills and leaks, and for making employees aware of the types of hazards that may occur as a result of the chemicals or the pipes carrying the chemicals.

The hazard assessment should be conducted by trained environmental health personnel under the direction of an industrial hygienist. The definition of hazardous chemicals will need to be clearly stated in order to properly identify the potential problems within the workplace. The hazardous chemical assessment will include all chemicals that may create a physical hazard or a health hazard. It will not include hazardous waste, tobacco products, wood products, and other substances that may be consumed by employees at the workplace, or food, drugs, or cosmetics.

A list of hazardous chemicals may be obtained from OSHA, the American Conference of Governmental Industrial Hygienists, the National Toxicology Program, and the International Agency for Research on Cancer. If there are questions concerning chemicals not listed by any of these groups, they may be called for further information.

If a facility or workplace uses common consumer products for housekeeping, such as soap, typing correction fluid, etc., they may be excluded from the hazard assessment, providing they are used in the same way and the same quantities as the typical consumer would. However, if a commercial cleaning solution, such as Clorox®, is used to disinfect surfaces, it should be included in the program. Only the chemicals known to be present in the workplace should be included in the program. It is then essential to determine which employees have a potential for exposure to the chemical. An exposure means that the employee comes in contact with the material by inhalation, ingestion, or direct contact with the skin during normal work activities, nonroutine work tasks, and unforeseeable emergencies.

The survey itself involves identifying the chemicals, how they are being used or produced, and in what quantities. Determine if warning labels are placed in readily observable places on drums or containers. Next set up a flow chart of each operation. The flow chart should start at the place where the raw materials enter the building, is taken to storage, and then is introduced into the actual manufacturing operation. From this point, each step along the way in the manufacturing of the product should be shown and the potential problems listed under the particular step. The flow chart will proceed to the final step of production packaging and storage, and also to byproducts, waste products, waste product storage, and disposal. By using the flow chart and diagraming the actual flow, it is easier to determine who may come in contact with the hazardous chemical, where it may occur, how it may happen, when it may happen, why it may happen, what the final results may be, and the kind of protective measures that are being utilized by the employee to avoid contact with the chemical. An important part of the study includes ventilation controls related to local exhaust ventilation as well as general ventilation. Since the chemicals vary so, the requirement for respirators and protective equipment varies considerably. NIOSH, in its Morgantown, West Virginia facility, is doing a considerable amount of research work on the use of respirators and protective equipment. They would be a good resource to contact in the event of problems. The environmental health practitioner, while conducting this study, should look for obvious signs of exposure as follows:

1. Is there airborne dust, smoke, or mist present?
2. Are there accumulations of dust, liquid, or oil on machines, on the floor or on the ledges?
3. Are there odors of solvent vapors or gases present?
4. Is there an unusual taste present in the mouth?
5. Are the eyes burning or is the throat and nose irritated?
6. Are hazardous operations being performed at times other than the survey time? What special tasks do maintenance people perform? Where do they perform them (such as closed vessels)? How do they perform them?
7. Are there procedures available for response to emergencies such as chemical spills, leaks, explosions, and fires?
8. Have employees complained of skin rashes or dermatitis, cough, tightness of the chest and difficulty in breathing, stuffy nose, and colds that won't go away? Do employees have headaches, dizziness, light-headedness, loss of appetite, fatigue, or nausea? Do employees feel numbness in their fingers, hands, arms, and legs? Do the symptoms go away when the individuals leave work? Also determine if the employees are given special medical tests such as blood, urine, lung function, and X-rays.

An air-sampling survey should be conducted in those areas that appear to have potential chemical hazard problems. The air-sampling survey should consist of a general air survey, specific air samples in high concentration areas, and the monitoring of personnel with personal monitors at the breathing level.

Once the survey has been concluded, a hazardous substances inventory should be developed and a hazardous chemical MSDS file should be provided. When hazardous chemicals are utilized, they should be clearly identified by chemical name, trade name and common name. All of the hazardous ingredients in the substance should be listed. The physical and chemical characteristics of the substance, such as boiling and freezing points, density, vapor pressure, specific gravity, solubility, volatility, and the product's general appearance and odor, should be included. All physical hazards, such as fire and explosion, should be described. All health hazards, such as carcinogenicity, corrosive, toxic, irritant, and other hazards, should be included. The route of entry should be clearly described. Special precautions and procedures should be established for spills, leaks, cleanup controls, emergencies and first aid. The program must provide access to the MSDSs for employees. All incoming labels and existent labels need to be checked.

The last part of the program is employee training. The employees need to clearly understand the operations of their work areas, the hazardous chemicals present, the means of prevention and control, the physical and health hazards associated with the chemicals, and the types of protective clothing or respirators that they should wear in order to avoid problems. Finally, they should be well aware of any symptoms of illness that may be attributed to the chemicals or other substances in the workplace and immediately seek help from trained medical personnel when needed. Along with this, they should recognize any potential physical problems that may occur from hazardous chemicals, and therefore, avoid explosions, fires, or asphyxiation from carbon monoxide or other gases.

SUMMARY

Occupational disease, injury, and death is of vital concern to our society because the costs and physical and emotional anguish are great. Occupational disease may be caused by physical, biological, or chemical factors. Although there are numerous techniques available for hazard reduction and the potential for intervention is excellent in certain areas, it is extremely poor in others, such as the chemical environment.

The two major routes of entry for toxic substances are the skin and respiratory tract. In addition, the liver and kidney show a significant amount of disease. Death is attributed not only to the dysfunctions of the liver and kidney, but also to the cardiovascular system. It is essential that the incidence of occupationally involved illness be recognized and the necessary changes made.

The basic principles of occupational disease control involve prevention primarily and treatment secondarily. To have proper prevention and treatment, it is necessary to not only study the individual employee and ensure that job placement is carried out in an effective manner, but the level of hazard associated with the various agents in the workplace must be determined. Cause and effect relationships and dose-response relationships should be understood. The occupational health specialist, physician, toxicologist, and chemist need to work together to further evaluate and eventually establish feasible controls for the given occupational environment. Once the controls are established, both the environment and the worker must be constantly monitored to ensure that exposure levels to a given set of occupational hazards will not exceed the appropriate established standards.

RESEARCH NEEDS

There is a need for better collection, analysis, and dissemination of morbidity and mortality data in the occupational environments. A national death index should be developed and combined with occupational information. Systematic studies should be conducted to determine the comparative value of existing tests used for screening potential mutagenic substances. A working committee of representatives of industry, labor, universities, and federal agencies should develop industry-wide uniform medical data reporting systems that can be coded in an appropriate manner for all to understand. These data systems should be available for all researchers in all areas. Efforts should be made to identify the early indications of carcinogenesis.

Efforts to develop new techniques for gathering the basic data for epidemiological studies should be made. Basic research is necessary to fully understand the mechanisms of respiratory stresses and host responses. Studies are needed to determine various organ responses to environmental stresses. Epidemiological studies should be made to determine the relationship between occupational exposure and fertility, miscarriages, congenital malformation, mental retardation, cancer, and the relationship of stress to mental health problems. Industries having excessive injury rates should be identified and the various risk factors analyzed. Centers need to be developed for the combined study of biologists, toxicologists, occupational medicine experts, environmental epidemiologists, industrial engineers, and environmental health practitioners. Additional research is needed in a variety of control technologies. Personal protective equipment should be further evaluated to ensure that the individuals are receiving the proper protection. All the research must be combined with necessary techniques to make available to all who are interested the results and possible application of the work completed.

Additional studies, epidemiological and toxicological, need to be conducted into the reproductive effects of occupational exposures. Studies should include the etiology of adverse reproductive outcomes, such as fetal chromosomal abnormalities or abnormal spermatogenesis. Improved methodologies are needed to evaluate the quality of semen, the outcomes of pregnancy, and the potential effects of a variety of chemicals on these areas.

Ongoing research is needed to establish an acceptable human exposure level for lowering toxic agents that have been identified. Epidemiological evaluations of suspected neurotoxicity should be conducted for a variety of identified chemicals. Simple screening tools are needed to test the asymptomatic populations exposed to known neurotoxic agents.

Studies should be implemented for determining the relationship between dermatological conditions and a variety of contact allergens and the immune response of people. Studies should be conducted to determine which chemicals can be used to substitute for those that are causing dermatological problems.

Computers and robots are expected to affect 7 million factory jobs and 39 million office jobs. This job displacement may cause severe psychological disorders. It is essential to study the potential problems before they occur.

Additional research should be carried out in the area of occupational lung disease and in the area of occupational cancer.

REFERENCES

Chapter 1

Abercrombie, M. *Dictionary of Biology* (Baltimore: Penguin Books, 1964).

Adams, J. "Clinical Relevance of Experimental Behavioral Teratology", *Neurotoxicology* 7(2):44–90 (1980).

American Cancer Society. *Cancer Facts and Figures, 1983* (New York: American Cancer Society, 1983).

American Gas Association. *Total Energy Resource Analysis* (Arlington, VA: January 1988).

Ames, B. N. "Identifying Environmental Chemicals Causing Mutation and Cancer", *Science* 204:587–592 (1979).

Anderson, L. E. *The Mosby Medical Encyclopedia* (New York: New American Library, 1985).

Archer, V. E. *Health Implications of New Energy Technologies* (Ann Arbor, MI: Ann Arbor Science, 1980).

Arthur Anderson & Company. *Oil & Gas Reserve Disclosures* (Houston, TX: 1988).

Atlas, R. M. *Basic and Practical Microbiology* (New York: MacMillan Publishing Co., 1986).

Austin, D. F. *Epidemiology for the Health Sciences* (Springfield, IL: Charles C Thomas, 1982).

Beckman, D. A., and R. L. Brent. "Mechanism of Known Environmental Teratogens: Drugs and Chemicals", *Clin. Perinatol.* 13(3):649–87 (1986).

Billings, C. H. "Health Risk Assessment", *Public Works* 117:107 (1986).

Bloomfield, M. M. *Chemistry and the Living Organism* (New York: John Wiley & Sons, 1984), p. 610.

Boehm, P. D., and D. L. Fiest. "Subsurface Distributions of Petroleum From an Offshore Well Blowout", *Environ. Sci. Technol.* 16(2):67–76 (1982).

Boyland, E. "The Metabolism of Foreign Compounds and the Induction of Cancer", *Xenobiotica* 16:901–915 (1986) .

Brenson, A. *Control of Communicable Diseases in Man* (Washington, DC: American Public Health Association, 1985).

Brent, R. L. "Evaluating the Alleged Teratogenicity of Environmental Agents", *Clin. Perinatol.* 13(3):609–613 (1986).

Broom, B. *Anatomy and Physiology* (New York: Cambridge Communications, 1985).

Brown, T. L. *Chemistry: The Central Science* (Englewood Cliffs, NJ: Prentice Hall, 1984), p. 840.

Butler, G. C. *Principles of Ecotoxicology* (New York: John Wiley & Sons, 1981), p. 350.

Carlo, G. L., and S. Hearn. "Practical Aspects of Conducting Reproductive Epidemiology Studies Within Industry", *Progr. Clin. Biol. Res.* 160:49–65 (1984).

Chen, R. "Detection and Investigation of Subtle Epidemics", *Prog. Clin. Biol. Res.* 163B:39–43 (1985).

Clemmesen, J. "International Commission for Protectoin Against Environmental Mutagens and Carcinogens", *Mutat. Res.* 98(1):97–100 (1982).

Committee on the Institutional Means for Assessment of Risks to Public Health, Commission of Life Sciences, National Research Council. *Risk Assessment in the Federal Government: Managing the Process* (Washington, DC: National Academy Press, 1983).

Coon, M. J., and A. V. Persson. *Enzymatic Basis of Detoxificaton* (New York: Academic Press, 1980).

Covello, V., Von Winterfeldet, Detlof, Slovic and Paul. *Risk Communication: An Assessment of the Literature on Communicating Information about Health, Safety and Environmental Risks,* draft (Los Angeles: Institute of Safety and Systems Management, 1986).

Darmstadter, J. *Energy in America's Future: The Choices Before Us* (Baltimore, MD: John Hopkins Press, 1980).

Data Resources, Inc. *Energy Review*, (Lexington, MA: Data Resources Inc., 1988).

Davies, D.M. *Textbook of Adverse Drug Reactions* (New York: Oxford Medical Publications, 1981).

Department of the Environment. *Heavy Metals in the Environment* (London: H.M.S.O., 1983).

Di Carlo, F. J., P. Bickart and C. M. Auer. "Structure—Metabolism Relationships (SMR) for the Prediction of Health Hazards by the EPA", *Drug Metab. Rev.* 17(3–4):187–220 (1986).

Doll, R. "Relevance of Epidemiology to Policies for the Prevention of Cancer", *Human Toxicol.* 4(1):81–96 (1985).

Doyle, D. K. "Teratology: A Primer", *Neonatal Network* 4:24–29 (1986).

Dufus, J. H. *Environmental Toxicology* (London: Edward Arnold Publishers, 1982), p. 164.

Ecker, M. D., and N. J. Bramesco. *Radiation* (New York: Vintage Books, 1981).

El-Hinnawi, E. E. *The Environmental Impacts of Production and Use of Energy* (Shannon, 1981).

Ellerman, V. "Cancer Carcinogenesis", *Science America* 243:98–105 (1984).

Energy Information Administration. *Annual Energy Outlook 1987* DOEEIA-0383(87) (Washington, DC: Dept. of Energy, 1988).

Energy Information Administration. *Manufacturing Energy Consumption Survey: Consumption of Energy, 1985* DOE/EIA-000512(85) (Washington, DC: Dept. of Energy, 1988).

Evans, D.W. "Effects of Ionizing Radiation", *Cana. Med. Assoc. J.* 129:10–11 (1983).

Evans, W. E. *Applied Pharmacokinetics* (San Francisco: Applied Therapeutics Inc., 1980).

Fawcett, H. H. *Hazardous and Toxic Materials* (New York: John Wiley & Sons, 1984).

Fein, G. G. et al. "Environmental Toxins and Behavioral Development", *Am. Psychol.* 38(11):1188–1197 (1983).

Fein, G. *Intrauterin Exposure of Humans to PCB's: Newborn Effects,* (Duluth, MN: U.S. EPA, 1984).

Fochtman, E. G. *Biodegration and Carbon Adsorption of Carcinogenic and Hazardous Organic Compounds* (Cincinnati, OH: U.S. EPA, 1981).

Frederick, G. L. "Assessment of Teratogenic Potential of Therapeutic Agents", *Clin. Invest. Med.* 8(4):323–27 (1985).

Fukino, H., and S. Mimura. "Mutagenicity of Airborne Particles", *Mutat. Res.* 102: 237–241 (1982).

Gas Resources Institute. *1987 GRI Baseline Projection of U.S. Energy Supply and Demand* (Chicago, IL: Gas Resources Institute, 1987).

Gofman, J. W. *Radiation and Human Health* (San Francisco: Sierra Club Books, 1981).

Goldsmith, J. R. "Epidemiology Monitoring in Environmental Health, Introduction, and Overview", *Sci. Total Environ.* 32:211–218 (1984).

Gorbaty, M. L. *Coal Science* (New York: Academic Press, 1982).

Grahame-Smith, D. G., and J. K. Aronson. *Oxford Textbook of Clinical Pharmacology and Drug Therapy* (Oxford: Oxford University Press, 1984).

Green, G. "Ionizing Radiation and its Biological Effects", *Occupat. Health,* 38:354–357 (1986).

Green, L. W. *Community Health* (St. Louis, MO: Times Mirror/Mosby College Publishing, 1986).

Heinonen, O. P., D. Slone and S. Shapiro. *Birth Defects and Drugs in Pregnancy* (Boston, MA: John Wright-PSG Inc., 1982).

Hendee, W. R. "Real and Perceived Risks of Radiation Exposure", *West. J. Med.* 138:380–386 (1983).

Henold, K. L., and F. Walmsley. *Chemical Principles, Properties, and Reactions* (Reading, MA: Addison-Wesley, 1984).

Hodgson, E. *Introduction to Biochemical Toxicology* (New York: Elsevier, 1980).

Inskip, H., V. Beral, P. Fraser and J. Haskey, "Epidemiological Monitoring: Methods for Analyzing Routinely-Collected Data", *Sci. Total Environ.* 32:219–232 (1984).

Jefrey, A. M. "DNA Modification by Chemical Carcinogens", *Pharmacol Terminol.,* 28:237–241 (1985).

Khoury, D. L. *Coal Cleaning Technology* (Parkridge, NJ: Noyes Data Corp., 1981).

Khoury, M. J., and N. A. Holtzman. "On the Ability of Birth Defects Monitoring to Detect New Teratogens", *Am. J. Epidemiol.* 126(1):136–143 (1987).

Kleinbaum, D. G., L. L. Kupper and H. Morgenstern. *Epidemiologic Research* (London: Lifetime Learning Publications, 1982).

Larson, T. E. *Cleaning Our Environment: A Chemical Perspective* (Washington DC: American Chemical Society, 1983).

Legator, M. S., M. Rosenberg and H. Zenick. *Environmental Influences on Fertility, Pregnancy, and Development* (New York: Alan R. Liss, 1984).

Lilienfeld, A. *Foundations of Epidemiology* (New York: Oxford University Press, 1980).

Lippmann, M., and R. B. Schlesinger. *Chemical Contamination in the Human Environment* (New York: Oxford University Press, 1980).

March, J. *Advanced Organic Chemistry* (New York: John Wiley & Sons, 1985).

Marcus, J., and S. L. Hans. "Neurobehavorial Toxicology and Teratology", Department of Psychiatry, University of Chicago, 1982.

Meyers, V. K. "Chemicals Which Cause Birth Defects—Teratogens A Special Concern of Research Chemists", *Sci. Total Environ.* 32:1–12 (1983).

Miller, G. T. *Living in the Environment* (Bellmont, CA: Wadsworth Publishing Company, 1985).

Morgenstern, H. "Uses of Ecologic Analysis in Epidemiological Research", *Am. J. Public Health* 72:1336–1344 (1982).

Mortimer, C. E. *Chemistry* (Belmont, CA: Wadsworth, 1981).

Naeye, R., and H. Tafari. *Risk Factors in Pregnancy and Diseases of the Fetus and Newborn* (Baltimore: Williams and Wilkins, 1983).

Nau, H. "Species Differences in Pharmacokinetics and Drug Teratogenesis", *Environ. Health Perspec.* 70:113–129 (1986).

Neubert, D., et al. "Principles and Problems in Assessing Prenatal Toxicity", *Arch. Toxicol.* 60:238–245 (1987).

NIH. *Cancer Risk and Rates.* NIH Publication No. 85–691 (Washington, DC: NIH, 1985).

Niswander, K. R. *Obstetrics* (Boston: Little Brown and Co., 1981).

Oftedal, P., and A. Brogger. *Risk and Reason—Risk Assessment In Relation to Environmental Mutagens and Carcinogens* (New York: Alan R. Liss, 1986).

Parks, C. F., Jr. *Earthbound: Minerals, Energy, and Man's Future* (San Francisco: Freeman, Cooper and Company, 1975).

Poch, D. I. *Radiation Alert* (Toronto/Ontario: An Energy Prove Project, 1985).

Pochin, E. E. *Nuclear Radiation: Risks and Benefits* (Oxford: Clarendon Press, 1983).

Porter, C. R. *Coal Processing Technology* (American Institute of Chemical Engineers, 1981).

Price, S. A. *Pathophysiology Clinical Concepts of Diseases Processes* (New York: McGraw-Hill, 1986).

Quilligan, E. J., and N. Kretchmer. *Fetal and Maternal Medicine* (New York: John Wiley & Sons, 1980).

Robertson, R., and P. M. MacLeod. "Accutane-Induced Teratogenesis", *Canadian Med. Assoc. J.* 133(11):1147–1148 (1985).

Rosa, F. W., A. L. Wilk and F. O. Kelsey. "Teratogen Update", *Teratology* 33(3):356–364 (1986).

Rosenber, M. J., and W. E. Halperin. "The Role of Surveillance in Monitoring Reproductive Health", *Teratogen. Carcinogen. Mutagen.* 4(1):15–24 (1984).

Rowland, M., and T. Tozer. *Clinical Pharmacokinetics: Concepts and Applications* (Philadelphia: Lea & Febiger, 1980).

SCEP. "Study of Critical Environmental Problems", in *Man's Impact on the Global Environment* (Cambridge, MA: MIT Press, 1983).

Schardein, J. L. "Current Status of Drugs as Teratogens in Man", *Progr. Clin. Biol. Res.* 163C:181–190 (1985).

Schlesselman, J. *Case-Control Studies* (New York: Oxford University Press, 1982).

Schottenfeld, D., and J. Fraumeni, Jr. *Cancer Epidemiology and Prevention* (Philadelphia: W. B. Saunders, 1982).

Shepard, T. H. "Human Teratogens: How Can We Sort Them Out?", *Ann. N. Y. Acad. Sci.* 477:105–115 (1986).

Sherwin, R. P. "What is an Adverse Health Effect?", *Environ. Health Perspec.* 52:177–182 (1983).

Sorensen, H. *Energy Conversion Systems* (New York: John Wiley & Sons, 1983).

Swaab, D. F., and M. Mirmiran. "Functional Teratogenic Effects of Chemicals on the Developing Brain", *Monogr. Neural Sci.* 12:45–57 (1986).

Ujino, Y. "Epidemiological Studies on Disturbances of Human Fetal Development", *Arch. Environ. Health* 40:181–184 (1985).

Ulsamer, A. G., P. D. White and P. W. Preuss. "Evaluation of Carcinogens: Perspective of the Consumer Product Safety Commission", in *Handbook of Carcinogen Testing*, H. A. Milman and E. K. Weisburger Eds. (Park Ridge, NJ: Noyes Publications, 1985).

U.S. Department of Health and Human Services. *Determining Risks to Health* (Dover, MA: Auburn Publishing Co., 1986).

U.S. Department of Health and Human Services, Food and Drug Administration. "Identifying and Regulating Carcinogens", Comments on unpublished OTA draft report (1987).

U.S. Department of Health and Human Services, National Toxicology Program. *NTP Tech. Bull.* 1(3):1–2, (1980).

U.S. Department of Health and Human Services, National Toxicology Program. "Explanation of Levels of Evidence of Carcinogenic Activity", 1986.

U.S. Department of Health and Human Services, National Toxicology Program. *Annual Plan, FY 1986*, DHHS Publ. No. NTP-86-086 (Research Triangle Park, NC:1986).

U.S. Department of Health and Human Services, National Toxicology Program. Board of Scientific Counselors *Report of the NTP Ad Hoc Panel on Chemical Carcinogensis Testing and Evaluation* (Research Triangle Park, NC: 1980).

U.S. Department of Health and Human Services, National Toxicology Program. *Fourth Annual Report on Carcinogens* (Research Triangle Park, NC: 1985).

U.S. Department of Health and Human Services, National Toxicology Program. *Review of Current DHHS, DOE, and EPA Research Related to Toxicology, Fiscal Year 1986*, Publ. No. NTP-86-087 (Research Triangle Park, NC: 1987).

U.S. Department of Health and Human Services, National Toxicology Program. *Second Annual Report on Carcinogens* (Research Triangle Park, NC: 1981).

U.S. Department of Health and Human Services, National Toxicology Program. *Third Annual Report on Carcinogens* (Research Triangle Park, NC: 1983).

U.S. Environmental Protection Agency. *Environmental Outlook 1980* (Washington, DC: 1980).

U.S. Environmental Protection Agency. "Guidelines for Carcinogen Risk Assessment", *Fed. Reg.* 51:33992 (1986).

U.S. Environmental Protection Agency. "Guidelines for Mutagenicity Risk Assessment", *Fed. Reg.* 51:34006 (1986).

U.S. Environmental Protection Agency, Office of Policy and Resource Management. *Environmental Mathematical Pollutant Fate Modeling Handbook/Catalogue* (draft), Contract no. 68-01-5146 (Washington, DC: 1982).

U.S. Environmental Protection Agency. "Proposed Guidelines for Carcinogen Risk Assessment", *Fed. Reg.* 49(227):46294–46301 (1984).

U.S. Environmental Protection Agency. "Proposed Guidelines for the Health Risk Assessment of Chemical Mixtures and Request for Comments", *Fed. Reg.* 50(6):1170–1176 (1985).

U.S. Environmental Protection Agency. Proposed Guidelines for the Health Assessment of Suspect Developmental Toxicants and Request for Comments", *Fed. Reg.* 49 (227):46324–46331 (1984).

U.S. Environmental Protection Agency. *Risk Assessment and Management: Framework for Decision Making* (Washington, DC: 1984).

Vainio, H. "Current Trends in the Biological Monitoring of Exposure to Carcinogens", *Scand. J. Environ. Health* 11:1–3 (1985).

Verschueren, K. *Handbook of Environmental Data on Organic Chemicals* (New York: Van Nostrand Reinhold Company 1983).

Vesilind, A. P., and J. Pierce. *Environmental Pollution and Control,* 2nd ed. (Ann Arbor, MI: Ann Arbor Science, 1983).

Vesiland, P. A., and J. J. Peirce. *Environmental Engineering* (Boston: Butterworth Publishers, 1982).

Wagner, P. *Coal Processing Technology* (American Institute of Chemical Engineers, 1980).

Wallace, R. A. *Biology, the Science of Life* (San Francisco: Scott, Foresman and Company, 1983), p. 1140.

Wartak, J. *Clinical Pharmacokinetics* (Baltimore: University Park Press, 1983).

Washington Policy and Analysis, Inc. *Natural Gas: A Strategic Resource for the Future* (Washington, DC: 1988).

Wehrle, P. *Communicable and Infectious Disease* (St. Louis, MO: C.V. Mosby Co., 1981).

Wharton Econometric Forecasting. Associates *Energy Analysis Quarterly* (Bala Cynwyd, PA: 1988).

Whitmore, R. W. *Methodology for Characterization of Uncertainty in Exposure Assessments* (Research Triangle Park, NC: Research Triangle Institute, 1985).

Winter, M. E., *Basic Clinical Pharmacokinetics* (San Francisco: Applied Therapeutics, Inc., 1980).

Yoyn, D., and Blkeslee, Eds. *Advances in Comparative Leukemia* (New York: Elsevier North Holland, 1981), p. 1–16.

Zitko, V. "Graphical Display of Environmental Quality Criteria", *Sci. Total Environ.* 72:217–220 (1988).

Chapter 2

Akinboade, O. A., J. O. Hassan and A. Adejinmi. "Public Health Importance of Market Meat Exposed to Refuse Flies and Air-borne Microorganisms", *Int. J. Zoonoses* 11 (1):111–114 (1984).

Archer, D. L., and F. E. Young. "Contemporary Issues: Diseases with a Food Vector", *Clin. Microbiol. Rev.* 1(4): 377–398 (1988).

Baddour, L. M., S. M. Gaia, R. Griffin and R. Hudson. "A Hospital Cafeteria-Related Food-borne Outbreak due to *Bacillus Cereus*: Unique Features", *Infect. Control* 7 (9):4652–4655 (1986).

Brown, P., D. Kidd, T. Riordan and A. Barrell. "An Outbreak of Food-Borne *Campylobacter jejuni* Infection and the Possible Role of Cross-Contamination", *J. Infect.* 17 (2):171–176 (1988).

Bull, P. *Prevention of Food-Borne Disease* 22(2):207–208 (1988).

Bush, M. F. "The Symptomless Salmonella Excretor Working in the Food Industry", *Community Med.* 7 (2):133–135 (1985).

Chia, J. K., J. B. Clark, C. A. Ryan and M. Pollack. "Botulism in an Adult Associated With Food-Borne Intestinal Infection With *Clostridium botulinum*", *N. Engl. J. Med.* 315(4):239–241 (1986).

Cohen, M. L., and R. V. Tauxe. "Drug-Resistant Salmonella in the United States: An Epidemiologic Perspective," *Science* 234(4779):964–969 (1986).

Corbe, S., P. Barton, R. C. Nair and C. Dulbert. "Evaluation of the Effect of Frequency of Inspection on the Sanitary Conditions of Eating Establishments ", *Can. J. Public Health* 75(6):434–438 (1984).

Cordle, M. K. "USDA Regulation of Residues in Meat and Poultry Products ", *J. Anim. Sci.* 66(2):413–433 (1988).

de-Wit, J. C., and E. H. Kampelmacher. "Some Aspects of Bacterial Contamination of Hands of Workers in Food Service Establishments", *Zentralbl. Bakteriol. Mikrobiol. Hyg. B.* 186(1):45–54 (1988).

Gay, C. T., W. A. Marks, H. D. Riley, J. B. Bodensteiner, M. Hamza, P. A. Noorani and G. B. Bobele. "Infantile Botulism ", *South. Med. J.* 81(4):457–460 (1988).

Greene, J. "FDA Talk Paper ", *Food and Drug Admin.* 25 Feb. 1983.

Greene, J. "FDA Talk Paper ", *Food and Drug Admin.* 5 Apr. 1983.

Greene, J. "FDA Talk Paper ", *Food and Drug Admin.* 26 Sept. 1984.

Guilfoyle, D. E., and J. F. Yager. "Survey of Infant Foods for *Clostridium botulinum* Spores", *J. Assoc. Off. Anal. Chem.* 66(5):1302–1304 (1983).

Harris, N. V., D. Thompson, D. C. Martin and C. M. Nolan. "A Survey of Campylobacter and Other Bacterial Contaminants of Pre-Market Chicken and Retail Poultry and Meats, King County, Washington", *Am. J. Public Health* 76(4):401–406 (1986).

Hayes, A. H. "The Food and Drug Administration's Role in the Canned Salmon Recalls of 1982 ", *Public Health Rep.* 98(4):144–149 (1983).

Hobbs, G. "Food Poisoning and Fish ", *J. R. Soc. Health* 103(4):144–149 (1983).

Hay, R. N. "Environmental Effects on Mastitis and Milk Quality ", *Vet. Clin. North Am. [Large Anim. Pract.]*, 6(2):371–375 (1984).

Kay, R. N. "Poultry Meat Inspection — A Service to the Community ", *J. R. Soc. Health* 108(2):39–40 (1988).

Kliks, M. M. "Anisakiasis in the Western United States: Four New Case Reports From California ", *Am. J. Trop. Med. Hyg.* 32(3):526–532 (1983).

Kwee, H. G. and R. L. Sautter. "Anisakiasis ", *Am. Fam. Phys.* 36(2):137–140 (1987).

Lim, Y. S., S. Y. Khor, M. Jegathesan and S. H. Kang. "The Isolation of Salmonella from Raw, Cooked and Dried Foods", *Med. J. Malaysia* 39(3):220–224 (1984).

Macdonald, K. L., and P. M. Griffin. "Foodborne Disease Outbreaks, Annual Summary, 1982", *MMSR. CDC. Surveill. Summ.* 35(1):775–1655 (1986).

Martin, D. L., T. L. Gustafson, J. W. Pelosi, L. Suarez and G. V. Pierce. "Contaminated Produce — A common Source for Two Outbreaks of Shigella Gastroenteritis ", *Am. J. Epidemiol.* 124(2):299–305 (1986).

Mascola, L., L. Lieb, J. Chiu, S. L. Fannin and M. J. Linnan. "Listeriosis: An Uncommon Opportunistic Infection in Patients with Acquired Immunodeficiecy Syndrome. A report of five cases and a review of the Literature ", *Am. J. Med.* 84(1):162–164 (1988).

Meitert, E., A. Vlad and F. Sima. "Food Poisoning Associated with *Pseudomonas aeruginosa*" *Arch. Roum. Pathol. Exp. Microbiol.* 43(2):115–119 (1984).

Miller, S. A. "Toxicology and Food Safety Regulations ", *Arch. Toxicol.* 60(1–3):212–216 (1987).

Nastasi, A., C. Mammina, M. R. Villafrate, G. Scarlata, M. F. Massenti and M. Diquattro. "Multiple Typing of *Salmonella typhimurium* Isolates: An Epidemiological Study", *Microbiologica* 11(3):173–178 (1988).

Reed, D. V. "The FDA Surveillance Index for Pesticides: Establishing Food Monitoring Priorities Based on Potential Health Risk", *J. Assoc. Off. Anal. Chem.* 68(1):122–124 (1985).

Rennels, M. B., and M. M. Levine. "Classical Bacterial Diarrhea: Perspectives and Updates—Salmonella, Shigella, Escherichia coli, Aeromonas, and Plesiomonas", *Pediatr. Infect. Dis.* 5(Suppl. 1):91–100 (1986).

Tartakow, I., and Jackson. *Foodborne and Waterborne Disease* (Westport, CT: AVI Publishing Company, 1981).

U. S. Department of Health and Human Services, Food and Drug Administration. "General Principles for Evaluating the Safety of Compounds Used in Food-Producing Animals", typescript (July 1983).

Wainwright, R. B., W. L. Heyward, J. P. Middaugh, C. L. Hatheway, A. P. Harpster and T. R. Bender. "Food-Borne Botulism in Alaska, 1947–1985: Epidemiology and Clinical Findings", *J. Infect. Dis.* 157(6):1158–1162 (1988).

Wilson, J. "Foodborne Illness: The Problem That Just Won't Go Away", *Can. Nurse* 82(2):14–15 (1986).

Wong, H. D., M. H. Chang and J. Y. Fan. "Incidence and Characteriziation of *Bacillus cereus* Isolates Contaminating Dairy Products", *Appl. Environ. Microbiol.* 54(3)669–702 (1988).

Yamagishi, T., K. Sakamoto, S. Sakurai, K. Konishi, Y. Daimon, M. Matsuda, Y. Gyobu, Y. Kubo and H. Kodama. "A Nosocomial Outbreak of Food Poisoning Caused by Enterotoxigenic *Clostridium perfringens*", *Microbiol. Immunol.* 27(3):291–296 (1983).

Chapter 3

Alfin-Slater, R. B., and F. X. Pi-Sunyer. "Sugar and Sugar Substitutes. Comparisions and Indications", *Postgrad. Med.* 82(23):46–50 (1987).

Anderson, J. A. "The Establishment of Common Language Concerning Adverse Reactions to Foods and Food Additives", *J. Allergy Clin. Immunol.* 78(1):140–144 (1986).

Angus, D. S. "Comparative Mutagenicity of Two Triarylmethane Food Dyes in Salmonella, Saccharamyces, and Drosophila", *Food Chem. Toxicol.* 19(4):419–424 (1981).

Anliker, R. "Ecotoxicology of Dyestuffs—A Joint Effort by Industry", *Ecotoxicol. Environ. Safety* 3:59–74 (1979).

Betina, V. *Citrinin and Related Substances*, Vol. 8 (Amsterdam: Elsevier/North Holland Publishing Co., 1984), p. 217–236.

Chung, K. T., G. E. Fulk and A. W. Andrews. "Mutagenicity Testing of Some Commonly Used Dyes", *Appl. Environ. Microbiol.* 42(4):641–648 (1981).

Combes, R. D., and R. B. Haveland. "A Review of the Genotoxity of Food, Drug, and Cosmetic Colors and Other Azo, Triphenylmethane, and Xanthene Dyes", *Mutat. Res.* 98:101–248 (1982).

Cordle, M. K. "USDA Regulation of Residues in Meat and Poultry Products", *J. Anim. Sci.* 66(2):413–433 (1988).

Coye, M. J. "The Health Effects of Agricultural Production: II The Health of the Community", *Public Health Policy* 7(3):340–354 (1986).

Diamond, S., J. Prager and F. G. Freitag. "Diet and Headache. Is There a Link?", *Postgrad. Med.* 79(4):279–286 (1986).

Dixon, E. A., and G. Renyk. "Isolation, Separation, and Identification of Synthetic Food Colors", *J. Chem. Educ.* 59(1):67–69 (1982).

Ecobichon, D. J. "Pesticide Residues in Foods", *Proc. West. Pharmacol. Soc.* 29:499–502 (1986).

Freydbert, N., and W. Gortner. *The Food Additives Book* (New York: Bantam Books, 1982).

"Hearings Before the Subcommittee on Health and the Environment of the Committee on Interstate and Foreign Commerce House of Representatives", *Antibiotics in Animal Feed*, (Washington, DC: U.S. Government Printing Office, 1980).

Hecht, A., and J. Willis. "Sulfites: Preservatives That Can Go Wrong?", *FDA Consumer* 11:(1983).

Janssen, P. J., and C. A. van-der-Heijden. "Aspartame: Review of Recent Experimental and Observational Data", *Toxicology* 50(1):1–26 (1988).

Kropf, W., and M. Houben. "Harmful Food Additives", in *The Eat Safe Guide* (New York: Ashley Books, 1980).

Lecos, C. "Food Preservatives: A Fresh Report", *FDA Consumer* April (1984), p. 23–25.

Lehmann, P. "More Than You Ever Thought You Would Know About Food Additives", *FDA Consumer* Feb. (1981), p.1–4.

Leist, K. H. "Subacute Toxicity Studies of Selected Organic Colorants", *Ecotoxicol. Environ. Safety* 6:457–463 (1982).

Lener, J., and B. Bibr. "Cadmium Content in Some Foodstuffs in Respect of its Biological Effects", *Vitalstoffe Zivilisations-Krankheiten* 15(4):139–141 (1980).

London, R. "Saccharin and Aspartame. Are They Safe to Consume During Pregnancy?", *J. Reprod. Med.* 33(1):17–21 (1988).

Longer, K. "Lead Toxicity and Metabolism From Lead Sulfate Fed to Holstein Calves", *J. Dairy Sci.* 67(5):1007–1013 (1984).

Longstaff, E. "An Assessment and Categorization of the Animal Carcinogenicity Data on Selected Dyestuffs and An Extrapolation of those Data to an Evaluation of the Relative Carcinogenic Risk to Man", *Dyes Pigm.* 4:243–304 (1983).

Mahaffy, K. R. "Heavy Metal Exposure From Foods", *Environ. Health Perspec.* 12:63 (1983).

Marmion, D. M. *Handbook of U.S. Colorants for Foods, Drugs, and Cosmetics,* (New York: John Wiley & Sons, 1979).

Morbidity and Mortality Weekly Report "Evaluation of Consumer Complaints Related to Aspartame Use", 33:605–607 (1984).

National Academy of Sciences. *The Effects of Human Health of Subtherapeutic Use of Antimicrobials in Animal Feeds* (Washington, DC: U.S. Government Printing Office, 1980).

Richmond, M. H., and D. I. Smith. "The Use of Tetracycline in the Community and its Possible Relation to the Excretion of Tetracycline-Resistant Bacteria", *J. Antimicrobial Chemother.* 6:33–41 (1980).

Simon, R. A. "Adverse Reactions to Food Additives", *N. Eng. Reg. Allergy Proc.* 7(6):533–546 (1986).

Stegink, L. D. "The Aspartame Story: A Model for the Clinical Testing of a Food Additive", *Am. J. Clin. Nutr.* 46(Suppl 1):204–215 (1987).

Stobbaerts, R. F., L. VanHarverbeke and M. A. Herman. "Qualitative and Quantitative Determination of Some Yellow, Orange, and Red Food Dyes by Resonance Raman Spectroscopy", *J. Food Sci.* 48:521–525 (1983).

Uno, Y. "The Toxicology of Mycotoxins", *CRC Crit. Rev. Toxicol.* 14(2):99–132 (1985).

U.S. Department of Health and Human Services, Food and Drug Administration. "Policy for Regulating Carcinogenic Chemicals in Food and Color Additives; Advance Notice of Proposed Rulemaking", *Fed. Reg.* 47:14464 (1982).

U.S. Department of Health and Human Service, Food and Drug Administration. "Sponsored Compounds in Food-Producing Animals; Proposed Rule and Notice", *Fed. Reg.* 50:45530 (1985).

U.S. Department of Health and Human Services, Food and Drug Administration, Bureau of Foods. *Toxicologic Principles for the Safety Assessment of Direct Food Additive and Color Additives Used in Food* (Washington, DC: 1982).

Wagner, G. J. "Cadmium in Wheat Grain", *J. Toxicol. Environ. Health* 13(4–6):979–989 (1984).

Walton, J. R. "Antibiotic Residues in Meat", *Br. Vet. J.* 143(6):485–486 (1987).

Watanabe, H. K., and Y. Hasimoto. "Carcinogenicities of 3-Methoxy-4-Aminoazobenzene and Related Axo Dyes in the Mouse", *Gann (Japanese Cancer Assoc.)* 73(136):136–140 (1982).

Watson, D. H. "Toxic Fungal Metabolites in Food", *CRC Crit. Rev. Food Sci. Nutr.* 22(3):177–198 (1985).

Winter, R. *A Consumers Dictionary of Food Additives* (New York: Crown Publishing, 1982).

Wolnik, K. "Cadmium in Lettuce, Peanuts, Potatoes, Sweet Corn and Wheat", *J. Agric. Food Chem.* 31(6):120–124 (1983).

Wurtman, R. "Neurochemical Changes Following High-Dose Aspartame with Dietary Carbohydrates", *N. Engl. J. Med.* 309(7):429–430 (1983).

Chapter 4

Adamis, Z., A. Antal, I. Fuzesi, J. Molnar, L. Nagy and M. Susan. "Occupational Exposure to Organophosphorus Insecticides and Synthetic Pyrethroid", *Int. Arch. Occup. Environ. Health* 56(4):299–305 (1985).

Ali, A., H. N. Nigg, J. H. Stamper, M. L. Kok-Yokomi and M. Weaver. "Diflubenzuron Application to Citrus and its Impact on Invertebrates in an Adjacent Pond", *Bull. Environ. Contam. Toxicol.* 41(5):781–790 (1988).

Babu, C. J., K. N. Panicker and P. K. Das. "Breeding of *Aedes aegypti* in Closed Septic Tanks", *Indian J. Med. Res.* 77:637 (1983).

Beausoleil, E. G. "A Review of Present Antimalaria Activities in Africa", *Bull. WHO.* 62(Suppl):13–17 (1984).

Carlson, D. B., and P. D. O'Bryan. "Mosquito Production in a Rotationally Managed Impoundment Compared to Other Management Techniques", *J. Am. Mosq. Control Assoc.* 4(2):146–151 (1988).

Cherniack, M. G. "Organophosphorus Esters and Polyneuropathy", *Ann. Intern. Med.* 104(2):264–266 (1986).

D'Ambrosio, S. M. Chlorinated Hydrocarbons: Insecticide Versus Carcinogenic Action, (Cincinnati, OH: U.S. EPA, 1983).

Flannigan, S. A., S. B. Tucker, M. M. Key, C. E. Ross, E. J. Fairchild, B. A. Grimes and R. B. Harris. "Synthetic Pyrethroid Insecticies: A Dermatologist Evaluation", *Br. J. Ind. Med.* 42(6):363–372 (1985).

Guzman, D. R., and R. C. Axtell. "Temperature and Water Quality Effects in Simulated Woodland Pools on the Infection of Culex Mosquito Larvae by *Lagenidium giganteum* (Domycetes: Lagenidiales) in North Carolina", *J. Am. Mosq. Control Assoc.*

Hogdson, E. "Development of Safer Insecticides", *Drug. Metab. Rev.* 15(5–6):881–895 (1985).

Kumari, K., R. P. Singh, and S. K. Saxena. "Movement of Carbufuran (nematicide) in Soil Columns", *Ecotoxicol. Environ. Safety* 16(1):36–44 (1988).

Liebowitz, D. P., and J. A. Kriz. "Collection and Determination of Counter (terbufos) Insecticide in Air", *Am. Ind. Hyg. Assoc. J.* 44(8):567–571 (1983).

Maloney, S. E., A. Maule and A. R. Smith. "Microbial Transformation of the Pyrethroid Insecticides: Permethrin, Deltamethrin, Fastac, Fenvalerate, and Fluvalinate", *Appl. Environ. Microbiol.* 54(11):2874–2876 (1988).

Matsumura, F. *Toxicology of Insecticides*, 2nd ed. (New York: Plenum Press, 1985).

Mulla, M. S., and H. A. Darwazeh. "Efficacy of New Insect Growth Regulators Against Mosquito Larvae in Dairy, Wastewater Lagoon", *J. Am. Mosq. Control Assoc.* 4(3):322–325 (1988).

Pandita, T. K. "Mutagenic Studies on the Insecticide Metasystox-R With Different Genetic Systems", *Mutat. Res.* 124(1):971–972 (1983).

Parry, J. E. "Control of *Aedes triseriatus* in LaCrosse, Wisconsin", *Prog. Clin. Biol. Res.* 123:355–363 (1983).

Reeves, R. R., and R. O. Pendarvis. "A New Method for the Differentiation of Naphthalene and Paradichlorobenzene Mothballs", *J. Am. Ostepath. Assoc.* 85(12):806–808 (1985).

Schaefer, C. H., and E. F. Dupres, Jr. "Recycling Suitability of Wash Waters From Mosquito Abatement Vehicles and Equipment Into Spray Diluent", *J. Am. Mosq. Control Assoc.* 4(1):1–3 (1988).

Sharma, V. P., and K. N. Mehrotra. "Malaria Resurgence in India: A Critical Study", *Soc. Sci. Med.* 22(8):835–845 (1986).

Tucker, S. B., S. A. Flannigan and C. E. Ross. "Inhibition of Cutaneous Parethesia Resulting From Synthetic Pyrethroid Exposure", *Int. J. Dermatol.* 23(10):686–689 (1984).

U. S. Department of Health and Human Services. *Insecticides: For the Control of Insects of Public Health Importance* (Washington, DC: Department of Health, 1982).

U. S. Department of Health, Education, and Welfare. *Insecticide Application Equipment* (Atlanta: Public Health Service, 1984).

Warren, M., H. C. Spencer, F. C. Churchill, V. J. Francois, R. Hippolyte and M. A. Staiger. "Assessment of Exposure to Organophosphate Insecticides During Spraying in Haiti: Monitoring of Urinary Metabolites and Blood Cholinesterases Levels", *Bull. WHO* 63(2):353–360 (1985).

Walliker, D. "The Genetic Basis of Diversity in Malaria Parasites", *Adv. Parasitol.* 22:217–259 (1983).

Chapter 5

Jones, E. C., G. H. Grow and S. C. Naiman. "Prolonged Anticoagulation in Rat Poisoning", *Pr. Med. J. Clin. Res.* 287(6385):118–119 (1983).

Kimsey, S. W., T. E. Carpenter, M. Pappaioanou and E. Lusk. "Benefit Cost Analysis of Bubonic Plague Surveillance and Control at Two Campgrounds in California, USA", *JAMA* 252(21):3005–3007 (1984).

Lipton, R. A., and E. M. Klass. "Human Infeston of a 'Superwarfarin' Rodenticide Resulting in a Prolonged Anticoagulant Effect", *J. Med. Entomol.* 22(5):499–506 (1985).

Paynter O. E., U. S. Environmental Protection Agency, Office of Pesticides Programs, Hazard Evaluation Division, Standard Evaluation Procedure. *Oncogenicity Potential: Guidance for Analysis and Evaluation of Long-Term Rodent Studies,* EPA-540/9-85-019 (Washington, DC: U.S. Environmental Protection Agency, 1985).

Schwar, T. G., D. Thompson and B. C. Nelson. "Fleas on Roof Rats in Six Areas of Los Angeles County, California: Their Potential Role in the Transmission of Plague and Murine Typhus to Human", *Am. J. Trop. Med. Hyg.* 34(2):372–379 (1985).

Chapter 6

Adamis, Z., A. Antal, I. Fuzesi, J. Molnar, L. Nagy and M. Susan. "Occupational Exposure to Organophosphorus Insecticides and Synthetic Pyrethroid", *Int. Arch. Occup. Environ. Health* 56(4):399–305 (1985).

Balding, H. "Inert Pesticide Ingredients Toxic", *Tech. Rev.* 80:181 (1985).

Bull, D. *Pesticides and the Third World* (New York: Hayes & Pinkle, 1982).

Chaboussov, F. "How Pesticides Increase Pests", in *The Ecologist* (Wadebridge: Ecosystems Ltd., 1982).

Corning, M. C. "Pesticide Eating Bacteria", *Science News* 127:329 (1985).

Dindal, D. L., Ed. *Soil Biology as Related to Land Use Practices* (Washington, DC: U.S. EPA, WASA, DD, Office of Pesticides and Toxic Substances, 1980).

Dover, M. "Getting Off the Pesticide Treadmill", *Tech. Rev.* 88:111 (1985).

Edgerton, S. *Pesticides Analysis in Select Soil Samples* (New York: John Wiley & Sons, 1985).

Fleming, J. *Organochlorine Pesticides and PCB's: A Continuing Problem for the 1980's* (Washington, DC: U.S. Fish and Wildlife Service, 1983).

Gough, M. *Dioxin Agent Orange* (New York: Plenum Press, 1986).

Hayes, W. J. *Pesticides Studied in Man* (Baltimore: Williams & Wilkins, 1982).

Hudson, R. H. *Toxicology of Pesticides to Wildlife* (Washington: Resource Publication, 1984).

Immerman, F. W. National Urban Pesticide Usage (Cincinnati, OH: U.S. EPA, Economic Analysis Branch, Benefits and Use Division, 1984).

Krueger, R. F. and D. J. Severn. *Regulation of Pesticide Disposal* Vol. 295 (Washington, DC: American Chemical Soc., 1984), p. 1–15.

Li, W. "The Role of Pesticides in Skin Disease", *Int. J. Dermatol.* 25(5):295–297 (1986).

Marquis, J. K., *Contemporary Issues in Pesticides Toxicology and Pharmacology* (Basel: S. Karger, 1986).

McEwen, F. L., and G. R. Stephenson. *The Use and Significance of Pesticides in the Environment* (New York: John Wiley & Sons, 1980).

Morrill, L. G., et al. *Organic Compounds in Soil* (Ann Arbor, MI: Butterworth Ltd., 1982).

Ninety-Eighth Congress. "Government Regulation of the Pesticides Ethelene Dibromide, Joint Hearing" (Washington, DC: U.S. Government Printing Office, 1984).

Ninety-Eighth Congress. "Hearings of Pesticides Tolerance Legislation" (Washington, DC: U.S. Government Printing Office, 1984).

Ninety-Ninth Congress. "Regulation of Pesticide Residues" (Washington DC: U.S. Government Printing Office, 1986).

Norris, R. *Pills, Pesticides, and Profits* (New York: Pergamon Press, 1982).

Saltzman, S., Ed. *Pesticides in Soil* (New York: Van Nostrand Reinhold, 1986).

Schmitt, C. J. *National Pesticide Monitoring Program: Organochlorine Residues in Freshwater Fish* (Washington, DC: U.S. EPA, 1983).

Schoor, J. L. *Verification of a Toxic Organic Transport and Bioaccumulation Model* (Washington DC: U.S. EPA, Environmental Research, 1986).

Sims, R. *Contaminated Surface Soils* (Park Ridge, NJ: Noyes, 1986).

Smith, D. C. *Pesticides Residues in Soil* (New York: John Wiley & Sons, 1981).

Soren, J. *Aldicarb—Pesticide Contamination of Ground Water in Eastern Suffolk County, Long Island, New York* (Syosset, NY: U.S. Geological Survey, 1984).

U.S. Environmental Protection Agency. "Data Requirements for Pesticide Registration; Final Rule", *Fed. Reg.* 49:42856 (1984).

U.S. Environmental Protection Agency. "Inert Ingredients in Pesticide Products; Policy Statement", *Fed. Reg.* 52:13305 (1987).

U.S. Environmental Protection Agency, Office of Pesticides and Toxic Substances. *Pesticide Assessment Guidelines, Subdivision F. Hazard Evaluation: Human and Domestic Animals,* Series 83: Chronic and Long-Term Studies EPA 540/9-82-025 (Washington, DC: U.S. EPA, 1982).

U.S. Environmental Protection Agency. *Pesticides & Toxic Substances Monitoring Report* (Washington, DC: U.S. EPA, 1980).

U.S. Environmental Protection Agency. *The Federal Insecticide, Fungicide and Rodenticide Act as Amended* (Washington, DC: U.S. EPA, Office of Public Affairs, 1985).

Walsh, J. "EPA, OSHA Act to Curb Pesticide Ethylene Dibromide", *Science* 222:400 (1983).

WHO. *Field Studies in Exposure to Pesticides* (Washington DC: U.S. Printing Office, 1983).

Whorton, M. D. and D. E. Foliart. "Mutagenicity, Carcinogenicity and Reproductive Effects of Dibromochloropropane (DBCP)", *Mutat. Res.* 123(1):13–30 (1983).

Chapter 7

Abu-Jarad, F., and J. H. Fremlin. "Effect of Internal Wall Covers on Radon Emanation Inside Houses", *Health Physics* 44(3):243–248 (1983).

Alter, H. W., and R. A. Oswald. "Results of Indoor Radon Measurements Using the Track Etch Method", *Health Physics* 45(2):425–428 (1983).

Anderson, I. "Showers Pose Risk to Health", *New Scientist* 11:23 (1986).

Angle, C. R. "Indoor Air Pollutants", *Adv. Pediatr.* 35:239–280 (1988).

Bare, J. C. "Indoor Air Pollution Source Database", *J. Air Pollut. Control Assoc.* 38(5):670–671 (1988).

Ballenger, J. J. "Some Effects of Formaldehyde on the Upper Respiratory Tract", *Laryngoscope* 94(11, pt. 1):1411–1413 (1984).

Balmot, J. L. "Monitoring of Formaldehyde in Air", *Am. Industr. Hyg. J.* 46:574–584 (1985).

Bartnett, S. G. "Wood Stove Design and Control Mode as Determinants of Efficiency, Creosote Accumulation, and Condensable Particulate Emissions", *Residential Wood and Coal Combustion (J. Air Pollution Control Assoc.,* 1982).

Beck, B. "Prediction of Pulmonary Toxicity of Respirable Combustion Products From Residential Wood and Coal Stoves", *Residential Wood and Coal Combustion (J. Air Pollut. Control Assoc.,* 1982).

Boleij, J. S. and B. Brunekreff. "Indoor Air Pollution", *Public Health Rev.* 10(2):169–198 (1982).

Brown, R. "Home and Office: Shelter or Threat", *EPA J.* 13:2–5 (1987).

Brutsaert, W. F., S. A. Norton, C. T. Hess and J. S. Williams. "Geologic and Hydrologic Factors Controlling Radon-222 in Ground Water in Maine", *Ground Water* 19(4):407–417 (1981).

Cohen, B. L. "Health Effects of Radon From Insulation of Buildings", *Health Physics* (39):937–941 (1980).

Committee on Aldehydes *Formaldehyde and Other Aldehydes* (Washington, DC: National Academy Press, 1981).

Committee on Indoor Pollutants, Board of Toxicology and Environmental Health Hazards Assembly of Life Science National Research Council. *Indoor Pollutants* (Washington, DC: National Academy Press, 1981).

Committee on Science and Technology, U.S. House of Representatives 98th Congress. *Indoor Air Quality Research* (Washington, DC: U.S. Government Printing Office, 1984).

"EPA Restricts Wood Preservatives", *J. Environ. Health* 47(2) (1984).

Esmen, N. A. "The Status of Indoor Air Pollution", *Environ. Health Perspect.* 62:259–265 (1985).

Fleischer, R. L., A. Mogro-Campero and L. G. Turner. "Indoor Radon Levels in the Northeastern U.S.: Effects of Energy-Efficiency in Homes", *Health Physics* 45 (2):407–412 (1983).

Fleischer, R. L. "Theory of Passive Measurement of Radon Daughters and Working Levels by the Nuclear Track Technique", *Health Physics* 47(2):263–270 (1984).

Gammage, R. B., and S. V. Kaye. *Indoor Air and Human Health* (Chelsea, MI: Lewis Publishers, 1985).

George, A. C. "Passive Integrated Measurement of Indoor Radon Using Activated Carbon", *Health Phys.* 46(4):867–872 (1984).

George, A. C., E. O. Knutson and H. Franklin. "Radon and Radon Daughter Measurements in Solar Building", *Health Physics* 45(2):413–420 (1983).

Gesell, T. F. "Background Atmospheric ^{222}Rn Concentrations Outdoors and Indoors: A Review", *Health Physics* 45(2):289–302 (1983).

Godish, T. *Air Quality* (Chelsea, MI: Lewis Publishers, 1985).

Godish, T. "Mitigation of Residential Formaldehyde Contamination by Indoor Climate Control", *Am. Indust. Hyg. Assoc.* 47:792–797 (1986).

Greenfield, E. J. *House Dangerous* (New York: Vintage Books, 1987).

Harley, N. "Radon and Lung Cancer in Mines and Homes", *N. Engl. J. Med.* 310(23):1525–1527 (1984).

Hayden, A. C., and R. W. Braaten. "Effects of Firing Rate on Domestic Wood Stove Performance", in *Residential Wood and Coal Combustion* (*J. Air Pollut. Control Assoc.*, 1982).

Hendee, W. R., and T. C. Doege. "Origin and Health Risks of Indoor Radon", *Semin. Nucl. Med.* 18(1):3–9 (1988).

Humble, C. "Marriage to a Smoker and Lung Cancer Risk", *Am. J. Public Health* 77:598–602 (1987).

Ingersol, J. G. "A Survey of Radionuclide Contents and Radon Emanation Rates in Building Materials Used in the U.S.", *Health Physics* 45(2):363–368 (1983).

Johnson, C. F., and D. L. Coury. "Bruising and Hemophilia: Accident or Child Abuse?", *Child Abuse Negl.* 12(3):409–415 (1988).

Kazantzis, G. "Lead: Sources, Exposure and Possible Carcinogenicity", *IARC Sci. Publ.* (71):103–111 (1986).

Klein, G. L., and R. W. Ziering. "Environmental Control of the Home", *Clin. Rev. Allergy* 6(1):3–22 (1988).

Lais, M. "Characterization of Emissions from Building Materials, Building Occupants and Consumer Products: Laboratory Studies", *Atmosph. Environ.* 21(2):313–314 (1987).

Landman, K. A. "Diffusion of Radon Through Cracks in a Concrete Slab", *Health Physics* 43(1):65–71 (1982).

Marshall, E. "EPA Indicts Formaldehyde, 7 Years Later", *Science* 236:381 (1987).

McKone, T. "Human Exposure to Volatile Organic Compounds in Household Tap Water: The Indoor Inhalation Pathway", *Environ. Sci. Technol.* 21:1194–1207 (1987).

Meyer, B. "Diurnal Variations of Formaldehyde Exposure in Mobile Homes", *J. Environ. Health* 48:57–61 (1985).

Meyer, B. *Indoor Air Quality* (Reading MA: Addison-Wesley, 1983).

Morawska, L. "Influence of Sealants on 222 Radon Emanation Rate From Building Materials", *Health Physics* 44(4):416–418 (1983).

Munford, J. "Indoor Air Sampling and Mutagenicity Studies of Emissions from Unvented Coal Combustion", *Environ. Sci. Technol.* 21:308–311 (1987).

National Research Council. *Indoor Pollutants* (Washington, DC: National Academcy Press, 1981).

Nazaroff, W. W. "Experiments of Pollutant Transport from Soil into Residential Basements by Pressure-Driven Airflow", *Environ. Sci. Technol.* 21:459–466 (1987).

Nazazrorr, W. "Mathematical Modeling of Chemically Reactive Pollutants in Indoor Air", *Environ. Sci. Technol.* 20:924–934 (1986).

Nero, A. V. "Airborne Radionuclides and Radiation in Buildings: A Review", *Health Physics* 45(2):303–322 (1983).

Nero, A. V., Jr. "Controlling Indoor Air Pollution", *Sci. Am.* 258:42–48 (1988).

Nero, A. V. "Distribution of Airborne Radon 222 Concentration of U.S. Homes", *Science* 234:992–997 (1986).

Nero, A. V. "Indoor Radiation Exposure From 222 and Its Daughters: A View of the Issue", *Health Physics* 45(2):277–288 (1983).

Nero, A. V., et al. "Radon Concentrations and Infiltration Rates Measured in Conventional and Energy-Efficient Houses", *Health Physics* 45(2):401–405 (1983).

Nishemera, H. "Reduction of NO^2 and NO by Rush and Other Plants", *Environ. Sci. Technol.* 20:413–416 (1986).

Prichard, H. M., T. F. Gesell, C. T. Hess, et al. "Integrated Radon Data From Dwellings in Maine and Texas", *Health Physics* 45(2):428–432 (1983).

Raloff, J. "Toxic Showers and Bath", *Science News* 130:190 (1986).

Reasor, J. "The Composition and Dynamics of Environmental Tobacco Smoke", *J. Environ. Health* 50:109 (1987).

Repace, J. L., and A. H. Lowrey, "Indoor Air Pollution, Tobacco Smoke, and Public Health", *Science* 208:464–472 (1980).

Samet, J. M., M. C. Marbury and J. D. Spengler. "Respiratory Effects of Indoor Air Pollution", *J. Allergy Clin. Immunol.* 79(5):685–700 (1987).

Samson, R. A. "Occurrence of Moulds in Modern Living and Working Environments", *Eur. J. Epidemiol.* 1(1):54–61 (1985).

Schneider, C. "The Indoor Air Pollution Burden", *EPA J.* 12:14–15 (1986).

Shephard, R. J. *Carbon Monoxide: The Silent Killer* (Springfield, IL: Charles C Thomas, 1983).

Smith, H. "Lung Cancer Risk From Indoor Exposure to Radon Daughters", *Radiology* 167:580 (1958).

Spengler, J. D. *Indoor Pollutants* (Washington, DC: National Academy Press, 1981).

Straden, E. "Radon in Dwellings and Lung Cancer—A Discussion", *Health Physics* 38:301–306 (1980).

Straden, E., and L. Berteig. "Radon in Dwellings and Influencing Factors", *Health Physics* 39:275–284 (1980).

Steinhausler, F., et al. "Radiation Exposure of the Respiratory Tract and Associated Carcinogenic Risk Due to Inhaled Radon Daughters", *Health Physics* 45(2):331–337 (1983).

Stowlwijk, J. *Health Effects of Indoor Air Contaminants* (Wilson, Inc., 1986).

Tartaglia, M., S. R. DiNardi and J. Ludwig. "Radon and Its Progeny in the Indoor Environment", *J. Environ. Health* 47(2):62–67(1984).

Traynor, G. "Indoor Air Pollution Due to Emissions From Wood Burning Stove", *Environ. Sci. Technol.* 21:691–697 (1987).

Turiel, I. *Indoor Air Quality and Human Health* (Stanford, CA: Stanford University Press, 1985).

U.S. Consumer Product Safety Commission and EPA. "Asbestos in the Home" (1986).

U.S. Consumer Product Safety Commission. "Labeling Asbestos-Containing Household Products; Enforcement Policy", *Fed. Reg.* 51-33910 (1986).

U.S. Consumer Product Safety Commission. no. 13, *"Carbon Monoxide"* (Sept. 1980).

U.S. Consumer Product Safety Commission. no. 97, *"Kerosene Heater"* (Dec. 1982).

U.S. Consumer Product Safety Commission. no. 34, *"Space Heaters"* (Nov. 1983).

U.S. Consumer Product Safety Commission. *Questions and Answers on Urea-Formalde-hyde Foam Insulation* (July 22, 1983).

U.S. Consumer Product Safety Commission. *Unvented Gas Space Heaters* (Oct. 10, 1980).

U.S. EPA. OPA 86–005, "Radiation Reduction Methods" (Aug. 1986).

U.S. EPA. OPA-86–005, "Radon Reduction Methods" (Aug. 1986).

U.S. EPA .520/1–86–09. "Interim Indoor Radon and Radon Decay Product Measurement Protocols" (April 1986).

U.S. EPA. Residential Wood Combustion Study, 1980–1982, Executive Summary.

U.S. EPA. "Wood Stove Features and Operation Guideline for Cleaner Air", EPA 600/D-83–12.

Wadden, R. A., and P. A. Schaff. *Indoor Air Pollution* (New York: John Wiley & Sons, 1983).

Wallance, L. "Surprising Results From a New Way of Measuring Pollutants", *EPA J.* 13:15–16 (1987).

Walsh, P. J. *Indoor Air Quality* (Boca Raton, FL: CRC Press, 1984).

Zirschy, J., and L. Witherell. "Cleanup of Mercury Contamination of Thermometer Workers' Homes", *Am. Ind. Hyd. Assoc. J.* 48(1):81–84 (1987).

Chapter 8

Ackerman, J. "Monitoring Waste Nitrous Oxide. One Medical Center's Experience", *AORN J.* 41(5):895,897,898 (1985).

Altree, W. S., and J. S. Preston. "Asbestos and Other Fibre Levels in Buildings", *Ann. Occup. Hyg.* 29(3):357–363 (1985).

American Dental Association. *Guideline for Infection Control in the Dental Office and the Commercial Laboratory* (American Dental Assocciation, 1985).

Angle, C. R. "Indoor Air Pollutants", *Adv. Pediatr* 35:239–280 (1988).

Bagga, B. S. R., R. A. Murphy, A. W. Anderson and I. Punwani. *Contamination of Dental Unit Cooling Water With Oral Microorganisms and Its Prevention* (American Dental Association, 1984).

Bardana, E. J., Jr., A. Montanaro and M. T. O'Hollaren. "Building-Related Illness. A Review of Available Scientific Data", *Clin. Rev. Allergy* 6(1):61–89 (1988).

Beebe, G. W. "Ionizing Radiation and Health", *Am. Sci.* 70:35–44 (1982).

Belger, A. "The Determination of Ambient Air Quality Within an Environmental Care Unit", *J. Environ. Health* 49:2880–2893 (1987).

Bennett, L. C., and S. Searl. *Communicable Disease Handbook* (New York: John Wiley & Sons, 1982).

Bennett, N. M. "Disposal of Medical Waste", *Med. J. Aust.* 149(8):400–402 (1988).

Birnbaum, D. "Analysis of Hospital Infection Surveillance Data", *Infect. Control* 5(7):332–338 (1984).

Block, S. S. *Disinfection, Sterilization, and Preservation,* 3rd ed. (Philadelphia: Lea and Febiger, 1983).

Campos, L. L. "Measurement of the Exposure Rate Due to Low Energy X-rays Emitted From Video Display Terminals", *Int. J. Rad. Appl. Instrum. A* 39(2):173–174 (1988).

Cotton, D. J. "The Impact of AIDS on the Medical Care System", *JAMA* 260(4):519–523 (1988).

Cottone, J. A., E. W. Mitchell and C. H. Baker. *Proceeding of the National Symposium on Hepatitis B and the Dental Profession* (American Dental Association 1985).

Council on Dental Therapeutics. *Accepted Dental Therapeutics,* 40th ed. (Chicago: American Dental Association, 1984).

Council on Dental Therapeutics, Council on Prosthetic Services and Dental Laboratory Relations. "Guidelines for Infection Control in the Dental Office and the Commercial Dental Laboratory", *JADA* 110 (1985).

Deaderer, B. P. "Characterization of Contaminant Emissions From Indoor Sources", *Atmosph. Environ.* 21(2):279–462 (1987).

Decke, J. L. "Herpes", *Clinical Center Office of Clinical Reports and Inquiries* (Bethesda, MD: 1985).

Donnell, H. D., J. R. Bagby, R. G. Harmon, J. R. Brellin, H. C. Chaski, M. F. Bright, M. Van-Tuinen and R. W. Metzger. "Report of an Illness Outbreak at the Harry S. Truman State Office Building", *Am. J. Epidemiol.* 129(3):550–558 (1989).

Doxsey, K. "Toxic Substances in the Hospital Environment", *J. Nurs. Staff. Dev.* 3(1):41–42 (1987).

Ember, L. "Toxic Chemicals Levels Higher Indoors Than Out", *Chem. Engin. News* 63:22 (1985).

Feldman, R. G., and G. L. Ridgway. "Database Handling for Infection Control and Hospital Epidemiology", *J. Hosp. Infect.* 11(Suppl.):37–42 (1988).

Fiehn, N. E., and K. Henriksen. "Methods of Disinfection of the Water System of Dental Units by Water Chlorination", *J. Dent. Res.* 57(12):1499–1504 (1988).

Fischer, P., and Brunekreff. *Indoor NO^2 Pollution and Personal Exposure* (Dordrecht: D. Reidel Publishing Company, 1981).

Friedman, H. M., M. R. Lewis, D. M. Nemerofsky and S. A. Plotkin. "Acquisition of Cytomegalovirus Infection Among Female Employees at a Pediatric Hospital", *Pediatr. Infect. Dis.* 3(3):233–235 (1984).

Gammage, S. V. *Indoor Air and Human Health* (Chelsea, MI: Lewis Publishers, 1984).

Garner, J. S., and B. P. Simmons. *CDC Guideline for Isolation Precautions in Hospitals* (Atlanta, GA: Centers for Disease Control, 1983).

Gofman, J. W. "Ionizing Radiation: Concepts for the Dental Assistant", *Dental Assistant* 51:17–22 (1982).

Guidotti, T. L. "Quantitative Risk Assessment of Exposure to Airborne Asbestos in an Office Building", *Can. J. Public Health* 79(4):249–254 (1988).

Haley, R. "Infection Control Strategies Save $250,000 Annually", *Hospitals* 59(22):63–65 (1985).

Haley, R. W., D. H. Culver, J. W. White, W. M. Morgan and T. G. Emori. "The Nationwide Nosocomial Infection Rate. A New Need for Vital Statistics", *Am. J. Epidemiol.* 121(2):159–167 (1985).

Haley, R. W., W. M. Morgan, D. H. Culver, J. W. White, T. G. Emori, J. Mosser and J. M. Hughes. "Update From the SENIC Project. Hospital Infection Control: Recent Progress and Opportunities Under Prospective Payment", *Am. J. Infect. Control* 13(3):971–978 (1985).

Harris, A. A., S. Levin and G. M. Trenholme. "Selected Aspects of Nosocomial Infection in the 1980's", *Am. J. Med.* 77(1B):3–10 (1984).

Hoffman, P. "Disinfection in Hospitals", *Nursing-Lond*, 3(3):106–108 (1986).

Holthaus, D. "Right-to-Know Rules Cover Hospital Employees", *Hospitals* 61(20):58 (1987).

Hughes, R. T., and D. M. O'Brien. "Evaluation of Building Ventilation Systems", *Am. Ind. Hyg. Assoc. J.* 47(4):207–213 (1986).

Jacobson, J. T., D. S. Johnson, C. A. Ross, M. T. Conti, R. S. Evans and J. P. Burke "Adapting Disease-Specific Isolation Guidelines to a Hospital Information System", *Infect. Control* 7(8):411–418 (1986).

Kneedler, J. A., and G. H. Dodge. *Perioperative Patient Care* (Boston: Blackwell Scientific, 1983).

Lewey, R. "Visits to a Hospital-Based Employee Health Service", *J. Occup. Med.* 28(3):241–242 (1986).

MacNeil, J. M., and G. M. Weisz. "Critical Care Nursing Stress: Another Look", *Heart-Lung* 16(3):274–277 (1987).

Mallison, G. F. "A Critical Review of Draft Manual For Infectious Waste Management", *Am. J. Infect. Control* 13(1):45–46 (1985).

Manzella, J. P., J. H. McConville, W. Valenti, M. A. Menegus, E. M. Swierkosz and M. Arens. "An Outbreak of Herpes Simples Virus Type I Gingivostomatitis in a Dental Hygiene Practice", *JAMA* (1984).

Marsh, L. A. "Quality Assurance Activities in a Small Community Hospital", *Quality Rev. Bull.* 9(3):77–80 (1983).

Mattia, M. A. "Hazards in the Hosptial Environment. The Sterilants: Ethylene Oxide and Formaldehyde", *Am. J. Nurs.* 83(2):240–243 (1983).

Mattila, M. K. "Job Load and Hazard Analysis: A Method for the Analysis of Workplace Conditions for Occupational Health Care", *Br. J. Ind. Med.* 41(1):656–666 (1985).

McCunney, R. J. "The Role of Building Construction and Ventilation in Indoor Air Pollution. Review of a Recurring Problem", *N.Y. State J. Med.* 87(4):203–209 (1987).

Michaels, D., D. Nagin and S. Zoloth. "Occupational Hazards in the Dental Office", *N.Y. State Dent. J.* 50(5):294–296 (1984).

Morgan, M. S., and S.W. Horstman. "Systematic Investigation Needed for Tight Building Syndrome Complaints", *Occup. Health Saf.* 56(5):79–82,85–88 (1987).

Morse, D. L., M. A. Gordon, T. Matte and G. Eadie. "An Outbreak of Histoplasmosis in a Prison", *Am. J. Epidemiol.* 122(2):253–261 (1985).

Nero, A. V. "Airborne Radionuclides and Radition in Buildings: A Review", *Health Phys.* 45(2):303–322 (1983).

Neu, H. C. "Unusual Nosocomial Infection", *Dis. Mon.* 30(13):1–68 (1984).

Occupational Safety and Health Administration. *Risk of Hepatitis B Infection for Workers in Health Care Delivery System and Suggested Methods for Risk Reduction* (Washington DC: U.S. Department of Labor, 1983).

Palmer, S. R., and B. Rowe. "Investigation of Outbreaks of Salmonella in Hospitals", *Br. Med. J. Clin. Res.* 287(6396):891–893 (1983).

Paz J. D., R. Milliken, W. T. Ingram, A. Frank and A. Atkin. "Potential Ocular Damage From Microwave Exposure During Electrosurgery: Dosimetric Survey", *J. Occup. Med.* 29(7):580–583 (1987).

Reed, C. M., T. L. Gusafson, J. Siegel and P. Duer. "Nosocomial Transmission of Hepatitis A From a Hospital-Acquired Case", *Pediatr. Infect. Dis.* 3(4):300–303 (1984).

Ringham, S., M. Worsley and J. Archer. "Journal of Infection Control Nursing. Building Standards", *Nurs. Times* 82(38):97–98 (1986).

Robinson, G. V., B. R. Tegtmeier and J. A. Zaia. "Brief Report: Nosocomial Infection Rates in a Cancer Treatment Center", *Infect. Control* 5(6):289–294 (1984).

Runnells, R. R. *Infection Control in the West Finger Environment* (Salt Lake City: Publishers Press, 1984).

Rutala, W. A., and F. A. Sarubbi. "Management of Infectious Waste From Hospitals", *Infect. Control* 4(4):198–204 (1983).

Rulander, R., and P. Haglind. "Airborne Endotoxins and Humidifier Disease", *Clin. Allergy* 14(1):109–112 (1984).

Samet, J. M., M. C. Marbury and J. D. Spenger. "Health Effects and Sources of Indoor Air Pollution, Part II", *Am. Rev. Respir. Dis.* 137(1)221–242 (1988).

Sanford, J. P., and J. P. Luby. *Infectious Diseases* (New York: Grune & Stratton, 1982).

Sawyer, D. L., and B. L. Braddock. "Infection Control: Consultation Provides New Frontier", *Am. J. Infect. Control* 14(5)229–233 (1986).

"Sick Building Syndrome Affects People Too", *Environment* 29:23 (1987).

Simpson, R. "Priorities For Hospital Cleaning, Disinfection, Sterilization, and Control of Infection", *Br. Med. J. Clin. Res.* 288(6434):1898–1900 (1984).

Standfast, S. J., P. B. Michelsen, A. L. Baltch, R. P. Smith, E. K. Latham, A. B. Spellacy, R. A. Venezia and M. H. Andritz. "A Prevalence Survey of Infections in a Combined Acute and Long-Term Care Hospital", *Infect. Control* 5(4):177–184 (1984).

Stellman, J. M., S. Klitzman, G. C. Gordon and B. R. Snow. "Air Quality and Ergonomics in the Office: Survey Results and Methodologic Issues", *Am. Ind. Hyg. Assoc. J.* 45(5):286–293 (1985).

Sterling, T. D., E. Sterling and W. Dimich. "Building Illness in the White-Collar Workplace", *Int. J. Health Serv.* 13(2):277–287 (1983).

SyWassink, J. M., and L. I. Lutwick. "Risk of Hepatitis B in Dental Care Providers: A Contact Study", *JADA* 106 (1983).

Tyzack, R. "The Management of Methicillin-Resistant *Staphylococcus Aureus* in a Major Hospital", *J. Hosp. Infect.* 6(Suppl. A):195–199 (1985).

U.S. Department of Labor, OSHA Office of Occupational Medicine. OSHA Instruction CPL 2-2.36, (Washington, DC: Nov 30, 1983).

Weinstein, R. A. "Resistant Bacteria and Infection Control in the Nursing Home and Hospital", *Bull. N.Y. Acad. Med.* 63(3):337–344 (1987).

Welliver, R. C., and S. McLaughlin. "Unique Epidemiology of Nosocomial Infection in a Children's Hospital", *Am. J. Dis. Child.* 138(2):131–135 (1984).

Welty, C., S. Burstin, S. Muspratt and I. B. Tager. "Epidemiology of Tuberculous Infection in a Chronic Care Population", *Am. Rev. Resp. Dis.* 132(1):133–136 (1985).

West, K. H. "Infection Control. Disposal of Hazardous Waste: Implications for Nurses", *J. Oper. Room. Res. Inst.* 3(2):5–7 (1983).

Wilkinson, W. E. "Occupational Injury at a Midwestern Health Science Center and Teaching Hospital", *AAOHN J.* 35(8):367–376 (1987).

Wilson, S. J., and H. J. Wilson. "Mercury Vapour Levels in a Dental Hospital Environment", *Br. Dent. J.* 159(7):233–234 (1985).

Yaffe Y., D. Jenkins, H. Mahon-Haft, W. Winkelstein, C. P. Flessel and J. J. Wesolowski. "Epidemiological Monitoring of Environmental Lead Exposures in California State Hospital", *Sci. Total. Environ.* 32(3):261–275 (1984).

Zautra, A. J., C. Eblen and K. D. Reynold. "Job Stress and Task Interest: Two Factors in Work Life Quality", *Am. J. Community Psychol.* 14(4):377–393 (1986).

Chapter 9

Ahonen, E., and U. Nousiainen. "The Sauna and Body Fluid Balance", *Ann. Clin. Res.* 20(4):257–261 (1988).

Cuddihy, R. G., W. C. Griffith and R. O. McClellan. "Potential Health and Environmental Effects of Light-Duty Vehicles", *Toxicology Rep.* LMF 89:3–25 (1981) .

Godsey, M. S., T. E. Amundson, E. C. Burgess, W. Schell, J. P. David, R. Kaslow and R. Eldeman. "Lyme Disease Ecology in Wisconsin: Distribution and Host Preference of *Ixodes dammini*, and Prevalence of Antibody to *Borrelia burgdorferi* in Small Mammals", *Am. J. Trop. Med. Hyg.* 37(1):180–187 (1987).

Gong, H. J., F. Bedi and S. M. Horvath. "Inhaled Albuterol Does Not Protect Against Ozone Toxicity in Nonasthmatic Athletes", *Arch. Environ. Health* 43(1):46–53 (1988).

Gregory, D. W., and W. Schaffner. "Pseudomonas Infections Associated with Hot Tubs and Other Environments", *Infect. Dis. Clin. North Am.* 1(3):635–648 (1987).

Grobler, S. R., L. S. Maresky and R. J. Rossouw. "Blood Lead Levels of South African Long-Distance Road-Runners", *Arch. Environ. Health* 41(3):155–158 (1986).

Holtan, N. R. "Giardiasis. A Crimp in the Life-Style of Campers, Travelers, and Others", *Postgrad. Med.* 83(5):54–56, 59–61 (1988).

Kauppihen, K., and I. Vuori. "Man in the Sauna", *Ann. Clin. Res.* 18(4):173–185 (1986).

Linn, W. S., E. L. Avol, D. A. Shamoo, R. C. Peng, C. E. Spier, M. N. Smith and J. D. Hackney. "Effect of Metaproterenol Sulfate on Mild Asthmatics' Response to Sulfur Dioxide Exposure and Exercise", *Arch. Environ. Health* 43(6):399–406 (1988).

Marwick, C. "Olympic Athletes May Face Extra Challenge — Pollution [news]", *JAMA* 251(19):2495–2497 (1984).

Morrison, A. B. "Recreational Water Quality and Human Health", *Can. J. Public Health* 75(1):13–14 (1984).

Shephard, R. J. "Athletic Performance and Urban Air Pollution", *Can. Med. Assoc. J.* 131(2):105–109 (1984).

Solomon, S. L. "Host Factors in Whirlpool-Associated *Pseudomonas aeruginosa* Skin Disease", *Infect. Control* 6(10):402–406 (1985).

Tunnicliff, B., and S. K. Brickler. "Recreational Water Quality Analyses of the Colorado River Corridor in Grand Canyon", *Appl. Environ. Microbiol.* 48(5):909–917 (1984).

Chapter 10

Abrams, H. K. "Credibility in the TLV Process [letter]", *Am. J. Ind. Med.* 13(5):609–610 (1988).

"Acute Occupational Exposure to Sulfur Dioxide—Missouri", *MMWR* 32(41):541–542 (1983).

Alessio, L. "Relationship Between 'Chelatable Lead' and the Indicator of Exposure and Effect in Current and Past Occupational Exposure", *Sci. Total Environ.* 71(3):292–299 (1988).

Akesson, B., M. Bengtsson and I. Floren. "Visual Disturbances After Industrial Triethyl-amine Exposure", *Int. Arch. Occup. Environ. Health* 57(4):297–302 (1986).

Allison, W. W. "Industrial Hygiene [letter]", *Am. Ind. Hyg. Assoc. J.* 48(4):A244, A246 (1987).

Angerer, J. "Occupational Chronic Exposure to Organic Solvents", *Int. Arch. Occup. Environ. Health* 56(4):307–321 (1985).

Ashford, N. A. "Policy Considerations for Human Monitoring in the Workplace", *J. Occup. Med.* 28(8):563–568 (1986).

Atuhaire, L. K., J. J. Campbell and M. Jones. "Specific Causes of Death in Miners and Ex-miners of the Rhondda Fach 1950–80", *Br. J. Ind. Med.* 43:497–499 (1986).

Austin, W. B., and C. F. Phillips. "Development and Implementation of a Health Surveil-lance System", *Am. Ind. Hyg. Assoc. J.* 44(9):6538–6542 (1983).

Baird, D. D. "Occupational Exposure to Heat or Noise and Reduced Fertility [letter]", *JAMA* 253(18):2643–2644 (1985).

Baker, D. G. "Radiology, Is There an Occupational Hazard?", *Am. Ind. Hyg. Assoc. J.* 49(1):17–20 (1988).

Baloch, U. K. "Problems Associated With the Use of Chemicals by Agricultural Workers", *Basic Life Sci.* 34:63–78 (1985).

Baram, M. S. "The Right to Know and the Duty to Disclose Hazard Information", *Am. J. Public Health* 74(4):385–390 (1984).

Becklake, M. R. "Use of Information: Environmental Standards, Assessments of Risk, Prevention and Clinical Implication", *Med. J. Aust.* 143(2):63–64 (1985).

Blackwell, R. P. "The Effects of Low-Level Radiofrequency and Microwave Radiation on Brain Tissue and Animal Behavior", *Int. J. Radiat. Biol. Relat. Stud. Phys., Chem. Med.* 50(5):761–787 (1986).

Blot, W. J., and J. F. Fraumeni. "Cancer Among Shipyard Workers", *Branbury Rep.* 9:37–49 (1981).

Borak, T. B. "Radomization of Grab-Sampling Strategies for Estimating the Annual Exposure of U Miners to Rn Daughters", *Health Phys.* 50(4):465–472 (1986).

Bransford, J. S. "Proliferation of New Chemicals Increasing Need for Databases", *Occup. Health Saf.* 53(4):32–35 (1984).

Brewster, M. A. "Biomarkers of Xenobiotic Exposures", *Ann. Clin. Lab. Sci.* 18(4):306–317 (1988).

Brix, K. "Environmental and Occupational Hazards to the Fetus", *Medicine* 27(9):577–583 (1982).

Bunn, W. B. "Right-to-Know Laws and Evaluation of Toxicologic Data", *Ann. Intern. Med.* 103(6 pt 1):947–949 (1985).

Burgess, J. E. "Occupational Health in Agriculture", *Curr. Approaches Occup. Health* 2:62–85 (1982).

Burgess, W. A. *Recognition of Health Hazards in Industry* (New York: Wiley-Interscience Publication, 1981).

Burton, J. *Industrial Ventilation* (Ive Inc. 1986).

Cabrera, I. T. *Encyclopedia of Occupational Health and Safety* (Washington, DC: Government Printing Office, 1983), p. 1712–1715.

Castleman, B. I., and V. Navarro. "International Mobility of Hazardous Products, Industries, and Wastes", *Annu. Rev. Public Health* 8:1–19 (1987).

Castren, J. "On the Microwave Exposure", *Acta Ophthalmol.* 60(4):647–654 (1982).

Checkoway, H. "Methods of Treatment of Exposure Data in Occupational Epidemiology", *Med. Lav.* 77(1):48–73 (1986).

Cheremisinoff, P. N., and F. Ellerbusch. *Guide for Industrial Noise Control* (Ann Arbor, MI: Ann Arbor Science Publishers, 1982).

Christenson, W. N., C. F. Brennan and J. DeVito. "Even 'Healthy' Environments Can Harbor Many Hidden Hazards", *Occup. Health Saf.* 54(3):40–42, 46 (1985).

Corn, M. "Assessment and Control of Environmental Exposure", *J. Allergy Clin. Immunol.* 72(3):231–241 (1983).

Costa, M. "Principles of Industrial Toxicology", *Clin. Podiatr. Med. Surg.* 4(3):559–570 (1987).

Cox, C., W. E. Murray and E. P. Foley. "Occupational Exposures to Radiofrequency Radiation", *Am. Industr. Hyg. Assoc. J.* 43:149–153 (1982).

Cullen, A. P. "Industrial Nonionizing Radiation and Contact Lenses", *Can. J. Pub. Health* 73:251–254 (1982).

Davies, J. E. "A Global Need: Farm Worker Safety", *Am. J. Ind. Med.* 13(6):725–729 (1988).

Delaine, J. "Lead Oxide Emission—A Case History", *Ann. Occup. Hyg.* 30(2):257–261 (1986).

DeSantis, V., and I. Londo. "Health Effects of Airborne Effluents Released From the Nuclear Cycle", *Environ. Res.* 37(2):300–312 (1985).

Division of Labor Statistics & Research. Occupational Injuries and Illnesses Survey (San Francisco, CA: 1984).

Dooms, A. E. Goossens, K. M. Debusschere, D. M. Gevers, K. M. Dupre, H. J. Degreef, J. P. Loncke and J. E. Snauwaert. "Contact Dermatitis Caused by Airborne Agents. A Review and Case Reports", *J. Am. Acad. Dermatol.* 15(1):1–10 (1986).

Dreyer, N., and E. Friedlander. "Identifying the Health Risks From Very Low-Dose Sparsely Ionizing Radiation", *Am. J. Pub. Health* 72:585–587 (1982).

Drury, C. G. "Injury Potential of Industrial Jobs", *Semin. Occup. Med.* 2:41–49 (1987).

Dye, and E. White. "Environmental Hazards in the Work Setting: Their Effect on Women of Child-Bearing Age", *AAOHN J.* 34(2):76–78 (1986).

Dukes, F. N., Dobos and A. Henschel. *Recommended Heat Stress Standards* (Washington DC: Government Printing Office, 1980).

Ecker, M. D., and N. J. Bramesco. *Radiation* (New York: Vintage Books, 1981).

Edgerton, T. R. *Method Development for the Assessment of Possible Human Exposure to Pesticides and Industrial Chemicals* (Research Triangle Park, NC: U.S. EPA, 1981).

Eisenbud, M. *Environmental Radioactivity from Natural Industrial and Military Sources*, 3rd ed. (Troy, MO: Academic Press, 1987).

Fischbein, A., and Y. Lerman. "Environmental and Occupational Factors in General Medical Practices", *Mt. Sinai J. Med. N.Y.* 50(6):468–475 (1983).

Fletcher, R. A. "A Review of Persoanl Portable Monitors and Samplers for Airborne Particles", *J. Air Pollut. Control Assoc.* 34(10):1014–1016 (1984).

Flodin, U., M. Fredribsson, B. Persson, L. Hardell and O. Axelson. "Background Radiation, Electrical Work, and Some Other Exposures Associated with Acute Myeloid Leukemia in a Case", *Arch. Environ. Health* 41:77–84 (1986).

Fomenko, B. S. "Effect of Microwave Radiation (340 and 900 MHz) on Different Structural Levels of Erthrocyte Membranes", *Bioelectromagnetics* 6(3):305–312 (1985).

Frazier, T. "NIOSH Occupational Health and Hazard Surveillance Systems", *J. Toxicol. Clin. Toxicol.* 21(2):201–209 (1983–1984).

Gardner, W. A. *Current Approaches to Occupational Health 2* (London: Weight, PSG, 1982).

Goldman, M. "Ionizing Radiation and its Risks", *West. J. Med.* 137:540–547 (1982).

Goldman, R. H. "General Occupational Health History and Examinations", *J. Occup. Med.* 28(10):967–974 (1986).

Gramicidin, A. "Effects of Ionizing Radiation on Artificial Lipid Membranes", *Int. J. Radiat. Biol. Rel. Studies in Phy. Chem. Med.* 51:265–286 (1987).

Grandjean, P. "Indirect Exposures: The Significance of Bystanders at Work and at Home", *Ame. Ind. Hyg. Assoc.* 47:819–824 (1986).

Grazer, R. E., and H. W. Meislin. "A Nine-Year Evaluation of Emergency Department Personnel Exposure to Ionizing Radiation", *Ann. Emerg. Med.* 16:340–342 (1987).

Green, J. "Detecting the Hypersusceptible Workers; Genetics and Politics in Industrial Medicine", *Int. J. Health Serv.* 13(2):247–264 (1983).

Gun, R. A. "National Institute of Occupational Health and Safety", *Community Health Stud.* 8(3 Suppl):3–7 (1984).

Halder, C. A., G. S. VanGorp, N.S . Hatoum and T. M. Warne. "Gasoline Vapor Exposures, Part I, Characteriziaton of Workplace Exposures", *Am. Ind. Hyg. Assoc. J.* 47(3):164–172 (1986).

Halperin, W. E., and T. M. Frazier. "Surveillance for the Effects of Workplace Exposure", *Annu. Rev. Public Health* 6:419–432 (1985).

Hamburger, S. "Occupational Exposure to Nonionizing Radiation and an Association with Heart Disease", *J. Chron. Dis.* 36:791–802 (1983).

Harm, W. *Biological Effects of Ultraviolet Radiation* (New Rochelle, NY: Cambridge University Press, 1980).

Hart, D. L., and M. Jaraidei. "Effect of Lumbar Posture on Lifting", *Spine* 12:138–145 (1987).

Hema, P., K. K. Pushpavathi and P. P. Reddy. "Cytogenetic Damage in Lymphocytes of Rubber Industry Workers", *Env. Res.* 40(1):199–201 (1986).

Henderson, J., H. W. Baker and P. J. Hanna. "Occupation-Related Male Infertility: A Review", *Clin. Reprod. Fertil.* 4(2):87–106 (1986).

Hughes, D. "Ionizing Radiation: A Guide to Regulations", *Occup. Health* 38:362–366 1986.

Hutton, W. C. "Diurnal Variations in Stresses on the Lumbar Spine", *Spine* 12:130–137 (1987).

Inoue, O., K. Seiji, M. Kashara, H. Nakatsuka, T. Watanabe, S. G. Yin, G. L. Li, C. Jin, S. X. Cai, X. Z. Wang, et al. "Quantitative Relation of Urinary Phenol Levels to Breathzone Benzene Concentratons: A Factory Survey", *Br. J. Ind. Med.* 42(1):692–697 (1986).

International Labour Organization. *Occupational Exposure Limits For Airborne Toxic Substances*, 2nd ed. (Geneva: International Labour Office, 1980).

Irwin, J. D., and E. R. Graff. *Industrial Noise and Vibration Control* (Englewood Cliffs, NJ: Prentice-Hall, 1982).

Jarost, D. "Health Complaints", *Can. J. Pub. Health* 77:132–135 (1986).

Johnson, M. L. "Human Effects Following Exposure to Ionizing Radiation", *Arch. Dermatol.* 122:1380–1382 (1986).

Leads from the MMWR "Cytotoxicity of Volcanic Ash: Assessing the Risk for Pneumonconiosis", *JAMA* 255(20):2727, 2731 (1986).

Kanna Pilly, G. M., R. K. Woolf and P. B. DeNee. "Generation Characteriziations and Inhalation of Ultrafine Aerosols", *Ann. Occup. Hyg.* 26:77–81 (1982).

Karrh, B. W. "Reproductive Hazards in the Workplace", *Del. Med. J.* 58(3):203–204 (1986).

Kauppinen, T. P., T. J. Partanen, M. M. Nurminen, J. I. Nickels, S. G. Hernberg, T. R. Hakulinen, E. I. Pukkala and E. T. Savonen. "Respiratory Cancers and Chemical Exposures in the Wood Indsutry: A Nested Case-Control Study", *Br. J. Ind. Med.* 43(2):84–90 (1986).

Koplin, A. N. "Right-to-Know: Implications of New Jersey's Law", *J. Public Health Policy* 5(4):538–549 (1984).

Korn, R. J., D. W. Dockery, F. E. Speizer, J. H. Ware and B. G. Ferris. "Occupational Exposures and Chronic Respiratory Symptoms. A Population-Based Study", *Am. Rev. Respir. Dis.* 136(2):298–304 (1987).

Lammintausta, K., and H. I. Maibach. "Dermatologic Considerations in Worker Fitness Evaluation", *State Art Rev. Occup. Med.* 3(2):341–350 (1988).

Laufer, A. "Assessment of Safety Performance Measures at Construction Sites", *J. Constr. Eng. Manage.* 112:530,542 (1986).

Lauwerys, R. *Industrial Chemical Exposure: Guidelines for Biological Monitoring* (Cal. Biomedical, 1983).

Laverghetta, T. *Practical Microwaves* (Indiana: Howard W. Sams and Company, Inc., 1984).

Lecea, J. "Injuries in Small Press Work", *Semin. Occup. Med.* 2:69–70 (1987).

Logan, D. C. "Reproduction and the Workplace: An Industry Perspective", *State of Art Rev.: Occup. Med.* 1(3):473–481 (1986).

Logue, N. J., and S. Hamburger. "Congenitial Anomalies and Paternal Occupational Exposure to Shortwave, Microwave, Infrared, and Acoustic Radition", *J. Occup. Med.* 27:(1985)451–452.

Lohman, W. *Low Back Pain in Nurses* (Washington, DC: Government Printing Office, 1987).

Loomis, T. A. *Essentials of Toxicology* (Philadelphia: Lea & Febriger, 1978).

Lowry, L. K. "Biological Exposure Index as a Complement to the TLV", *J. Occup. Med.* 28(8):578–582 (1986).

Malek, R. F., J. M. Daise, and B. S. Cohen. "The Effect of Aerosol on Estimates of Inhalation Exposure to Airborne Styrene", *Am. Ind. Hyg. Assoc. J.* 47(9):524–529 (1986).

McAtt, F. L. "Reducing Repetitive Motions", *Semin. Occup. Med.* 2:73–74 (1987).

McDonald, J. C. *Recent Advances in Occupational Health* (London: Churchill Livingston, 1981).

McMichael, A. J. "Carcinogenicity of Benzene, Toluene and Xylene: Epidemiological and Experimental Evidence", *IARC Sci. Publ.* 85:3–18 (1988).

Merz, B. "Multiple Efforts Directed at Defining, Eliminating Excess Radiation", *Med. News Perspect.* 258:577 (1987).

Moson, R. R. *Occupational Epidemiology* (Boca Raton, FL: CRC Press, 1980).

Morgan, W., and P. Ravens. "Prediction of Distress for Individuals Wearing Industrial Respirators", *Am. Ind. Hyg. Assoc. J.* 46:363–368 (1985).

Morrow, P. E. "Toxicological Data on Nox: An Overview", *J. Toxicol. Environ. Health* 13(2–3):205–257 (1984).

Murphy, M. J. "Environmental Risk Assessment of Industrial Facilities: Techniques, Regulatory Initiatives and Insurance", *Sci. Total Environ.* 51:185–196 (1986).

NIOSH Criteria Document. Occupational Exposure to Hot Environments, NIOSH Publ. No. 72-10269 (1972).

National Safety Council. *Fundamentals of Industrial Hygiene*, (6th Printing 1983).

National Toxicology Program. *Technical Report on the Toxicology and Carcinogenesis of Benzene* (Bethesda, MD: National Institutes of Health Publications, 1986).

National Toxicology Program. *Technical Report on the Toxicology and Carcinogenesis of Tetrachloroethylene* (Bethesda, MD: National Institutes of Health Publication, 1986).

National Toxicology Program. *Technical Report on the Toxicology and Carcinogenesis of Trichloroethylene* (Bethesda, MD: National Institutes of Health Publication, 1986).

National Toxicology Program. *Technical Report on the Toxicology and Carcinogenesis of 1,1,1-Trichloroethane* (Bethesda, MD: National Institutes of Health Publication, 1986).

National Toxicology Program. *Technical Report on the Toxicology and Carcinogenesis of Vinyl Chloride* (Betheseda, MD: National Institutes of Health Publication, 1986).

Nickson, K. "Occupation: Nurse Occupational Hazard: Radiation", *Can. Nurse* 80:30–31 (1984).

Nussey, C., and B. Worthington. "The Use of Systematic Reliability Assessment Techniques for the Evaluation and Design of Mining Electronic Systems", *J. Occup. Accidents* 7:1–17 (1985).

Occupational Exposure to Ionizing Radiation in the United States (Washington, DC: Office of Radiation Programs, U.S. Environmental Protection Agency, 1984).

"Occupational Fatality Following Exposure to Hydrogen Sulfide—Nebraska", *MMWR* 35(33):533–535 (1986).

Olishifski, J. B., and E. R. Hartford. *Industrial Noise and Hearing Conservation* (Chicago: National Safety Council, 1975).

Oser, L. J. "Extent of Industrial Exposure to Epichlorohydrin, Vinyl Fluoride, Vinyl Bromide and Ethylene Dibromide", *Am. Indust. Hygiene Assoc. J.* 41:461–468 (1980).

Ott, M. G., H. G. Scharnweber and R. R. Langner. "Mortality Experience of 161 Employees Exposed to Ethylene Dibromide in Two Production Units", *Br. J. Indust. Med.* 37:163–168 (1979).

Parkinson, D. K., and M. J. Greenan. "Establishment for Medical Surveillance in Industry: Problems and Procedures", *J. Occup. Med.* 28(8):772–777 (1986).

Paull, J. M. "The Origin and Basis of Threshold Limit Values", *Am. J. Ind. Med.* 5(3):227–238 (1984).

Pellizzari, E. D., R. A. Zweidinger and L. S. Sheldon. "Breath Sampling", *IARC Sci. Publ.* (68):399–411 (1985).

Peters, J. M., D. H. Garabrant, W. E. Wright, L. Bernstein and T. M. Mack. "Uses of a Cancer Registry in the Assessment of Occupational Cancer Risks", *Natl. Cancer Inst. Monogr.* 69:157–161 (1985).

Petreas, M. "A Laboratory Evaluation of Two Methods of Measuring Low Levels of Formaldehyde in Indoor Air", *Am. Indust. Hygi. Assoc. J.* 47:276–280 (1986).

Phillip, R. "Cadmium Risk Assessment of an Exposed Occupational Population", *J. Soc. Med.* 78(4):328–333 (1985).

Poch, D. I. *Radiation Alert* (New York: Doubleday & Company, 1985).

Pollack, H. "Medical Aspects of Exposure to Radiofrequency Radiation Including Microwaves", *South. Med. J.* 76:(6)759 (1983).

"Polychlorinated Biphenyl Transformer Incident—New Mexico", *MMWR* 34(36):557–559 (1985).

"Radiation", *Occup. Health Nurs.* 32:263–264 (1984).

Ratoff, J. "Microwaves: Hints of Low-Dose Hazards", *Sci. News* 126:(7)103 (1984).

Rice, C., R. L. Harris, J. C. Lumsden and M. J. Symons. "Reconstruction of Silica Exposure in the North Carolina Dusty Trades", *Am. Ind. Hyg. Assoc. J.* 45(10):689–696 (1984).

Riggin, R. M. "Determination of Volatile Organic Compounds in Ambient Air Using Tenax Adsorption and Gas Chromatography/Mass Spectrometry", *IARC Sci. Publ.* (68):269–289 (1985).

Robertson, J. M., J. F. Diaz, I. M. Fyfe and T. H. Ingalls. "A Cross-sectional Study of Pulmonary Function in Carbon Black Workers in the United States", *Am. Ind. Hyg. Assoc. J.* 49(4):161–166 (1988).

Ross, D. "Case History: An Acid Fumes Hazard", *Occup. Health Lond.* 37(4):187–188 (1985).

Rycroft, R. J. "Environmental Aspects of Occupational Dermatology", *Derm. Beruf. Umwelt.* 34(6):157–159 (1986).

Sagripanti, J. L. "DNA Structural Changes Caused by Microwave Radiation", *Int. J. Rad. Biol. Related Stud. Phys. Chem. Med.* 50(1)47–50 (1986).

Saltzman, B. E. "Variability and Bias in the Analyses of Industrial Hygiene Samples", *Am. Ind. Hyg. Assoc. J.* 46(3):134–141 (1985).

Samuels, S. W. "Medical Surveillance: Biological, Social, and Ethical Parameters", *J. Occup. Med.* 28(8):572–577 (1986).

Sass, R. "What's in a Name? The Occupational Hygienists Problem With Threshold Limit Values", *Am. J. Ind. Med.* 14(3):355–363 (1988).

Schilling, R. S. "The Role of Medical Examination in Protecting Worker Health", *J. Occup. Med.* 28(8):553–537 (1986).

Schwartz, D. A., and J. P. LoGerfo. "Congenital Limb Reduction Defects in the Agricultural Setting", *Am. J. Public Health* 78(6):645–648 (1988).

Sever, L. E. "Epidemiologic Approaches to Reproductive Hazards of the Workplace", *Birth Defects: Original Article Series* 18(3A):33–38 (1982).

Singh, N. P., M. E. Bennett, D. B. Wrenn, and G. Saccomanno. "Concentrations of ^{210}Pb and Its States of Equilibrium With ^{238}U, ^{234}U, and ^{2309}Th in U Miners' Lungs", *Health Physics*, 51:501–507 (1986).

Sinks, T. and S. Newman. "Surveillance of Work-Related Cold Injuries Using Workers", *J. Occup. Med.* 29:504–509 (1987).

Sittig, M. *Hazardous and Toxic Effects of Industrial Chemicals* (New Jersey: Noyes Data, 1979).

Smith, J. M., and R. D. Costlow. "Recognition, Evaluation, and Control of Chemical Embryotoxins in the Workplace", *Fundament. Appl. Toxicol.* 5(4):626–633 (1985).

Smith, R. G. "Cummings Award Address. Occupational Health Standard Setting in the United States", *Am. Ind. Hyg. Assoc. J.* 46(1):541–546 (1985).

Solomon, C. J. "Occupational Health Problems in High Tech Industries", *Occup. Health Nurs.* 32(5):262–265 (1984).

Spear, R. C., S. Selvin and M. Francis. "The Influence of Averaging Time on the Distribution of Exposures", *Am. Ind. Hyg. Assoc. J.* 47(6):365–368 (1986).

Stallard, C. W. "Use of SOHIO's Health Information System", *J. Occup. Med.* 28(8):684–686 (1986).

Stinnett, S. *Mutagenistic Testing of Industrial Wastes From Representative Organic Chemical, Industries* (Ada, OK: U.S. EPA, 1981).

Thorpe, J. J. "Occupational Medicine in the Petroleum Industry: An Historical Perspective", *State Art Rev. Occup. Med.* 3(3):371–390 (1988).

Thrane, K. E., and H. Stray. "Organic Air Pollutants in an Aluminum Reduction Plant", *Sci. Total Environ.* 53(1–2):111–131 (1986).

Ungers, L. J., and J. H. Jones. "Industrial Hygiene and Control Technology Assessment of Ion Implantation Operations", *Am. Ind. Hyg. Assoc. J.* 47(10):607–614 (1986).

U.S. Congress, Office of Technology Assessment. *Preventing Illness and Injury in the Workplace* (Washington, DC: U.S. Government Printing Office, 1985).

U.S. Congress, Office of Technology Assessment. *Reproductive Health Hazards in the Workplace* (Washington, DC: U.S. Government Printing Office, 1985).

U.S. Department of Health and Human Services. *Handbook for Industrial Noise Control* (Cincinnati, OH: 1981).

U.S. Department of Health and Human Services. *Hot Environment* (Washington, DC: U.S. Government Printing Office, 1986).

U.S. Department of Health and Human Services, National Institute of Occupational Safety and Health, Division of Standards Development and Technology Transfer. "Evaluation of Epidemiologic Studies Examining the Lung Cancer Mortality of Underground Uranium Miners", report prepared for U.S. Mine Safety and Health Administration (May 9, 1985).

U.S. Department of Health and Human Services. *Occupational Exposure to Hot Environments* (Washington, DC: U.S. Government Printing Office, 1986).

U.S. Department of Health and Human Services. *Recognition of Occupational Health Hazards* (Washington, DC: Government Pritning Office, 1982).

U.S. Department of Health and Human Services. *Survey of Hearing Conservation Programs in Industry* (Washington, DC: U.S. Government Printing Office, 1982).

U.S. Department of Labor, Mine Safety and Health Administration. "Ionizing Radiation Standards for Underground Metal and Nonmetal Mines; Proposed Rule", *Fed. Reg.* 51:45678 (1986).

U.S. Department of Labor, Occupational Safety and Health Administration. "Identification, Classification and Regulation of Potential Occupational Carcinogens", *Fed. Reg.* 45:5002, (1980.)

U.S. Department of Labor. *OSHA Programs* (Washington, DC: U.S. Government Printing Office, 1985).

U.S. Department of Labor. *Job Hazard Analysis* (Washington, DC: U.S. Government Printing Office, 1986).

U.S. Environmental Protection Agency. "Asbestos; Proposed Mining and Import Restrictions and Proposed Manufacturing Importation and Processing Prohibitions", *Fed. Reg.* 51:3738, (1986).

U.S. Environmental Protection Agency. *Mortality and Tumor Incidence Among Ferochromium Workers* (1983).

Upton, A. C. "The Biological Effects of Low-Level Ionizing Radiation", *Sci. Am.* 236:41–49 (1982).

Upton, A. C. "Environmental Standards for Ionizing Radiation: Theoretical Basis for Dose-Respoxe Curves", *Environmental Health Perspectives* 52:31–39 (1983).

Utidjian, W. O., and H. M. Karten. "Retrospective Evaluation of Reproductive Performance of Workers Exposed to Ethylene Dibromide", *J. Occup. Med.* 21: 98–102 (1979).

Voelz, G. L. "Ionizing Radiation", *Occup. Health Saf.* 51:34–37 (1982).

Weil, C. S. "1984 Stokinger Lecture. Some Questions and Opinions on Issues in Toxicology and Risk Assessment", *Am. Ind. Hyg. Assoc. J.* 45(1):663–670 (1984).

Weisburger, J. H., and G. M. Williams. "Chemical Carcinogens", in *Toxicology: The Basic Science of Poisons,* 2nd ed. J. Doull, C. D. Klassen, and M. O. Amdur, Eds. (New York: Macmillan, 1980).

Wick, J. L. "Workplace Design Changes", *Sem. Occup. Med.* 2:75–77 (1987).

Williams, P. L., and J. Burson. *Industrial Toxicology: Safety and Health Applications in the Workplace* (New York: Van Nostrand Reinhold, 1985).

Witt, M. *White Book, Noise Control, A Guide For Workers and Employers* (New York: Holt, Reinehart & Winston, 1985).

Wood, J. L. "Right-to-Know Laws: Maintaining Compliance", *Occup. Health Saf.* 56(3):20–21,24,27–29 (1987).

WHO Offset Publ. "Evaluation of Exposure to Airborne Particles in the Work Environment", (80):1–75 (1984).

Yodaiken, R.E. "Medical Screening and Biological Monitoring for the Effects of Exposure in the Workplace. Surveillance, Monitoring, and Regulatory Concerns", *J. Occu. Med.* 28(8):569–571 (1986).

Index

VOLUME I

Heat, *see also* Temperature
  hospital environment, 333—334, 365
  internal housing environment, 278
    accidents and injuries, 296
    disease and, 283
    formaldehyde and, 288—289, 319
    standards, 303
  occupational environment, 417—418, 437—
    438, 482
  prison environment, 353
  school environment, 355, 377
Heat cramps, 437—438
Heat exhaustion, 437
Heat fatigue, transient, 438
Heating, in food preservation, 154
Heat rash, 438
Heat recovery ventilation, radon gas control,
  318
Heat stroke, 437
Heat surveys, 469
Heavy-duty cleaners, 98
Heavy metals, *see also* specific metals
  food chain and, 147—148, 171
  occupational environment, 457
Helper T-cells, 31
Hemodialysis machines, 341
Hemopoiesis, 32
Hepatitis A virus, food-borne, 104, 136
Hepatitis B virus
  health-care workers and, 338, 374, 396—
    397
Hepatotoxicity, 428
Hepatoxins, 111, 447
Heptachlor, 240, 246
Herbicides, 252—253, 457
Herpes simplex, 397
Heterotrophs, 2
HHS, *see* Department of Health and Human
  Services
Highly toxic chemical, 446
High-temperature short-time (HTST)
    pasteurization, 158—160, 168—169
HIV, *see* Human immunodeficiency virus
Home environment, *see* Housing environment
Homeostasis, 30, 438
Hopewell, Virginia, 264
Hormones, 35, 144
Horseback-type survey, in community rodent
    control, 233
Horseflies, 179, 182
Hospitals, 332—338
  accidents, 346, 378—379, 386

air, 339—340
electrical facilities, 333, 363—365
emergency medical services, 336, 368
environmental control, 336—338
equipment, 337
  design and construction, 340—341
  electrical, 363—365
explosive materials, 366
fire safety, 336, 368—369
flammable materials, 366
food-service program, 341—342, 386—389
hand washing, 336—337, 373—374
hazardous chemicals, 342—343
hazardous wastes, 347
housekeeping, 335, 337, 343, 367
infections acquired in, 336—338, 348—349
  routes of transmission, 349—350
  surveillance and evaluation, 377—379,
    380—385
infectious waste management, 371—373,
  394—396
insect control, 343—344
interior structural requirements, 359—363
laboratories, 344
laundry, 338, 344—345
lighting, 345, 364, 368
maintenance, 335
noise, 345—346
nursing homes and, 351—352
occupational health and safety, 350
organization, 332—333
patient accidents, 346
patient safety, 370
physical plant, 358—359
plumbing, 335, 346, 368
problems throughout, 349
radiation safety, 370—371
radioactive materials, 335, 346—347, 442—
  443
rodent control, 343—344
site selection, 358—359
solid wastes, 347
structural features, 366—367
surgical suites, 347—348, 362
thermal environment, 333—335, 365
water supply, 336, 368
Host, defined, 101
Host factors, susceptibility and, 90
Host-parasite relationship, 41
Hotels, 292
Hot water, food service facilities, 117
House fly, 179—182

Index

VOLUME II

567

Refractory substances, 416
Refractory wall furnaces, 134
Refrigerators, toxic chemicals, 79
Refuse, 103, 113
Refuse-derived fuel (RDF), 79, 114
Regulated air pollutants, 16
Regulation, *see also* Legislation; Standards
  air pollution control, 2, 3, 59—60
    components of, 59—61
    results of, 62—64
    standards, 60—61
  economic impacts of pollution control, 502
  solid and hazardous waste laws, 191—193
  water quality, 269
Relative humidity, 12
Remote optical sensing of emissions (ROSE), 48
Remote sensing, 47—48
Reoviruses, 485, 486, 488
Reovirus-like agent, 485
Reports, water pollution surveys, 520
Research facilities
  health hazards, 146
  laboratory waste, 72
Research needs
  air pollution effects, 65—66
  ocean dumping, 174
  plumbing, 353
  sewage disposal systems, 455—456
  solid and hazardous waste management, 206—207
  swimming areas, 320
  water quality, 278—279, 536—537
Reservoirs, water treatment, 221—222
Reservoir system, 91
Residential areas, solid waste storage, 160
Residential waste, *see* Domestic waste
Residual chlorine test, 444
Residue analysis, 117
Resins, 181
Resource Conservation and Recovery Act (RCRA), 85, 168, 191, 192
  definition of solid and hazardous waste, 69—70
  groundwater contamination, 271
  incinerator performance standards, 174—175
  programs, 195—198
  underground storage tanks, 197
  waste minimization, 80
Resource Conservation and Recovery Act Hotline, 160

Resource-recovery facilities, 68
Resources, natural, *see* Water suppies
Resources (information sources), *see* Scientific and technical resources
Respirator, 184
Respiratory system, air pollution and, 33—35
Return beam vidicaon (RBV), 48
Reverse air baghouse, 51
Reverse osmosis, 178—179, 273, 392
Reynolds number, 325
Ringlemann Scale, 44, 47
Rio Linda Chemical Company, 188
Rivers, 282, 292
Road salting, 211, 229
Rocket fuels, 181
Rodenticides, air pollution, 26
Rodents
  disaster plans, 569
  disease transmission, 141
ROSE (remote optical sensing of emissions), 48
Rotary cycle engines, 58
Rotary kiln incinerator, 116
Rotary screens, wastewater treatment, 388
Rotating biological contactors, 376
Rotaviruses, 485, 486
Roundworms, 490—491, 492
Routing, solid waste collection, 91
RRT, *see* Regional Response Team
Rubber, 75, 96
  classification of solid waste for incineration, 113
  recycling, 132
Runoff
  and bacterial content of water, 484
  beach water quality, 292
  feedlot, 481, 530
  water quality controls, 530

Safe Drinking Water Act (SDWA), 192, 210—211, 241
Safety hazards
  hazardous waste sites, 146—147
  swimming areas, 294—295, 319
Salinity, water, 484
Salmonella, 235, 413, 483
  paratyphi, 482—484
  typhi, 235, 413, 482—484
  typhimurium, 413
Salmonella infections, 155
Salmonellosis, 292